Representative Points of Statistical Distributions

Statistical simulation has become a cornerstone in statistical research and applications. The aim of *Representative Points of Statistical Distributions: Applications in Statistical Inference* is to present a comprehensive exploration of various methods for statistical simulation and resampling, focusing on consistency and efficiency. It covers a range of representative points (RPs)—Monte Carlo (MC) RPs, quasi-Monte Carlo (QMC) RPs, and mean square error (MSE) RPs—and their applications, and includes a collection of recent developments in the field. It also explores other types of representative points and the corresponding approximate distributions, and delves into the realm of statistical simulation by exploring the limitations of traditional MC methods and the innovations brought about by the bootstrap method. In addition, the text introduces other kinds of representative points and the corresponding approximate distributions, such as QMC and MSE methods.

Features

- Comprehensive exploration of statistical simulation methods: provides a deep dive into MC methods and bootstrap methods, and introduces other kinds of RPs and the corresponding approximate distributions, such as QMC and MSE methods.
- Emphasis on consistency and efficiency: highlights the advantages of these methods in terms of consistency and efficiency, addressing the slow convergence rate of classical statistical simulation.
- Collection of recent developments: brings together the latest advancements in the field of statistical simulation, keeping readers up to date with the most current research.
- Practical applications: includes numerous practical applications of various types of RPs (MC-RPs, QMC-RPs, and MSE-RPs) in statistical inference and simulation.
- Educational resource: can serve as a textbook for postgraduate students, a reference book for undergraduate students, and a valuable resource for professionals in various fields.

The book serves as a valuable resource for postgraduate students, researchers, and practitioners in statistics, mathematics, and other quantitative fields.

Representative Points of Statistical Distributions

Applications in Statistical Inference

Kai-Tai Fang, Huajun Ye, and Yongdao Zhou

CRC Press
Taylor & Francis Group
Boca Raton London New York

CRC Press is an imprint of the
Taylor & Francis Group, an **informa** business
A CHAPMAN & HALL BOOK

First edition published 2025
by CRC Press
2385 NW Executive Center Drive, Suite 320, Boca Raton FL 33431

and by CRC Press
4 Park Square, Milton Park, Abingdon, Oxon, OX14 4RN

CRC Press is an imprint of Taylor & Francis Group, LLC

ISBN: 978-1-032-96411-9 (hbk)
ISBN: 978-1-032-96415-7 (pbk)
ISBN: 978-1-003-58938-9 (ebk)

DOI: 10.1201/9781003589389

Typeset in CMR10 font
by KnowledgeWorks Global Ltd.

Publisher's note: This book has been prepared from camera-ready copy provided by the authors.

Contents

Foreword

The classical statistical simulation by the Monte Carlo (MC) methods is helpful, but the convergence rate is slow. Using computer software, MC methods take random samples from a given population $F(x)$. Then, we calculate the statistics of the interest for statistical inference based on the empirical distribution of the random sample. The bootstrap method proposed by B. Efron (1979) is the first resampling technique and has been popularly used. The bootstrap takes random samples from the empirical distribution and finds a numerical estimator from the bootstrap sample. However, the empirical distribution can be regarded as approximate to $F(x)$, and the random sample as representative points (MC-RPs). The book concerns other kinds of RPs and the corresponding approximate distributions, i.e., via quasi-Monte Carlo and mean square error methods. The statistical simulation and resampling by these methods have advantages in consistency and efficiency.

The book collects recent developments and reviews several kinds of representative points (MC-RPs, QMC-RPs, and MSE-RPs) and their applications in statistical inference and simulation. Many applications are included.

The text can be used as a textbook for postgraduate students and a reference book for undergraduate students and users in various fields.

Representative Points of Statistical Distributions: Applications in Statistical Inference collects recent developments and provides a comprehensive review of various types of representative points (MC-RPs, QMC-RPs, and MSE-RPs) and their applications in statistical inference and simulation. It includes numerous practical applications.

This book can serve as a textbook for postgraduate students, a reference book for undergraduate students, and a valuable resource for professionals in various fields.

Preface

Statistical simulation has become a cornerstone in statistical research and applications, mainly through the Monte Carlo (MC) method. This approach involves drawing random samples from a specified population $F(x)$ using computer software, then calculating the statistic of interest based on the empirical distribution of these random samples, facilitating statistical inference.

One of the pioneering resampling techniques is the bootstrap method, introduced by B. Efron in 1979. This method involves taking random (bootstrap) samples from the empirical distribution and deriving numerical estimators from these samples. For example, finding the confidence interval of a statistic directly from the p-quantile of $F(x)$ can be challenging. However, the bootstrap method simplifies this process by providing an easy-to-compute numerical estimator. As a computer-based method of statistical inference, the bootstrap method effectively addresses many real-world statistical problems without relying on complex formulas.

Many theoretical analyses contribute to finding the asymptotic justification of the bootstrap method, such as Bickel and Freedman (1981), Singh (1981), Hall (1990, 1992), Efron and Tibshirani (1993), and Shao and Tu (1995). They mentioned that the consistency results of the bootstrap distribution of the sample mean hold not only for resampling from the empirical distribution but also from other approximating discrete distributions. The empirical distribution can be regarded as an approximation distribution of $F(x)$, but they did not provide an example of another kind of approximating distribution.

Fang and Wang (1994) were the first to explore statistical simulation using number-theoretic methods (NTM) or quasi-Monte Carlo (QMC) methods. They constructed approximate distributions to $F(x)$ via QMC-based representative points (QMC-RPs). Later, Fang et al. (2014) further refined these methods, considering additional approximate distributions and employing the mean square error (MSE) criterion. Their research demonstrated that resampling from these approximate distributions offers benefits in terms of consistency and efficiency. Building on this foundation, significant research has been conducted in the past decade. This monograph compiles recent advances and reviews various types of representative points (MC-RPs, QMC-RPs, and MSE-RPs) and their applications in statistical inference simulation and simulation, experimental design, geometric probability problems, and finance.

Cox (1957) explored the classification of observations from a continuous univariate distribution into k groups, associating individuals in the jth group with a value ξ_j to measure the "representativeness" of selected points to the distribution using Mean Square Error-Representative Points (MSE-RP). In engineering, the search for MSE-RPs is framed as an optimal quantization problem, originally used to enhance signal transmission through suitable discretization methods. Max (1960) delved into minimizing signal distortion through a quantizer with a fixed number of output levels. Furthermore, Fang and He (1982) proposed a computational approach for selecting RPs with minimized MSE to establish clothing standards. Flury (1990) investigated MSE-RPs with elliptical contour distributions to determine optimal sizes and shapes of protection masks for the Swiss Army. Tarpey et al. (1995) referred to MSE-RPs as "principal points" due to their connection to the principal components in multivariate statistical analysis. Yang et al. (2021) examined the mean vector and covariance matrix using three approximation distribution methods based on MC, QMC-RPs, and MSE-RPs for the elliptical and skew-normal distributions. Furthermore, Xu et al. (2024) demonstrated the consistency of a novel method that relies on MSE-RPs for various univariate distributions. Fang and Pan (2023) provided an overview of recent advances in representative point theory and their practical applications.

There are several books on the theory and applications of RPs, for example, Graf and Luschgy (2000) *Foundations of Quantization for Probability Distributions* and Pagès (2018) *Numerical Probability, An Introduction with Applications to Finance*. However, this new book *Representative Points of Statistical Distributions: Applications in Statistical Inference* emphasizes statistical inference by statistical simulation and resampling. It is different from all existing books.

Comprising nine chapters, this book begins with a review of fundamental concepts of probability and statistics in the first chapter, covering random variables, statistical distributions, estimation methods, and resampling techniques. The second chapter introduces discretization methods based on various RPs, including MC-RPs, RMC-RPs, QMC-RPs, and MSE-RPs, to construct approximate discrete distributions from continuous distributions. The subsequent chapters delve into the properties and applications of MSE-RPs for univariate and multivariate distributions, statistical simulation, and estimation under unknown parameters. The text concludes with discussions on geometric probability problems, meta-distributions, and subsampling methods in scenarios where sample size exceeds variable count.

There are nine chapters in this book. The first chapter reviews fundamental concepts in probability and statistics, including random variables, random vectors, statistical distributions, and their characteristics (such as mean, variance, and higher-order moments). The content also involves statistical estimation methods, including moment estimation, maximum likelihood estimation, nonparametric estimation, and resampling methods. Numerical optimization techniques are also discussed, including the "sequential number-theoretic method for optimization" (SNTO). A list of notation used throughout is provided for the reader's convenience.

The second chapter introduces approximated discretization methods based on four types of representative points (RPs): Monte Carlo RPs (MC-RPs), revised Monte Carlo RPs (RMC-RPs), quasi-Monte Carlo RPs (QMC-RPs), and mean square error RPs (MSE-RPs). The objective is to construct approximate discrete distributions corresponding to a given continuous distribution. This chapter also provides the motivation and literature review for MSE-RPs and investigates the asymptotic behavior of MSE-RPs in univariate distributions.

The third chapter focuses on the properties of the MSE-RPs of a continuous univariate distribution $F(x)$ with finite variance. There are three methods for generating MSE-RPs: theoretical, numerical, and a mixture of theoretic and numerical. The chapter introduces the theory and algorithms for the generation of MSE-RPs. The uniqueness of MSE-RPs will be studied.

The purpose of the fourth chapter is to provide an overview of statistical simulation based on approximate distributions constructed using RPs, with a particular emphasis on MSE-RPs. This chapter examines several types of distribution, including the gamma distribution, mixtures of normal distributions, the extreme value distribution, and the arcsine distribution. It introduces various measures for assessing the closeness of two distributions and providing numerical comparisons to illustrate these concepts.

The fifth chapter deals with statistical simulation under the distribution with some unknown parameters. Assume that the type of $F(x, \boldsymbol{\theta})$ is known but with some unknown parameters $\boldsymbol{\theta}$. By a random sample, one can find the estimates of the unknown parameters, denoted by $\hat{\boldsymbol{\theta}}$, then we can generate the MSE-RP of $F(x, \hat{\boldsymbol{\theta}})$ and construction approximation distributions using a set of RPs to $F(x, \boldsymbol{\theta})$. Statistical estimation via resampling from these approximation distributions is considered. A general theory for the consistency of statistical estimation is introduced. In particular, the estimation of MSE-RPs for location-scale distributions with unknown parameters is discussed. Moment and maximum likelihood estimations and their bias correction are studied.

The sixth and seventh chapters focus on three types of RPs and the corresponding approximation distributions for continuous multivariate distributions. Chapter 6 discusses various algorithms for generating these three types of RPs based on the stochastic representation of the random vector. The elliptical distribution, which includes the multivariate normal distribution, has a simple stochastic representation, allowing it to generate its MSE-RPs and QMC-RPs. Relevant algorithms are presented. Other multivariate distributions, such as the skew-normal and strongly symmetric distributions, are also discussed. Chapter 7 focuses on the properties of MSE-RPs and the related discrete approximation distribution to $F(\boldsymbol{x})$. However, self-consistent points (SC-RPs) have a close relationship with MSE-RPs. Many properties of self-consistent points can proceed to those of MSE-RPs. An exciting aspect is the relationship between the RP and the principal components.

Chapter 8 shows some applications of MRs in geometric probability problems, meta-distributions via copula, applications in finance, experiments with mixtures, goodness-of-fit test hypotheses, and various applications.

The last chapter considers representative points for discrete data, especially for discrete massive data. The technique for finding representative points for discrete data differs from continuous distributions. This chapter considers cases where the sample size N is significantly larger than the number of variables p and introduces various model-based and model-free subsampling methods.

We put the most detailed proofs of theorems into the last section of the corresponding chapter. If the proof is short, the proof follows the theorem.

The first author, Kai-Tai Fang, of this book proposed the mathematical problem based on the national Chinese garment standard in Fang and He (1982) and Fang and He (1984). Later, Fang et al. (1994b) proposed computational algorithms for generating MSE-RPs of the elliptical distributions. Fang et al. (2014) considered applications of the three kinds of RPs in statistical estimation and resampling. The other two authors, Professors Huajun Ye and Yongdao Zhou, contributed significantly and published several articles on this research topic.

This text is a valuable resource for statisticians, researchers, and postgraduate students, offering insights into the theory and applications of representative points. The authors have made significant contributions to the field with ongoing seminars and collaborations at BNU-HKBU United International College since 2019. Grants from the Guangdong Provincial Key Laboratory and BNU-HKBU United International College have supported the work.

Since 2019, the BNU-HKBU United International College has held a regular seminar on this exciting and challenging topic. The working team includes our colleagues: Kai-Tai Fang, Jianxin Pan, Jiajun Liang, Huajun Ye, Ping He, Xiaoling Peng, and Min Zhou; and many postgraduate or undergraduate students, Xiao Ke, Yinan Li, Jun Yang, Wenjun Wang, Sirao Wang, Jia-Jian Jiang, Yangyi Zhang, Pingan He, and Longhao Xu. Several participants were from the Ocean University of China and The Renmin University of China. Significantly, Dr. Xiaoling Peng contributed to writing Section 8.6, and Dr. Yinan Li and Dr. Sirao Wang helped us with the numerical analysis of the examples in the monograph. We sincerely thank Dr. Changxing Ma for establishing the webpage on the generation of MSE-RPs: https://fst.uic.edu.cn/isci/Representative/Introduction.htm.

Our work was supported in part by the Guangdong Provincial Key Laboratory of Interdisciplinary Research and Application for Data Science, BNU-HKBU United International

College, grants (UIC-R201409 and UIC-R201314) from BNU-HKBU United International College (2022B1212010006), and the National Natural Science Foundation of China (12131001).

Kai-Tai Fang
Department of Statistics and Data Science
BNU-HKBU United International College, Zhuhai, China;
The Key Lab of Random Complex Structures and Data Analysis,
The Chinese Academy of Sciences, Beijing, China

Huajun Ye
Department of Statistics and Data Science
BNU-HKBU United International College, Zhuhai, China;
Guangdong Provincial Key Laboratory of Interdisciplinary Research and Application for Data Science

Yongdao Zhou
School of Statistics and Data Science
Nankai University, Tianjin, China

Authors

Kai-Tai Fang is a reputable statistician. He was educated at Peking University and The Chinese Academy of Sciences for undergraduate and postgraduate studies. He was elected as a fellow by the Institute of Mathematical Statistics (IMS) in 1992 and a fellow by the American Statistical Association (ASA) in 2001 as well as an elective member of the International Statistical Institute (ISI) in 1985. Professor Fang visited Yale University and Stanford University for two years and was invited as a guest professor at the Swiss Federal Institute of Technology and a visiting professor at the University of North Carolina at Chapel Hill. He was chair professor of the Department of Mathematics at Hong Kong Baptist University from 1993–2006. Currently, he is a chair professor at Beijing Normal-Hong Kong Baptist University (BNBU). His research interests are in statistics and mathematics, specifically experimental design, multivariate analysis, and data mining. He published 24 books (including six monographs in English) and more than 300 referred papers.

Huajun Ye received both bachelor's and master's degrees in probability and mathematical Statistics from Peking University in 1999 and 2002, respectively. He received a PhD in statistics from Manchester University, UK in 2005, and he did his PhD research on covariance structures modeling of longitudinal data. In 2007, he joined Beijing Normal-Hong Kong Baptist University (BNBU) as an assistant professor. Now, he is a full professor in the Department of Statistics and Data Science at Beijing Normal-Hong Kong Baptist University (BNBU). His research interests include statistical modeling, inference, financial risk management, and statistical representative points. More than ten research papers have been published in international journals and conferences, including *Biometrika*, *Mathematics*, *Journal of Complexity*, *Journal of Statistical Computation and Simulation*, and more.

Yongdao Zhou earned a BS in pure mathematics in 2002 and MS and PhD degrees in statistics in 2005 and 2008, respectively, from Sichuan University, China. He was a postdoctoral fellow at HKBU-UIC Joint Institute of Research Studies. Then, he joined Sichuan University and became a full professor in 2015. In 2017, he joined Nankai University, where he is presently a full professor in statistics. He visited UCLA, the University of Manchester, the National University of Singapore, and Simon Fraser University as a visiting scholar. His research agenda focuses on experimental design and big data analysis. He published over 70 papers, such as in *JRSSB*, *JASA*, *Biometrika*, and *IEEE TKDE*, as well as eight monographs and textbooks. His research publications have won two best paper awards.

1

Statistical Distributions and Preliminary Distributions

In this chapter, we introduce some basic knowledge in probability and statistics, including random variables, random vectors, statistical distributions, and their characteristics (e.g., mean, variance and high-order moments, quantiles); statistical estimation methods include the moment, maximum likelihood, and nonparametric estimations; and statistical simulation and resampling methods including the bootstrap method. Readers can refer to some textbooks, e.g., Hogg and Craig (2005) and Serfling (1980). Some numerical optimization techniques are discussed, including a powerful method, the "sequential number-theoretic method for optimization (SNTO)." Several tables of notations used in this text are given for convenience in reading.

1.1 Random Variables and Their Distributions

1.1.1 Definition

A *random variable* is a variable in $\mathbb{R}$, whose value is uncertain but follows some probability rule used by a statistical distribution. Let X be a random variable, and its domain or support region be $\mathcal{D}_X \subseteq \mathbb{R}$. The probability rule of X is defined by its *cumulative distribution function (cdf)*

$$F(x) = P(X \leq x), \quad -\infty \leq x \leq \infty. \tag{1.1}$$

When $\mathcal{D}_X = \{x_1, x_2, \ldots, x_k, \ldots\}$ is either finite or countable, X is called as *a discrete random variable.* Its *probability mass function (pmf)* $P(X = x_i) = p_i > 0, i = 1, \ldots, k, \ldots$ with $\sum_{i=1} p_i = 1$ is often expressed as

$$\begin{array}{c|ccccc} X & x_1 & x_2 & \cdots & x_k & \cdots \\ \hline p & p_1 & p_2 & \cdots & p_k & \cdots \end{array} \tag{1.2}$$

where $x_1, x_2, \ldots, x_k$ are called the *support points* of X. The binomial distribution and Poisson distribution are two commonly used discrete distributions in statistics.

When the domain of X is an interval, and there exists a function $p(x) \geq 0$ such that

$$F(x) = \int_{-\infty}^{x} p(t)dt, \quad p(x) = \frac{d}{dx}F(x), x \in \mathcal{D}_X, \tag{1.3}$$

X is called a *continuous random variable* and $p(x)$ is called *probability density function (pdf)* of X. There are many proper continuous distributions including the uniform distribution $U(a,b)$ in a finite interval (a,b), the exponential distribution $EP(\lambda)$ in $(0,\infty)$, the normal distribution $N(\mu, \sigma^2)$ and the gamma distribution $Gamma(\alpha, \beta)$ in $\mathbb{R}$, the beta distribution $Beta(\alpha, \beta)$ in $(0,1)$, extreme distributions in $(0,\infty)$, etc. See Table 1.4 for more distributions.

DOI: 10.1201/9781003589389-1

1.1.2 Classification of random variables

We have defined two classes of random variables: discrete random variables and continuous random variables. Other classifications are also discussed below.

A. Symmetric distribution

The distribution of $X \sim F(x)$ is symmetric about 0 if $X \stackrel{d}{=} -X$ that is equivalent to $F(x) = 1 - F(x)$. Here, the notation $X \stackrel{d}{=} Y$ means that the two variables X and Y have the same distribution. When X has a density $p(x)$, X is symmetric about 0 if $p(x) = p(-x)$. If $(X - \mu)$ is symmetric about 0, we say that X is symmetric about μ.

B. Mixture distribution

A *finite mixture distribution* is defined

$$F(x) = \sum_{j=1}^{m} p_j F_j(x), \ p_1, \ldots, p_m \geq 0, \ p_1 + \cdots + p_m = 1, \tag{1.4}$$

where p_j is the weight of the jth population $F_j(x)$. Various mixture distributions are useful in practice; for example, F_j are normal distributions (Subsection 3.9.1) or Weibull distributions (Section 3.9.5). When all $F_j(x)$ are continuous and have pdf $p_j(x)$, the pdf of X is

$$p(x) = \sum_{j=1}^{m} p_j p_j(x). \tag{1.5}$$

C. Location-scale distribution A distribution is called a *location-scale distribution* if its pdf has the form

$$p(x) = \frac{1}{\sigma} f\left(\frac{x - \mu}{\sigma}\right), \tag{1.6}$$

where μ is the location parameter and $\sigma > 0$ is the scale parameter. The normal distribution $N(\mu, \sigma^2)$ has the pdf

$$\phi(x; \mu, \sigma) = \frac{1}{\sqrt{2\pi}\sigma} \exp\left(-\frac{(x - \mu)^2}{2\sigma^2}\right) \tag{1.7}$$

and is a location-scale distribution. In the literature, $\Phi(x)$ and $\phi(x)$ denote the cdf and pdf of $N(0, 1)$, respectively.

D. Distribution with unimodal, bimodal, or multi-model

From the shape of the pdf, there are several types such as *unimodal, bimodal, or multimodel.* The tail of a pdf distribution (or cdf) significantly affects the behavior of statistical inference. There are *light-tailed, heavy-tailed, or long-tailed* that will be discussed in Subsection 2.4.4.

E. Univariate and multivariate distributions

The concepts in the above univariate distribution can be extended to the multivariate case. Let $\boldsymbol{X} = (X_1, \ldots, X_p)^T$ be a column random vector in $\mathbb{R}^p$, where $X_1, \ldots, X_p$ be p random variables, then $\boldsymbol{X}$ is called a random vector. The cdf of $\boldsymbol{X}$ is defined as

$$F(x_1, \ldots, x_p) = P(X_1 \leq x_1, \ldots, X_p \leq x_p), \ \ \boldsymbol{x} = (x_1, \ldots, x_p)^T \in \mathbb{R}^p. \tag{1.8}$$

The pdf $p(\boldsymbol{x}) = p((x_1, \ldots, x_p)$ of a continuous random vector $\boldsymbol{X}$ satisfies

$$F(\boldsymbol{x}) = \int_{-\infty}^{x_1} \cdots \int_{-\infty}^{x_p} p(x_1, \ldots, x_p) dx_1 \cdots dx_p, \ \boldsymbol{x} \in \mathbb{R}^p. \tag{1.9}$$

1.2 Characteristics of Statistical Distributions

Both the cdf and pdf of a statistical distribution are more complicated functions in practical uses and theoretical approaches. Therefore, statisticians have defined several types of characteristics of a statistical distribution. The expectation of a function $g(\boldsymbol{X})$

$$E(g(\boldsymbol{X})) = \int_{-\infty}^{\infty} \cdots \int_{-\infty}^{\infty} g(x_1, \dots, x_p)p(x_1, \dots, x_p)dx_1 \cdots dx_p, \tag{1.10}$$

has been used for defining many characteristics of a distribution.

For the univariate case the *moments at the origin* and the *central moments* are defined by

$$\mu_r = E(X^r), r = 1, 2, \dots, \quad \mu_r^c = E(X - \mu_1)^r, r = 1, 2, \dots, \tag{1.11}$$

respectively. It is well known that $\mu \equiv \mu_1 = E(X)$ is the *mean of* X, and $\mu_2^c = \mathrm{Var}(X) = \sigma^2$ is the *variance of* X. Other useful high-order moments are the *skewness* and *kurtosis*, which are defined as

$$Sk(X) = \frac{E(X-\mu)^3}{\sigma^3}, \quad Ku(X) = \frac{E(X-\mu)^4}{\sigma^4} - 3, \tag{1.12}$$

respectively. Skewness refers to a distortion or asymmetry that deviates from the symmetric bell curve or normal distribution, where the normal distribution has a zero skew. If the curve is shifted to the left or right, it is said to be skewed. Skewness can be quantified as a representation of how much a given distribution varies from the normal distribution. Kurtosis measures the probability distribution's "tailedness." For example, the normal distribution has zero kurtosis.

The question of the level within a population below which a random value from that population can be attained with a given probability is ubiquitous. Such levels are known as *distribution quantiles* or *percentiles* and are used very frequently in many scientific fields, such as biology, medicine, economics, etc. For given $0 < q < 1$, the *q-quantile* of $F(x)$ is denoted by $Q_q(X)$ or simply Q_q that satisfies

$$Q_q = F^{-1}(q) = \inf\{x : F(x) \geq q\}. \tag{1.13}$$

When Q_q is the continuous point of $F(x)$, we have $F(Q_q) = q$. The following quartiles are useful

$q = \frac{1}{4}, Q_{1/4}$ is the *first quartile* of $F(x)$;
$q = \frac{1}{2}, Q_{1/2}$ is the *median* of $F(x)$;
$q = \frac{3}{4}, Q_{3/4}$ is the *third quartile* of $F(x)$.

For a multivariate distribution $F(\boldsymbol{x}) = F(x_1, \dots, x_p)$, the mean extends into the *mean vector* of $\boldsymbol{X}$, the variance extends into the *covariance matrix*, that is,

$$\boldsymbol{\mu} = E(\boldsymbol{X}) = (E(X_1), \dots, E(X_p))^T, \tag{1.14}$$

$$\boldsymbol{\Sigma} = \mathrm{Cov}(\boldsymbol{X}) = E(\boldsymbol{X} - \boldsymbol{\mu})(\boldsymbol{X} - \boldsymbol{\mu})^T). \tag{1.15}$$

The *correlation matrix* of X is defined as a $p \times p$ matrix

$$\mathrm{corr}(\boldsymbol{X}) = \boldsymbol{R} = (\mathrm{corr}(X_i, X_j), i, j = 1, \dots, p), \tag{1.16}$$

where

$$\mathrm{corr}(X_i, X_j) = \frac{E(X_i - \mu_i)(X_j - \mu_j)}{\sqrt{\mathrm{Var}(X_i)\mathrm{Var}(X_j)}}, i, j = 1, \dots, p. \tag{1.17}$$

1.3 Statistical Estimation

Most useful statistical distributions involve few parameters. For example, the normal distribution $N(\alpha, \sigma^2)$ has two parameters μ and σ^2 and the gamma distribution $Be(\alpha, \beta)$ has parameters α and β.

Let X be a random variable with cdf $F(x; \boldsymbol{\theta})$, where $\boldsymbol{\theta} = (\theta_1, \ldots, \theta_m)$ are the parameters and their domain is indicated by Ω. Let $p(x; \boldsymbol{\theta})$ be the pdf of $F(x; \boldsymbol{\theta})$ if X is a continuous variable. A set of *random samples*, $X_1, \ldots, X_n$, from $F(x; \boldsymbol{\theta})$, i.e., they are independently identical distributed (iid) to $F(x; \boldsymbol{\theta})$. When the parameters $\boldsymbol{\theta}$ are unknown, a function of a random sample is used to estimate the parameters $\boldsymbol{\theta}$. A function of the sample $T = T(X_1, \ldots, X_n)$ is called the *statistic*. One often needs to estimate some characteristics such as the mean, variance, median, q-quartile, etc. Two strategies are commonly used to estimate these characteristics.

The first strategy is the plug-in method: we first estimate the unknown parameters $\boldsymbol{\theta}$ and plug-in their estimator $\hat{\boldsymbol{\theta}}$ to form the estimator distribution $\hat{F} = F(x, \hat{\boldsymbol{\theta}})$, then the mean, variance, median, q-quartile of $\hat{F}$ can be used to estimate these characteristics.

The second strategy is that one can find some statistics of $T = T(X_1, \ldots, X_n)$ to directly estimate these characteristics.

There are various methods for estimating $\boldsymbol{\theta}$ and many criteria for evaluating different estimation methods. Denote an estimate of $\boldsymbol{\theta}$ by $\hat{\boldsymbol{\theta}}$. There are three types of estimators of $\boldsymbol{\theta}$:

Point estimator: The statistic T outputs a vector $\hat{\boldsymbol{\theta}}$ as an estimate of $\boldsymbol{\theta}$.

Confidence region: The statistic T produces a confidence region $G \in \mathbb{R}^m$ for a given significant level α such that

$$P(T \in G) = 1 - \alpha.$$

When $m = 1$ (θ is only one unknown parameter), the confidence region reduces to a confidence interval (l, u), for a given significant level α such that

$$P(l \leq T \leq u) = 1 - \alpha.$$

Estimation of density: The statistic T_n produces a density curve $\hat{p}(x)$ as the estimate of $p(x)$. The detailed method will be introduced in Subsection 1.6.1.

There are four popular criteria for evaluating estimators in statistical theory (Serfling (1980)): (a) unbiasedness, (b) efficiency, (c) consistency, and (d) L_2-distance or mean square error (MSE). We will introduce them in the following subsections.

1.3.1 Unbiasedness

Quite a lot of statistics can be found for estimating the parameters $\boldsymbol{\theta}$, and it needs some criteria to evaluate them. The most important criterion in traditional statistics is unbiasedness.

Definition 1.1 *(Unbiasedness)*

Let X be a random variable with cdf $F(x; \boldsymbol{\theta}), \boldsymbol{\theta} \in \Omega$ with finite mean μ. Let $X_1, X_2, \cdots, X_n$ be a random sample from the distribution of X, and T denotes an estimator of $\boldsymbol{\theta}$. We say that T is an unbiased estimator of $\boldsymbol{\theta}$ if

$$E(T) = \boldsymbol{\theta}, \quad \textit{for all } \boldsymbol{\theta} \in \Omega;$$

otherwise T is a biased estimator of $\boldsymbol{\theta}$.

Example 1.1 *(Sample mean)*
Let $X_1, X_2, \cdots, X_n$ be a random sample from the population of $X \sim F(x)$ with the mean $\mu < \infty$. The sample mean $\bar{X} = \frac{1}{n}\sum_{i=1}^n X_i$ is the most widely used estimator of μ if $\mu < \infty$. It is obviously $E(X) = \mu$. Hence, $\bar{X}$ is an unbiased estimator of μ.

We can find many unbiased estimators of μ. Let $\boldsymbol{a} = (a_1, a_2, \cdots, a_n)^T$, where $a_i \geq 0, i = 1, \ldots, n$ and $a_1 + \ldots + a_n = 1$. The statistic

$$\hat{\mu}_{\boldsymbol{a}} = \boldsymbol{a}^T \boldsymbol{x} = \sum_{i=1}^n a_i X_i$$

is an unbiased estimator of μ, since $E(\hat{\mu}_{\boldsymbol{a}}) = \sum_{i=1}^n a_i E(X_i) = \mu$. Three examples are: for $\boldsymbol{a} = \boldsymbol{a}_1 = (1, 0, \ldots, 0)^T$ we have $\hat{\mu}_1 = X_1$; for $\boldsymbol{a} = \boldsymbol{a}_2 = (0.5, 0.5, 0, \ldots, 0)^T$ we have $\hat{\mu}_2 = (X_1 + X_2)/2$; and for $\boldsymbol{a} = \boldsymbol{a}_3 = (\frac{1}{n}, \ldots, \frac{1}{n})^T$ we have $\hat{\mu}_3 = \bar{X}$.

1.3.2 Minimum variance unbiased estimator

How to compare different $\hat{\mu}_{\boldsymbol{a}}$ in Example 1.1? A natural idea is to compare the variance of $\hat{\mu}_{\boldsymbol{a}}$. Suppose the variance $\text{Var}(X) = \sigma^2$ exists in Example 1.1. It is easy to find

$$\text{Var}(\hat{\mu}_{\boldsymbol{a}}) = \sigma^2 \boldsymbol{a}^T \boldsymbol{a}.$$

Note that $\text{Var}(\hat{\mu}_{\boldsymbol{a}_1}) = \sigma^2$, $\text{Var}(\hat{\mu}_{\boldsymbol{a}_2}) = \sigma^2/2$, and $\text{Var}(\hat{\mu}_{\boldsymbol{a}_3}) = \sigma^2/n$. It can be proven that $\hat{\mu}_{\boldsymbol{a}_3}$ has the minimum variance among all the unbiased estimates of $\hat{\mu}_{\boldsymbol{a}}$.

Definition 1.2 *(Minimum variance unbiased estimate (MVUE))*
Let $X_1, \cdots, X_n$ be a random sample from the distribution of X with finite mean and variance, an unbiased estimate of θ is called the minimum variance unbiased estimator of θ if it has the smallest variance among all the unbiased estimates of θ.

Example 1.2 *(Sample Variance) Let $X_1, \cdots, X_n$ be a random sample from the population of X which has finite mean μ and finite variance σ^2. Define the sample variance by*

$$s^2 = \frac{1}{n-1}\sum_{i=1}^n (X_i - \overline{X})^2 = \frac{1}{n-1}\left(\sum_{i=1}^n X_i^2 - n\overline{X}^2\right). \tag{1.18}$$

It can be proved that $E(s^2) = \sigma^2$ and s^2 is an unbiased estimator of σ^2. We can directly show this unbiasedness by the fact that $E(X^2) = \sigma^2 + \mu^2$, then

$$\begin{aligned} E(s^2) &= \frac{1}{n-1}\left(\sum_{i=1}^n E(X_i^2) - nE(\overline{X}^2)\right) \\ &= \frac{1}{n-1}\{n\sigma^2 + n\mu^2 - n[(\sigma^2/n) + \mu^2]\} = \sigma^2. \end{aligned}$$

In the literature it is shown that $\bar{X}$ and s^2 are MVUEs of μ and σ^2 of $N(\mu, \sigma^2)$, respectively. See Serfling (1980) for the details. The following example gives the variance of s^2.

Example 1.3 *(Variance of the Sample Variance)*
The sample variance can be expressed as

$$s^2 = \frac{1}{n(n-1)}\sum_{i \neq j} \frac{1}{2}(X_i - X_j)^2. \tag{1.19}$$

Suppose $E(X^4) < \infty$. From the definition of variance, we have

$$\mathrm{Var}(s^2) = \frac{1}{n^2(n-1)^2} \sum_{i \neq j, k \neq l} E\left[(\frac{1}{2}(X_i - X_j)^2 - \sigma^2)(\frac{1}{2}(X_k - X_l)^2 - \sigma^2) \right].$$

Denote

$$S_{ij} = \frac{1}{2}(X_i - X_j)^2 - \sigma^2, \; S_{kl} = \frac{1}{2}(X_k - X_l)^2 - \sigma^2.$$

Let v be the number of $\{i, j\} \cap \{k, l\}$ and consider the following situations:

1. $v = 0$: $E(S_{ij}S_{kl}) = 0$;
2. $v = 1$: *There are $4n(n-1)(n-2)$ such items each is equal to*

$$E\left[(\frac{1}{2}(X_1 - X_2)^2 - \sigma^2)(\frac{1}{2}(X_1 - X_3)^2 - \sigma^2) \right] = \frac{1}{4}(\mu_4 - \sigma^4),$$

where $\mu_4 = E(X - \mu)^4$.

3. $v = 2$: *There are $2n(n-1)$ such terms each is equal to*

$$\begin{aligned} & E\left[(\frac{1}{2}(X_1 - X_2)^2 - \sigma^2)^2 \right] \\ &= \frac{1}{4} E\left[(X_1 - \mu) - (X_2 - \mu) \right]^4 \sigma^4 = \frac{1}{2}(\mu_4 + \sigma^4). \end{aligned}$$

Now we have

$$\mathrm{Var}(s^2) = \frac{1}{n^2(n-1)^2} \left[4n(n-1)(n-2)\frac{1}{4}(\mu_4 - \sigma^4) + 2n(n-1)\frac{1}{2}(\mu_4 + \sigma^4) \right],$$

thus

$$\mathrm{Var}(s^2) = \frac{\mu_4}{n} - \frac{n-3}{n(n-1)}\sigma^4 = \frac{\mu_4 - \sigma^4}{n} + \frac{2\sigma^4}{n(n-1)}. \tag{1.20}$$

1.3.3 Consistency

Many statistics are biased but converge to the estimated parameter when the number of samples $n \to \infty$. It motivates to have a criterion for the estimator, i.e., *consistent estimator.* Various kinds of stochastic convergence are needed.

Definition 1.3 *(Stochastic Convergence)*

Let $\{X_n\}$ be a sequence of random variables, and let X be a random variable. Let F_{X_n} and F_X be the cdfs of X_n and X, respectively. In the following sense, X_n converges to X as $n \to \infty$.

1. *Convergence almost surely,* if

$$P(X_n \to X) = 1, \text{ denoted by } X_n \xrightarrow{a.s.} X.$$

2. *Convergence in probability,* if for every $\epsilon > 0$,

$$P(|X_n - X| > \epsilon) \to 0, \text{ denoted by } X_n \xrightarrow{P} X.$$

3. *Convergence in distribution,* if for each continuity point x of F,

$$F_{X_n}(x) \to F_X(x), x \in C(F_X), \text{denoted by } X_n \xrightarrow{D} X,$$

where $C(F_X)$ denotes the set of all points x where F_X is continuous at x.

4. *Converges in rth mean,* if $|X|^r < \infty$, we say that X_n converges in rth mean to X if

$$\lim_n \to \infty E|X_n - X|r = 0,$$

and write $X_n \xrightarrow{rth} X$.

Relationships among these different modes of convergence:

$$X_n \xrightarrow{a.s.} X \Rightarrow X_n \xrightarrow{P} X \Rightarrow X_n \xrightarrow{D} X, \tag{1.21}$$

$$X_n \xrightarrow{rth} X \Rightarrow X_n \xrightarrow{P} X, \ X_n \xrightarrow{rth} X \Rightarrow X_n \xrightarrow{sth} X, 0 < s < r. \tag{1.22}$$

More Properties:

1. $X_n \xrightarrow{D} X \nRightarrow X_n \xrightarrow{P} X.$
2. $X_n \xrightarrow{D} b$ (b is a constant) $\Rightarrow X_n \xrightarrow{P} b.$
3. $X_n \xrightarrow{D} X$ and $Y_n \xrightarrow{P} 0 \Rightarrow X_n + Y_n \xrightarrow{D} X.$
4. $X_n \xrightarrow{D} X \Rightarrow g(X_n) \xrightarrow{D} g(X)$, where g is a continuous function on the support of X.
5. Let X_n, X, A_n, B_n be the random variables, $X_n \xrightarrow{D} X$, $A_n \xrightarrow{P} a$, and $B_n \xrightarrow{P} b$, where a and b are constants, then $A_n + B_n X_n \xrightarrow{D} a + bX$.

The above definitions of convergence can be extended to multivariate distributions.

Definition 1.4 *(Consistent estimator)*
Let X be a random variable with cdf $F(x;\boldsymbol{\theta})$, $\boldsymbol{\theta} \in \Omega$. Let $X_1, \cdots, X_n$ be a sample of $F(x;\boldsymbol{\theta})$ and let T_n be a statistic to estimate $\boldsymbol{\theta}$. We say T_n is a weakly consistent estimator of $\boldsymbol{\theta}$ as $n \to \infty$ if

$$T_n \xrightarrow{P} \boldsymbol{\theta}, \ \text{or } T_n \xrightarrow{D} \boldsymbol{\theta},$$

and say T_n is a strongly consistent estimator of $\boldsymbol{\theta}$ as $n \to \infty$ if

$$T_n \xrightarrow{a.s.} \boldsymbol{\theta}.$$

The law of large numbers is useful to check whether a statistic T_n is a consistent estimator.

Theorem 1.1 *(The Law of Large Numbers)*

(A) The weak law of large numbers: If the random variables $X_1, \ldots, X_n$ are independent with finite $\mu_i = E(X_i)$ and as $n \to \infty$

$$\frac{1}{n^{1+\delta}} \sum_{i=1}^{n} |x_i|^{1+\delta} \to 0, \quad \text{for a positive } \delta \leq 1,$$

then

$$\frac{1}{n}\sum_{i=1}^{n} c_i(X_i - \mu_i) \xrightarrow{p} 0.$$

(B) The strong law of large numbers: Let $X_1, \ldots, X_n$ be an independent identically distribution (iid) with $E|X_1| < \infty$, then as $n \to \infty$

$$\frac{1}{n}\sum_{i=1}^{n} c_i(X_i - E(X_1)) \xrightarrow{a.s.} 0,$$

where c_i is a sequence of bounded numbers.

(C) If $X_1, \ldots, X_n$ are independent (not necessarily identically distributed) with finite $\mu_i = E(X_i)$, $\sigma_i^2 = \mathrm{Var}(X_i)$ and $\sum_{i=1}^{n} \sigma_i^2/i^2 < \infty$, then as $n \to \infty$

$$\frac{1}{n}\sum_{i=1}^{n} c_i(X_i - E(X_i)) \xrightarrow{a.s.} 0.$$

(D) The Marcinkiewicz strong law of large numbers: If $X_1 \ldots, X_n$ are iid and $E|X_1|^\delta < \infty$ for $\delta \in (0,1)$, then

$$\frac{1}{n^{1/\delta}}\sum_{i=1}^{n} |X_i| \xrightarrow{a.s.} 0.$$

(E) The Law of the Iterated Logarithm (LIL) (see P.29, Pagès (2018)) provides an a.s. rate of convergence, namely

$$\overline{\lim}_n \sqrt{\frac{n}{2\log(\log(n))}}(\bar{x}_n - \mu) = \sigma_X;$$
$$\underline{\lim}_n \sqrt{\frac{n}{2\log(\log(n))}}(\bar{x}_n - \mu) = -\sigma_X.$$

Theorem 1.2 *(Berry-Esseen Theorem)*
Let $\{X, X_1, \ldots, X_n\}$ be the iid with mean μ and variance σ^2 and $E|X|^3 < \infty$. Let $G_n(x)$ be the cdf of $\sqrt{n}\bar{X}_n$ and $\Phi(x)$ be the cdf of $N(0,1)$, Then

$$G_n(x) = \Phi(x) + O\left(\frac{1}{\sqrt{n}}E(|X|^3)\right), \ \sup_x |G_n(x) - \Phi(x)| \le \frac{33}{4}\frac{E|X-\mu|^3}{\sigma^3\sqrt{n}}.$$

See Serfling (1980), p. 33.

Theorem 1.3 *Let $\{\hat{g}_1(T_n), \ldots, \hat{g}_m(T_n)\}$ be a consistent estimator of $\{g_1(\boldsymbol{\theta}), \ldots, g_m(\boldsymbol{\theta})\}$ based on a statistic T_n. Let the function $h(v_1, \ldots, v_m)$ continue at $(g_1(\boldsymbol{\theta}), \ldots, g_m(\boldsymbol{\theta}))$. Then $h(\hat{g}_1(T_n), \ldots, \hat{g}_m(T_n))$ is a consistent estimator of $(g_1(\boldsymbol{\theta}), \ldots, g_m(\boldsymbol{\theta}))$.*

Example 1.4 *Let $X_1, \cdots, X_n$ be a random sample from $F(x)$ with a finite mean μ and a finite variance σ^2.*

1) The sample mean $\bar{X}_n = \frac{1}{n}\sum_{i=1}^n X_i$ is a weakly consistent estimator of μ, that is,

$$\bar{X}_n \xrightarrow{P} \mu.$$

2) Assume $E(X_1^4) < \infty$. By the law of large numbers, we have

$$\begin{aligned} s_n^2 = \frac{1}{n-1}\sum_{i=1}^n (X_i - \bar{X}_n)^2 &= \frac{n}{n-1}\left(\frac{1}{n}\sum_{i=1}^n X_i^2 - \bar{X}_n^2\right) \\ &\xrightarrow{P} 1 \cdot [E(X_1^2) - \mu^2] = \sigma^2. \end{aligned}$$

Hence, the sample variance is the weak consistency of σ^2.

1.4 Moment Estimators

Suppose that a random variable $X \sim F(x;\boldsymbol{\theta})$, where $\boldsymbol{\theta} = (\theta_1, \ldots, \theta_m) \in \Omega$ are unknown parameters. There are many construction methods of statistics for the estimation of *vgv*. The moment estimation (ME) by Pearson (1894) is one of the oldest methods. If X has m different moments (for example, moment about zero or moment about the mean) that have analytic expressions with the m unknown parameters, we can use sample estimators for these m moments and plug the analytic expressions to form a system of equations m. The solution of the unknown parameters is called their *moment estimators* and denotes them by $\hat{\boldsymbol{\theta}}_{ME}$.

Suppose that a random variable $X \sim F(x;\boldsymbol{\theta})$ has finite m moments. Let $x_1 \ldots, x_n$ be a random sample from F and denote its *empirical rth moments* or *sample rth central moments* by

$$m_r(x) = \frac{1}{n}\sum_{i=1}^n x_i^r, \; m_r^c(x) = \frac{1}{n}\sum_{i=1}^n (x_i - \bar{X})^r, r = 1, \ldots, m. \tag{1.23}$$

From the law of large numbers, it is known

$$m_r(x) \xrightarrow{P} \mu_r = E(X^r), \quad m_r^c(x) \xrightarrow{P} \mu_r^c = E(X-\mu)^r, \text{ as } n \to \infty.$$

If there are some relationships

$$g(\boldsymbol{\theta}) = h_{ij}(\mu_i, \mu_j^c), \text{ for some } i, j,$$

we can replace μ_i by $m_i(x)$ and μ_j^c by $m_j^c(x)$ to form a system of equations. The solution of the equations for $g(\boldsymbol{\theta})$ is a moment estimator, denoted by $\hat{g}(\boldsymbol{\theta})_{ME}$.

Example 1.5 *Let X have the density function*

$$p(x;\theta_1,\theta_2) = \frac{\theta_2}{\Gamma(\frac{1+\theta_1}{\theta_2})} x^{\theta_1} e^{-x^{\theta_2}}, x > 0,$$

where $-1 < \theta_1 < \infty, 0 < \theta_2 < \infty$. By a straightforward calculation, one can find

$$\mu_1 = E(X) = \frac{\Gamma(\frac{2+\theta_1}{\theta_2})}{\Gamma(\frac{1+\theta_1}{\theta_2})}, \quad \mu_2 = E(X^2) = \frac{\Gamma(\frac{3+\theta_1}{\theta_2})}{\Gamma(\frac{1+\theta_1}{\theta_2})}.$$

Replace μ_1 *by* $m_1(x)$ *and* μ_2 *by* $m_2(x)$ *in the above formulas, we have a system of two equations*

$$\begin{cases} m_1(x) = \dfrac{\Gamma(\frac{2+\theta_1}{\theta_2})}{\Gamma(\frac{1+\theta_1}{\theta_2})}, \\ m_2(x) = \dfrac{\Gamma(\frac{3+\theta_1}{\theta_2})}{\Gamma(\frac{1+\theta_1}{\theta_2})}. \end{cases} \tag{1.24}$$

The solutions $\hat{\theta}_1$ *and* $\hat{\theta}_2$ *of the equations are the moment estimates of* θ_1 *and* θ_2*, respectively.*

One advantage of moment estimation is that it is easy to implement, but there are some limitations/weaknesses:

(1) The population distribution $F(x)$ must have m finite moments;

(2) Moment estimators are not unique since they depend on the chosen moments;

(3) The system of equations for solving the moment estimators is usually nonlinear. An efficient algorithm is needed to find the solution. A robust numerical algorithm (SNTO) to solve a system of nonlinear equations will be introduced in Section 1.13.

The plug-in principle

The plug-in principle is a simple method to estimate parameters from samples. The plug-in estimate of a parameter $\boldsymbol{\eta} = t(F)$ is defined to be

$$\hat{\boldsymbol{\eta}} = t(\hat{F}).$$

The moment estimator is a particular application of the plug-in method. Plug-in methods are familiar friends in classical statistics (Efron (2003)). An obvious way to estimate some interesting aspect of F, like its mean, median, or quantiles, is to use the corresponding aspect of $\hat{F}$. This is the "plug-in principle".

1.5 Maximum Likelihood Estimators

The maximum likelihood estimation method is a philosophy. The maximum likelihood estimator (MLE) identifies the population parameter values that are most likely to have produced a particular sample of data. Let $X_1, X_2, \cdots, X_n$ be iid with pdf $p(x;\boldsymbol{\theta}),\ \boldsymbol{\theta} \in \Omega$. The parameter $\boldsymbol{\theta}$ is unknown and must be estimated. The following procedure can obtain the maximum likelihood estimation of $\boldsymbol{\theta}$ based on the random sample.

Step 1. Write the likelihood function $L(\boldsymbol{\theta}; X) = \prod_{i=1}^n p(x_i; \boldsymbol{\theta})$.

Step 2. Confirm the parameter space (domain) Ω.

Step 3. Maximization: to find a $\boldsymbol{\theta}^* \in \Omega$ such that $L(\boldsymbol{\theta}^*; X) = \max_{\boldsymbol{\theta}\in\Omega} L(\boldsymbol{\theta}; X)$.

The value $\boldsymbol{\theta}^*$ is called the *maximum likelihood estimator* of $\boldsymbol{\theta}$ (denote it as $\hat{\boldsymbol{\theta}}_{MLE}$ and $L(\hat{\boldsymbol{\theta}}_{MLE}; X)$ is called the *maximum likelihood.* In reality, the logarithm of the likelihood function is more convenient in the maximization step

$$l(\boldsymbol{\theta}; X) = \log L(\boldsymbol{\theta}; X) = \sum_{i=1}^n \log p(x_i, \boldsymbol{\theta}),\ \boldsymbol{\theta} \in \Omega.$$

Step 3 needs some optimization algorithms. SNTO (refer to Section 1.13) is recommended for this purpose.

An advantage of the maximum likelihood estimation is given in the following theorem. We omit "MLE" in the lowercase of $\hat{\boldsymbol{\theta}}_{MLE}$ for short.

Theorem 1.4 *Let $X_1, X_2, \cdots, X_n$ be iid with the pdf $p(x; \boldsymbol{\theta}), \boldsymbol{\theta} \in \Omega$. For a one-to-one function g, let $\eta = g(\boldsymbol{\theta})$ be a parameter of interest. Suppose $\hat{\boldsymbol{\theta}}$ is the MLE of $\boldsymbol{\theta}$. Then $g(\hat{\boldsymbol{\theta}})$ is the MLE of $\eta = g(\boldsymbol{\theta})$.*

Proof: First suppose g is a one-to-one function. The likelihood of interest is $L(g(\theta))$. Because g is one-to-one, we have

$$\max L(g(\boldsymbol{\theta})) = \max_{\eta = g(\boldsymbol{\theta})} L(\eta) = \max_{\eta} L(g^{-1}(\eta)).$$

But the maximum occurs when $g^{-1}(\eta) = \hat{\boldsymbol{\theta}}$; i.e., take $\hat{\eta} = g(\hat{\boldsymbol{\theta}})$. When g is more general, the proof can refer to Zhang and Fang (1982).

The Fisher Information Matrix

The Fisher Information Matrix (FIM) is a good measure of the amount of information on parameters from the sample data. Suppose $p(x; \boldsymbol{\theta})$ is the density function of the object model and $l(\boldsymbol{\theta}; X)$ is the log-likelihood function. The score function is defined as

$$F(\boldsymbol{\theta}|X) = \frac{\partial l(\boldsymbol{\theta}; X)}{\partial \boldsymbol{\theta}} = \sum_{i=1}^{n} \frac{\partial log\ p(x_i; \boldsymbol{\theta})}{\partial \boldsymbol{\theta}}.$$

The expected FIM can be defined as a $m \times m$ matrix:

$$I(\boldsymbol{\theta}|X) = E\left[\frac{\partial l(\boldsymbol{\theta}; X)}{\partial \boldsymbol{\theta} \partial \boldsymbol{\theta}^T}\right]. \tag{1.25}$$

Under some regular conditions, we can prove (Meng (2016))

$$E\left[\frac{\partial l(\boldsymbol{\theta}; X)}{\partial \boldsymbol{\theta}}\right] = \mathbf{0}.$$

And we have

$$I(\boldsymbol{\theta}|X) = \text{Cov}\left[\frac{\partial l(\boldsymbol{\theta}; X)}{\partial \boldsymbol{\theta}}\right].$$

Escobar and Meeker Jr (1994) discussed the Fisher information matrix for normal, extreme value, logistic distributions, and censored data.

1.6 Nonparametric Estimation

In the previous sections, all estimators depend on the type of population distribution. When an estimator is not based on any specific population distribution, it is called *a nonparametric estimator*. In this section, we introduce two important nonparametric estimators: density and quantile estimators.

1.6.1 Density estimation

When the input population has a continuous distribution with an unknown density function $p(x)$, the Parzen-Rosenblatt window method proposed by Rosenblatt (1956) and Parzen (1962) provides a way to estimate the density function based on a set of samples, $x_1, \ldots, x_n$. The estimate function

$$\hat{p}_h(x) = \frac{1}{n}\sum_{i=1}^{n} k_h(x - x_i) = \frac{1}{nh}\sum_{i=1}^{n} k\left(\frac{x - x_i}{h}\right), \tag{1.26}$$

is called the *kernel density estimator*, where $k(\cdot)$ is the kernel function, h is the bandwidth and $k_h(x) = \frac{1}{h}k(x/h)$. One of the most popular kernels is the standard density function $k(x) = \phi(x) = \frac{1}{\sqrt{2\pi}}e^{-\frac{1}{2}x^2}$. In this section, we choose it as the kernel. The n iid samples can be replaced by a set of representative points (RPs), $b_1, \ldots, b_n$ with related probabilities $p_1, \ldots, p_n$. Chapter 2 will introduce several kinds of RPs. In this case formula (1.26) becomes

$$\hat{p}_h(x) = \sum_{i=1}^{n} k_h(x - x_i)p_i = \frac{1}{h}\sum_{i=1}^{n} k(\frac{x - x_i}{h})p_i. \tag{1.27}$$

The choice of the bandwidth is critical. There are many studies on this topic. The user can try several h and choose one such that the corresponding estimate performs well in a certain sense. Some software, like MATLAB, has a function that helps the user find the optimal bandwidth. In our study, we choose h to have a minimum L_2-distance between $\hat{p}_h(x)$ and $p(x)$.

The L_2-distance between two densities $\hat{p}_p(x)$ and $p(x)$, denoted by L2.pdf, is defined as:

$$L_2(p, \hat{p}) = \left\{\int_{-\infty}^{\infty} (p(x) - \hat{p}(x))^2 dx\right\}^{1/2}. \tag{1.28}$$

Sometimes, one can consider the L_2-distance between two cdfs $F(x)$ and $\hat{F}(x)$, denoted by L2.cdf, as follows:

$$L_2(F, \hat{F}) = \left\{\int_{-\infty}^{\infty} (F(x)(x) - \hat{F}(x))^2 dx\right\}^{1/2}. \tag{1.29}$$

Section 2.2.1 will show some examples of density estimation. We find that L2.pdf and L2.cdf may have different performances. As soon as given an estimated pdf $\hat{p}(x)$, one can calculate various estimated moments of X and statistics of the interest.

1.6.2 Order statistics

Let $X_1, \ldots, X_n$ be a random sample from $X \sim F(x)$. Sort them as

$$X_{(1)} \leq X_{(2)} \leq \cdots \leq X_{(n)}, \tag{1.30}$$

where $X_{(1)}$ is the smallest of $\{X_i\}$, $X_{(2)}$ is the second smallest of $\{X_i\}$, ..., $X_{(n)}$ is the largest of $\{X_i\}$. The order statistics are handy in statistical inference, including parametric and nonparametric estimation and hypothesis testing. David (1981) and David and Nagaraja (2003) gave a comprehensive review of the order statistics.

The joint pdf of $\{Y_i = X_{(i)}, i = 1, \ldots, n\}$ is given by

$$g(y_1, \cdots, y_n) = \begin{cases} n!f(y_1)f(y_2)\cdots f(y_n), & -\infty < y_1 \leq \cdots \leq y_n < \infty, \\ 0, & \text{elsewhere.} \end{cases}$$

The cdf of $Y_k = X_{(k)}$ is

$$F_{Y_k}(y) = \sum_{i=k}^{n} \binom{n}{i} F^i(y)[1 - F(y)]^{n-i}.$$

And the pdf of Y_k is given by

$$g_k(y) = \begin{cases} \frac{1}{B(k,n-k+1)}[F(y)]^{k-1}[1 - F(y)]^{n-k}p(y), & -\infty < y < \infty, \\ 0, & \text{elsewhere,} \end{cases}$$

where $B(\alpha, \beta)$ is the beta function with the parameters α and β and $p(\cdot)$ is the pdf of F. It can be shown that

$$\begin{aligned} E(X_{(k)}) &= \frac{1}{B(k, n-k+1)} \int_{-\infty}^{\infty} xF(x)^{k-1}(1 - F(x))^{n-k} dF(x) \\ &= \frac{1}{B(k, n-k+1)} \int_{0}^{1} F^{-1}(y)y^{k-1}(1 - y)^{n-k} dy. \end{aligned} \quad (1.31)$$

The order statistics plays an important role in statistical inference. For example, the order statistics can estimate the quartile Q_q. Especially estimate the median $Q_{1/2}$ of F. Define the sample median of X by

$$\text{med}\{X_1, \cdots, X_n\} = \begin{cases} X_{((n+1)/2)}, & \text{if } n \text{ is odd,} \\ (X_{(n/2)} + X_{(n/2+1)})/2, & \text{if } n \text{ is even.} \end{cases}$$

1.7 Quantile Estimation

The quantile of $F(x)$ has been defined in (1.13). Estimation of population quantiles or percentiles is of great interest. Many authors have studied a q-quantile estimator based on a random sample from F. There are two kinds of estimators: parametric and non-parametric.

1.7.1 Parametric estimate

When the population distribution type $F(x; \boldsymbol{\theta})$ is known with unknown parameters $\boldsymbol{\theta}$, we can obtain an estimate $\hat{\boldsymbol{\theta}}$ of $\boldsymbol{\theta}$ and find the q-quantile estimator of $F(x; \hat{\boldsymbol{\theta}})$ by theoretic or numerical methods.

1.7.2 Nonparametric estimate

A nonparametric estimate of q-quantile is not based on any assumptions for the distribution. In the literature, many methods for nonparametric quantile estimators have been proposed. For example, Harrell and Davis (1982), Kalgh and Lachenbruch (1982), Yang (1985), and Sfakianakis and Verginis (2008).

Sample quantiles have many desirable properties. The most common estimator of Q_q is the sample quantile $S_q = X_{\lfloor nq \rfloor + 1}$, where $\lfloor nq \rfloor$ is the integer part of nq. Several similar quantile estimators are based on one or two-order statistics; see Hyndman and Fan (1996) and Parrish (1990). The main drawback of this kind of quantile estimators is that their variance is relatively large.

Harrell and Davis (1982) proposed a nonparametric estimation based on a linear combination of the order statistics. One traditional estimator of the qth population quantile $F^{-1}(q)$ is a learn combination of $(1-g)X_{(j)} + X_{(j+1)}$, where $(n+1)q = j+g$ and $j = \lfloor (n+1)q \rfloor$. From $E(X_{(k)})$ in (1.31), and $E(X(n+l)q)$ converges to $F^{-1}(q)$ for $q \in (0,1)$, we take as an estimator of $F^{-1}(q)$ something which estimates $E(X(n+1)q)$ whether $(n+1))q$ is an integer or not, i.e.,

$$SQ_p = \frac{1}{B\{(n+1)p,(n+1)(1-p)\}} \int_0^1 F_n^{-1}(y) y^{(n+1)p-1}(1-y)^{(n+1)(1-p)-1} dy,$$

where $F_n(x)$ is the empirical distribution function of the sample. Harrell and Davis (1982) suggested the following estimator:

Step 1. Generate n quasi-random numbers $q_i = (2i-1)/2n$, for $i = 1, \ldots, n$, which are uniformly scattered on the interval $(0,1)$.

Step 2. The quantile estimator of $F^{-1}(p)$ based on a random sample is proposed to be $SQ(p)$ that can be expressed as

$$SQ(p) = \sum_{i=1}^{n} W_{n,i} X_{(i)}, \tag{1.32}$$

where

$$\begin{aligned} W_{n,i} =& \frac{1}{\beta\{(n+1)p,(n+1)(1-p)\}} \int_{(i-1)/n}^{i/n} y^{(n+1)p-1}(1-y)^{(n+1)(1-p)-1} dy \\ =& I_{i/n}\{p(n+1),(1-p)(n+1)\} - I_{(i-1)/n}\{p(n+1),(1-p)(n+1)\}, \end{aligned}$$

and $I_x(a,b)$ denotes the incomplete beta function.

The Harrell-Davis quantile (HD-quantile) estimator offers a significant gain in efficiency with emphasis on small sample results. This estimator is more efficient than the traditional quantile estimator over a wide variety of distributions in general, especially for light-tailed distributions (such as the beta, uniform, normal, exponential, Gumbel (0, 1) distributions). However, it could be less efficient in areas close to the low-density regions like tail areas of heavy-tailed distributions and middle areas between modes of the given bimodal distributions. A review of various estimators of quantiles was conducted, and some new estimates of quantiles were proposed by Navruz and Ozdemir (2020).

1.7.3 SV estimators

Sfakianakis and Verginis (2008) proposed other quantile estimators that again use all of the order statistics and have advantages in certain situations. Let $S_0 = \left(L, X_{(1)}\right), S_1 = \left[X_{(1)}, X_{(2)}\right), \ldots, S_{n-1} = \left[X_{(n-1)}, X_{(n)}\right), S_n = \left[X_{(n)}, U\right)$ are constructed, where $X_{(1)}, \ldots, X_{(n)}$ are order statistics of the sample, U and L are upper and lower bounds for X(which may be $-\infty$ or ∞). The qth quantile of the population, $Q(q)$, is in one of the S_i. Define the random variables:

$$\delta_i = \begin{cases} 1, & X_i \leq Q(q), \\ 0, & X_i > Q(q), \end{cases}$$

where $\delta_i \sim$ Bernoulli(q), i.e., the Bernoulli distribution. Then, their sum $N = \delta_1 + \delta_2 + \cdots + \delta_n$ has a binomial distribution with probability q, supposing δ_i are independent.

So, $P(Q(q) \in S_i) = P(N = i) = B(i; n, q)$, $i = 0, 1, \ldots, n$ Let the random variables $W(q) = Q(q)$ and $W_i(q) = Q_i(q)^\star, i = 0, 1, \ldots, n$, where $Q_i(q)^\star$ is a point estimator of $Q(q)$ conditioned on the event $Q(q) \in S_i, i = 0, 1, \ldots, n$. An estimator of $Q(q)$ is obtained by calculating $Q(q) = \mathrm{E}(W(q))$.

- $Q_{SV1}(q)$ is constructed by setting $Q_i(q)^\star = (X_{(i)} + X_{(i+1)})/2$, $i = 1, \ldots, n-1$, proxy $\mathrm{Q}_n(q)^\star$ by assuming that $Q_n(q)^\star - Q_{n-1}(q)^\star = Q_{n-1}(q)^\star - Q_{n-2}(q)^\star$.
- $Q_{SV2}(q)$ is constructed by setting $Q_i(q)^\star = \sup(Y \in S_i) = X_{(i+1)}$, $i = 0, 1, \ldots, n-1$, proxy $\mathrm{Q}_0(q)^\star$ and $\mathrm{Q}_n(q)^\star$ by assuming that $\mathrm{Q}_0(q)^\star - \mathrm{Q}_1(q)^\star = \mathrm{Q}_1(q)^\star - \mathrm{Q}_2(q)^\star$ and $Q_n(q)^\star - Q_{n-1}(q)^\star = Q_{n-1}(q)^\star - Q_{n-2}(q)^\star$.
- $Q_{SV3}(q)$ is constructed by setting $Q_i(q)^\star = \inf(Y \in \mathrm{S}_i) = X_{(i)}$, $i = 1, 2, \ldots, n$, proxy $\mathrm{Q}_0(q)^\star$ by assuming that $\mathrm{Q}_0(q)^\star - \mathrm{Q}_1(q)^\star = \mathrm{Q}_1(q)^\star - \mathrm{Q}_2(q)^\star$.

Substitute the above equation into $W_i(q) = Q_i(q)^\star$, and find the expectation $Q(q) = \mathrm{E}(W(q))$. After simplification, the following three SV estimators are obtained:

$$\begin{aligned} Q_{SV1}(q) =& \frac{2B(0;n,q)+B(1;n,q)}{2}X_{(1)} + \frac{B(0;n,q)}{2}X_{(2)} - \frac{B(0;n,q)}{2}X_{(3)} \\ &+ \sum_{i=2}^{n-1}\left[\frac{B(i;n,q)+B(i-1;n,q)}{2}X_{(i)}\right] - \frac{B(n;n,q)}{2}X_{(n-2)} \\ &+ \frac{B(n;n,q)}{2}X_{(n-1)} + \frac{2B(n;n,q)+B(n-1;n,q)}{2}X_{(n)}; \end{aligned} \tag{1.33}$$

$$Q_{SV2}(q) = \sum_{i=0}^{n-1} B(i;n,q)X_{(i+1)} + \left(2X_{(n)} - X_{(n-1)}\right)B(n;n,q); \tag{1.34}$$

$$Q_{SV3}(q) = \sum_{i=1}^{n} B(i;n,q)X_{(i)} + \left(2X_{(1)} - X_{(2)}\right)B(0;n,q). \tag{1.35}$$

1.7.4 Revised random sample – QMC-data

The performances of sample moments and likelihood functions highly depend on the quality of the random sample. Can we revise a set of random samples to another set of "data" with a higher representativeness for the underlying distribution $F(x)$? The Quasi-Monte Carlo (QMC) method is one of the powerful methods, which points out that the discrete distribution for the uniform distribution $U(0,1)$ is given by Y_{QMC}. Let $p_1 = \frac{1}{2n}, p_2 = \frac{3}{2n}, \ldots, p_i = \frac{2i-1}{2n}, i = 1, \ldots, n$,

$$\begin{array}{c|cccc} Y_{QMC} & F^{-1}(p_1) & F^{-1}(p_2) & \cdots & F^{-1}(p_n) \\ \hline p & \frac{1}{n} & \frac{1}{n} & \cdots & \frac{1}{n} \end{array} \tag{1.36}$$

In fact, $F^{-1}(p_i)$ is the quantile $Q_{(p_i)}$ and can be estimated by $SQ(p_i)$. Therefore, Y_{QMC} based on a random sample $\boldsymbol{x} = (X_1, \ldots, X_n)$ is estimated by $Y_{QMC|\boldsymbol{x}}$ that is defined as follows

$$\begin{array}{c|cccc} Y_{QMC|\boldsymbol{x}} & z_1 & z_2 & \cdots & z_n \\ \hline p & \frac{1}{n} & \frac{1}{n} & \cdots & \frac{1}{n} \end{array} \tag{1.37}$$

where $\{z_i = SQ(p_i), i = 1, ..., n\}$ are called as *QMC-data* proposed by Li and Fang (2022). They proposed z_i instead of X_i in estimating $\boldsymbol{\theta}$ in the moment estimation and maximum likelihood estimation, which will be introduced in Section 5.7. Yan et al. (2024) employed both HD quantile estimators and three SV estimators to construct QMC-data for the bimodal mixture Weibull distributions and compare their performance in estimation for several statistics.

1.8 Statistical Simulation, Bootstrap, and Resampling

1.8.1 Statistical simulation

The primary objective of statistical analysis is "extracting all the information from the data" to deduce the properties of the population. Statistical simulation plays an essential role in statistical research and applications. The statistical simulation is also called the Monte Carlo method. It takes random samples from a given population using computer software, and then we calculate the statistic T_n of the interest based on a random sample generated by a computer.

Let $X \sim F(x)$ be the population random variable and its cdf $F(x)$ with mean μ and variance σ^2. Let $x_1, \ldots, x_n$ be a random sample from the population, and let $T_n = T(x_1, \ldots, x_n)$ be a *statistic of the interest*, a function of the sample. We are very interested in the distribution, mean, median, variance, quantiles of T_n, etc. Many textbooks in this direction, such as Ross (2013), Johnson (1987), and Zhou et al. (2020), are good references.

Most statistical procedures require some knowledge of the sampling distribution of T_n. Statistical inference in estimation and testing hypothesis is based on a random sample $X_1, \ldots, X_n$ from the population $F(x)$. The *empirical distribution of the sample* is defined as

$$F_n(x) = \frac{1}{n}\sum_{i=1}^{n} I\{X_i \le x\}. \tag{1.38}$$

All the traditional statistical inferences are based on $F_n(x)$. Statisticians try to find the analytic expression of T_n, like the distribution of T_n, the mean and variance of T_n, and the confidence interval of some model parameter $\boldsymbol{\theta}$. Unfortunately, the analytic formula for many statistics of interest is not easy to find. Currently, Monte Carlo simulation methods are efficient ways for statistical inferences.

1.8.2 Bootstrap

Efron (1979) first proposed the so-called *Bootstrap methods* that take random samples (bootstrap sample) from the empirical distribution and find numerical estimators from the bootstrap sample. For example, if one is interested in finding the confidence interval of q-quantile of $X \sim F(x)$. It is not easy to find the corresponding formula, but it is easy to find some numerical estimators using the bootstrap method. The bootstrap is a computer-based method of statistical inference and can answer many real statistical questions without formulas. Hall (1992) commented the bootstrap "Efron's contribution was to marry the bootstrap to modern computational power". Efron (2003), Hall (1992), and Shao and Tu (1995) gave a system of research on the theory of the bootstrap. Some material in this subsection is from their books.

Efron and Tibshirani (1993) pointed out that one of the bootstrap's charms is its direct appreciation of variance, bias, coverage, and other probabilistic phenomena. Since any estimator may have an estimation error, the knowledge of accuracy measures such as the variance, bias, and mean squared error are required when the bootstrap estimator accuracy is considered. These accuracy measures are characteristics of the estimator's sampling distribution. Accuracy measures can also be used to select the best estimator from a class of appropriate estimators.

In the traditional approach, an empirical analog estimates an accuracy measure. The formula for the accuracy measure (or its approximation) is derived from a postulated model. Some weaknesses of the traditional approach:

- Large n: Sometimes it requires a huge sample size n to have accurate variance estimators or another accuracy measure.

- Not Robustness: The theoretical formula or approximation is based on the postulated model. When the model is slightly wrong, the obtained accuracy estimator may not be valid anymore.

- Lack of theory: We have to derive a theoretical formula such as σ_F^2 in (1.40) for different problems. These derivations may be difficult or tedious.

Denote $\mu = E(X)$ and the centered r moment of X by $\mu_r^c = E(X-\mu)^r$, see (1.11). Let us see an example from Shao and Tu (1995).

Example 1.6 *Consider the statistic* $T_n = \bar{x}$ *to estimate the mean* μ. *It is known that* $\text{Var}(T_n) = \frac{1}{n}\sigma^2$. *For another statistic* $T_n = \bar{x}^2$, *we have*

$$\text{Var}(\bar{x}^2) = \frac{4\mu^2\mu_2^c}{n} + \frac{4\mu\mu_3^c}{n^2} + \frac{\mu_4^c}{n^3}. \tag{1.39}$$

For most statistics, obtaining an exact and explicit formula for $\text{Var}(T_n)$. Thus, the traditional approach is to simplify the problem by considering the approximations or asymptotic expansions of $\text{Var}(T_n)$. Under some regularity conditions, we can often establish

$$\lim_{n\to\infty} n[\text{Var}(T_n)] = \sigma_F^2, \tag{1.40}$$

where $\sigma_F^2 = \sigma^2(F)$ is a simple function of F. It is clear that for $T_n = \bar{x}^2$

$$\lim_{n\to\infty} n[\text{Var}(T_n)] = 4\mu^2\mu_2^c,$$

i.e., $\sigma_F^2 = 4\mu^2\mu_2^c$. Definition 1.5 will give another kind of consistency from the discussion in this example.

Efron (2003) asked how far the plug-in principle can be trusted. "Pretty far" is a reasonable summary of current bootstrap research. Simple bootstrap ideas, like resampling from the empirical distribution, work surprisingly well in a surprisingly large catalog of cases, yet there are situations where the plug-in method starts to get worrisome.

- It is usually quite good if the only available information about F comes from the sample $\boldsymbol{x}$.

- Under this circumstance $\hat{\boldsymbol{\theta}} = t(F_n)$ cannot be improved upon as an estimator of $\boldsymbol{\theta} = t(F)$, at least not in the usual asymptotic $(n \to \infty)$ sense of statistical theory.

- The bootstrap, as an application of the plug-in principle, can be used to study the bias and standard error of the plug-in estimate $\hat{\boldsymbol{\theta}} = t(F_n)$.

- The bootstrap's virtue is that it automatically produces biases and standard errors, no matter how complicated the functional mapping $\boldsymbol{\theta} = t(F)$ may be.
- The plug-in principle is less good when there is information about F other than that provided by the sample $\boldsymbol{x}$.
- However, the plug-in principle and the bootstrap can be adapted to parametric families and regression models.

1.8.3 The bootstrap-variance estimation

Bootstrap is a very convenient and appealing tool for statistical analysis, as it includes variance estimation, estimation of the confidence region, and estimation of sampling distribution. A general theory for the bootstrap distribution and variance estimation for the given statistic is presented in this subsection.

Suppose that the data $X_1, \ldots, X_n$ are iid from F, and F is estimated by F_n. Substituting F_n for F, we obtain the *bootstrap variance estimator.*

$$V_{\mathrm{BOOT}} = \mathrm{Var}_* \left[T_n\left(X_1^*, \ldots, X_n^*\right) \mid X_1, \ldots, X_n\right], \tag{1.41}$$

where $\{X_1^*, \ldots, X_n^*\}$ are an iid samples from F_n, i.e., a bootstrap sample; and $\mathrm{Var}_*\left[\cdot \mid X_1, \ldots, X_n\right]$ denotes the conditional variance for given $X_1, \ldots, X_n$.

Equation (1.41) is the theoretical form of the bootstrap variance estimator for T_n. It may not be used directly for practical applications when V_{BOOT} is not an explicit function of $X_1, \ldots, X_n$. If V_{BOOT} is an explicit function of $X_1, \ldots, X_n$, then it is actually a substitution estimator of $\mathrm{Var}\left(T_n\right)$. Here are two examples of V_{BOOT}.

Example 1.7 *Consider $T_n = \bar{X}_n$ and $\hat{F} = F_n$. It is well known that*

$$\mathrm{Var}(\bar{X}_n) = \frac{\sigma^2}{n},$$

and

$$V_{BOOT} = \frac{\sum_{i=1}^{n}\left(X_i - \bar{X}_n\right)^2}{n^2}.$$

Consider $T_n = \bar{X}_n^2$, we have

$$V_{\mathrm{BOOT}} = \frac{4\bar{X}_n^2 \tilde{m}_2^c}{n} + \frac{4\bar{X}_n \tilde{m}_3^c}{n^2} + \frac{\tilde{m}_4^c}{n^3},$$

where $\tilde{m}_k^c = n^{-1}\sum_{i=1}^{n}\left(X_i - \bar{X}_n\right)^k$, the maximum likelihood estimate of $E(X-\mu)^k$.

Example 1.8 *Consider estimation for the sample median of X, i.e., $Q_{1/2}$. Let $X_1, \ldots, X_n$ be a random sample from X and $n = 2m-1$ is odd. Let $T_n = X_{(m)}$ be the sample median.*

Let $X_1^*, \ldots, X_n^*$ be a bootstrap sample from F_n. Then

$$\begin{aligned} p_k &= P_*\left\{X_{(m)}^* = X_{(k)} \mid X_1, \ldots, X_n\right\} \\ &= \sum_{j=0}^{m-1}\binom{n}{j}\frac{(k-1)^j(n-k+1)^{n-j} - k^j(n-k)^{n-j}}{n^n}. \end{aligned} \tag{1.42}$$

When the expression (1.41) is usually complicated and V_{BOOT} *is not an explicit function of* $X_1, \ldots, X_n$*, we can draw* $\{X_{1b}^*, \ldots, X_{nb}^*\}, b = 1, \ldots, B$*, independently from* $\hat{F}$*, conditioned on* $X_1, \ldots, X_n$*. Compute* $T_{n,b}^* = T_n(X_{1b}^*, \ldots, X_{nb}^*)$ *and approximate* V_{BOOT} *by the following Monte Carlo approximation:*

$$V_{\text{BOOT}}^{(B)} = \frac{1}{B} \sum_{b=1}^{B} \left(T_{n,b}^* - \frac{1}{B} \sum_{l=1}^{B} T_{n,l}^* \right)^2 .$$

From the law of large numbers, $V_{BOOT}^{(B)} \xrightarrow{a.s.} V_{BOOT}$ *as* $B \to \infty$*. It should be mentioned that in Efron and Tibshirani (1993)*

$$V_{\text{BOOT}}^{(B)} = \frac{1}{B-1} \sum_{b=1}^{B} \left(T_{n,b}^* - \frac{1}{B} \sum_{l=1}^{B} T_{n,l}^* \right)^2 ,$$

for decreasing the bias.

In essence, the bootstrap method is a mixture of two techniques, that is, the *substitution principle* and the *numerical approximation.*

- If V_{BOOT} is an explicit function of $X_1, \ldots, X_n$, the bootstrap coincides with the traditional substitution approach.
- Otherwise, the bootstrap approximates V_{BOOT} numerically by re-sampling.
- In this book, we name $V_{\text{BOOT}}^{(B)}$ as "re-sampling estimator".

The distribution $\hat{F}$ is the empirical distribution F_n used in the bootstrap method, but we can extend to any approximate discrete distribution as

$$\begin{array}{c|cccc} X & x_1 & x_2 & \cdots & x_n \\ \hline p & p_1 & p_2 & \cdots & p_n \end{array} \tag{1.43}$$

The bootstrap chooses the equal probabilities $p_i = 1/n, i = 1, \ldots, n$. In the general case, the bootstrap sampling is called resampling from an approximate discrete distribution. The next chapter will introduce several resampling methods.

For estimating the variance of T_n, usually $\text{Var}(T_n)$ depends on n and $\text{Var}(T_n) \to 0$ as $n \to \infty$, then the definition of consistency has to be modified slightly.

Definition 1.5 *A variance estimator* V_{BOOT} *of* $\text{Var}(T_n)$ *is said to be*

- weakly consistent *if*

$$V_{BOOT}/\text{Var}(T_n) \xrightarrow{P} 1, \quad \textit{or } V_{BOOT}/\text{Var}(T_n) \xrightarrow{D} 1.$$

- strongly consistent *if*

$$V_{BOOT}/\text{Var}(T_n) \xrightarrow{a.s.} 1.$$

1.9 Bootstrap and Elementary Asymptotic Theory

This section introduces the asymptotic theory for the bootstrap. More details can refer to Bickel and Freedman (1981). Let $X_1, \ldots, X_n$ be iid from $F(x)$ with finite mean μ and variance σ^2. Denote the sample mean by $\bar{x}_n$ and sample variance by

$$s_n^2 = \frac{1}{n-1}\sum_{i=1}^{n}(X_i - \bar{x}_n)^2. \tag{1.44}$$

By the classical Central Limit Theorem, the distribution of the pivotal quantity

$$T_n = \sqrt{n}(\bar{x}_n - \mu)/s_n \xrightarrow{D} N(0,1).$$

Let F_n be the empirical distribution of the sample, and $X_1^*, \ldots, X_m^*$ be the resampling sample from F_n, where m can be different from n. Let

$$T_m^* = \sqrt{m}(\overline{x^*}_m - \bar{x}_n)/s_m^*,$$

where $\overline{x^*}_m = \frac{1}{m}\sum_{j=1} X_j^*$ and $(s_m^*)^2 = \frac{1}{m-1}\sum_{j=1}^{m}(X_j^* - \overline{x^*}_m)^2$.

The resample size m is statistically important. For instance, a sample of size n can be bootstrapped to see what would happen with a sample of size n^2, or $\sqrt{n}$. Furthermore, with m and n free to vary separately, the second-moment condition in Theorem 1.5 becomes quite natural. If m goes to infinity first, then the conditional law of $\sqrt{m}(\overline{x^*}_m - \bar{x}_n)$ tends to $N(0, \sigma^2)$. As n tends to infinity, this converges only if S_n does.

Theorem 1.5 *Suppose $X_1, X_2, \ldots$ are iid from $F(x)$ with finite mean μ and variance σ^2. Along almost all sample sequences $X_l, X_2, \ldots$ given $(X_1, \ldots, X_n)$, as n and m tend to ∞:*

(a) The conditional distribution of $T_m^* = \sqrt{m}(\overline{x^*}_m - \bar{x}_n)$ converges weakly to $N(0, \sigma^2)$ in distribution;

(b) $s_m^* \xrightarrow{P} \sigma$: that is for $\epsilon > 0$,

$$P(|s_m^* - \sigma| > \epsilon | X_1 \ldots, X_n) \xrightarrow{a.s.} 0, \quad \text{as } m \to \infty.$$

Conclusions (a) and (b) imply that the asymptotic distribution of the bootstrap pivot T_n^* coincides with the classical one: convergence to the standard normal distribution holds. The proof will be given in Section 1.14.

Efron considered the possibility of resampling not from F_n, but from some other estimator, called it $\tilde{F}_n$, of F. The argument for Theorem 1.5 shows that this works too, provided $\tilde{F}_n \Rightarrow F$ in Γ_2, that is, $\tilde{F}_n$ gets F almost right in the weak topology and gets the second moment almost right. If $E(X^4) < \infty$, the estimated variance-covariance matrix can be bootstrapped, and so on.

1.10 *U*-Statistics

The U-statistic is a beneficial class of statistics. Let $\mathcal{F}$ be a class of statistical distributions and $F \in \mathcal{F}$. Consider a parametric function of $\theta = \theta(F)$,

$$\theta(F) = E_F[h(X_1, \ldots, X_m)] = \int \cdots \int h(x_1, \ldots, x_m)dF(x_1) \cdots dF(x_m),$$

for some exchangeable function $h = h(x_1, \ldots, x_m)$ with respect to $x_1, \ldots, x_m$, where $X_1, X_2, \ldots, X_m$ be iid from F. The function h is called a "kernel". We want an unbiased estimator for $\theta(F)$.

For any kernel h the corresponding U-statistic for θ on the sample $X_1, \ldots, X_n, n \geq m$, is

$$U_n = U(X_1, \ldots, X_n) = \frac{1}{\binom{n}{m}} \sum_c h(X_{i_1}, \ldots, X_{i_m}), \quad (1.45)$$

where $\sum_c$ denotes the summation over the $\binom{n}{m}$ combinations of m distinct elements from $\{1, \ldots, n\}$. We choose h such that U_n is an unbiased estimate of θ. Table 1.1 lists some useful U-statistics.

TABLE 1.1
U-statistics with their kernels

$\theta(F)$	Kernel	U-statistics
$E(X)$	$h(x) = x$	$U = \bar{x}$
$E(X_1X_2)$	$h(x_1, x_2) = x_1x_2$	$U = \frac{2}{n(n-1)} \sum_{1\leq i<j\leq n} X_iX_j$
$\mathrm{Var}(X)$	$h = \frac{1}{2}(x_1 - x_2)^2$	$U = \frac{2}{n(n-1)} \sum_{1\leq i<j\leq n} h(X_i, X_j) = s^2$
$F(t_0)$	$h(x) = I(x \leq t_0)$	$U = F_n(t_0)$
$E(X^k)$	$h(x) = x^k$	$U = \frac{1}{n} \sum_{i=1}^n x_i^k$
$E_F\|X_1 - X_2\|$	$h(x_1 - x_2) = \|x_1 - x_2\|$	$U = \frac{2}{n(n-1)} \sum_{1\leq i<j\leq n} \|X_i - X_j\|$
$P_F(X_1 + X_2 \leq 0)$	$h = I(x_1 + x_2 \leq 0)$	$U = \frac{2}{n(n-1)} \sum_{1\leq i<j\leq n} I(X_i + X_j \leq 0)$

Example 1.9 *The Wilcoxon 2-sample statistic. Let $\{X_1, \ldots, X_{n_1}\}$ and $\{Y_1, \ldots, Y_{n_2}\}$ be independent samples from F and G, respectively. Then for the probability $\theta(F, G) = P(X \leq Y)$, an unbiased estimator is*

$$U = \frac{1}{n_1n_2} \sum_{i=1}^{n_1} \sum_{j=1}^{n_2} I(X_i \leq Y_j).$$

1.10.1 An optimality property of U-statistics

Let $X_{(n)}$ be the largest order statistics of a sample $X_1, \ldots, X_n$. For a kernel $h(x_1, \ldots, x_m), m \leq n$, the corresponding U-statistics may be expressed as

$$U_n = E[h(X_1, \ldots, X_m)|F_{(n)}],$$

where $F_{(n)}$ is the distribution of $X_{(n)}$.

Theorem 1.6 *Let $S = S(X_1, \ldots, X_n)$ be an unbiased estimator of $\theta(F)$ based on a sample $X_1, \ldots, X_n$. Then the corresponding U-statistic is also unbiased and*

$$\mathrm{Var}_F(U) \leq \mathrm{Var}_F(S),$$

with equality if and only if $P_F(U = S) = 1$.

Proof. Take the kernel associated with S by

$$h(x_1, \ldots, x_n) = \frac{1}{n!} \sum_c S(x_{i_1}, \ldots, x_{i_n}),$$

and the corresponding U-statistic can be expressed as $U = E(S|X_{(n)})$. Therefore,

$$E_F(U^2) = E_F(E^2(S|F_{(n)})) \leq E_F(E(S^2)|F_{(n)})) = E_F(S^2),$$

with equality if and only if $E(S|X_{(n)})$ is degenerate and equals S with probability 1. Since $E_F(U) = E_F(S)$, the proof is completed.

1.10.2 Bootstrapping for U-statistics

Let a function h on R and define

$$\Gamma_h = \left\{ h : \int h^2(x) dG(x) < \infty \right\}, \tag{1.46}$$

where G is a distribution function. If a random sample X_i from G, then the estimator $(1/n)\sum_{i=1}^n h(X_i)$ can be bootstrapped. The relevant notion of convergence is

$$G_\alpha \Rightarrow G \text{ in } \Gamma_h, \text{ iff } \int h^2 dG_\alpha \to \int h^2 dG, \int \theta(h) dG_\alpha \to \int \theta(h) dG \tag{1.47}$$

for all bounded continuous functions θ on the line.

Let $\tilde{F}_n$ be an estimator of F, where $F \in \Gamma_h$. Consider bootstrapping $(1/n)\sum_{i=1}^n h(X_i)$ resampling from $\tilde{F}_n$. What is needed is the analog of the strong law of large numbers,

$$\int v(x) d\tilde{F}_n(x) \to \int v(x) dF(x), \ a.s. \tag{1.48}$$

whenever $\int |v(x)| dF(x) < \infty$. For the parametric bootstrap, $\tilde{F}_n = F_{\hat{\theta}_n}$, where θ is the unknown parameter in the distribution $F(x, \theta)$ and $\hat{\theta}$ is an estimator of θ based on the random sample. Then (1.48) holds when $F = F_{\theta_0}$ is strongly consistent and the map $\theta \to \int v(x) dF_\theta(x)$ is continuous at θ_0 whenever $\int |v(x)| dF_{\theta_0}(x) < \infty$.

1.10.3 Bootstrapping for U-statistics with two-dimensional kernel

Considered a U-statistic with a two-dimensional kernel

$$U_n = \frac{1}{\binom{n}{2}} \sum_{1 \le i < j \le n} h(X_i, X_j)$$

under the condition

$$Eh^2(X_1, X_2) < \infty, \ Eh^2(X_1, X_1) < \infty,$$

where $X_1, \ldots, X_n$ are iid from $F(x)$. Shi (1986) considered a statistic

$$T(X_1, \ldots, X_n; F) = \sqrt{n}(U_n - \theta),$$

where $\theta = E[h(X_1, X_2)] = E(U_n)$, and he gave the following theorem.

Theorem 1.7 *Under the above notation and assumptions:*

$$E|h(X_1, X_1)| < \infty, \ E[h^2(X_1, X_2)] < \infty, \tag{1.49}$$

$$\sigma_g^2 = E(g^2(X_1)) > 0, g(X_1) = E[h(X_1, X_2) - \theta | X_1], \tag{1.50}$$

then

$$\sup_x |P^*(\sqrt{n}(U_n^* - \theta^*) \le x) - P(\sqrt{n}(U_n - \theta) \le x)| \to 0 \ \ a.s., \tag{1.51}$$

where

$$U_n^* = \frac{1}{\binom{n}{2}} \sum_{1 \le i < j \le n} h(Y_i, Y_j), \ \theta^* = E^* U_n^* = \frac{1}{n^2} \sum_{i,j=1}^n E^* h^2(X_1, X_2),$$

and the notation * *means that the operator is under* F_n.

1.11 Quasi-Monte Carlo Methods

Many statistical problems, such as in statistical inference, statistical computation, geometric probability, experimental design, and optimization need low-discrepancy sequences/sets on a specific domain. These problems may involve the following characteristics:

(a) multiple variables,
(b) non-normality,
(c) non-linearity,
(d) computation complexity to be an NP-Hard complex problem,
(e) the objective function to be multi-extremal, and
(f) underlying model to be complicated.

The treatment of such problems requires more powerful methods. The number-theoretic (NTM) or Quasi-Monte Carlo (QMC) methods are such methods. These methods have been popularly used in high-dimension integration (Hua and Wang, 1981; Niederreiter, 1992; Pagès, 2018), experimental design (Fang and Wang, 1994; Fang et al., 2018), geometric probability, and others. These methods define low-discrepancy sequences on a unit cube $C^p = [0,1]^p$ instead of iid random samples on C^p in simulation and high-dimensional numerical integration. This section introduces generation algorithms of low-discrepancy sequences on C^p and other domains.

1.11.1 Brief introduction to QMC

Definition 1.6 *Let $\mathcal{P} = \{\boldsymbol{x}_1, \ldots, \boldsymbol{x}_n\}$ be a set of points in $\mathbb{R}^p$ and $F_n(\boldsymbol{x})$ or $F_{\mathcal{P}}(\boldsymbol{x})$ be its empirical distribution. Then,*

$$D_F(n, \mathcal{P}) = \sup_{\boldsymbol{x} \in \mathbb{R}^p} |F_n(\boldsymbol{x}) - F(\boldsymbol{x})| \tag{1.52}$$

is said to be the F-discrepancy of $\mathcal{P}$ with respect to $F(\boldsymbol{x})$, where $F(\boldsymbol{x})$ is the cdf of the underlying distribution on $\mathbb{R}^p$.

The star L_p-discrepancy for the uniform distribution on $C^p = [0,1]^p$ is defined as

$$D_p^*(\mathcal{P}) = \left\{ \int_{C^p} |F_{\mathcal{P}}(\boldsymbol{x}) - F(\boldsymbol{x})|^p d\boldsymbol{x} \right\}^{1/p}, \; p > 0, \tag{1.53}$$

where $F(\boldsymbol{x})$ is the cdf of $U(C^p)$ and $F_{\mathcal{P}}$ is the empirical distribution of $\mathcal{P}$.

Under the F-discrepancy, Fang et al. (1994b) defined the F-optimum set of points with respect to $F(\boldsymbol{x})$ as follows.

Definition 1.7 *For given $F(\boldsymbol{x})$ and n, a set $\mathcal{P}^* = \{\boldsymbol{x}_1^*, \ldots, \boldsymbol{x}_n^*\}$ is called the set of optimum representative points with respect to $F(\boldsymbol{x})$ if*

$$D_F^*(n, \mathcal{P}^*) = \min_{\mathcal{P}} D_F^*(n, \mathcal{P}), \tag{1.54}$$

where $\mathcal{P} = \{\boldsymbol{x}_j, j = 1, \cdots, n\}$.

Fang et al. (1994a) gave the following lemma, which shows that there always exists a set of F-optimum set of points with respect to $F(X)$ where X has a continuous univariate distribution.

Lemma 1.1 *Let $F(x)$ be a continuous distribution function and let $F^{-1}(X)$ be its inverse function. Then the set $\{F(^{-1}((2i-1)/(2n)), i = 1, \cdots, n\}$ with F-discrepancy $\frac{1}{2n}$ is the F-optimum set of points with respect to $F(X)$.*

Definition 1.8 *Let $\mathcal{P}_n$, $n \in \mathbb{N}^+$, where $\mathbb{N}^+$ is an infinite subset of nonnegative integers, be a sequence of sets of points in $\mathbb{R}^p$ with a certain structure, and let $F(\boldsymbol{x})$ be a p-dimensional distribution function. If $D_F(n, \mathcal{P}_n) = O(n^{-1+\varepsilon})$ as $n \to \infty$, where $0 < \varepsilon < 1/2$, the points of $\mathcal{P}_n$ are called representative points (RPs) or quasi-random F-numbers of $F(\boldsymbol{x})$.*

The set of quasi-random F-numbers on the distribution $F(\boldsymbol{x})$ implies the set of points is representative of $F(\boldsymbol{x})$. When the random vector $\boldsymbol{X} \sim U(C^p)$, uniformly distributed on $C^p = [0,1]^p$, the quasi-random F-numbers reduce to the common *quasi-random numbers.* There are some methods to generate the quasi-random numbers sequence on C^p, such as the good lattice point (*glp*) method and the good point (*gp*) method. More methods can refer to Niederreiter (1992) and Pagès (2018). Among all those methods, the *glp* method has been widely applied to generate the quasi-random numbers on C^p. Let us briefly introduce the *glp* set obtained by good lattice points modulo n.

Definition 1.9 *Let $(n; h_1, \cdots, h_p)$ be a vector of positive integer satisfying $1 \le h_i \le n$, $h_i \neq h_j$ for $i \neq j$, and $p < n$. Let*

$$x_{ki} = \left\lfloor \frac{2kh_i - 1}{2n} \right\rfloor, \quad i = 1, \cdots, p, \; k = 1, \cdots, n \tag{1.55}$$

where $\lfloor x \rfloor$ is the fractional part of x. Then, the set $\mathcal{P} = \{\boldsymbol{x}_k = (x_{k1}, \cdots, x_{kp}), k = 1, \ldots, n\}$ is called the set of lattice points of the generating vector $(n; h_1, \cdots, h_p)$. If the sequence $\mathcal{P}$ has discrepancy $D(n, \mathcal{P}) = O(n^{-1+\varepsilon})$, the set $\mathcal{P}$ is called a glp set.

Hence, by (1.55), we can easily generate the quasi-random number for $U(C^p)$ if we can find a suitable generating vector $(n; h_1, \cdots, h_p)$. For $2 \le p \le 18$, the reader can find many choices of generating vectors for different values of n's in Fang and Wang (1994).

Qi et al. (2018) used the linear level permutation of the glp sets to define the *generalized glp* (GGLP) that performs better in statistical simulation. Lin et al. (2024) further studied GGLP under more measures, including the Kullback-Leibler (KL) divergence and Frobenius distance between two sample matrices $\boldsymbol{\Sigma}_1$ and $\boldsymbol{\Sigma}_2$ by

$$\|\boldsymbol{\Sigma}_1 - \boldsymbol{\Sigma}_2\|_F = trace(\boldsymbol{\Sigma}_1 - \boldsymbol{\Sigma}_2). \tag{1.56}$$

They applied a modified threshold accepting algorithm proposed by Fang et al. (2017) for generating more glp sets by the GGLP method with better performance.

1.11.2 Measure of uniformity

If $F(\boldsymbol{x})$ is the uniform distribution on the unit cube C^p, QMC methods have various uniformity measures. In the past, the most popular used measure was the star discrepancy defined in (1.52) proposed by Weyl (1916) and star L_p-discrepancy defined in (1.53). The star discrepancy is the star L_p-discrepancy with $p = \infty$ and has been used in goodness of fit tests as the Kolmogorov-Smirnov statistic. Any set $\mathcal{P} = \{\boldsymbol{x}_1, \ldots, \boldsymbol{x}_n\} \subset C^p$ is associated with an $n \times p$ matrix, $X_{\mathcal{P}} = (x_{ki})$. The L_p-star discrepancy is invariant to permutation of rows and columns of $X_{\mathcal{P}}$, but it is not invariant if the cube $\mathcal{P}$ is rotated by mapping x_{ki} to $1 - x_{ki}$. When n is small, the star discrepancy is not sensitive enough, while the L_2-star discrepancy ignores the differences between $F_{\mathcal{P}}(\boldsymbol{x})$ and $F(\boldsymbol{x})$ in any low-dimensional manifold. It is easy to find many sets of points whose L_2-star discrepancies have unreasonable results.

Due to the above weakness of the star L_p-discrepancy, Hickernell (1998) and Hickernell (2000) suggested several modified versions of the star L_p-discrepancy by the reproducing kernel Hilbert space, such as the centered L_p-discrepancy (CD) and the wrap-around L_p-discrepancy (WD). Many authors have found that CD and WD have some attractive properties and can be used in several kinds of experimental designs. For a set of points $\mathcal{P} = \{\boldsymbol{x}_1, \ldots, \boldsymbol{x}_n\} = (x_{ij})_{n\times s}$, the computational formulas for CD and WD are given as follows,

$$(CD(\mathcal{P}))^2 = \left(\frac{13}{12}\right)^s - \frac{2}{n}\sum_{k=1}^{n}\prod_{j=1}^{s}\left[1 + \frac{1}{2}|x_{kj} - 0.5| - \frac{1}{2}|x_{kj} - 0.5|^2\right]$$
$$+\frac{1}{n^2}\sum_{k=1}^{n}\sum_{j=1}^{n}\prod_{i=1}^{s}\left[1 + \frac{1}{2}|x_{ki} - 0.5| + \frac{1}{2}|x_{ji} - 0.5| - \frac{1}{2}|x_{ki} - x_{ji}|\right],$$

and

$$(WD(\mathcal{P}))^2 = \left(\frac{4}{3}\right)^s + \frac{1}{n^2}\sum_{k=1}^{n}\sum_{j=1}^{n}\prod_{i=1}^{s}\left[\frac{3}{2} - |x_{ki} - x_{ji}|(1 - |x_{ki} - x_{ji}|)\right].$$

Zhou et al. (2013) proposed the mixture discrepancy (MD), which combines the advantages of CD and WD to overcome their shortcomings. The computational formula of MD is given as follows:

$$\mathrm{MD}^2(\mathcal{P}) = \left(\frac{19}{12}\right)^s - \frac{2}{n}\sum_{i=1}^{n}\prod_{j=1}^{s}\left[\frac{5}{3} - \frac{1}{4}|x_{ij} - \frac{1}{2}| - \frac{1}{4}|x_{ij} - \frac{1}{2}|^2\right]$$
$$+\frac{1}{n^2}\sum_{i=1}^{n}\sum_{k=1}^{n}\prod_{j=1}^{s}\left[\frac{15}{8} - \frac{1}{4}|x_{ij} - \frac{1}{2}| - \frac{1}{4}|x_{kj} - \frac{1}{2}| - \frac{3}{4}|x_{ij} - x_{kj}| + \frac{1}{2}|x_{ij} - x_{kj}|^2\right].$$

1.12 NT-Net on a Finite Domain

Many statistical problems need low-discrepancy sequences/sets on a finite region $G \subset \mathbb{R}^p$, as NT-net on G. To solve the difficulty in generating an NT-net on any G, Wang and Fang (1990a,b) proposed a way for generating NT-nets on several useful G by *transformation methods via stochastic representation.*

1.12.1 Transformation methods via stochastic representation

Let us see an example first to help you understand this method.

Example 1.10 *Let* $\boldsymbol{X} \sim N_2(\boldsymbol{0}, \boldsymbol{I}_2)$. *The Box-Muller method for the generation of this distribution is based on the following stochastic representation (SR):*

$$\begin{bmatrix} X_1 \\ X_2 \end{bmatrix} \stackrel{d}{=} (-2\log(U_1))^{1/2}\begin{bmatrix} \cos(2\pi U_2) \\ \sin(2\pi U_2) \end{bmatrix} = \boldsymbol{T}(\boldsymbol{U}), \tag{1.57}$$

where U_1 *and* U_2 *are iid* $U(0,1)$, $\boldsymbol{U} = (U_1, U_2)$ *and*

$$\boldsymbol{T}(\boldsymbol{U}) = \begin{bmatrix} T_1(U_1, U_2) \\ T_2(U_1, U_2) \end{bmatrix} = \begin{bmatrix} (-2\log(U_1))^{1/2}\cos(2\pi U_2) \\ (-2\log(U_1))^{1/2}\sin(2\pi U_2) \end{bmatrix}.$$

Suppose the random vector $\boldsymbol{X} \in \mathbb{R}^p$ has a stochastic representation

$$\boldsymbol{X} \stackrel{d}{=} \boldsymbol{g}(\boldsymbol{Y}) = \begin{bmatrix} g_1(Y_1, \ldots, Y_q) \\ \vdots \\ g_p(Y_1, \ldots, Y_q) \end{bmatrix},$$

for some random vector $\boldsymbol{Y} \in \mathbb{R}^q$ and the measurable functions $g_1 \ldots, g_p$, and suppose that there is a way to find random samples from $\boldsymbol{Y}$. Let $\boldsymbol{y}_1, \ldots, \boldsymbol{y}_n$ be a random samples from $\boldsymbol{Y}$, then $\boldsymbol{x}_i = \boldsymbol{g}(\boldsymbol{y}_i), i = 1, \ldots, n$ are random samples from $\boldsymbol{X}$.

Suppose $\boldsymbol{Y} = (Y_1, \cdots, Y_q)^T$ has independent components $Y_1, \cdots, Y_q$, where each Y_i has a continuous distribution function F_{Y_i}. By the inverse transformation method, $F_{Y_i}^{-1}(u)$ is a random sample from Y_i where u is a random number, i.e., $u \sim U(0, 1)$. For this case, Wang and Fang (1990a,b) proposed the SR algorithm as follows.

Algorithm 1.1 (SR algorithm)

Step 1: Generate q random numbers $u_1, \ldots, u_q$.

Step 2: Calculate y_i by $F_{y_i}^{-1}(u_i)$ for $i = 1, \ldots, q$.

Step 3: Produce and deliver $\boldsymbol{x} = \boldsymbol{g}(y_1, \ldots, y_q)$ as a random sample from the population of $\boldsymbol{X}$.

Step 4: Repeat the above steps k times, and we can output a set of k random samples from $\boldsymbol{X}$.

Following the above SR algorithm, we can generate another kind of representative points of $F(\boldsymbol{x})$. The following are denoted by number-theoretic method representative points (NTM-RPs) or quasi-Monte Carlo representative points (QMC-RPs).

Suppose that random vector $\boldsymbol{X} \stackrel{d}{=} \boldsymbol{g}(\boldsymbol{Y}) = \boldsymbol{g}(Y_1, \ldots, Y_q)$, where $Y_1, \cdots, Y_q$ are independent. Let $\{\boldsymbol{c}_i, i = 1, \ldots, k\}$ be a set of quasi-random numbers, and let $\boldsymbol{x}_i = \boldsymbol{g}(\boldsymbol{c}_i), i = 1, \ldots, k$. Then $\{\boldsymbol{x}_1, \ldots, \boldsymbol{x}_k\}$ is called a set of quasi-random F-numbers to $\boldsymbol{X}$. Especially when $F(\boldsymbol{x})$ is the uniform distribution on a region G in $\mathbb{R}^p$, its set of NTM-RPs is called NT-net on G and is useful in a study on the geometric probability.

The SR algorithm can be extended to generate multivariate distributions.

Algorithm 1.2 (QMC-SR algorithm)

Step 1: Generate a set of quasi-random numbers on C^q denoted as $\{\boldsymbol{c}_i, i = 1, \ldots, k\}$ by some algorithm;

Step 2: Find $\{\boldsymbol{y}_i, i = 1, \ldots, k\}$ by the transformation method based on $\{\boldsymbol{c}_i, i = 1, \ldots, k\}$;

Step 3: Let $\boldsymbol{x}_i = \boldsymbol{g}(\boldsymbol{y}_i), i = 1, \ldots, k$;

Step 4: Deliver $\{\boldsymbol{x}_i\}$ as a set of quasi-random F-numbers of $\boldsymbol{X} = g(Y_1, \ldots, Y_q)$.

1.12.2 Some NT-nets

When $F(\boldsymbol{x})$ is the uniform distribution on a finite region G in $\mathbb{R}^p$, a set of NTM-RPs of F is called NT-net on G that are useful in the geometric probability study. When G is a rectangle including a unit cube

$$[\boldsymbol{a}, \boldsymbol{b}] = [a_1, b_1] \times [a_2, b_2] \times \cdots \times [a_p, b_p], \text{ a unit cube } C^p = [0, 1]^p. \tag{1.58}$$

Wang and Fang (1990a,b) and Fang and Wang (1994) found the transformations for the following domains with $r = 1$:

(A) a ball or a sphere in projection pursuit and geometric probability

$$B^p(r) = \{\boldsymbol{x} : \boldsymbol{x}^T\boldsymbol{x} \le r^2\}, S^p(r) = \{\boldsymbol{x} : \boldsymbol{x}^T\boldsymbol{x} = r^2\}, \tag{1.59}$$

(B) a simplex with different shapes in order statistics, experiments with mixtures and reliability,

$$\begin{aligned} A^p(r) &= \{\boldsymbol{x} : 0 \le x_1 \le x_2 \cdots \le x_p \le r\}, \\ V^p(r) &= \{\boldsymbol{x} : \boldsymbol{x} \in R^p_+, x_1 + \ldots + x_p \le r, r > 0\}, \\ T^p(r) &= \{\boldsymbol{x} : \boldsymbol{x} \in R^p_+, x_1 + \ldots + x_p = r, r > 0\}. \end{aligned} \tag{1.60}$$

The notation A^p is used for $A^p(1)$ and similarly for others.

From now on, we denote the uniform distribution on a finite manifold $G \in \mathbb{R}^p$ by $U(G)$. How can we generate a low-discrepancy set on these domains? The stochastic representation method is useful (Wang and Fang (1990a,b)). Most methods in Monte Carlo simulation can be applied to QMC to generate a low-discrepancy set. Fang and Wang (1994) proposed a way to generate NT-nets on these domains based on the following theory.

(A) Generation for NT-net on C^p or a rectangle.

There are several ways, such as the good lattice point method (Definition 1.9), the good point method, the Halton sequence, and its various versions, (t, s)-sequences and (t, m, s)-nets, etc., to generate low-discrepancy sequences/sets on a unit cube (see, e.g., Hua and Wang (1981) and Niederreiter (1992)). For simplicity, a low-discrepancy set on a finite domain G is called as a net on G in the literature, for example, the (t, m, s)-net by Niederreiter (1992) and the NT-net by Fang and Wang (1994). By linear transformation, it is straightforward to find an NT-net on $[\boldsymbol{a}, \boldsymbol{b}]$ from an NT-net on C^p.

(B) Generation NT-net on A^p.

The domain A^p defined in (1.60) is related to the order statistics with many applications. Let $\boldsymbol{X} = (X_1, \ldots, X_p)^T \sim U(A^p)$. Consider the transformation

$$\begin{cases} X_1 = \phi_1\phi_2\cdots\phi_p \\ X_2 = \phi_2\cdots\phi_p \\ \cdots \\ X_p = \phi_p \end{cases} \tag{1.61}$$

where $(\phi_1, \ldots, \phi_p) \in C^p$ are independent with the respective cdf and pdf of ϕ_i as ϕ^i and $i\phi^{i-1}$, for $i = 1, \ldots, p$. From the transformation (1.61) and the inverse transformation methods, we can obtain

$$\begin{cases} x_{i1} = c_{i1}c_{i2}^{1/2}\cdots c_{ip}^{1/p}, \\ x_{i2} = c_{i2}^{1/2}\cdots c_{ip}^{1/p}, \\ \cdots \\ x_{ip} = c_{ip}^{1/p}. \end{cases} \tag{1.62}$$

Now the set $\{\boldsymbol{x}_i = (x_{i1}, \ldots, x_{ip}), i = 1, \ldots, n\}$ is an NT-net on A^p.

(C) Generation for NT-net on B^p.

Let $\boldsymbol{X} = (X_1, \ldots, X_p)^T \sim U(B^p)$, the uniform distribution on the unit ball in $\mathbb{R}^p$. The use of the spherical coordinate transformation

$$\begin{cases} X_j = \phi_1 S_2\cdots S_j C_{j+1}, j = 1, \ldots, p-1 \\ X_p = \phi_1 S_2\cdots S_{p-1}S^p, \end{cases} \tag{1.63}$$

where

$$S_i = \sin(\pi\phi_i), C_i = \cos(\pi\phi_i), i = 1, \ldots, p-1,$$
$$S_p = \sin(2\pi\phi_p), C_p = \cos(2\pi\phi_p),$$

in which $(\phi_1, \ldots, \phi_p) \in C^p$ are independent with the respective pdf of ϕ_j

$$p_j(\phi) = \begin{cases} p\phi^{p-1} & \text{if } j = 1, \\ \frac{\pi}{B(\frac{1}{2}, \frac{p-j+1}{2})}(\sin(\pi\phi))^{p-j} & \text{if } j = 2, \ldots, p, \end{cases}$$

where $B(a, b)$ is the beta function.

From the transformation (1.63) and the inverse transformation methods, we have

$$\begin{cases} x_{i1} = b_{i1}C_{i2}, \\ x_{ij} = b_{i1}\prod_{k=2}^{j} S_{ik}C_{i,j+1}, j = 2, \ldots, p-1, \\ x_{ip} = b_{ip}\prod_{k=2}^{p} S_{ik}, \end{cases} \tag{1.64}$$

where $b_{i1} = c_{i1}^{1/p}, b_{ik} = F_i^{-1}(c_{ik}), i = 1, \ldots, n, k = 2, \ldots, p$, and the cdf F_i^{-1} is the distribution function corresponding to the pdf in the following

$$F_i(\phi) = \begin{cases} \phi^p, & i = 1, \\ \frac{\pi}{B(1/2, (p-i+1)/2)} \int_0^\phi (\sin(\pi x))^{p-i} dx, & i = 2, \ldots, p. \end{cases}$$

Now the set $\{\boldsymbol{x}_i = (x_{i1}, \ldots, x_{ip}), i = 1, \ldots, n\}$ is an NT-net on B^p.

In general, we need to use numerical calculation for obtaining $b_{ik} = F_i^{-1}(c_{ik})$. When $p = 2, 3$, we can find the simple formulas as follows. For example, for $p = 3$, the three cdfs are

$$\begin{cases} F_1(\phi) = \phi^3, \\ F_2(\phi) = \frac{1}{2}(1 - \cos(\pi\phi), \\ F_3(\phi) = \phi. \end{cases}$$

Let $\{\boldsymbol{c}_i = (c_{i1}, c_{i2}, c_{i3}), i = 1, \ldots, n\}$ be an NT-net on C^3, we can find

$$b_{i1} = c_{i1}^{1/3}, b_{i2} = \frac{1}{\pi}\arccos(1 - 2c_{i2}), b_{i3} = c_{i3}.$$

And we have

$$\begin{cases} x_{i1} = c_{i1}^{1/3}\cos(\pi\frac{1}{\pi}\arccos(1 - 2c_{i2})) = c_{i1}^{1/3}(1 - 2c_{i2}), \\ x_{i2} = b_{i1}\sin(\pi b_{i2})\cos(2\pi b_{i3}) = 2\sqrt{c_{i2}(1 - c_{i2})}c_{i1}^{1/3}\cos(2\pi c_{i3}), \\ x_{i3} = b_{i1}\sin(\pi b_{i2})\sin(2\pi b_{i3}) = 2\sqrt{c_{i2}(1 - c_{i2})}c_{i1}^{1/3}\sin(2\pi c_{i3}). \end{cases} \tag{1.65}$$

Now $\{\boldsymbol{x}_i = (x_{i1}, x_{i2}, x_{i3}), i = 1, \ldots, n\}$ is an NT-net on B_3.

(D) Generation for NT-net on $U(S^p)$.

Let $\boldsymbol{X} = (X_1, \ldots, X_p)^T \sim U(S^p)$, the uniform distribution on the sphere of the unit ball in $\mathbb{R}^p$. The distribution $U(S^p)$ has played an important role in the elliptical distributions. (6.7). Similarly, consider the spherical coordinate transformation

$$\begin{cases} X_j = S_1 \cdots S_{j-1}C_j, j = 1, \ldots, p-1 \\ X_p = S_1 \cdots S_{p-1}, \end{cases} \tag{1.66}$$

where

$$S_i = \sin(\pi\phi_i), C_i = \cos(\pi\phi_i), i = 1, \ldots, p-2,$$
$$S_{p-1} = \sin(2\pi\phi_{p-1}), C_{p-1} = \cos(2\pi\phi_{p-1}),$$

in which $(\phi_1, \ldots, \phi_{p-1}) \in C^{p-1}$ are independent with the respective pdf of ϕ_j

$$p_j(\phi) = \frac{\pi}{B(\frac{1}{2}, \frac{p-j}{2})}(\sin(\pi\phi))^{p-j-1} \quad \text{if } j = 1, \ldots, p-1. \tag{1.67}$$

It is easy to generate an NT-net on a rectangle $[\boldsymbol{a}, \boldsymbol{b}] \in R^{p-1}$ from an NT-net on C^{p-1}. Section 6.3.1 will introduce an efficient algorithm for NT-net on S^p.

(E) Generation for NT-net on V^p.

Let $\boldsymbol{X} = (X_1, \ldots, X_p)^T \sim U(V^p)$, where V^p is the unit L_1-ball in $\mathbb{R}^p$ defined in (1.60). Consider the transformation

$$\begin{cases} X_i = \phi_1 \prod_{j=2}^{i} S_j^2 C_{i+1}^2, i = 1, \ldots, p-1, \\ X_p = \phi_1 \prod_{j=2}^{p} S_j^2, \end{cases} \tag{1.68}$$

where

$$S_i = \sin\left(\frac{1}{2}\pi\phi_i\right), \; C_i = \cos\left(\frac{1}{2}\pi\phi_i\right), i = 1, \ldots, p.$$

$\phi_1, \ldots, \phi_p$ are mutually independent with pdf

$$p_j(\phi) = \begin{cases} p\phi^{p-1}, & j = 1, \\ \pi(s-j+1)\left(\sin(\frac{\pi\phi}{2})\right)^{2p-2j+1} \cos(\frac{\pi\phi}{2}), & j = 2, \ldots, p. \end{cases} \tag{1.69}$$

From the transformation (1.68) and the inverse transformation method, we can find a similar way to generate an NT-net on V^p based on an NT-net on C^p as follows:

$$\begin{cases} x_{ij} = c_{i1}^{1/p} \prod_{k=2}^{j} c_{ik}^{1/(p-k+1)}(1 - c_{i,j+1}^{1/(p-j)}), j = 1, \ldots, p-1, \\ x_{ip} = c_{i1}^{1/p} \prod_{k=2}^{p} c_{ik}^{1/(p-k+1)}, \quad i = 1, \ldots, n. \end{cases} \tag{1.70}$$

When $p = 2$ we have

$$\begin{cases} x_{i1} = \sqrt{c_{i1}}(1 - c_{i2}), \\ x_{i2} = \sqrt{c_{i1}} c_{i2}, \quad i = 1, \ldots, n. \end{cases} \tag{1.71}$$

(F) Generation for NT-net on T^p.

Let $\boldsymbol{X} = (X_1, \ldots, X_p)^T \sim U(T^p)$. The domain T^p defined in (1.60) is the base of the experiments with mixtures (see Wang and Fang (1996)). Consider the transformation

$$\begin{cases} X_i = \prod_{j=1}^{i-1} S_j^2 C_i^2, i = 1, \ldots, p-1 \\ X_p = \prod_{j=1}^{p-1} S_j^2, \end{cases} \tag{1.72}$$

where

$$S_i = \sin\left(\frac{1}{2}\pi\phi_i\right), \; C_i = \cos\left(\frac{1}{2}\pi\phi_i\right), i = 1, \ldots, p.$$

$\phi_1, \ldots, \phi_p$ are mutually independent with cdf

$$F_j(\phi) = \left(\sin\left(\frac{1}{2}\pi\phi_i\right)\right)^{2(p-j)}, j = 1, \ldots, p-1.$$

From the transformation (1.72) and the inverse transformation method, we can find a similar way to generate an NT-net on T^p based on an NT-net on C^p as follows:

$$\begin{cases} x_{ij} = \prod_{k=1}^{j-1} c_{ik}^{1/(p-k)}(1 - c_{ij}^{1/(p-j)}), j = 1, \ldots, p-1, \\ x_{ip} = \prod_{k=1}^{p-1} c_{ik}^{1/(p-k)}, \quad i = 1, \ldots, n. \end{cases} \tag{1.73}$$

where

$$S_i = \sin\left(\frac{1}{2}\pi\phi_i\right), \quad C_i = \cos\left(\frac{1}{2}\pi\phi_i\right), i = 1, \ldots, p.$$

$\phi_1, \ldots, \phi_p$ are mutually independent with cdf

$$F_j(\phi) = \left(\sin\left(\frac{1}{2}\pi\phi_i\right)\right)^{2(p-j)}, j = 1, \ldots, p-1.$$

1.13 Sequential Number-Theoretic Methods in Optimization

Statistics need powerful algorithms in estimation (the maximum likelihood estimation), experimental design (the optimal design, the uniform design), the non-linear regression analysis, etc. However, there are some difficulties in optimization problems requested in the above studies.

1. The objective function is a multi-variable function with more local optimal points (local extrema);
2. The partial differentials of the objective function don't exist or are difficult to calculate;
3. The domain of the multi-variables is broad.

Besides the quasi-Newton methods and the EM algorithm, the sequential number-theoretic algorithm for optimization (SNTO) proposed by Fang and Wang (1991) is an efficient optimization method. Suppose the objective function $f(\boldsymbol{x})$ is a continuous function defined on a closed and bounded domain D. For example, D is a rectangle (1.58). One wishes to find the global maximum M of f on D and the corresponding maximum point $\boldsymbol{x}^* \in D$, i.e.,

$$M = f(\boldsymbol{x}^*) = \max_{\boldsymbol{x} \in D} f(\boldsymbol{x}). \tag{1.74}$$

Assume a unique maximum point of $f(\boldsymbol{x})$ exists. Let $\mathcal{P}_N = \{\boldsymbol{x}_1, \ldots, \boldsymbol{x}_N\}$ be a set of points uniformly scattered on D in a certain sense, i.e., the F-discrepancy of $\mathcal{P}_N$ in (1.52) has the order of $O(N^{-1}(\log N)^s)$ as $N \to \infty$. Let b_N be the dispersion of $\mathcal{P}_N$ on D defined by

$$b_N(D, \mathcal{P}_N) = \max_{\boldsymbol{x} \in D} \min_{1 \le k \le N} d(\boldsymbol{x}, \boldsymbol{x}_k),$$

where $d(\boldsymbol{x}, \boldsymbol{x}_k)$ is the L_2-distance between $\boldsymbol{x}$ and $\boldsymbol{x}_k$. Fang and Wang (1994) pointed out that $b_N(D, \mathcal{P}_N) \to 0$ with the convergence rate $O(N^{-1/s} \log N)$ that is very slow. Denote the maximum of $f(\boldsymbol{x})$ on $\mathcal{P}_N$ by

$$M_N = \max_{1 \le i \le N} f(\boldsymbol{x}_i) = f(\boldsymbol{x}_N^*).$$

It is known that M_N will converge to M and the corresponding maximum points $\boldsymbol{x}_N^*$ converge to $\boldsymbol{x}^*$ as $N \to \infty$ if $\boldsymbol{x}^*$ is unique. To solve the slow convergence problem, Fang and Wang (1991) proposed a sequential method SNTO. In each step, SNTO reduces the searching domain and uses a set of points uniformly scattered on the reduced domain to search the global maximum. That improves the efficiency of convergence significantly. That is, the convergence speed will be much faster. The following theorem (see Theorems 3 and 4 in Fang et al. (1994a)) supports the main idea of SNTO.

Theorem 1.8 *Suppose that $f(\boldsymbol{x})$ is a continuous function defined on a closed and bounded domain $D = [\boldsymbol{a}, \boldsymbol{b}] \in \mathbb{R}^s$ and has a unique maximum point $\boldsymbol{x}^* \in D$. Let $\{\mathcal{P}_{N_i}\}, N_1 < N_2 < \cdots$, be a sequence of sets on D which are uniformly scattered on D with F-discrepancy $D_F(N_i, \mathcal{P}_{N_i}) = o(1)$ as $i \to \infty$. Let $\boldsymbol{x}_{N_i}^* \in \mathcal{P}_{N_i}$ be a point satisfying*

$$M_{N_i} = f\left(\boldsymbol{x}_{N_i}^*\right) = \max_{\boldsymbol{x} \in \mathcal{P}_{N_i}} f(\boldsymbol{x}).$$

Then $M_{N_i} \to M$ as $i \to \infty$. More precisely, we have $M_{N_i} = M + o\left(D_F(N_i, \mathcal{P}_{N_i})^{1/s}\right)$ and $x_{N_i}^* \to x^*$ as $i \to \infty$ as $\boldsymbol{x}^*$ is unique.

Furthermore, suppose $\Delta f = (\partial f/\partial x_1, \ldots, \partial f/\partial x_s)$ is continuous and bounded on D in the sense of the L_2-norm $||\Delta f|| < C$ with some positive number C. Then

$$M_N \leq M \leq M_N + sCb_N\left(D, \mathcal{P}_N\right),$$

where b_N is the dispersion of $\mathcal{P}_N$ on D.

In literature, many methods have been proposed to generate a set of points uniformly scattered in a high-dimensional rectangle; see Hua and Wang (1981) and Niederreiter (1992). The good lattice points (GLP) method (see Fang and Wang (1994)) and the generalized good lattice (GGLP), (Qi et al., 2018) are among them.

Algorithm 1.3 (SNTO algorithm, Fang and Wang (1991))

Step 0. Initialization: *Set $t = 0, D(0) = D, \boldsymbol{a}(0) = \boldsymbol{a}$, and $\boldsymbol{b}(0) = \boldsymbol{b}$.*

Step 1. Generate an NT-net: *Use a number-theoretic method to generate a set of N_t points $\mathcal{P}^{(t)}$ uniformly scattered on $D(t) = [\boldsymbol{a}(t), \boldsymbol{b}(t)]$.*

Step 2. Compute a new approximation: *Find $\boldsymbol{x}^{(t)} \in \mathcal{P}^{(t)} \cup \{\boldsymbol{x}^{(t-1)}\}$ and $M^{(t)}$ such that*

$$M^{(t)} = f(\boldsymbol{x}^{(t)}) \geq f(\boldsymbol{y}), \ \boldsymbol{y} \in \mathcal{P}^{(t)} \cup \{\boldsymbol{x}^{(t-1)}\},$$

where $\{\boldsymbol{x}^{(-1)}\}$ is the empty set, $\boldsymbol{x}^{(t)}$ and $M^{(t)}$ are the best approximations to $\boldsymbol{x}^*$ and M so far.

Step 3. Termination criterion: *Let $\mathbf{c}^{(t)} = (\boldsymbol{b}(t) - \boldsymbol{a}(t))$. If the largest element of $\mathbf{c}^{(t)}$ is smaller than a pre-decided small positive number δ, terminate the algorithm and output $\boldsymbol{x}^{(t)}$ and $M^{(t)}$, otherwise, proceed to the next step.*

Step 4. Contract domain: *Construct new domain $D(t+1) = [\boldsymbol{a}(t+1), \boldsymbol{b}(t+1)]$ as follows:*

$$a_i(t+1) = \max(x_i^{(t)} - \gamma c_i^{(t)}, a_i), \ b_i(t+1) = \min(x_i^{(t)} + \gamma c_i^{(t)}, b_i),$$

where γ is a predefined contraction ratio. Set $t := t+1$ and go to Step 1.

Fang and Wang (1994) suggested to choosing a much larger number N_1 compared with N_2 and $N_2 = N_3 = \cdots$. For example, γ can be chosen as $1/2$ or $\gamma_i = \gamma^i$ for facter contraction.

The SNTO algorithm requires a weak assumption compared to traditional quasi-Newton methods. It can be used to maximize or minimize any system of non-linear equations with a flexible form of continuous functions $f_i(x)$ in the system equations; for instance, max, min, and $|x|$ are allowed in the expressions of f_i's. Fang and Wang (1991) and Fang et al. (1994a) gave some comparisons between SNTO and BFGS (which is a quasi-Newton method) for various optimization problems, and the advantages of SNTO are shown, especially when the objective function $f(\boldsymbol{x})$ has many local extrema. Another attractive feature of SNTO is that it always produces meaningful solutions, as the prior information and experience could be "plug-in" the algorithm by specifying the range of searching per parameter dimension. Moreover, the initial searching region D and the contraction rate γ can fully control the convergence rate.

SNTO in Moment Estimators

Consider to estimate m unknown parameters $\boldsymbol{\theta} = (\theta_1, \ldots, \theta_m)$ of the population $F(x)$. At the moment, the estimation method (Section 1.4) allows us to construct a system of m equations

$$\begin{cases} f_1(\boldsymbol{\theta}) = 0, \\ \cdots \\ f_m(\boldsymbol{\theta}) = 0, \end{cases} \quad \boldsymbol{\theta} \in \Omega,$$

where Ω is the domain of $\boldsymbol{\theta}$. Let the objective function be

$$f(\boldsymbol{\theta}) = f_1^2(\boldsymbol{\theta}) + \cdots + f_m^2(\boldsymbol{\theta}), \boldsymbol{\theta} \in \Omega. \tag{1.75}$$

The minimum point of $f(\boldsymbol{\theta})$ is a moment estimator of $\boldsymbol{\theta}$. Finding the moment estimator $\hat{\boldsymbol{\theta}}_{ME}$ is equivalent to finding a vector $\boldsymbol{\theta}^*$ such that $f(\boldsymbol{\theta}^*) = 0$. Section 5.6 gives a study of SNTO for ME and MLE.

1.14 Proof of Theorem 1.5

There are several equivalent ways to prove these results. We choose a qualitative argument that requires some machinery.

Let $\Gamma_2 = \{F : X \sim F, E(X^2) < \infty\}$. For $G_\alpha, G \in \Gamma_2$, denote

$$G_\alpha \Rightarrow G, \ \text{ iff } \ G_\alpha \xrightarrow{D} G \ \text{ and } \int x^2 dG_\alpha \to \int x^2 dG.$$

The strong law implies

$$F_n \Rightarrow F, \quad \text{along almost all sample sequences.} \tag{1.76}$$

The conclusions of the theorem hold along any such sample sequence.

The D_2-distance between G and H in Γ_2 is defined as follows: $D_2(G, H)^2$ is the infimum of $E(X - Y)^2$ over all joint distributions for the pair of random variables X and Y whose fixed marginal distributions are G and H, respectively. Introduced this metric citeMCL72.

Let $Z_1(G), \ldots, Z_m(G)$ be iid random samples from G. Let $G^{(m)}$ be the distribution of

$$S_m(G) = m^{-1/2} \sum_{j=1}^{m} [Z_j(G) - E(Z_j(G))].$$

If $G \in \Gamma_2$, so is $G^{(m)}$. By Lemma 3 of Mallows (1972),

$$D_2[G^{(m)}, H^{(m)}] < D_2[G, H]. \tag{1.77}$$

Proof of Theorem 1.5 For part (a), the bootstrap construction can be put into present notation as follows: conditionally, the law of $\sqrt{m}(\bar{x^*}_m - \bar{x}_n)$ is $F_n^{(m)}$. But F_n is close to F in the D_2-metric on Γ_2, by (1.76). Then, $F_n^{(m)}$ is close to $F^{(m)}$ by (1.77), that is,

$$D_2[F_n^{(m)}, F^{(m)}] < D_2[F^{(m)}, F].$$

Now, use the ordinary Central Limit Theorem on $F^{((m)}$.

For part (b), similarly define Γ_1 to Γ_2 under the condition $\int |x| G(dx) < \infty$. Define the metric D_1 on Γ_1 as the infimum of $E|X - Y|$ over all pairs of random variables X and Y with marginal distributions F and G, respectively. Let $G^{(m)}$ be the distribution of $\frac{1}{m} \sum_{j=1}^{m} Z_j(G)$. The requisite analog of (1.77) is

$$D_1[G^{(m)}, H^{(m)}] < D_1[G, H]. \tag{1.78}$$

1.15 Notation in This Book

The following tables list the notations used in the book.

TABLE 1.2
List of abbreviations

Abbreviation	Complete words
ABI	absolute bias index
cdf	cumulative distribution function
cf	characteristic function
DBM	design of experiments with mixtures
DI	disparity index
ECD	elliptically contoured distribution
ESD	elliptically symmetric distribution
EMSD	expected mean squared distance
FIM	Fisher Information Matrix
GGLP or gglp	generalized good lattice point
GLP or glp	good lattice point
GLS	global likelihood sampler
iid	independently identical distributed
GLS	global likelihood sampler
ITM	inverse transformation methods
KL	Kullback-Leibler divergence
Lp.cdf	L_p-distance between two cdfs
Lp.pdf	L_p-distance between two densities
MC	Monte Carlo method
ME	moment estimation
MLE	maximum likelihood estimator
MSE	mean square error
MVUE	minimum variance unbiased estimator
NTM	number-theoretic method
OVL	overlapping coefficient
pmf	probability mass function
pdf	probability density function
QMC	quasi-Monte Carlo
QMC-data	revised random samples
SSD or SD	spherically symmetric distribution
SR	stochastic representation
SRM	stochastic representation method
TG	transformation group
VDR	vertical density representation

TABLE 1.3
Notation for random variables

Notation	Definition
$\mathbb{R}$	Real number space
$\mathbb{R}^+$	Positive real number space
$\mathbb{N}^+$	Positive integer space
$\lfloor x \rfloor$	the largest integer not exceeding x
$X \sim F$	Random variable X follows the distribution F
$\mathcal{D}_X$	Domain of X
$X \overset{d}{=} Y$	X and Y have the same distribution
$\boldsymbol{X} \overset{iid}{=} \boldsymbol{Y}$	X and Y are iid
$F(x; \boldsymbol{\theta})$	cumulative distribution function (cdf) of X with parameters $\boldsymbol{\theta}$
$p(x; \boldsymbol{\theta})$	probability density function (pdf) of X with parameters $\boldsymbol{\theta}$
Ω	the domain of $\boldsymbol{\theta} = (\theta_1, \ldots, \theta_m) \in \Omega$
$X \overset{d}{=} Y$	X and Y have the same distribution
$\boldsymbol{X} \overset{iid}{=} \boldsymbol{Y}$ X and Y are iid	
$C(F_X)$	the set of all continuous points x of F_X
$p(x)$	probability density function of a continuous random variable X
μ, and σ^2	mean of X, $\mu = E(X)$ and variance of X, $\sigma^2 = \text{Var}(X)$
μ_r and μ_r^c	rth moment $E(X^r)$ and central rth moment $E(X-\mu)^r$
$m_r(x)$ and $m_r^c(x)$	sample rth moment at zero and central sample rth moment
$Sk(X)$	skewness of X, $\frac{1}{\sigma^3}E(X-\mu)^3$
$Ku(X)$	kurtosis of X, $\frac{1}{\sigma^4}E(X-\mu)^4 - 3$
Q_q	q-quantile of $F(x)$, $Q_q = F^{-1}(q) = \inf\{x : F(x) \geq q\}$
$Q_{1/4}, Q_{1/2}$ and $Q_{3/4}$	first quartile, median, and third quartile of $F(x)$
$\Phi(x)$ and $\varphi(x)$	cdf and pdf of $N(0,1)$
$B(a,b)$	the beta function with parameters a and b
$\Gamma(a)$	the gamma function with parameter a
$X_n \overset{a.s.}{\to} X$	Convergence almost surely
$X_n \overset{P}{\to} X$	Convergence in probability
$X_n \overset{D}{\to} X$	Convergence in distribution
$X_n \overset{rth}{\to} X$	Converges in rth mean

TABLE 1.4
Some univariate continuous distributions

Distribution	Notation	Density $p(x)$
Uniform	$U(0,1)$	$p(x) = 1, 0 \leq x \leq 1$
Normal	$N(\mu, \sigma^2)$	$\frac{1}{\sqrt{2\pi}\sigma} \exp\left\{-\frac{(x-\mu)^2}{2\sigma^2}\right\}, \sigma > 0$
Lognormal	$LN(\mu, \sigma^2)$	$\frac{\exp(-(x-\mu)/\sigma)}{\beta(1+\exp(-(x-\mu)/\sigma))^2}, \sigma > 0$
Student-t	t_ν	$\frac{\Gamma(\frac{\nu+1}{2})}{\sqrt{\nu\pi}\Gamma(\frac{\nu}{2})}\left(1+\frac{x^2}{\nu}\right)^{-\frac{\nu+1}{2}}, \nu \in \mathbb{N}^+$
Exponential	$EP(\lambda)$	$\lambda e^{-\lambda x}, x \geq 0, \lambda > 0$
Laplace	$Laplace(\mu, b)$	$\frac{1}{2b} e^{-\frac{\lvert x-\mu\rvert}{b}}, b > 0$
Gamma	$Gamma(\alpha, \beta)$	$\frac{\beta^\alpha}{\Gamma(\alpha)} x^{\alpha-1} e^{-\beta x}, \alpha > 0, \beta > 0$
Generalized gamma	$Ggamma(a, \lambda)$	$\frac{1}{2\Gamma(a)}(\lambda\lvert x\rvert)^{a-1} e^{-\lambda\lvert x\rvert}, a > 0, \lambda > 0$
Arcsine	Arcsine	$\frac{1}{\pi\sqrt{x(1-x)}}, 0 \leq x \leq 1$
F	$F(d_1, d_2)$	$\frac{1}{\mathrm{B}(\frac{d_1}{2}, \frac{d_2}{2})}\left(\frac{d_1}{d_2}\right)^{\frac{d_1}{2}} x^{\frac{d_1}{2}-1}\left(1+\frac{d_1}{d_2}x\right)^{-\frac{d_1+d_2}{2}}$
χ^2	χ^2_ν	$[2^{\nu/2}\Gamma(\frac{\nu}{2})]^{-1} e^{-x/2} x^{\nu/2-1}, x > 0, \nu \in \mathbb{N}^+$
χ	χ_ν	$[2^{\nu/2-1}\Gamma(\frac{\nu}{2})]^{-1} x^{\nu-1} e^{-x^2/2}, x > 0, \nu \in \mathbb{N}^+$
Beta	$Beta(a, b)$	$\frac{1}{B(a,b)} x^{a-1}(1-x)^{b-1}, 0 \leq x \leq 1, a, b > 0$
Beta prime	$BP(\alpha, \beta)$	$\frac{1}{\mathrm{B}(\alpha,\beta)}\frac{x^{\alpha-1}}{(1+x)^{\alpha+\beta}}, x > 0, \alpha, \beta > 0$
General Beta prime	$BP2((p, q, \alpha, \beta)$	$\frac{\alpha(1+x/\beta)^\alpha)^{-p-q}(x/\beta)^{p\alpha-1}}{\beta B(p,q)}, p, q, \alpha, \beta > 0, x > 0.$
Logistic	$L(\mu, \sigma)$	$\frac{1}{\sigma} e^{-\frac{x-\mu}{\sigma}}/[1+e^{-\frac{x-\mu}{\sigma}}]^2, \sigma > 0$
Gumbel	$Gu(\mu, \sigma)$	$e^{-\frac{x-\mu}{\sigma}} \exp\{-e^{-\frac{x-\mu}{\sigma}}\}, \sigma > 0$
Frèchet	$Fr(m, \sigma)$	$m\sigma^m x^{-m+1} e^{-(x/\sigma)^{-m}}$
Weibull	$We(m, \sigma)$	$\frac{m}{\sigma^m} x^{m-1} e^{-(x/\sigma)^m}$
Inverse Gaussian	$IG(\mu, \sigma)$	$\sqrt{\frac{\sigma}{2\pi x^3}} \exp\left\{-\frac{\sigma(x-\mu)^2}{2\mu^2 x}\right\}, x > 0, \mu > 0, \sigma > 0$

TABLE 1.5
Notation for sampling and resampling

Notation	Definition
$X_1, \ldots, X_n$	a random sample from $F(x, \boldsymbol{\theta})$
$X_{(1)} \leq \cdots \leq X_{(n)}$	order statistics of a random sample
$F_n(x)$	the empirical distribution of a random sample
T_n	statistics of the interest
$\hat{F}$	approximate distribution to F
$\hat{\boldsymbol{\theta}}_{ME}$	estimator of $\boldsymbol{\theta}$ by the moment estimation
$\hat{\boldsymbol{\theta}}_{MLE}$	estimator of $\boldsymbol{\theta}$ by the maximum likelihood estimation
$\hat{\boldsymbol{\eta}} = t(\hat{F})$	the plug-in estimate of $\boldsymbol{\eta} = t(F)$
$SE_F(\hat{\boldsymbol{\theta}})$	standard error of the estimate $\hat{\boldsymbol{\theta}}$
$\widehat{SE}_{F_n}(\hat{\boldsymbol{\theta}}^*)$	ideal bootstrap estimate of $SE_F(\hat{\boldsymbol{\theta}})$
$X_1^*, \ldots, X_n^*$	a bootstrap sample from F_n
$\hat{\boldsymbol{\theta}}^*$	$q(\boldsymbol{x}^*)$, bootstrap of replication of $\hat{\boldsymbol{\theta}}$
V_{BOOT}	bootstrap variance estimator
$U_n = U(X_1, \ldots, X_n)$	U-statistic: $\theta(F) = E_F[h(X_1, \ldots, X_m)]$
$Y_{MC} \sim F_n$	random variable follows the empirical distribution F_n
Y_{RMC}	random variable formed by MC-RPs
$Y_{QMC} \sim F_{QMC}$	random variable formed by QMC-RPs
$Y_{MSE} \sim F_{MSE}$	random variable formed by MSE-RPs
Y_{SC}	a discrete random variable is self-consistent to X
$\mathcal{Y}^{(k)}$	a set of k MSE-RPs of F: $\mathcal{Y}^{(k)} = \{\xi_1^{(k)}, \ldots, \xi_k^{(k)}\}$

TABLE 1.6
Notation for random vector

Notation	Definition
$\boldsymbol{X} = (X_1, \ldots, X_p)^T$	random vector in $\mathbb{R}^p$
$\mathcal{D}(\boldsymbol{X})$	the domain of random vector $\boldsymbol{X}$
$F(x_1, \ldots, x_p)$	cumulative distribution function of $\boldsymbol{X}$
$p(x_1, \ldots, x_p)$	pdf of a continuous random vector $\boldsymbol{X}$
$\boldsymbol{\mu} = E(\boldsymbol{X})$	mean vector of $\boldsymbol{X}$
$\boldsymbol{\Sigma} = \text{Cov}(\boldsymbol{X})$	covariance matrix of $\boldsymbol{X}$
$\boldsymbol{R} = \text{corr}(\boldsymbol{X})$	the correlation matrix of $\boldsymbol{X}$
$tr(\boldsymbol{A})$	the trace of the square matrix $\boldsymbol{A}$
$\boldsymbol{Y}_{SC} = \{\boldsymbol{y}_1, \ldots, \boldsymbol{y}_k\}$	self-consistent vector to $\boldsymbol{X}$
$\boldsymbol{\xi} = \boldsymbol{y}_{MSE}$	defined by MSE-RPs of $\boldsymbol{\xi}_1, \ldots, \boldsymbol{\xi}_k$

2

Approximated Discretization Methods to a Given Continuous Distribution

This chapter introduces approximated discretization methods based on the four kinds of representative points (RPs), that is, Monte Carlo RPs (MC-RPs), revised Monte Carlo RPs (RMC-RPs), quasi-Monte Carlo RPs (QMC-RPs), and mean square. The motivation and literature review for MSE-RPs are given, and the asymptotic behavior of MSE-RPs of univariate distributions is also investigated in this chapter.

2.1 Motivation

In the literature, researchers pay special attention to the problem of selecting a given number of representative points (RPs) to retain as much information about the given probability distribution as possible. How to discrete a given continuous distribution arises more attention in many situations.

Let a continuous random variable X follow the probability distribution $F(x)$ with the density function $p(x)$, mean $\mu = E(X)$ and variance $\sigma^2 = \text{Var}(X)$. One wishes to find a discrete random variable Y to approximate X in a certain sense, that is, to determine the support points $x_1, \ldots, x_k$, and the corresponding probabilities $p_1, \ldots, p_k$.

Many methods have been discussed to construct RPs and related approximation distributions in the literature, such as the Monte Carlo method, revised Monte Carlo method, quasi-Monte Carlo method, and mean square error method.

2.1.1 MC-RPs

Random variables and their distribution, denoted by $X \sim F(x)$, are the key concepts in traditional statistics. One often employs *random samples* to infer the population $F(x)$. A set of random samples, denoted by $X_1, \ldots, X_k$, that are independent and identically distributed (iid) samples from the distribution $F(x)$. When a set of $x_1, \ldots, x_k$ random samples has been observed, they are used for statistical inference, including
(1) parameter estimation (point estimation and confidence interval),
(2) density estimation, and
(3) testing hypothesis.

In the literature, many authors use the capital letter (like $X_1, \ldots, X_k$) for random samples and the small letter (like $x_1, \ldots, x_k$) for observations. A random sample $\{X_1, \ldots, X_k\}$ from X can be used to define a discrete random variable $Y_{MC}(X)$, or Y_{MC} for short, as

DOI: 10.1201/9781003589389-2

follows

$$
\begin{array}{c|cccc}
Y_{MC} & X_1 & X_2 & \cdots & X_k \\
\hline
p & \frac{1}{k} & \frac{1}{k} & \cdots & \frac{1}{k}
\end{array} \tag{2.1}
$$

The distribution of Y_{MC} has the well-known *empirical distribution function*

$$F_k(x) = \frac{1}{k}\sum_{i=1}^{k} I(X_i \le x), \tag{2.2}$$

where $I(\cdot)$ is the indicator function. In the book, we use F_{MC} for the empirical distribution, i.e., $Y_{MC} \sim F_{MC}$. Most of the statistical inferences are based on Y_{MC} as most statisticians believe that Y_{MC} is a good representative points of X. That is Y_{MC} is close to X under the theory of probability, especially when $k \to \infty$, for example,

$$
\begin{aligned}
E(Y_{MC}) &= \bar{X} = \frac{1}{n}\sum_{i=1}^{k} X_i, \text{ and } E_X(\bar{X}) = \mu, \\
\mathrm{Var}(Y_{MC}) &= \frac{1}{k}\sum_{i=1}^{k}(x_i - \bar{x})^2, \text{ and } E_X(\mathrm{Var}(Y_{MC})) = \frac{k-1}{k}\sigma^2 \to \sigma^2, \\
F_k(x) &\xrightarrow{P} F(x), \text{ as } k \to \infty.
\end{aligned}
$$

There are many different measurements for choosing a discrete random variable Y to approximate X. For example, the L_p-distance between the distribution functions of X and Y, that includes

$$D_p(F_X, F_Y) = \left[\int_{\mathbb{R}} |F_X(t) - F_Y(t)|^p dt\right]^{1/p}, p > 0, \tag{2.3}$$

where $F_X(t)$ and $F_Y(t)$ are the cdfs of X and Y, respectively. When both X and Y are continuous, the L_p-distance between their pdfs $p_X(x)$ and $p_Y(x)$ is

$$D_p(p_X, p_Y) = \left[\int_{\mathbb{R}} |p_X(x) - p_Y(x)|^p dx\right]^{1/p}, p > 0. \tag{2.4}$$

Some conditions are required for the existence of $D_p(p_X, p_Y)$. Note that these two distances in (2.3) and (2.4) may have different performance.

However, the approach based on random samples is not always satisfactory in many areas, such as numerical integration, geometric probability, and density estimations. Alternative approaches are discussed in the literature as follows.

2.1.2 RMC-RPs

Considering the unstable and unsatisfied performance of the MC method, Fang et al. (2020) proposed the revised Monte Carlo (RMC) method, where a k-point discrete random variable Y_{RMC} and its distribution F_{RMC} are defined for representing a continuous random variable X with pdf $p(x)$. The main idea of RMC-RPs is to give unequal weights to the support points of Y_{RMC}, compared with the Monte Carlo method. Suppose $\boldsymbol{x}_1, \ldots, \boldsymbol{x}_k$ are the k support points of the random vector $\boldsymbol{X}$ in $\mathbb{R}^p$. Then, through partitioning $\mathbb{R}^p$ based on those k support points, k subregions are obtained and denoted by $D_1, \ldots, D_k$, respectively. Then, the probability of each support point is $p(Y = \boldsymbol{x}_i) = \int_{D_i} p(\boldsymbol{x})d\boldsymbol{x}$, $i = 1, \ldots, k$.

Especially for one-dimensional cases, the support points of Y_{RMC} are sorted random samples $x_1 \leq \ldots \leq x_k$, which are essentially the order statistics, and the related probabilities are can be calculated as follows,

$$\begin{aligned} p_1^{(RMC)} &= \int_{-\infty}^{\frac{x_1+x_2}{2}} p(x)dx,\ p_2^{(RMC)} = \int_{\frac{x_1+x_2}{2}}^{\frac{x_2+x_3}{2}} p(x)dx, \\ &\cdots \\ p_i^{(RMC)} &= \int_{\frac{x_{i-1}+x_i}{2}}^{\frac{x_i+x_{i+1}}{2}} p(x)dx, i = 2, \ldots, k-1,\ p_k^{(RMC)} = \int_{\frac{x_{k-1}+x_k}{2}}^{\infty} p(x)dx. \end{aligned}$$

Fang et al. (2020) showed that RMC-RPs perform better than MC-RPs in statistical inference. The probabilities of RMC-RPs modify the equality probability of MC-RPs to form a set of self-consistent RPs that will be defined in Definition 2.4. Therefore, we might to use Y_{SC} for Y_{RMC}.

2.1.3 QMC-RPs

This approach was motivated by the numerical integration in $\mathbb{R}^p$ in the 1950s. Consider a canonical form

$$I(f) = \int_0^1 \cdots \int_0^1 f(x_1, \ldots, x_p) dx_1 \cdots dx_p, \tag{2.5}$$

where f is a continuous function on $C^p = [0,1]^p$. Let $\mathcal{Y} = \{\boldsymbol{y}_1, \ldots, \boldsymbol{y}_k\}$ be a set of k points on C^p. The mean of $f(\boldsymbol{y}_i), \boldsymbol{y}_i \in \mathcal{Y}$ can be used to approximate $I(f)$. By the MC method, we can employ a random sample from the uniform distribution $U(C^p)$ as $\mathcal{Y}$ and related $\boldsymbol{y}_{MC}$ similar to (2.1), we have

$$f(\bar{\boldsymbol{y}}_{MC}) = \frac{1}{k} \sum_{i=1}^{k} f(\boldsymbol{y}_i) \xrightarrow{P} I(f), \text{ as } k \to \infty.$$

The rate of convergence by the MC approach is $O_p(1/\sqrt{k})$, which is independent of p, but it is slow when p is not large, for example, $p \leq 18$. The Number-theoretic methods (NTM) or Quasi-Monte Carlo methods (QMC) (see Subsection 1.11.1) are proposed for construction of $\mathcal{Y}$ such that $\boldsymbol{y}_1 \ldots, \boldsymbol{y}_k$ are uniformly scattered on C^p under a certain sense, and it can increase the rate of convergence up to $O(k^{-1}(\log k)^p)$ compared with MC methods.

To find a good set of $\mathcal{Y}$ under a certain measure of uniformity, many uniformity measures are proposed (see Subsection1.11.2 and Fang et al. (2018)).

Quasi-Monte Carlo methods have been used in high-dimensional integration, finance, statistics, experimental design, geometric probability, and optimization. See Hua and Wang (1981), Niederreiter (1992), Fang and Wang (1994), and Pagès (2018) for many constructions methods for low-discrepancy sequences and their applications.

When $p = 1$, the set of QMC-RPs for the uniform distribution $U(0,1)$ is given by

$$C^{(k)} = \left\{ \frac{1}{2k}, \frac{3}{2k}, \ldots, \frac{2k-1}{2k} \right\}, \tag{2.6}$$

under some measures (see Theorem 4.1, Fang and Wang (1994)). For other distributions, the *inverse transformation method (ITM)* based on the following theorem is often used, which gives the link from $U(0,1)$ to any continuous distribution $F(x)$.

Theorem 2.1 *Let X have a continuous cdf $F(x)$ and $U \sim U(0,1)$. Then*
(a) $F^{-1}(U) \sim F(\cdot)$;
(b) $F(X) \sim U(0,1)$.

Proof *Since $F(x)$ is continuous and nondecreasing, $F(x)$ must be strictly increasing and has an inverse function F^{-1}. Let $Y = F^{-1}(U)$ be a random variable. Then,*

$$P(Y \leq y) = P(F^{-1}(U) \leq y) = P(U \leq F(y)) = F(y), \forall y \in \mathbb{R},$$

that is, $F^{-1}(U) \sim F(\cdot)$. A random variable on [0,1]. For any $z \in [0,1]$, we have

$$P(Z \leq z) = P(F(X) \leq z) = P(X \leq F^{-1}(z)) = F(F^{-1}(z)) = z,$$

and $P(Z \leq z) = 0$ for $z < 0$ and $P(Z \leq z) = 1$ for $z > 1$, therefore, $Z \sim U(0,1)$.

A set of QMC-RPs for the continuous distribution $F(x)$ is suggested by

$$\mathcal{Y}_F^{(k)} = \left\{F^{-1}\left(\frac{1}{2k}\right), F^{-1}\left(\frac{3}{2k}\right), \ldots, F^{-1}\left(\frac{2k-1}{2k}\right)\right\}. \tag{2.7}$$

Through the inverse of the cumulative distribution function (CDF), a random variable $Y_{QMC}(X)$ (or Y_{QMC} for short) can be defined as follows

Y_{QMC}	$F^{-1}(\frac{1}{2k})$	$F^{-1}(\frac{3}{2k})$	$\cdots$	$F^{-1}(\frac{2k-1}{2k})$
p^{QMC}	$\frac{1}{k}$	$\frac{1}{k}$	$\cdots$	$\frac{1}{k}$

The points in (2.7) are $\frac{1}{2k}, \frac{3}{2k}, \ldots, \frac{2k-1}{2k}$-quantiles of $F(x)$, respectively. Those points should have a good representation for $F(x)$. The distribution is denoted by $F_{QMC}^{(k)}$ or simply by F_{QMC}.

For a multivariate distribution $F(\boldsymbol{x})$, denote $\boldsymbol{y}_1 \ldots, \boldsymbol{y}_k$ as the corresponding QMC set. Let each weigh $1/k$; we obtain the QMC-RPs for $F(\boldsymbol{x})$. Fang and Wang (1994) and Fang et al. (1994a) show that QMC performs better than MC in many applications, especially for problems in high dimension scenarios.

2.1.4 MSE-RPs

Suppose that it is required to condense observations of a variate into a small number of groups, the grouping intervals to be chosen to retain as much information as possible. In this subsection, we consider the one-dimensional scenario. Let $X \sim F(x)$ have a density function $p(x)$, consider a set of k points $\mathcal{Y} = \{(\xi_1^{(k)}, \ldots, \xi_k^{(k)}) : -\infty < \xi_1^{(k)} < \xi_2^{(k)} < \cdots < \xi_k^{(k)} < \infty\}$ to minimize the mean square error (MSE)

$$\text{MSE}(\mathcal{Y}) = \int_{-\infty}^{+\infty} \min_i (x - \xi_i^{(k)})^2 p(x) dx \tag{2.8}$$

on the domain $\mathbb{R}$, the corresponding k intervals are $(a_1, a_2], (a_2, a_3], \ldots, (a_k, a_{k+1})$, where endpoints are

$$a_1 = -\infty, a_i = \left(\xi_{i-1}^{(k)} + \xi_i^{(k)}\right)/2, \ i = 2, \ \cdots, \ k, a_{k+1} = \infty. \tag{2.9}$$

Definition 2.1 *Suppose that a random variable $X \sim F(x)$ has a density function $p(x)$ with finite mean $\mu = E(x)$ and variance σ^2. The k points $\mathcal{Y}_{MSE}^{(k)} = \{\xi_1^{(k)}, \xi_2^{(k)}, \ldots, \xi_k^{(k)}\}$ obtained by minimizing the mean square error in (2.8) are called the MSE-representation points of*

$F(x)$ (or X) and denoted as MSE-RPs. Define a discrete random variable $Y_{MSE}^{(k)}$, or Y_{MSE} for short, with the support points $\mathcal{Y}_{MSE}^{(k)}$ and $P(Y_{MSE} = \xi_i^{(k)}) \equiv p_i^{MSE} \equiv p_i^{(k)}$, where

$$\begin{aligned} p_1^{(k)} &= \int_{-\infty}^{(\xi_1^{(k)}+\xi_2^{(k)})/2} p(x)dx = \int_{a_1}^{a_2} p(x)dx, \\ p_i^{(k)} &= \int_{(\xi_{i-1}^{(k)}+\xi_i^{(k)})/2}^{(\xi_i^{(k)}+\xi_{i+1}^{(k)})/2} p(x)dx = \int_{a_i}^{a_{i+1}} p(x)dx, 2 \le i \le k-1, \\ p_k^{(k)} &= \int_{(\xi_{k-1}^{(k)}+\xi_k^{(k)}/2)}^{\infty} p(x)dx = \int_{a_k}^{a_{k+1}} p(x)dx. \end{aligned} \tag{2.10}$$

A random variable Y_{MSE} can be defined as follows,

$$\begin{array}{c|cccc} Y_{MSE} & \xi_1^{(k)} & \xi_2^{(k)} & \cdots & \xi_k^{(k)} \\ \hline p^{MSE} & p_1^{(k)} & p_2^{(k)} & \cdots & p_k^{(k)} \end{array} \tag{2.11}$$

Obviously, Y_{MSE} is a function of X and we denote it as $Y_{MSE}(X)$, where

$$Y_{MSE}(X) = \begin{cases} \xi_1^{(k)}, & X \le (\xi_1^{(k)} + \xi_2^{(k)})/2, \\ \xi_i^{(k)}, & (\xi_{i-1}^{(k)} + \xi_i^{(k)})/2 < X \le (\xi_i^{(k)} + \xi_{i+1}^{(k)})/2, \\ & \text{for } i = 2, \ldots, k-1, \\ \xi_k^{(k)}, & (\xi_{k-1}^{(k)} + \xi_k^{(k)})/2 < X. \end{cases} \tag{2.12}$$

The mean square error of Y_{MSE} is

$$\text{MSE}(Y_{MSE}) = E(X - Y_{MSE})^2 = \sum_{i=1}^{k} \int_{a_i}^{a_{i+1}} (x - \xi_i^{(k)})^2 p(x)dx, \tag{2.13}$$

that measures the bias between X and Y_{MSE}, where $\{a_i\}_{i=1}^{k+1}$ are defined in (2.9). Given following notations

$$I_1 = (a_1, a_2], I_i = (a_i, a_{i+1}], i = 2, \ldots, k-1, I_k = (a_k, a_{k+1}),$$

the MSE of Y_{MSE} can be written as

$$\text{MSE}(Y_{MSE}) = \sum_{i=1}^{k} \int_{I_i} (x - \xi_i^{(k)})^2 p(x)dx. \tag{2.14}$$

Theorem 3.5 will show out $\text{Var}(Y_{MSE}^{(k)}) = \text{Var}(X) - MSE(Y_{MSE}^{(k)})$. The MSE measure depends on σ^2, the variance of X. The loss function of Y_{MSE} is defined by

$$L(Y_{MSE}) = \frac{1}{\sigma^2} MSE(Y_{MSE}). \tag{2.15}$$

In Subsection 3.2, we shall show that:
1) $E(Y_{MSE}) = E(X)$;
2) $0 \le L(Y_{MSE}) \le 1$;
3) $L(Y_{MSE}^{(k)}) \to 0$ as $k \to \infty$.

2.1.5 Motivation of MSE-RPs

The concept of MSE-RPs of statistical distributions is motivated by many real-world problems as follows.

A. Grouping problem

Cox (1957) considered the grouping problem. Let X denote a continuous random variable and suppose that it is desired to classify the individuals in the population into k groups for the convenience of exposition. For example, X may be some property connected with the age status of an individual, and we may wish to classify the individuals as "young," "middle," or "old", where $k = 3$.

B. Canonical Correlation Analysis

Bofinger (1970) studied grouping a continuous bivariate distribution by intervals on the margins, thereby obtaining a discrete bivariate distribution. The grouping was provided by maximizing the possible correlation between these marginal variables. The solution is approximate when both margins are grouped and exact when only one margin is grouped. For the bivariate normal distribution with one margin grouped, tables of interval endpoints that maximize the correlation are provided at up to ten intervals, as suggested by the author.

C. Data-transmission problem

In many data transmission systems, analog input signals are converted to digital form at the transmitter, transmitted in digital form, and reconstituted at the receiver as analog signals. The problem of optimum quantization of a continuous random variable with a fixed number of levels was first precisely formulated by Max (1960). It seeks to quantize the univariate normal distribution so that k points can represent it. Employing a mean square error loss function, Max (1960) provided the tables for $k = 1, 2, \ldots, 36$, which yield optimal quantization for digitizing the univariate normal distribution. Treatments of the quantization problem with applications in information theory, such as the analog-to-digital conversion, signal compression, coding theory, etc., are contained in the Special Issue of *IEEE Transactions on Information Theory* (Vol. 28, pp. 127–202).

D. Garment standard and Swiss army protection masks

To obtain the national standard for the garment in China, p measurements of the body of every n individuals (in general, n is sufficiently large) are measured, and these p-dimensional data are projected onto a q ($q < p$) dimensional space by principal components analysis or by some other methods. They wish to select k points that best represent the data in the q-dimensional space. Fang (1976) proposed a way to analyze the body measurements for $q = 2$ dimensional normal distributions where the intervals are of equal lengths and the k points are centered in each interval. Fang and He (1982) proposed the mathematical problem based on the garment standard of choosing the k points under the mean square error criterion. Iyengar and Solomon (1983) pointed out the mathematical problems that arise in the theory of representing a distribution by optimally chosen points. Flury (1990, 1993) studied a project of the Swiss army to develop newly designed protection masks.

2.2 Four Methods in Statistical Estimation

Statistical simulation has been widely used in statistical research and data analysis. Monte Carlo methods are a traditional approach, but it is well known that the convergence rate of MC methods is slow. This section shows that other methods can improve the statistical estimation. In this section, we will show that the four types of approximation distributions for $F(x)$ that are MC-RPs, RMC-RPs, QMC-RPs, and MSE-RPs mentioned in the previous section, have different performances in density estimation and parameter estimation.

2.2.1 Four methods in density estimation

Density estimation is a nonparametric approach. Subsection 1.6.1 has introduced the Parzen-Rosenblatt window method. Let us see one example in a signal transaction by the four methods introduced in the previous section. In the signal transaction, one must estimate a density function and related statistics for the input without any prior information. When the population of the input has a continuous distribution with an unknown density function $p(x)$, one can use some density estimation methods to estimate the density function based on a set of samples, $x_1, \ldots, x_m$.

Assume $x_1, \cdots, x_m$ are a sample from the population with an unknown density function $p(x)$. The Parzen-Rosenblatt window method suggests to use

$$\hat{f}_h(x) = \frac{1}{m}\sum_{i=1}^{m} k_h(x - x_i) = \frac{1}{mh}\sum_{i=1}^{m} k\left(\frac{x - x_i}{h}\right), \quad x \in \mathbb{R},$$

as an estimator of $p(x)$, where $k(\cdot)$ is the kernel function, h is the bandwidth and $k_h(y) = \frac{1}{h}k(y/h)$. In practice, the most popular kernel function is $k(x) = \phi(x) = \frac{1}{\sqrt{2\pi}}e^{-\frac{1}{2}x^2}$.

Consider the mixture of two normal distributions $N(-8, 3^2)$ and $N(3, 4^2)$ with $\alpha = 0.6$, that is, the pdf

$$p(x) = \alpha\varphi(x; \mu_1, \sigma_1^2) + (1 - \alpha)\varphi(x; \mu_2, \sigma_2^2), \tag{2.16}$$

where $\varphi(x; 0, 1)$ is the pdf of $N(0, 1)$. When $k = 15$, the density estimation by the four methods is given in Figure 2.1, and the L_2-distances between $p(x)$ and $p_Y(x)$ by the four methods are given in Table 2.1. Comparing their performances, we know that MSE > QMC > RMC > MC, i.e., the MSE method has the best performance, and the MC method has the poorest results.

TABLE 2.1
The L_2-distance between $p(x)$ and $p_Y(x)$ by the four methods

k	MC	RMC	QMC	MSE
15	0.0676	0.0453	0.0171	0.0151
30	0.0674	0.0316	0.0129	0.0126
35	0.0615	0.0284	0.0124	0.0122

2.2.2 Four methods in moments estimation

Let $X \sim F(x)$ with pdf $p(x)$ and finite the first four moments. Let Y be an approximate discrete random variable to X with distribution $F_Y(x)$. Then, the mean, variance, skewness,

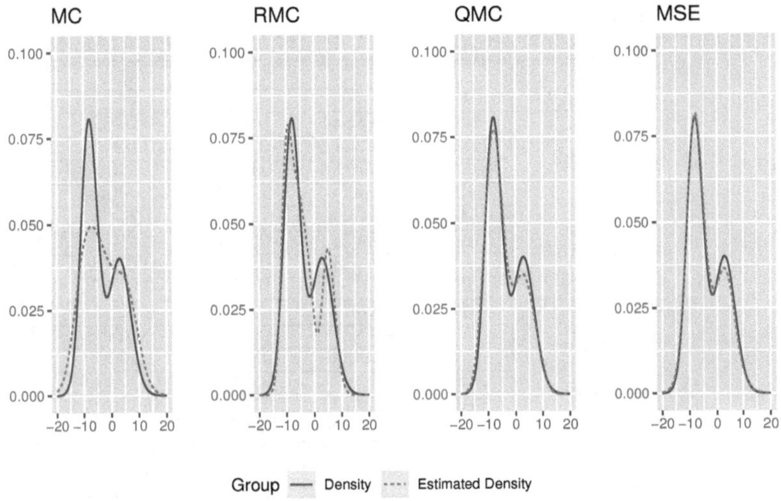

FIGURE 2.1
Density estimators of a mixture of two normal distributions by the four methods, $k = 15$.

and kurtosis of Y are given by

$$\begin{aligned} E(Y) &= \sum_{i=1}^{k} y_i p_i \equiv \mu_y, \ Var(Y) = \sum_{i=1}^{k} (y_i - \mu_y)^2 p_i \equiv \sigma_Y^2, \\ Sk(Y) &= \frac{1}{\sigma_Y^3} \sum_{i=1}^{k} (y_i - \mu_y)^3 p_i, \ Ku(Y) = \frac{1}{\sigma_Y^4} \sum_{i=1}^{k} (y_i - \mu_y)^4 p_i - 3. \end{aligned} \tag{2.17}$$

For the standard normal distribution $Z \sim N(0, 1)$, it is known that $E(Z) = 0, \quad \sigma^2 = 1$, and $Sk(Z) = Ku(Z) = 0$. Fang et al. (2014) gave the numerical comparisons of these statistics among MC, QMC, and MSE methods. Fang et al. (2020) added RMC in the comparison. Table 2.2 compares the four methods for $k = 5, 15, 21, 30, 35$. It is shown that.

- (1) MC has the worst estimation, and MSE gives the best results in most cases;
- (2) RMC improves MC significantly;
- (3) QMC gives a better estimation than RMC except for kurtosis estimation.

2.2.3 Resampling method

Let $X \sim F(x)$ be a continuous random variable and $Y \sim G(x)$ be an approximate random variable to $F(x)$ in a certain sense. The random samples $v_1, \ldots, v_N$ are from $G(x)$ and called as *resampling samples*. The bootstrap method introduced in Subsection 1.8.2 is one resampling technique. Efron and Tibshirani (1993) said: "The explanations that we will give for the bootstrap, and other computer-based methods, involve explanations of traditional ideas in statistical inference. The basic ideas of statistics haven't changed, but their implementation has. The modern computer lets us apply these ideas flexibly, quickly, easily, and with a minimum of mathematical assumptions. Our primary purpose in the book is to explain when and why bootstrap methods work and how they can be applied in various real data-analytic situations."

TABLE 2.2
Estimation errors by MC, RMC, QMC and MSE for $N(0,1)$

statistics	Y	k = 5	15	21	30	35
Mean	MC	-0.8534	-0.1328	-0.0862	-0.0177	0.0654
	RMC	-0.2511	-0.0200	-0.0045	-0.0010	0.0023
	QMC	0.0000	0.0000	0.0000	0.0000	0.0000
	MSE	0.0000	0.0000	0.0000	0.0000	0.0000
Variance	MC	0.4134	0.2640	0.2147	0.2534	0.3931
	RMC	-0.4733	-0.0935	-0.0279	-0.0241	-0.0095
	QMC	-0.2331	-0.0812	-0.0586	-0.0414	-0.0356
	MSE	-0.0799	-0.0107	-0.0057	-0.0028	-0.0021
Skewness	MC	-0.6610	-1.1892	-0.6055	-0.4466	-0.1385
	RMC	-1.5711	-0.6695	-0.0709	-0.0932	-0.0455
	QMC	0.0000	0.0000	0.0000	0.0000	0.0000
	MSE	0.0000	0.0000	0.0000	0.0000	0.0000
Kurtosis	MC	-1.2000	0.3546	0.6469	0.2836	0.1383
	RMC	1.3308	-0.1736	-0.0464	-0.1633	-0.0699
	QMC	-1.1143	-0.5903	-0.4794	-0.3813	-0.3445
	MSE	-0.5753	-0.1033	-0.0575	-0.0302	-0.0227

Let $T(x_1,\ldots,x_n)$ be a statistic for estimation of an unknown parameter θ of the interest, where $\{x_1,\ldots,x_n\}$ is a set of random samples from $F(x)$. A random simulation is useful when the analytic formula for the distribution of T is challenging to find. Fang et al. (2014) first applied RPs in the resampling technique. Let Y be a discrete random variable to approximate X in a certain sense. Denote the support points of Y by $\xi_1^{(k)} < \xi_2^{(k)} < \ldots < \xi_k^{(k)}$ with probabilities $P(Y=\xi_j^{(k)}) = p_j^{(k)}, j=1,\ldots,k$. Taking N sets of n-random samples from Y to calculate values of T, denote them by $T_1,\ldots,T_N$. Then the statistical inference on T can be based on the empirical distribution of $T_1,\ldots,T_N$. For example, we can estimate the expectation of T by $\bar{T} = (\sum_{j=1}^N T_j)/N$. The four methods based on MC-RPs, RMC-RPs, QMC-RPs, and MSE-RPs can construct four approximation distributions to $F(x)$.

The following algorithm is the traditional one for taking a random sample from the discrete distribution of Y, randomly generating a set of N random samples from the statistic T and an estimator of $E(T)$.

Algorithm 2.1 (Resampling on a discrete distribution formed by RPs:)

Step 1 Generate a random number $U \sim U(0,1)$.

Step 2 Construct a random sample from Y *by*

$$Y = \begin{cases} \xi_1^{(k)}, & \text{when } U < p_1^{(k)}, \\ \xi_2^{(k)}, & \text{when } p_1 \le U < p_1^{(k)} + p_2^{(k)}, \\ \vdots & \vdots \\ \xi_k^{(k)}, & \text{when } \sum_{i=1}^{k-1} p_i^{(k)} \le U. \end{cases}$$

Step 3 For each replicate, $j = 1, ..., N$:
(a) Repeat Step 1 and Step 2 n *times and obtain the* jth *sample from* Y.
(b) Compute the given statistic T_j *based on the* jth *sample from* Y.

Step 4 Use $\bar{T} = (\sum_{j=1}^N T_j)/N$ *to infer the statistic* T *of the population.*

The above four methods allow for the construction of approximate distributions for the population distribution $F(x)$. The bootstrap is one of the popular resampling techniques. Fang et al. (2014) and Fang et al. (2020) extended the bootstrap technique to RMC, QMC, and MSE methods. They showed the comparisons in terms of estimation for the mean, variance, skewness, and kurtosis for the mixture of the normal distributions $N(-8, 3^2)$ and $N(3, 4^2)$ with $\alpha = 0.6$. In the same way, our simulation results in Table 2.3 show that the performance of the four methods in estimation comparisons can be shown as follows,

$$\text{MSE} > \text{QMC} > \text{RMC} > \text{MC}.$$

It indicates an excellent potential for MSE and QMC methods in statistical estimation and inference.

TABLE 2.3
Estimation error by MC, RMC, QMC, and MSE for mixture normal with $k = 25$ by resampling

n	Category	1000	2000	5000	10000
Mean	MC	-0.8166	-0.8123	-0.8210	-0.8312
	RMC	-0.0087	-0.0516	-0.0360	-0.0422
	QMC	-0.0489	-0.0361	-0.0213	0.0122
	MSE	-0.0493	-0.0503	0.0142	-0.0081
Variance	MC	-13.8159	-14.2544	-14.5410	-14.4436
	RMC	-3.7956	-4.2640	-4.4867	-4.5306
	QMC	-2.5794	-2.5570	-2.7826	-2.4634
	MSE	-1.8174	-1.8934	-1.7205	-1.7869
Skewness	MC	0.6068	0.5830	0.5766	0.5837
	RMC	0.1707	0.1552	0.1406	0.1386
	QMC	-0.0213	-0.0093	-0.0232	-0.0264
	MSE	-0.0251	-0.0182	-0.0155	-0.0186
Kurtosis	MC	1.7145	1.6689	1.6419	1.6556
	RMC	0.3957	0.3662	0.3363	0.3251
	QMC	-0.0756	-0.0391	-0.0643	-0.0762
	MSE	-0.0048	0.0177	0.0012	0.0149

2.3 Representative Points of Multivariate Distributions

We have discussed four types of RPs and related approximation distributions for univariate distributions. Extending these types of RPs into multivariate distributions is necessary and challenging. There are different ways to define the MSE-RPs.

Definition 2.2 Optimal vector quantizer
Let $\boldsymbol{X} = (X_1, \ldots, X_p)^T$ *be a random vector in* $\mathbb{R}^p$ *with pdf* $p(\boldsymbol{x})$ *and finite second moments. A* k*-level quantizer* $\mathcal{Q} = \{\mathcal{Y}, \mathcal{S}\}$ *of* $\boldsymbol{X}$ *consists of*

(1) a set of output vector $\mathcal{Y} = \{\boldsymbol{y}_1, \ldots, \boldsymbol{y}_k\}$*;*

(2) a partition $\mathcal{S} = \{S_1, \ldots, S_k\}$ *of* $\mathbb{R}^p$ *with* k *disjoint and exhaustive regions, and*

(3) a mapping $Q: \mathbb{R}^p \to \mathcal{S} : Q(\boldsymbol{x}) = \boldsymbol{y}_i$ *if* $\boldsymbol{x} \in S_i$*.*

The optimal k*-level quantizer, denoted by* $Q(\boldsymbol{x})$*, minimizes*

$$MSE(Q(\boldsymbol{x})) = \min_{Q} E(||\boldsymbol{X} - Q(\boldsymbol{x})||^2), \tag{2.18}$$

the corresponding $\{\boldsymbol{y}_1, \ldots, \boldsymbol{y}_k\}$*, denoted by* $\{\boldsymbol{\xi}_1, \cdots, \boldsymbol{\xi}_k\}$ *are called representative points (MSE-RPs for short) of* $F(\boldsymbol{x})$ *and denote the corresponding* $Q(\boldsymbol{x})$ *by* $\boldsymbol{\xi}(\boldsymbol{x})$*, where* $\boldsymbol{\xi}(\boldsymbol{x})$

is a discrete random vector $\boldsymbol{\xi} = (\boldsymbol{\xi}_1, \cdots, \boldsymbol{\xi}_k)^T$ *with* $\boldsymbol{\xi} = \boldsymbol{\xi}_i$ *if* $\boldsymbol{x} \in S_i$ *with the probability* $p_i = P(\boldsymbol{x} \in S_i)$. *The vector* $\boldsymbol{\xi}$ *is also denoted by* $\boldsymbol{y}_{MSE}$.

The partition $\mathcal{S}$ of k-MSE-RPs is *a set of Voronoi regions.* Note that the notation $Q(\boldsymbol{x})$ may have a confusion with the quantile $Q(p)$. Here, we use $Q(\boldsymbol{x})$ to show our respect for the literature.

The term *principal points* introduced by Flury (1990, 1993) was motivated from a practical project of the Swiss Army on designing new protection masks (see Subsection 2.1.5). To put the construction of the new protection masks on reasonable empirical grounds, a group of anthropologists was hired to measure the heads of 900 Swiss soldiers. Twenty-five variables were regarded as potentially necessary for the fit of the masks. Since human heads differ in size and shape, several types of masks are needed for adequate fit. They define the representative points for a multivariate distribution as follows.

Definition 2.3 Principal points
Let $\mathcal{Y} = \{\boldsymbol{y}_j \in \mathbb{R}^p, 1 \leq j \leq k\}$ *denote a set of* k *points in* $\mathbb{R}^p$. *Then the minimal distance of* $\boldsymbol{z} \in \mathbb{R}^p$ *to* $\mathcal{Y}$ *is defined as*

$$d(\boldsymbol{z}|\mathcal{Y}) = \min_{1 \leq h \leq k} [(\boldsymbol{z} - \boldsymbol{y}_h))^T (\boldsymbol{z} - \boldsymbol{y}_h)]^{1/2}. \tag{2.19}$$

Let random vector $\boldsymbol{X} \sim F(\boldsymbol{x})$. *The set of vectors* $\boldsymbol{\xi} = \{\boldsymbol{\xi}_j \in \mathbb{R}^p, 1 \leq j \leq k\}$ *is called* k *principal points of* $\boldsymbol{X}$ *if*

$$\text{MSE}(\boldsymbol{\xi}) = E_F\{d^2(\boldsymbol{X}|\boldsymbol{\xi})\} = \min_{\boldsymbol{y}_i \in \mathbb{R}^p, 1 \leq i \leq k} E\{d^2(\boldsymbol{X}|\mathcal{Y})\}. \tag{2.20}$$

This definition uses the Euclidean distance, but alternative distances could also be considered. The term "principal points was introduced to stress their similarity with the least-squares definition of principal components by Pearson (1901), and distinguishing them from cluster means.

Flury (1993) introduced the concept of self-consistency as follows, which is more general than the concept of representative points.

Definition 2.4 *A set of* k *points* $\{\boldsymbol{y}_1, \ldots, \boldsymbol{y}_k\}$ *in* $\mathbb{R}^p$ *is called self-consistent with respect to the p-variate random vector* $\boldsymbol{X}$ *and the partition* $\{S_j, j = 1, \ldots, k\}$ *of* $\mathbb{R}^p$ *if*

$$E[\boldsymbol{X}|\boldsymbol{X} \in S_j] = \boldsymbol{y}_j, j = 1, ..., k. \tag{2.21}$$

For a given set of points $\{\boldsymbol{y}_1, \ldots, \boldsymbol{y}_k\}$ in $\mathbb{R}^p$, we can find a partition $\{S_j, j = 1, \ldots, k\}$ by

$$S_j = \{\boldsymbol{x} : (\boldsymbol{x} - \boldsymbol{y}_j)^T (\boldsymbol{x} - \boldsymbol{y}_j) \leq (\boldsymbol{x} - \boldsymbol{y}_h)^T (\boldsymbol{x} - \boldsymbol{y}_h), \forall h \neq j\} \tag{2.22}$$

that is called the domain of attraction of $\boldsymbol{y}_j$ or Voronoi region associated with $\boldsymbol{y}_j, j = 1.\ldots, k$.

Tarpey and Flury (1996) gave a comprehensive study on self-consistency. Sections 7.1 will introduce their theory. Here we only highlight two issues for your attention:

(1) MSE-RPs are self-consistent with respect to $\boldsymbol{X}$;
(2) MSE-RPs has the minimum mean square error among all the self-consistent of $\boldsymbol{X}$.

Chapters 6 and 7 will detail the study of RPs of multivariate distributions. Abaya and Wise (1984) justified the optimal quantizer as follows. "Suppose that a sequence of probability distribution functions F_j converges to the distribution function F. Does the sequence of optimal quantizers for the F_j's converge quantizer for F? If so, do the respective distortions converge to the optimal distortion for F? Sufficient conditions are given to guarantee the convergence for scalar and vector quantizers with a general class of distortion measures." Therefore, we have to pay attention to the asymptotic behavior of RPs.

2.4 Asymptotic Behavior of MSE-RPs

The asymptotic behavior of MSE-RPs (or short RPs in this section) has raised more and more attention in the past decades; for example, see Zador (1982), Su (2000), Graf and Luschgy (2000), and Pagès (2018). This section introduces only some results obtained by Xu et al. (2023) on the distribution tail for univariate distributions. Let random variable X have a density function $p(x)$ with finite mean μ and variance σ^2. Its RPs are sorted by increasing order: $\xi_1^{(k)} < ... < \xi_k^{(k)}$. Denote

$$\rho_k = \xi_k^{(k)}, \Delta_i^{(k)} = \xi_i^{(k)} - \xi_{i-1}^{(k)}, i = 2, \ldots, k. \tag{2.23}$$

There are many nice results on RPs when k converges to infinity. Theorem 3.5 will give some properties. Due to the need in this section, we list some properties in the next chapter as follows.

Lemma 2.1 *Let random variable X have a density function $p(x)$ with finite mean $\mu = E(X)$ and variance $\sigma^2 = \mathrm{Var}(X)$. We have*
(i) $E(Y_{MSE}) = E(X)$.
(ii) $\lim_{k\to\infty} \mathrm{MSE}(\xi_1^{(k)}, ..., \xi_k^{(k)}) = 0$.
(iii) $\mathrm{Var}(Y_{MSE}) = \mathrm{Var}(X) - \mathrm{MSE}(\xi_1^{(k)}, ..., \xi_k^{(k)}) \to \mathrm{Var}(X)$ *as* $k \to \infty$.
(iv) $\lim_{k\to\infty} \xi_k^{(k)} = \sup\{x, x \in \mathrm{supp}(X)\}$, *where* $\mathrm{supp}(X)$ *is the upper support of* X.

Xu et al. (2023) gave a comprehensive study on $\Delta_i^{(k)}$ as $k \to \infty$. For further discussion, we introduce some basic knowledge in survival analysis.

Definition 2.5 *The survival function of X is defined as*

$$S(x) = P(X > x) = \int_x^\infty p(u)\mathrm{d}u,$$

and the hazard function is defined as $h(x) = \frac{p(x)}{S(x)}$.

Definition 2.6 *Let X be a nonnegative random variable with $E(X) < \infty$. Define*

$$\rho(x) = E(X - x|X \geq x), \tag{2.24}$$

for all x such that $P(X \geq x) > 0$. The function $\rho(x)$ is the mean residual life (MRL). The doubly truncated mean residual life (DMRL) is defined by

$$\tilde{\rho}(x, y) = E(X - x|x \leq X \leq y) = \frac{\int_x^y (u - x)p(u)\mathrm{d}u}{\int_x^y p(u)\mathrm{d}u}. \tag{2.25}$$

The MRL expresses the expected remaining life of a component at time x given that the element has survived to time x, and MRL can uniquely determine the distribution. The DMRL measures the expected remaining life of a component with the time x and is broken before time y; see the details in Sankaran and Sunoj (2004).

It is an essential and common problem to study survival analysis based on conditional random variables because most of the observed data are censored.

In the literature, the expected remaining life and the mean residual life are interchangeably used, since

$$\begin{aligned}\int_x^\infty S(w)\mathrm{d}w &= \int_x^\infty \int_w^\infty p(u)\mathrm{d}u\mathrm{d}w = \int_x^\infty \int_x^u p(u)\mathrm{d}w\mathrm{d}u \\ &= \int_x^\infty \Big(\int_x^u \mathrm{d}w\Big)p(u)\mathrm{d}u = \int_x^\infty (u-x)p(u)\mathrm{d}u.\end{aligned}$$

See more details in Cox (1962), Kotz and Shanbhag (1980), Hall and Wellner (1981), and Bhattacharjee (1982). Therefore, we can rewrite $\rho(x)$ as

$$\rho(x) = E(X - x|X \geq x) = \frac{\int_x^\infty (u-x)p(u)\mathrm{d}u}{\int_x^\infty p(u)\mathrm{d}u} = \frac{\int_x^\infty S(w)\mathrm{d}w}{S(x)}.$$

Because $\rho(x)$ plays an important role in later analysis, we list some useful properties as follows.

Lemma 2.2 *Under the above notation, we have*

$$\rho(0) = E(X), \tag{2.26}$$

$$1 + \rho'(x) = h(x)\rho(x), \tag{2.27}$$

$$\lim_{x\to\infty} \rho(x) = \lim_{x\to\infty} \frac{S(x)}{p(x)} = \lim_{x\to\infty} \frac{1}{h(x)}, \tag{2.28}$$

$$\lim_{x\to\infty} \rho(x) = \lim_{x\to\infty} \frac{-1}{\frac{\mathrm{d}}{\mathrm{d}x}\ln p(x)}. \tag{2.29}$$

The property (2.26) is from Calabria and Pulcini (1987), (2.27) from Lai and Xie (2006), (2.28) from Calabria and Pulcini (1987), and (2.29) from Lawless (2011).

The ratio $\frac{S(x)}{p(x)}$ is called as *Mills ratio* in probability theory. It is well known that Mills ratio is reciprocal to the hazard function, namely $\frac{S(x)}{p(x)} = \frac{1}{h(x)}$. In Lemma 2.2, (2.28) relates the asymptotic behavior of MRL to the asymptotic behavior of hazard function $h(x)$, while (2.29) provides a useful relationship between the asymptotic behavior of MRL and the logarithm of $p(x)$.

The following theorem provides a simple way to compare the relative wideness of $\Delta_i^{(k)}$ and $\Delta_{i+1}^{(k)}$ based on the monotonicity of pdf $p(x)$.

Theorem 2.2 *Let a random variable X have a continuous pdf $p(x)$ and finite variance. If $p(x)$ is strictly decreasing (increasing) on the interval $((\xi_{i-1}^{(k)} + \xi_i^{(k)})/2, (\xi_i^{(k)} + \xi_{i+1}^{(k)})/2)$ for some $i \in \{2, ..., k-1\}$, then*

$$\Delta_{i+1}^{(k)} > (<)\Delta_i^{(k)},$$

where $\xi_1^{(k)} < ... < \xi_k^{(k)}$ are k RPs of X and $\Delta_i^{(k)}$ is defined in (2.23).

Subsection 2.6.1 gives a proof of the theorem. The following result is based on Theorem 2.2.

Corollary 2.1 *Let X be a random variable with a continuous density function $p(x)$ and finite variance. If $p(x)$ is strictly decreasing (increasing) on its domain, then $\{\Delta_i^{(k)}\}_{i=2}^k$ is an increasing (decreasing) sequence.*

The relative wideness of $\Delta_i^{(k)}$ and $\Delta_{i+1}^{(k)}$ is illustrated in two examples as follows.

Example 2.1 *For the exponential distribution $EP(\lambda)$ with pdf*

$$p(x;\lambda) = \lambda e^{-\lambda x}, x \geq 0, \lambda > 0,$$

we have

$$\Delta_2^{(k)} < ... < \Delta_k^{(k)} = \frac{2}{\lambda}, \text{ for any } k \geq 2,$$

which is consistent with the theoretical results in Theorem 3.14.

Example 2.2 *For the normal distribution $N(\mu, \sigma^2)$ with pdf*

$$p(x;\mu,\sigma^2) = \frac{1}{\sqrt{2\pi}\sigma} \exp\Big\{-\frac{1}{2}\Big(\frac{x-\mu}{\sigma}\Big)^2\Big\}, \ \mu \in \mathbb{R}, \sigma^2 > 0,$$

we note that $p(x)$ is strictly increasing on $(-\infty, \mu)$ and strictly decreasing on $(\mu, +\infty)$. It is noticeable that MSE-RPs of the normal distribution are symmetric about μ (see Theorem 2.2 in Li and Flury (1995)), that is

$$\Delta_i^{(k)} = \Delta_{k+2-i}^{(k)}, \ i = 2, ..., \lfloor (k+1)/2 \rfloor.$$

Therefore, if $k = 2l$, we have

$$\Delta_2^{(k)} > ... > \Delta_{l+1}^{(k)} < ... < \Delta_k^{(k)},$$

if $k = 2l+1$, we have

$$\Delta_2^{(k)} > ... > \Delta_{l+1}^{(k)} = \Delta_{l+2}^{(k)} < ... < \Delta_k^{(k)},$$

which is consistent with the numerical results in Fang et al. (2014).

Xu et al. (2023) divided the continuous probability distributions into three categories based on their support set and discuss the limiting behavior of $\Delta_k^{(k)}$ case by case.

2.4.1 Continuous distributions supported on $\mathbb{R}^+$

For a nonnegative random variable X, the following theorems provide a new explanation of the gap between RPs indicates that RPs may have an inner connection with survival analysis.

Theorem 2.3 *Let $\xi_1^{(k)} < ... < \xi_k^{(k)}$ $(k > 2)$ be k RPs of a non-negative random variable X with pdf $p(x)$ and finite variance, then*

$$\Delta_i^{(k)} = \xi_i^{(k)} - \xi_{i-1}^{(k)} = 2\tilde{\rho}\Big(\frac{\xi_{i-1}^{(k)} + \xi_i^{(k)}}{2}, \frac{\xi_i^{(k)} + \xi_{i+1}^{(k)}}{2}\Big), \quad i = 2, ..., k-1,$$

where $\tilde{\rho}(\cdot,\cdot)$ is defined in (2.25). Therefore, $\Delta_i^{(k)}$ can be interpreted as twice the doubly truncated mean residual life at times $(\xi_{i-1}^{(k)} + \xi_i^{(k)})/2$ and $(\xi_i^{(k)} + \xi_{i+1}^{(k)})/2$, $i = 2, ..., k-1$.

The proof of the theorem is presented in Subsection 2.6.2. The following theorem gives a relationship between $\Delta_k^{(k)}$ and MRL.

Theorem 2.4 *Under the assumption of Theorem 2.3, we have*

$$\Delta_k^{(k)} = \xi_k^{(k)} - \xi_{k-1}^{(k)} = 2\rho\Big(\frac{\xi_{k-1}^{(k)} + \xi_k^{(k)}}{2}\Big), \tag{2.30}$$

which says that $\Delta_k^{(k)}$ can be interpreted as twice the mean residual life at time $(\xi_{k-1}^{(k)}+\xi_k^{(k)})/2$, that is, $\rho((\xi_{k-1}^{(k)} + \xi_k^{(k)})/2)$.

The proof of the theorem is presented in Subsection 2.6.3. Theorems 2.3 and 2.4 offer a new point of view on $\Delta_i^{(k)}$, $i = 2, ..., k$, and Theorem 2.4 provides the powerful tool MRL to analyze the limiting behavior of $\Delta_k^{(k)}$. Taking the limit for both sides of (2.30), we have

$$\lim_{k\to\infty} \Delta_k^{(k)} = 2 \lim_{k\to\infty} \rho\Big(\frac{\xi_{k-1}^{(k)} + \xi_k^{(k)}}{2}\Big),$$

which shows that the limiting behavior of $\Delta_k^{(k)}$ is related to the limiting behavior of MRL, $\rho(x)$. In addition, (2.28) and (2.29) provide two different methods to calculate the limiting value of $\rho(x)$. From the information about the hazard function rather than the probability density function, we can compute the limiting value of $\rho(x)$ based on (2.28). The following example illustrates the two ways for the limiting value of $\rho(x)$.

Example 2.3 *Consider the Weibull distribution with pdf*

$$p(x;\sigma,m) = \frac{m}{\sigma^m}x^{m-1}e^{-(x/\sigma)^m}, \quad \sigma > 0, m > 0 \textit{ and } m \neq 1, x > 0.$$

Its hazard function $h(x)$ is $h(x) = \frac{m}{\sigma^m}x^{m-1}$. From (2.28), the limiting value of $\rho(x)$ as $x \to \infty$ is

$$\lim_{x\to\infty} \rho(x) = \lim_{x\to\infty} \frac{1}{h(x)} = \lim_{x\to\infty} \frac{\sigma^m}{m}\frac{1}{x^{m-1}} = \begin{cases} +\infty, & 0 < m < 1, \\ 0, & m > 1. \end{cases}$$

Moreover, we have

$$\ln p(x) = (m-1)\ln\Big(\frac{x}{\sigma}\Big) - \Big(\frac{x}{\sigma}\Big)^m + \ln\Big(\frac{m}{\sigma}\Big),$$

and

$$\frac{\mathrm{d}}{\mathrm{d}x}\ln p(x) = \frac{m-1}{x} - \frac{m}{\sigma^m}x^{m-1}.$$

From (2.29), the limiting value of $\rho(x)$ as $x \to \infty$ is

$$\lim_{x\to\infty} \rho(x) = \lim_{x\to\infty} \frac{-1}{\frac{\mathrm{d}}{\mathrm{d}x}\ln p(x)} = \lim_{x\to\infty} \frac{1}{\frac{m}{\sigma^m}x^{m-1} - \frac{m-1}{x}} = \begin{cases} +\infty, & 0 < m < 1, \\ 0, & m > 1. \end{cases}$$

which is consistent with the previous results.

The limiting behavior of $\Delta_k^{(k)}$ for some useful distributions on $\mathbb{R}^+$ is summarized in Table 2.4, where the distributions are given in Table 1.4.

TABLE 2.4
Limit behavior of $\Delta_k^{(k)}$ for some distributions on $\mathbb{R}^+$

Distribution	Notation	Limiting value of $\Delta_k^{(k)}$
Weibull	$We(m,\sigma), m > 1, \sigma > 0,$	0
Weibull	$We(m,\sigma), 0 < m < 1, \sigma > 0,$	$+\infty$
Gamma	$Gamma(\alpha,\beta), \alpha > 0, \beta > 0,$	2β
Exponential	$EP(\lambda), \lambda > 0,$	2β for all $k \geq 2$
F	$F(d_1,d_2), d_1, d_2 \in \mathbb{N}^+$	$+\infty, d_1 > 0, d_2 > 4$
Beta prime	$BP(\alpha,\beta), \alpha > 0, \beta > 2.$	$+\infty$
Lognormal	$LN(\mu,\sigma^2), \mu \in \mathbb{R}, \sigma > 0,$	$+\infty$
Inverse	$IG(\mu,\sigma), \mu > 0, \sigma > 0$	$\frac{2\mu^2}{\sigma}$

Example 2.3 shows that the parameter m takes effect on the limiting behavior of $\Delta_k^{(k)}$ in the Weibull distribution. Notice that the Weibull distribution for $m = 1$ degenerates into the exponential distribution and $\Delta_n^{(n)}$ is exactly equal to 2β for all $n \geq 2$ in the exponential distribution according to Theorem 2 in Xu et al. (2021) or Theorem 3.13 in this monograph.

To satisfy the condition $E(X^2) < \infty$, the parameter d_2 in the F distribution should be larger than 4, and the parameter β in the beta prime distribution should be larger than 2. It is well known that the limit value of $\Delta_k^{(k)}$ for the gamma distribution is independent of α, and the limit value of $\Delta_k^{(k)}$ for some distributions is independent of μ. Meanwhile some of them converge to the finite values, and some converge to ∞.

2.4.2 Continuous distributions supported on $\mathbb{R}$

Extending the MRL into a real-valued random variable on $\mathbb{R}$ because negative survival time makes no sense. Although MRL cannot be explained with the knowledge of survival analysis, Lemma 2.2 can still be used for calculating the limiting value of $\Delta_k^{(k)}$ (cf. Govil and Aggarwal (1983) and Calabria and Pulcini (1987)).

Table 2.5 lists the limiting behavior of $\Delta_k^{(k)}$ for some proper distributions on $\mathbb{R}$. To satisfy the condition $E(X^2) < \infty$, the parameter ν in the Student's t distribution should be larger than 2.

TABLE 2.5
Limit behavior of $\Delta_k^{(k)}$ for some distributions on $\mathbb{R}$

Distribution	Notation	Limiting value of $\Delta_k^{(k)}$
Normal	$N(\mu,\sigma^2), \mu \in \mathbb{R}, \sigma^2 > 0$	0
Logistic	$L(\mu,\sigma), \alpha \in \mathbb{R}, \beta > 0$	2β
Student's t	$t_\nu, \nu \in \mathbb{N}^+$	$+\infty, \nu > 2$

Remark:

1. If the set of MSE-RPs is unique and the corresponding probability density function is symmetric about zero, then the MSE-RPs must be symmetric about zero. Notice that the MSE-RPs of the normal distribution, logistic distribution, and t distribution are unique; see Theorem 2.2 in Li and Flury (1995). Thus, their MSE-RPs are symmetric, and

$$\lim_{k\to\infty} \Delta_2^{(k)} = \lim_{k\to\infty} \Delta_k^{(k)}.$$

2. The standard normal distribution is the limiting distribution of the t distribution as the degree of freedom $\nu \to \infty$, but their limiting behavior of $\Delta_k^{(k)}$ is quite different.
3. For the logistic distribution with $\alpha = 0$ and $\beta = 1$, the asymptotic behavior of MSE-RPs of the logistic distribution may be relevant to that of the exponential distribution since $\lim\limits_{x\to\infty} \frac{p(x)}{e^{-x}} = 1$.

2.4.3 Continuous distributions supported on a bounded interval

For a random variable defined on a bounded interval $[m, M]$, the limiting value of $\Delta_k^{(k)}$ as $k \to \infty$ is always 0. Xu et al. (2023) stated this fact in the following theorem.

Theorem 2.5 *Suppose that a random variable X is on a bounded interval $[m, M]$ with continuous pdf $p(x)$ and $E(X^2) < +\infty$. Denote $m \le \xi_1^{(k)} < ... < \xi_k^{(k)} \le M$ are the k MSE-RPs of X, and $p(\xi_k^{(k)} - y)$ converges uniformly to $p(M - y)$, $y \in [0, M - m]$, then*

$$\lim_{k\to\infty} \Delta_k^{(k)} = 0.$$

The proof of this theorem will be shown in the Subsection 2.6.4. The conclusion of Theorem 2.5 is illustrated in following example.

Example 2.4 *Consider the uniform distribution $U[a, b]$, its pdf is*

$$p(x) = \begin{cases} \frac{1}{b-a}, & a \le x \le b, \\ 0, & \text{otherwise}. \end{cases}$$

The MSE-RPs of $U[a, b]$ is $\xi_i^{(k)} = a + (b - a)\frac{2i-1}{2k}$, $i = 1, ..., k$ and the corresponding MSE is $\frac{(b-a)^2}{12k^2}$ (refer to Theorem 3.12). Obviously, $\Delta_i^{(k)} = \frac{b-a}{k}$ and $\lim\limits_{k\to\infty} \Delta_i^{(k)} = 0$ for all $i = 2, ..., k$.

Next, we provide an upper bound of MSE for any k points in the domain of X.

Theorem 2.6 *Suppose a random variable X has a continuous pdf $p(x)$ on $[m, M]$ and $E(X^2) < \infty$. Let k points set $\{x_i\}_{i=1}^k$ satisfy $m \le x_1 < ... < x_k \le M$, then the MSE of $\{x_i\}_{i=1}^k$ for X is bounded, that is,*

$$\mathrm{MSE}(x_1, ..., x_k) = E(\min_{i=1,...,k} (X - x_i)^2) \le \frac{h^2}{4},$$

where $h = \max\limits_{i=1,...,k+1} \{x_i - x_{i-1}\}$ provided that $x_0 = 2m - x_1$ and $x_{k+1} = 2M - x_k$.

The proof of this theorem is given in Subsection 2.6.5.

It is noticeable that the upper bound $h^2/4$ does not depend on the random variable X but on the dispersion of the points $\{x_i\}_{i=1}^n$ on $[m, M]$. Obviously, the range of h is $(M - m)/k \le h < M - m$. In order to show that h can be equal to $(M - m)/k$, we construct the following $\{b_i\}_{i=1}^k$ below,

Corollary 2.2 *If a continuous random variable X is defined on a bounded interval $[m, M]$ and*

$$b_i = m + (M - m)\frac{2i - 1}{2k}, i = 1, ..., k,$$

then

$$E(\min_{i=1,...,k} (X - b_i)^2) \le \frac{(M - m)^2}{4k^2}.$$

Based on the definition of MSE-RPs, we can obtain the following result.

Corollary 2.3 *If a continuous random variable X is defined on a bounded interval $[m, M]$ and $\xi_1^{(k)} < ... < \xi_k^{(k)}$ are the k RPs of X, then*

$$E(\min_{i=1,\dots,k}(X - \xi_i^{(k)})^2) \le \frac{(M-m)^2}{4k^2}. \quad (2.31)$$

Remark: In fact, (2.31) provides us an upper bound for the optimal MSE when X is defined on a bounded interval $[m, M]$.

2.4.4 Tail and MSE-RPs of statistical distributions

The statistical simulation results in Section 2.2 show that different types of RPs in the distribution may affect the estimation of mean, variance, skewness, and kurtosis performance. The tail of the distribution may play some role in the performance.

In the literature, many kinds of classification methods for statistical distribution are proposed based on different standards. Pagès and Sagna (2012) classified the distribution based on the maximal radius of RPs, and they proposed two kinds of distributions: *distributions with polynomial tails* and *distributions with hyper-exponential tails*. In addition, Embrechts et al. (1997) classified the distributions based on the convergence rate of pdf $p(x)$ as $x \to \infty$, and they defined the so-called *heavy-tailed distribution* and *light-tailed distribution*, in which the exponential distribution is used as a standard for comparison. The following formal definitions are from Foss et al. (2013).

Definition 2.7 *The univariate random variable X with the distribution function F is said to have a heavy-tailed if*

$$\int_{-\infty}^{+\infty} e^{\lambda x} \mathrm{d}F(x) = \infty, \quad \textit{for all } \lambda > 0;$$

otherwise, F is said to have a light-tailed if

$$\int_{-\infty}^{+\infty} e^{\lambda x} \mathrm{d}F(x) < \infty, \quad \textit{for some } \lambda > 0.$$

From Table 2.4, we have

1. **Heavy-tailed distribution**: the lognormal, Weibull with shape parameter $0 < m < 1$, and the Student's t distributions; many subclasses, such as the subexponential, long-tailed, and the regularly varying distributions.
2. **Light-tailed distribution**: the gamma, exponential, normal distributions, and Weibull distribution with the shape parameter $m > 1$; any univariate distributions supported on a bounded interval.

The equivalent version of Definition 2.7 is following Definition 2.8 (cf. Theorem 2.6. in Foss et al. (2013)). From this definition, it is intuitively known that the tail of the heavy-tailed distribution is heavier than the tail of the exponential distribution.

Definition 2.8 *The univariate random variable X with distribution function F is said to have a heavy-tailed if*

$$\lim_{x \to \infty} e^{\lambda x} P(X > x) = \infty \quad \textit{for all } \lambda > 0.$$

Moreover, long-tailed distribution is discussed here as an essential subclass of heavy-tailed distribution, which is more commonly used in applications. The formal definition of long-tailed distribution was given by Foss et al. (2013) as follows.

Definition 2.9 *The univariate random variable X with distribution function F is said to have a long-tailed if*

$$\lim_{x\to\infty} P(X > x + t|X > x) = 1 \quad \text{for all } t > 0,$$

or equivalently

$$\lim_{x\to\infty} \frac{\bar{F}(x+t)}{\bar{F}(x)} = 1. \quad \text{for all } t > 0,$$

where $\bar{F}(x) = P(X > x)$.

An interesting interpretation is that given a univariate random variable follows a long-tailed distribution and exceeds a threshold, it is likely to exceed a higher threshold. That is to say, if the situation is terrible, it will likely be worse than you think.

Similar to further analysis, we rewrite MRL as

$$\rho(x) = \frac{\int_x^\infty \bar{F}(w)\mathrm{d}w}{\bar{F}(x)} = \frac{\int_0^\infty \bar{F}(w+x)\mathrm{d}w}{\bar{F}(x)} = \int_0^\infty \frac{\bar{F}(w+x)}{\bar{F}(x)}\mathrm{d}w.$$

$\rho(x) \to \infty$ is $x \to \infty$ for long-tailed distribution. This conjecture is true, and we state this as the following lemma (Example 6.2.5 in Embrechts et al. (1997)).

Lemma 2.3 *Assume that X is a random variable with support unbounded to the right and cdf F. If for all $y \in \mathbb{R}$, we have*

$$\lim_{x\to\infty} \frac{\bar{F}(x+y)}{\bar{F}(x)} = 1,$$

then

$$\lim_{x\to\infty} \rho(x) = +\infty.$$

Now, the relationship between a long-tailed distribution and MSE-RPs is presented, and it is well shown that the long-tailed distribution can imply $\Delta_k^{(k)} \to \infty$ as $k \to \infty$.

Theorem 2.7 *If the univariate random variable X supported on $\mathbb{R}^+$ is long-tailed, then*

$$\lim_{k\to\infty} \Delta_k^{(k)} = +\infty.$$

Proof. Based on Lemma 2.3 and Theorem 2.4, the assertion is obvious.

□

2.5 Discrete Analogues

A continuous random variable may be characterized either by its probability density function (pdf), moment-generating function (mgf), moments, hazard rate function, etc. Construction of a discrete analog from a continuous distribution is based on the principle of preserving one or more characteristic properties of the continuous one.

The various methods by which discrete analog Y with pmf of a continuous random variable X with density $p(x)$ may be constructed can be classified as follows, summarized by Chakraborty (2015):

1. Difference equation analogs of Pearsonian differential equation.
2. Probability mass function (pmf) of Y retains the form of the pdf of X, and the support of Y is determined from the full range of X.

   $$P(Y = k) = \frac{f(k)}{\sum_{j=-\infty}^{\infty} f(j)}, \; k = 0, \pm 1, \pm 2, \ldots$$

   The distribution generated by this technique may not always have a compact form due to the normalizing constant. This approach is a discrete analog of the gamma distribution, general Dirichlet distribution, discrete normal, lognormal, exponential, Laplace, generalized exponential, and gamma distributions.
3. The pmf of Y retains the form of the pdf of X, and the support of Y is determined from a subset of the range of X.

   This is a modification of method 2. Here, the discrete analog is derived from having finite support. Barbiero (2010) applied this method to estimate the reliability of systems for which stress and strength are defined as complex functions whose reliability is not derivable through analytic techniques.
4. Survival function (SF) of Y retains the form of the survival function of X, and the support of Y is determined from the full range of X.

   The discrete SF is defined as $S_Y(k) = P(Y \geq k)$, and accordingly the cdf $F_Y(k) = P(Y \leq k)$ is related to the SF as $S_Y(k) = 1 - F_Y(k-1)$. This method proposes many discrete distributions, such as the discrete exponential, Weibull, geometric Weibull, normal, Rayleigh, Maxwell, extended exponential, and Lindley distributions.
5. Cumulative distribution function (cdf) of Y retains the form of the cdf of X and support of Y is determined from a subset of the range of X.

   The discrete Y is defined by

   $$P(Y = k) = F_X(k + \delta) - F_X(k - [1 - \delta]), \; 0 < \delta < 1, \tag{2.32}$$

   where the parameter δ is well chosen so that the first two raw moments of X and Y remains close (Roy and Dasgupta 2001). Except for a shift in the location by δ, the pdf in (2.32) preserves the form of the original cdf. The choice of the number of discretization points is derived from the trade-off between the accuracy and computational load of the results. Hence, to reduce the computational load, the number of points should be small, say 3. Meanwhile, for increasing accuracy, the number of points should be significant, say 9.
6. Hazard (failure) rate function of Y retains the form of the hazard (failure) rate function of X.

   This method preserves the hazard rate function. If the underlying continuous random variable X has the SF $S_X(x) = P(X \geq x)$ and hazard rate function $\lambda_X(x) = f_X(x)/S_X(x)$ then the SF of the discrete analog Y is given by

   $$P(Y \geq k) = \prod_{i=1}^{k-1} (1 - \lambda_X(i)), \; k = 1, \ldots, m,$$

and pmf of Y is

$$P(Y = k) = \lambda_X(k) \prod_{i=1}^{k-1} (1 - \lambda_X(i)).$$

This way, we propose new discrete Weibull, Rayleigh, and Lomax distributions.

7. Moments of Y and X up to a certain order coincide.

   This is a process proposed by Luceno (1999) of approximating a continuous random variable X having pdf $f(x)$, $a \leq x \leq b$ by a discrete random variable Y taking values $y_1, y_2, \ldots, y_{MN}$ having pmf $P(Y = y_j) = p_j; j = 1, \ldots, MN$ such that both X and Y have the same finite r^{th} moment for $r = 0, 1, \ldots, 2N - 1$ and their cdf coincides at least at $M + 1$ points. Here the support of random variable Y is $\{y_1, \ldots, y_{MN}\}$ is roots of polynomial equation of Nth degree and not necessarily the integers. As such, the derived distribution is not discrete in the sense of having integer support. So this is rather a way of approximating a pdf $f_X(x)$ by a pmf $\{p_j, j = 1, \ldots, MN\}$ that retains common moments and cdf value at the points of discretization. This method may require a solution for the system of non-linear equations in addition to the requirement for the existence of moments of the continuous distribution. Some new uniform, normal, gamma, and beta distributions are defined.

8. Any interval domain, any theoretically possible mean-variance pair for Y. Hagmark (2008) presented a method for constructing nonnegative integer-valued random variables with any interval domain, any theoretically possible mean and covariance pair, and different shapes using the basic tool of a mean-preserving discretization method in which the discretization of a nonnegative initial random variable X with cdf $F_X(x)$ is defined as the count variable Y with cdf

   $$F_Y(Y \leq y) = \int_y^{y+1} F_X(x)dx,$$

   By this construction, it was shown that (i) $E(X) = E(Y)$ and (ii) $\text{Var}(X) \leq \text{Var}(Y) \leq \text{Var}(X) + \min\{E(X), 1/4\}$.

9. Two-stage composite methods.

   A discrete random variable Y is defined by its pmf

   $$P(Y = k) = S_X(k)^\alpha - S_X(k+1)^\alpha, k = 0, 1, \ldots; \alpha > 0,$$

   where $S_X(x)$ is the SF of X.

Thus, the continuous distribution function is first exponentiated, and the resulting continuous distribution is then discretized using methodology 4. Gupta and Kundu (1999) derived generalized exponential distribution having pdf and cdf as follows.

$$f(x) = \alpha\lambda(1 - e^{-\lambda x})^{\alpha-1} e^{-\lambda x}, \ F(x) = (1 - e^{-\lambda x})^\alpha, x, \alpha, \lambda > 0.$$

A discrete analogue of this distribution can be obtained with pmf

$$P(Y = k) = (1 - e^{-\lambda(k+1)})^\alpha - (1 - e^{-\lambda k})^\alpha, k = 0, 1, \ldots$$

The above review shows that the basic purpose of two kinds of discrete distributions is different. Discrete analogs want to define new discrete distributions that preserve one or more characteristic properties of the continuous one. A discrete distribution $\hat{F}(x)$ is called an approximation distribution to a given continuous distribution $F(x)$ if it keeps information with $F(x)$ as much as possible. The information is a wide concept here. It can be some distance between $\hat{F}(x)$ and $F(x)$, or between moments of $\hat{F}(x)$ and $F(x)$, or simulation from $\hat{F}(x)$ and $F(x)$ or resampling from $\hat{F}(x)$ and $F(x)$. This monograph focuses on constructing approximate distributions and their applications in statistical simulation and inference.

2.6 Proofs of Some Theorems in This Chapter

The proofs in this section are from Xu et al. (2023).

2.6.1 Proof of Theorem 2.2

Proof *Take the derivative of* $\mathrm{MSE}(\xi_1, ..., \xi_n)$ *with respect to* $\xi_1, ..., \xi_n$ *and set them equal to 0, we obtain the following equations:*

$$\begin{cases} \int_{-\infty}^{(\xi_1^{(k)}+\xi_2^{(k)})/2} (x-\xi_1^{(k)})p(x)\mathrm{d}x = 0, \\ \int_{(\xi_{i-1}^{(k)}+\xi_i^{(k)})/2}^{(\xi_i^{(k)}+\xi_{i+1}^{(k)})/2} (x-\xi_i^{(k)})p(x)\mathrm{d}x = 0, \quad i=2,\ldots,k-1, \\ \int_{(\xi_{k-1}^{(k)}+\xi_k^{(k)})/2}^{\infty} (x-\xi_k^{(k)})p(x)\mathrm{d}x = 0. \end{cases} \tag{2.33}$$

From (2.33), we have

$$\int_{(\xi_{i-1}^{(k)}+\xi_i^{(k)})/2}^{(\xi_i^{(k)}+\xi_{i+1}^{(k)})/2} (x-\xi_i^{(k)})p(x)\mathrm{d}x = 0, \ i=2,...,k-1.$$

Consequently,

$$\int_{(\xi_{i-1}^{(k)}+\xi_i^{(k)})/2}^{\xi_i^{(k)}} (x-\xi_i^{(k)})p(x)\mathrm{d}x + \int_{\xi_i^{(k)}}^{(\xi_i^{(k)}+\xi_{i+1}^{(k)})/2} (x-\xi_i^{(k)})p(x)\mathrm{d}x = 0,$$

where $i = 2, ..., k-1$, *then,*

$$\int_{\xi_i^{(k)}}^{(\xi_i^{(k)}+\xi_{i+1}^{(k)})/2} (x-\xi_i^{(k)})p(x)\mathrm{d}x = -\int_{(\xi_{i-1}^{(k)}+\xi_i^{(k)})/2}^{\xi_i^{(k)}} (x-\xi_i^{(k)})p(x)\mathrm{d}x.$$

Since $(x-\xi_i^{(k)})$ *is an integrable function that does not change the sign on the interval* $[\xi_i^{(k)}, (\xi_i^{(k)}+\xi_{i+1}^{(k)})/2]$, *it follows the mean value theorem that there exists* u_i *in* $(\xi_i^{(k)}, (\xi_i^{(k)} +$

$\xi_{i+1}^{(k)})/2)$ *such that*

$$\int_{\xi_i^{(k)}}^{(\xi_i^{(k)}+\xi_{i+1}^{(k)})/2}(x-\xi_i^{(k)})p(x)\mathrm{d}x = p(u_i)\int_{\xi_i^{(k)}}^{(\xi_i^{(k)}+\xi_{i+1}^{(k)})/2}(x-\xi_i^{(k)})\mathrm{d}x$$

$$= p(u_i)\left[\frac{1}{2}(x-\xi_i^{(k)})^2\right]_{\xi_i^{(k)}}^{(\xi_i^{(k)}+\xi_{i+1}^{(k)})/2} = p(u_i)\frac{1}{2}\left(\frac{\xi_i^{(k)}+\xi_{i+1}^{(k)}}{2}-\xi_i^{(k)}\right)^2$$

$$= \frac{1}{8}p(u_i)(\Delta_{i+1}^{(k)})^2 \quad i = 2, ..., k-1.$$

In a similar way, since $(x-\xi_i^{(k)})$ *is an integrable function that does not change sign on* $[(\xi_{i-1}^{(k)}+\xi_i^{(k)})/2, \xi_i^{(k)}]$*, it follows that there exists* w_i *in* $((\xi_{i-1}^{(k)}+\xi_i^{(k)})/2, \xi_i^{(k)})$ *such that*

$$-\int_{(\xi_{i-1}^{(k)}+\xi_i^{(k)})/2}^{\xi_i^{(k)}}(x-\xi_i^{(k)})p(x)\mathrm{d}x = -p(w_i)\int_{(\xi_{i-1}^{(k)}+\xi_i^{(k)})/2}^{\xi_i^{(k)}}(x-\xi_i^{(k)})\mathrm{d}x$$

$$= -p(w_i)\left[\frac{1}{2}(x-\xi_i^{(k)})^2\right]_{(\xi_{i-1}^{(k)}+\xi_i^{(k)})/2}^{\xi_i^{(k)}} = p(w_i)\frac{1}{2}\left(\frac{\xi_{i-1}^{(k)}+\xi_i^{(k)}}{2}-\xi_i^{(k)}\right)^2$$

$$= \frac{1}{8}p(w_i)(\Delta_i^{(k)})^2 \quad i = 2, ..., k-1.$$

Thus,

$$\frac{1}{8}p(u_i)(\Delta_{i+1}^{(k)})^2 = \frac{1}{8}p(w_i)(\Delta_i^{(k)})^2, \quad i = 2, ..., k-1.$$

Further,

$$\left(\frac{\Delta_{i+1}^{(k)}}{\Delta_i^{(k)}}\right)^2 = \frac{p(w_i)}{p(u_i)}, \quad i = 2, ..., k-1.$$

Since $(\xi_{i-1}^{(k)}+\xi_i^{(k)})/2 < w_i < \xi_i^{(k)} < u_i < (\xi_i^{(k)}+\xi_{i+1}^{(k)})/2$ *and* $p(x)$ *is strictly decreasing on the interval* $((\xi_{i-1}^{(k)}+\xi_i^{(k)})/2, (\xi_i^{(k)}+\xi_{i+1}^{(k)})/2)$ *for some* $i \in \{2, ..., k-1\}$*, it follows that* $p(w_i) > p(u_i)$*, and hence* $\Delta_{i+1}^{(k)} > \Delta_i^{(k)}$.

2.6.2 Proof of Theorem 2.3

Proof *Let* $s_i = (\xi_{i-1}^{(k)}+\xi_i^{(k)})/2,\ i = 2, ..., k-1$. *From*

$$\int_{s_i}^{s_{i+1}}(x-\xi_i^{(k)})p(x)\mathrm{d}x = 0,$$

in (2.33), it follows that the RPs can be expressed as

$$\xi_i^{(k)} = \frac{\int_{s_i}^{s_{i+1}} xp(x)\mathrm{d}x}{\int_{s_i}^{s_{i+1}} p(x)\mathrm{d}x}.$$

Recall DMRL in Definition 2.25,

$$\tilde{\rho}(x, y) = E(X-x|x \le X \le y) = \frac{\int_x^y (u-x)p(u)\mathrm{d}u}{\int_x^y p(u)\mathrm{d}u}.$$

Thus, the RPs-gap can be expressed as

$$\begin{aligned}\Delta_i^{(k)} &= \xi_i^{(k)} - \xi_{i-1}^{(k)} = 2(\xi_i^{(k)} - s_i) \\ &= 2\Big(\frac{\int_{s_i}^{s_{i+1}} xp(x)\mathrm{d}x}{\int_{s_i}^{s_{i+1}} p(x)\mathrm{d}x} - s_i\Big) \\ &= 2\frac{\int_{s_i}^{s_{i+1}} (x - s_i)p(x)\mathrm{d}x}{\int_{s_i}^{s_{i+1}} p(x)\mathrm{d}x} = 2\tilde{\rho}(s_i, s_{i+1}).\end{aligned}$$

2.6.3 Proof of Theorem 2.4

Proof *For convenience, let* $s_k = (\xi_{k-1}^{(k)} + \xi_k^{(k)})/2$. *By the last equation in (2.33)*

$$\int_{s_k}^{\infty} (x - \xi_k^{(k)})p(x)\mathrm{d}x = 0,$$

it follows that the largest RPs can be expressed as

$$a_n^{(n)} = \frac{\int_{s_n}^{\infty} xp(x)\mathrm{d}x}{\int_{s_n}^{\infty} p(x)\mathrm{d}x}.$$

Recall the definition of MRL,

$$\rho(x) = E(X - x|X \geq x) = \frac{\int_x^{\infty} (u - x)p(u)\mathrm{d}u}{\int_x^{\infty} p(u)\mathrm{d}u}.$$

Thus, the gap between the largest two MSE RPs can be expressed as

$$\begin{aligned}\Delta_k^{(k)} &= \xi_k^{(k)} - \xi_{k-1}^{(k)} = 2(\xi_k^{(k)} - s_k) \\ &= 2\Big(\frac{\int_{s_k}^{\infty} xp(x)\mathrm{d}x}{\int_{s_k}^{\infty} p(x)\mathrm{d}x} - s_k\Big) = 2\rho(s_k).\end{aligned}$$

2.6.4 Proof of Theorem 2.5

Proof *If* $\lim_{k\to\infty} \Delta_k^{(k)}$ *exists, then there are only two kinds of possible values for this limit, namely, 0 and a positive real number, according to the definition of* $\Delta_k^{(k)}$. *Next, we will show that this limit value can not be a positive real number by using the mathematical contradiction method.*

Suppose that

$$\lim_{k\to\infty} \Delta_k^{(k)} = c > 0$$

and we have

$$
\begin{aligned}
E(\min_{j=1,\dots,k}(X-\xi_j^{(k)})^2) &\geq \int_{(\xi_{k-1}^{(k)}+\xi_k^{(k)})/2}^{M}(x-\xi_k^{(k)})^2 p(x)\mathrm{d}x \\
&= \int_{(\xi_{k-1}^{(k)}+y_k^{(k)})/2}^{\xi_k^{(k)}}(\xi_k^{(k)}-x)^2 p(x)\mathrm{d}x + \int_{\xi_k^{(k)}}^{M}(x-\xi_k^{(k)})^2 p(x)\mathrm{d}x \\
&\geq \int_{(\xi_{k-1}^{(k)}+\xi_k^{(k)})/2}^{\xi_k^{(k)}}(\xi_k^{(k)}-x)^2 p(x)\mathrm{d}x \\
&= -\int_{\Delta_k^{(k)}/2}^{0} y^2 p(\xi_k^{(k)}-y)\mathrm{d}y = \int_0^{\Delta_k^{(k)}/2} y^2 p(\xi_k^{(k)}-y)\mathrm{d}y.
\end{aligned}
$$

For any $\varepsilon > 0$, we have

$$
\begin{aligned}
&\Big|\int_0^{\Delta_k^{(k)}/2} y^2 p(\xi_k^{(k)}-y)\mathrm{d}y - \int_0^{c/2} y^2 p(M-y)\mathrm{d}y\Big| \\
&= \Big|\int_0^{\Delta_k^{(k)}/2} y^2 p(\xi_k^{(k)}-y)\mathrm{d}y - \Big(\int_0^{\Delta_k^{(k)}/2} y^2 p(M-y)\mathrm{d}y + \int_{\Delta_k^{(k)}/2}^{c/2} y^2 p(M-y)\mathrm{d}y\Big)\Big| \\
&= \Big|\int_0^{\Delta_k^{(k)}/2} y^2 (p(\xi_k^{(k)}-y)-p(M-y))\mathrm{d}y - \int_{\Delta_k^{(k)}/2}^{c/2} y^2 p(M-y)\mathrm{d}y\Big| \\
&\leq \Big|\int_0^{\Delta_k^{(k)}/2} y^2 (p(\xi_k^{(k)}-y)-p(M-y))\mathrm{d}y\Big| + \Big|\int_{\Delta_k^{(k)}/2}^{c/2} y^2 p(M-y)\mathrm{d}y\Big| \\
&\leq \int_0^{\Delta_k^{(k)}/2} y^2 |p(\xi_k^{(k)}-y)-p(M-y)|\mathrm{d}y + \Big|\int_{\Delta_k^{(k)}/2}^{c/2} y^2 p(M-y)\mathrm{d}y\Big| \\
&\leq \varepsilon \int_0^{\Delta_k^{(k)}/2} y^2\mathrm{d}y + \frac{1}{2}\Big(\max_{y\in[0,M-m]} y^2 p(M-y)\Big)|c-\Delta_k^{(k)}| \\
&= \frac{(\Delta_k^{(k)})^3}{24}\varepsilon + \frac{1}{2}\Big(\max_{y\in[0,M-m]} y^2 p(M-y)\Big)|c-\Delta_k^{(k)}|.
\end{aligned}
$$

Since $p(x)$ is supported on $[m, M]$, it follows that $p(x) > 0$ on $[m, M]$. When $0 \leq y \leq c/2$, the value $p(M-y)$ can not always be zero. Thus,

$$
\lim_{k\to\infty}\int_0^{\Delta_k^{(k)}/2} y^2 p(\xi_k^{(k)}-y)\mathrm{d}y = \int_0^{c/2} y^2 p(M-y)\mathrm{d}y > 0.
$$

It is noticeable that the integral $\int_0^{c/2} y^2 p(M-y)\mathrm{d}y$ does not depend on k. Then, we have

$$
\begin{aligned}
\lim_{k\to\infty} E(\min_{j=1,\dots,k}(X-\xi_j^{(k)})^2) &\geq \lim_{k\to\infty}\int_0^{\Delta_k^{(k)}/2} y^2 p(\xi_k^{(k)}-y)\mathrm{d}y \\
&= \int_0^{c/2} y^2 p(M-y)\mathrm{d}y > 0.
\end{aligned}
$$

Recalling Lemma 2.1, it is a contradiction. That is, $\lim_{k\to\infty}\Delta_k^{(k)}$ cannot be a positive real number. In fact, we have conclusion $\lim\limits_{k\to\infty}\Delta_k^{(k)} = 0$.

2.6.5 Proof of Theorem 2.6

Proof

$$(x - x_i)^2 \leq \max\Big\{\Big(\frac{x_{i-1} + x_i}{2} - x_i\Big)^2, \Big(\frac{x_i + x_{i+1}}{2} - x_i\Big)^2\Big\}.$$

Because $p(x) > 0$, it yields

$$(x - x_i)^2 p(x) \leq \max\Big\{\Big(\frac{x_{i-1} + x_i}{2} - x_i\Big)^2, \Big(\frac{x_i + x_{i+1}}{2} - x_i\Big)^2\Big\} p(x).$$

After taking integration from both sides with respect to x on the interval $[(x_{i-1}+x_i)/2, (x_i+x_{i+1})/2]$, we have

$$\int_{(x_{i-1}+x_i)/2}^{(x_i+x_{i+1})/2} (x - x_i)^2 p(x) dx \leq$$
$$\int_{(x_{i-1}+x_i)/2}^{(x_i+x_{i+1})/2} \max\Big\{\Big(\frac{x_{i-1} + x_i}{2} - x_i\Big)^2, \Big(\frac{x_i + x_{i+1}}{2} - x_i\Big)^2\Big\} p(x) dx.$$

The following is straightforward,

$$\begin{aligned}
& E(\min_{i=1,\ldots,k} (X - x_i)^2) \\
&= \sum_{i=1}^{k} \int_{(x_{i-1}+x_i)/2}^{(x_i+x_{i+1})/2} (x - x_i)^2 p(x) \mathrm{d}x \\
&\leq \sum_{i=1}^{k} \int_{(x_{i-1}+x_i)/2}^{(x_i+x_{i+1})/2} \max\Big\{\Big(\frac{x_{i-1} + x_i}{2} - x_i\Big)^2, \Big(\frac{x_i + x_{i+1}}{2} - x_i\Big)^2\Big\} p(x) \mathrm{d}x \\
&= \sum_{i=1}^{k} \int_{(x_{i-1}+x_i)/2}^{(x_i+x_{i+1})/2} \max\Big\{\Big(\frac{x_i - x_{i-1}}{2}\Big)^2, \Big(\frac{x_{i+1} - x_i}{2}\Big)^2\Big\} p(x) \mathrm{d}x \\
&\leq \sum_{i=1}^{k} \int_{(x_{i-1}+x_i)/2}^{(x_i+x_{i+1})/2} \frac{h^2}{4} p(x) \mathrm{d}x \\
&= \frac{h^2}{4} \sum_{i=1}^{k} \int_{(x_{i-1}+x_i)/2}^{(x_i+x_{i+1})/2} p(x) \mathrm{d}x = \frac{h^2}{4} \int_m^M p(x) \mathrm{d}x = \frac{h^2}{4}.
\end{aligned}$$

3

Property and Generation of MSE-RPs of Univariate Distributions

This chapter focuses on the properties of the MSE-RPs of a continuous univariate distribution with finite variance. There are three kinds of methods for generating MSE-RPs: Theoretical, numerical, and a mixture of theoretical and numerical. The chapter introduces the theory and algorithms for generating MSE-RPs. The uniqueness of MSE-RPs will be studied. In this chapter, we always assume that the random variable X follows the distribution $F(x)$ with pdf $p(x)$ on $\mathcal{D}_X$ having finite mean $\mu = E(X)$ and variance $\sigma^2 = \text{Var}(X)$. The most detailed proofs of theorems are in the last section of the chapter. If the proof is short, the proof immediately follows the theorem.

3.1 MSE-RPs of $F(x)$ for $k = 1, 2$

A set of MSE-RPs of $F(x)$ is denoted by $\mathcal{Y} = \{\xi_1, \dots, \xi_k\}$ or $\mathcal{Y}^{(k)} = \{\xi_1^{(k)}, \dots, \xi_k^{(k)}\}$ if we need to emphasize the size k. The corresponding discrete random variable is Y_{MSE} or $Y_{MSE}^{(k)}$.

Theorem 3.1 *For $k = 1$ the MSE-RP(ξ) of X is the mean $E(X) = \mu$ with $MSE(\xi) = \sigma^2$.*

Proof *Denote ξ as the MSE-RP of F. Its MSE*

$$\begin{aligned}\text{MSE}(Y_{MSE}) &= E(X-\xi)^2 = E(X-\mu+\mu-\xi)^2 \\ &= E(X-\mu)^2 + (\mu-\xi)^2 + 2E(X-\mu)(\mu-\xi) \\ &= \sigma^2 + (\mu-\xi)^2 \geq \sigma^2,\end{aligned}$$

where the equal holds if and only if $\xi = \mu$, i.e., $Y_{MSE} = \xi$ with probability one. Obviously, $\text{MSE}(Y_{MSE}) = \sigma^2$.

The conclusion of Theorem 3.1 indicates that the MSE-RPs can be regarded as an extension of the mean. For the case $k = 2$, we can not find a theoretical solution for MSE-RPs for all distributions, but we can discover a theoretical solution for the symmetric distributions.

Theorem 3.2 *Let $\mathcal{Y} = \{\xi_1, \dots, \xi_k\}$ be a set of MSE-RPs for a random variable $X \sim F(x)$ which is symmetrical about the origin, then the set $\{-\xi_l, \dots, -\xi_k\}$ is also a set of MSE-RPs. Furthermore, if the set of MSE-RPs for $F(x)$ is unique, then*

$$-\xi_i = \xi_{k-i+1}, i = 1, \dots, \lfloor k/2 \rfloor, \tag{3.1}$$

DOI: 10.1201/9781003589389-3

where $\lfloor x \rfloor$ is the largest integer not exceeding x.

Proof *As $X \stackrel{d}{=} -X$ we have*

$$\begin{aligned} MSE(\mathcal{Y}) &= E(X - Y_{MSE})^2 = E(-X + Y_{MSE})^2 \\ &= E(X - (-Y_{MSE}))^2 \end{aligned}$$

that shows that the set of $\{-\xi_1, \ldots, -\xi_k\}$ are also a set of MSE-RPs of X. If the set of MSE-RPs for $F(x)$ is unique,; two sets, $\{\xi_1, \ldots, \xi_k\}$ and $\{-\xi_1, \ldots, -\xi_k\}$ are identified, and (3.1) holds.

A set of RPs $\{\xi_1, \ldots, \xi_k\}$ is called *symmetric about* 0 if $\{\xi_1, \ldots, \xi_k\} = \{-\xi_1, \ldots, -\xi_k\}$. As the standard normal distribution $X \sim N(0, 1)$ is symmetric, from Theorem 3.2, we have $-\xi_1 = \xi_2$ for $k = 2$. Fang and He (1982) gave two-point MSE-RPs, that is, $\xi_1 = -\sqrt{\frac{2}{\pi}}$, $\xi_2 = \sqrt{\frac{2}{\pi}}$ with MSE=0.2634. Flury (1990) gave a more general result. Here, we may use a different proof.

Theorem 3.3 *Let X be a continuous random variable with pdf $p(x)$, finite mean μ, and variance σ^2, and the distribution of X is symmetric about μ. Its two MSE-RPs are*

$$\xi_1 = \mu - E(|X - \mu|), \ \text{and}\ \xi_2 = \mu + E(|X - \mu|),$$

and the corresponding MSE is

$$\mathrm{MSE}(\xi_1, \xi_2) = \mathrm{Var}(|X - \mu|). \tag{3.2}$$

Proof *Denote $Z = X - \mu$. It is clear that Z has mean 0 and variance σ^2, and is symmetric about 0. From Theorem 3.2, Z has two RPs, $-\xi$ and ξ, where $\xi = E(|Z|) > 0$. Denote the pdf of Z by $p_Z(z)$. The MSE of $\{-\xi, \xi\}$ for Z is*

$$\begin{aligned} \mathrm{MSE}(-\xi, \xi) &= \int_{-\infty}^{0} (z + \xi)^2 p_Z(z) dz + \int_{0}^{\infty} (z - \xi)^2 p_Z(z) dz \\ &= 2 \int_{0}^{\infty} (z - \xi)^2 p_Z(z) dz. \end{aligned}$$

Note that $|Z|$ has a density function $2p_Z(z), z \geq 0$. The above equation's right-hand side arrives at the minimum at $\xi = E(|Z|)$. So, the MSE-RPs of X are $-\xi$ and ξ with related $\mathrm{MSE}(\xi_1, \xi_2)$ *as the variance of $|Z|$.*

Furthermore, for the random variable satisfying the conditions in Theorem 3.3, Flury (1990) claimed that the two points $\{\xi_1 = \mu - E(|X - \mu|),\ \ \xi_2 = \mu + E(|X - \mu|)\}$ provide a local minimum of $\mathrm{MSE}(\xi_1, \xi_2)$ if and only if

$$p(\mu)E(|X - \mu|) < \frac{1}{2}. \tag{3.3}$$

However, Gu and Mathew (2001) pointed out that the condition (3.3) is not a sufficient condition and gave a counterexample about the standard symmetric exponential distribution with pdf $p(x) = \frac{1}{2}e^{-|x|}$ and $\mu = 1$. It is easy to find that $p(\mu)E(|X - \mu|) = \frac{1}{2}$, but $\xi = \pm 1$ are MSE-RPs. Moreover, when $p(\mu)E(|X - \mu|) = \frac{1}{2}$, the two MSE-RPs of a symmetric random variable may not be symmetric. For example, let X have the density function

$$p(x) = \frac{1}{2E(|X|)} - \beta |x|^{\alpha} + o(|x|^{\alpha}), \alpha > 0,$$

which is symmetric about $\mu = 0$ and $p(\mu)E(|X - \mu|) = \frac{1}{2}$. Gu and Mathew (2001) indicated that it does not exist two symmetric MSE-RPs in the following cases: (i) $\beta < 0$;

(ii) $\beta > 0, \alpha > 2$; and

(iii) $0 < \beta < (3/2)(1/E(|X|)^3, \alpha = 2$.

While $\xi_1 = \mu - E(|X|)$ and $\xi_2 = \mu + E(|X|)$ are two MSE-RPs if $\beta > 0, \alpha < 2$ or $\beta > (3/2)(1/E(|X|)^3, \alpha = 2$.

Furthermore, for some symmetric discrete distributions, the corresponding MSE-RPs still are not symmetric. For example, let $p(i) = P(X = i) = \frac{1}{3}, i = -1, 0, 1$. It is clear $\mu = 0, E|X| = \frac{2}{3}$ and $p(0) = \frac{1}{3} > 0$. The two MSE-RPS are $\{\frac{1}{2}, -1]$ or $\{1, -\frac{1}{2}\}$ are not symmetric about $\mu = 0$. This fact is against the result $\xi_1 = \mu - E(|X|) = -\frac{2}{3}$ and $\xi_2 = \mu + E(|X|) = \frac{2}{3}$.

Then, Gu and Mathew (2001) provided the following necessary and sufficient condition for a local minimum of MSE(c_1, c_2).

Theorem 3.4 *Let a random variable X have a density $p(x)$ symmetric about the mean μ and continuous at μ, and have a finite variance. Then*

(a) When $p(\mu)E(|X - \mu|) < \frac{1}{2}$, $\xi_1 = \mu - E(|X - \mu|)$ and $\xi_2 = \mu + E(|X - \mu|)$ provide a local minimum of MSE(c_1, c_2).

(b) $p(\mu)E(|X - \mu|) > \frac{1}{2}$ implies for any small number $\epsilon > 0$ there always exists $c \in N_\epsilon$ such that

$$2E|X - \mu| \int_{\mu-|c|}^{\mu+|c|} (|c| - |x - \mu|)dF(x) + \int_{\mu-|c|}^{\mu+|c|} (|c| - |x - \mu|)^2 dF(x) > c^2,$$

i.e., $\xi_1 = \mu - E(|X - \mu|)$ and $\xi_2 = \mu + E(|X - \mu|)$ do not provide a local minimum of MSE(c_1, c_2).

The proof The proof can refer to Gu and Mathew (2001).

3.2 Basic Properties of MSE-RPs

Many authors have studied some basic properties of MSE-RPs for $X \sim F$. Let $Y_{MSE}^{(k)}$ be the random variable defined in Definition 2.1. Fei (1991) gave the following result.

Theorem 3.5 *Let X be a continuous random variable with pdf $p(x)$, finite mean μ, and variance σ^2. Then we have*

(A) $E(Y_{MSE}^{(k)}) = E(X) = \mu$;

(B) $E((Y_{MSE}^{(k)})^t) = E(X(Y_{MSE}^{(k)})^{t-1}), t = 2, 3, \ldots, m-1,$ *if* $E(|X|^m) < \infty$;

(C) $\mathrm{Var}(Y_{MSE}^{(k)}) = \mathrm{Var}(X) - MSE(Y_{MSE}^{(k)})$;

(D) $\mathrm{MSE}(Y_{MSE}^{(k)}) \geq \mathrm{MSE}(Y_{MSE}^{(k+1)}), k > 1$.

The proof of the theorem can refer to Subsection 3.13.1.

In Theorem 3.5, the property (A) can be regarded as "unbiased mean". The property (C) gives a decomposition of variance of X as

$$\mathrm{Var}(X) = \mathrm{Var}(Y_{MSE}^{(k)}) + MSE(Y_{MSE}^{(k)}).$$

The loss function $L(Y_{MSE}^{(k)})$ defined in (2.15) shows how much percentage of $\mathrm{Var}(X)$ lost by use $Y_{MSE}^{(k)}$ replacing X.

Fei (1991) showed that as $k \to \infty$,

$$\mathrm{MSE}(Y_{MSE}^{(k)}) \to 0, \text{ and } L(Y_{MSE}^{(k)}) \to 0. \tag{3.4}$$

We get the following result by combining this fact and Theorem 3.5.

Corollary 3.1 *Under the assumption in Theorem 3.5 we have*
(1) $\mathrm{Var}(Y_{MSE}^{(k)}) \leq \mathrm{Var}(Y_{MSE}^{(k+1)})$;
(2) $\mathrm{Var}(Y_{MSE}^{(k)}) \to \mathrm{Var}(X)$ *as* $k \to \infty$;
(3) $\mathrm{Cov}(X, Y_{MSE}^{(k)}) = \mathrm{Var}(Y_{MSE}^{(k)})$;
(4) $\mathrm{Corr}(X, Y_{MSE}^{(k)}) = [1 - L(Y_{MSE}^{(k)})]^{1/2}$;
(5) $Y_{MSE}^{(k)} \overset{2ed}{\to} F$ *and* $F_{MSE}^{(k)} \to F(x)$ *as* $k \to \infty$, *for the converges in* r*th mean can refer to Subsection 1.3.3.*

Proof *The fact (1) follows from Theorem 3.5(C), and the fact (2) is a consequence of (3.4). Moreover, from Theorem 3.5 we have*

$$\begin{aligned}
\mathrm{Cov}(X, Y_{MSE}^{(k)}) &= E(XY_{MSE}^{(k)}) - E(X)E(Y_{MSE}^{(k)}) \\
&= E[(Y_{MSE}^{(k)})^2] - E^2(Y_{MSE}^{(k)}) = \mathrm{Var}(Y_{MSE}^{(k)}),
\end{aligned}$$

which proves the fact (3). As a consequence

$$\begin{aligned}
Corr(X, Y_{MSE}^{(k)}) &= \frac{Cov(X, Y_{MSE}^{(k)})}{\sqrt{\mathrm{Var}(X)\,\mathrm{Var}(Y_{MSE}^{(k)})}} = \sqrt{\frac{\mathrm{Var}(Y_{MSE}^{(k)})}{\mathrm{Var}(X)}} \\
&= \sqrt{1 - \frac{\mathrm{MSE}(Y_{MSE}^{(k)})}{\mathrm{Var}(X)}} = \sqrt{1 - L(Y_{MSE}^{(k)})}.
\end{aligned}$$

The fact (5) follows from the theory of the convergence in r*th mean implies the convergence in the distribution in Subsection 1.3.3 (*$r = 2$ *in our case). The proof is complete.*

Let $Z = a + bX$. It is easy to see

$$E[Z - (a + bY_{MSE})]^2 = E[a + bX - (a + bY_{MSE})]^2 = b^2 E[X - Y_{MSE}]^2,$$

which means that the MSE-RPs of X have the linearity. Then, we have the following result.

Theorem 3.6 *Let* $\mathcal{Y} = \{\xi_1 < \ldots < \xi_k\}$ *be the MSE-RPs of* X*, then* $Z = a + bX$ *has MSE-RPs to be* $a + b\mathcal{Y} = \{a + b\xi_1, \ldots, a + b\xi_k\}$ *with MSE to be* $b^2 MSE(\mathcal{Y})$.

Example 3.1 *Uniform distributions on an interval. Let* $X \sim U(-\theta, \theta), \theta > 0$*. The mean and variance of* X *are 0 and* $\theta^2/3$*, respectively.*

For $k = 1$*, the MSE-RP is* $\mu = 0$ *with MSE* $\sigma^2 = \theta^2/3$.

For $k = 2$, it is easy to find $E(|X|) = \theta/2$. The two MSE-PPs of X are given by $\pm\theta/2$. From Theorem 3.6 the result is easy to extend to $X \sim U(\mu - \theta, \mu + \theta)$, that is, the two MSE-RPs are $\mu \pm \theta/2$.

In many applications one meets the uniform distribution on (a, b), i.e., $X \sim U(a, b), a < b$, or $U \sim U(0, 1)$. Subsection 3.5.1 shows that the MSE-RPs of $U(0, 1)$ are

$$\xi_{U(0,1)} = \left\{ \frac{1}{2k}, \frac{3}{2k}, \ldots, \frac{2k-1}{2k} \right\}. \tag{3.5}$$

From Theorem 3.6, the k MSE-RPs of $X \sim U(\mu - \theta, \mu + \theta)$ are $\xi_i = \mu - \frac{\theta}{k}(k - 2i + 1), i = 1, \ldots, k$. Zoppè (1995) gave more results for those MSE-RPs. More discussion can refer to Subsection 3.5.1.

Example 3.2 *Let $X \sim N(\mu, \sigma^2)$. If we can find MSE-RPs for the standard normal distribution $Z \sim N(0, 1)$, from Theorem 3.6 we can obtain the MSE-RPs of $N(\mu, \sigma^2)$.*

It is known that $E(|Z|) = \sqrt{\frac{2}{\pi}}$ and two MSE-PPs of Z are $\xi_1 = -\sqrt{\frac{2}{\pi}}$, $\xi_2 = \sqrt{\frac{2}{\pi}}$. Consequently, the two MSE-RPs of X are $\mu \pm \sigma\sqrt{2/\pi}$. This result has been pointed out by different approaches, for example, Cox (1957), Max (1960), Fang and He (1982), Flury (1990), and Fang et al. (2014).

Moreover, we have the following result for the MSE-RPs of a univariate random variable X supported on a bounded interval $[m, M]$.

Theorem 3.7 *Suppose that the univariate random variable X is supported on a bounded interval $[m, M]$ with continuous probability density function $p(x)$ and $\mathbb{E}(X^2) < +\infty$. If $\xi_1 < \ldots < \xi_k (k \geq 2)$ are k RPs of X, then*

$$m < \xi_1 < \ldots < \xi_k < M.$$

The proof of the theorem is given in Subsection 3.13.2.

3.3 Uniqueness of RPs of Univariate Distributions

For a continuous random variable $X \sim F(x)$ with pdf $p(x)$ and finite mean μ and variance σ^2, the existence of MSE-RPs is no problem. A natural question is whether they are unique or not? If MSE-RPs of a distribution are unique for each size n, the corresponding theory and applications can be much easier. In this section, we will introduce the theory and examples of the uniqueness of the MSE-RPs of distributions. This is a challenging problem. Fleisher (1964) gave a sufficient condition "log-concavity" for the uniqueness of the MSE-RPs; that is, the pdf $p(x)$ is differentiable, and the derivative $(\log p(x))'$ is strictly decreasing. Trushkin (1982) proved that a log-concave probability density function has a unique set of MSE-RPs. Let us introduce the definition of log-concavity.

Definition 3.1 *A function $f : \mathbb{R}^p \to \mathbb{R} \cup \{+\infty\}$ is said to be convex if*

$$\lambda f(\boldsymbol{x}) + (1 - \lambda) f(\boldsymbol{y}) \geq f(\lambda \boldsymbol{x} + (1 - \lambda)\boldsymbol{y}),$$

for all $\boldsymbol{x}, \boldsymbol{y} \in \mathbb{R}^p$ and for all λ with $0 \leq \lambda \leq 1$, where it is understood that the inequality is satisfied if $f(\boldsymbol{x})$ or $f(\boldsymbol{y})$ is equal to $+\infty$. This inequality implies that the set

$$ED_f = \{\boldsymbol{x} \in \mathbb{R}^p \mid f(\boldsymbol{x}) < +\infty\}$$

is a convex set called the effective domain of f.

A point (or vector) $\boldsymbol{x}$ is said to be a global minimum (global optimum) of f if the inequality

$$f(\boldsymbol{x}) \leq f(\boldsymbol{y})$$

holds for every $\boldsymbol{y}$, and $\boldsymbol{x}$ is a local minimum (local optimum) if this inequality holds for every $\boldsymbol{y}$ in some neighborhood of $\boldsymbol{x}$. Global optimality implies local optimality. The converse is not generally valid. This is true for convex functions.

Definition 3.2 *A continuous random variable X is said to have a log-concave density $p(x)$ if it satisfies*

$$\log(p(\lambda x + (1-\lambda)y)) \geq \lambda \log p(x) + (1-\lambda)\log p(y),$$

for all $\lambda \in (0,1)$ and all x, y in the support of X.

Log-concavity of the density is a well-known property, which is satisfied by many remarkable distributions, including the normal distribution. Table 3.1 lists some log-concave densities, where the kernel of $p(x)$ ignores some constant $p(x)$ and $\nabla^2 \log p(x)$ also ignores some constant so that the condition for log-concavity of $p(x)$ stays the same (see Table 1.4 for reference). The exponential distribution is the case of gamma distribution with $\alpha = 1$, and the uniform distribution $U(0,1)$ is the special case of beta distribution with $a = 1$ and $b = 1$.

Based on the log-concavity property, Trushkin (1982) gave the following result.

Theorem 3.8 *Assume that there exists an open interval $I = (V, W)$, $-\infty \leq V < W \leq +\infty$, such that the pdf $p(x)$ is continuous and positive inside this interval and $p(x) = 0$ outside this interval. Then, the concavity of $\log p(x)$ on the interval I is sufficient for the uniqueness of the MSE-RPs of $p(x)$.*

In fact, $\log p(x)$ is concave is equivalent to $(\log p(x))'' < 0$. By a straightforward calculation, we can derive that

$$(\log p(x))' = \frac{p(x)'}{p(x)}, \quad (\log p(x))'' = \frac{p(x)''p(x) - (p(x)')^2}{p(x)^2},$$

then

$$(\log p(x))'' < 0 \Leftrightarrow \frac{p(x)''p(x) - (p(x)')^2}{p(x)^2} < 0.$$

TABLE 3.1
Some log-concave densities

Distribution	Kernel of $p(x)$	$\nabla^2 \log p(x)$	Condition
Normal	$\exp\left\{-\frac{(x-\mu)^2}{2\sigma^2}\right\}$	$-\frac{x-\mu}{\sigma^2}$	$\sigma > 0$
student t-	$(1+\frac{x^2}{v})^{-(v+1)/2}$	$v\frac{(v+1)(x^2+1)}{(v+x^2)^2}$	$v \geq 3$
Gamma	$x^{\alpha-1}e^{-\beta x}$	$-\frac{\alpha-1}{x^2}$	$\alpha \geq 1, \beta > 0$
Beta	$x^{\alpha-1}(1-x)^{b-1}$	$-[\frac{\alpha-1}{x^2}+\frac{\beta-1}{1-x^2}$	$\alpha \geq 1, \beta \geq 1$
Logistic	$e^{-\frac{x-\mu}{\sigma}}/[1+e^{-\frac{x-\mu}{\sigma}}]^2$	$\frac{-2}{\sigma^2}e^{-\frac{x-\mu}{\sigma}}/[1+e^{-\frac{x-\mu}{\sigma}}]^2$	$\sigma > 0$
Gumbel	$e^{-\frac{x-\mu}{\sigma}}\exp\{-e^{-\frac{x-\mu}{\sigma}}\}$	$\frac{-1}{\sigma^2}e^{-\frac{x-\mu}{\sigma}}$	$\sigma > 0$
Weibull	$x^{m-1}e^{-(x/\sigma)^m}$	$-(m-1)[\frac{1}{x^2}+\frac{m}{\sigma^m}x^{m-2}]$	$m \geq 1, \sigma > 0$

Thus, we can also say that $p(x)$ is log-concave if

$$p(x)''p(x) - (p(x)')^2 < 0. \tag{3.6}$$

A mathematical proof can refer to Trushkin (1982).

When the distribution of X is symmetric about 0, Li and Flury (1995) gave a weaker sufficient condition in the following theorem.

Theorem 3.9 *For a univariate symmetric distribution about 0, if* $p(x)''p(x) - 2(p(x)')^2 < 0$, *then there exists a unique set of MSE-RPs for any size* k.

The proof is picked up from Li and Flury (1995) and given in Subsection 3.13.3. The condition in the theorem is weaker than the strongly unimodal condition. Tarpey (1994) defined that a distribution with density $p(x)$ is *strongly unimodal* if $\log p(x)$ is concave. That is, if it is strongly unimodal,

$$\frac{p(x)''p(x) - (p(x)')^2}{p^2(x)} = [\log(p(x)]'' < 0$$

then

$$\frac{p(x)''p(x) - 2(p(x)')^2}{p^2(x)} < 0.$$

Example 3.3 *The student t-distribution with degrees of freedom* v *has a density function*

$$t(x) = \frac{\Gamma((v+1)/2)}{\sqrt{v\pi}\Gamma(v/2)}(1 + \frac{x^2}{v})^{-(v+1)/2}.$$

By a straightforward calculation, we have

$$t'(x) = -\frac{(v+1)x}{v+x^2}t(x), \ t''(x) = \frac{(v+1)[(v+2)x^2 - v]}{(v+x^2)^2}t(x),$$

$$[t \cdot t'' - 2(t')^2](x) = -v\frac{(v+1)(x^2+1)}{(v+x^2)^2}t^2(x) < 0,$$

if $v \geq 3$. *Yamamoto and Shinozaki (2000) proved that 2 MSE-RPs of the t-distributions with degrees of freedom* $v \geq 3$ *are unique. Zhou and Wang (2016) further showed that the* k *MSE-RPs of t-distributions are unique when* $v \geq 3$.

Tarpey (1994) showed that the uniqueness and symmetry of MSE-RPs do not necessarily hold when the pdf is unimodal and symmetric but not log-concave. Strong unimodality is shown to be a sufficient condition for the optimal cutpoint to occur at the mean. Tarpey (1994) gave the following result.

Theorem 3.10 *Suppose* $p(x)$ *is a symmetric, strongly unimodal density. Then, there exists a unique set of two self-consistent points.*

It is known that MSE-RPs are self-consistent points, while the self-consistent points are not necessarily MSE-RPs.

Example 3.4 *For a single cutpoint, Flury (1990) showed that if the loss function* L, *has a local minimum at* $\mu = E[X]$ *for a random variable* X *which is symmetric about* μ, *then*

$$p(\mu)E|X - \mu| \leq \frac{1}{2}. \tag{3.7}$$

On the other hand, if $p(\mu)E|X-\mu| < \frac{1}{2}$*, then the loss function* L_2 *has a local minimum at* $x = \mu$*. This result is inconclusive when* $p(\mu)E|X-\mu| = \frac{1}{2}$*, which is the case for the double exponential distribution with density* $c\exp|x|^{\alpha}, \alpha > 0$ *and the t-distribution on two degrees of freedom.*

Example 3.5 *Consider distributions with densities of the form*

$$p(x) = c \cdot \exp|x|^{\alpha}, \alpha > 0.$$

For $\alpha > 1$ *the distributions are strongly unimodal, and there exists a unique set of two self-consistent points, which must be symmetric MSE-RPs.*

Example 3.6 *The set of the k MSE-RPs of a univariate normal distribution is unique. Consider a location mixture of two normal distributions with density*

$$p(x) = \frac{1}{2\sqrt{2\pi}} e^{-(x-\mu)^2/2} + \frac{1}{2\sqrt{2\pi}} e^{-(x+\mu)^2/2}$$

Let $g(x) = 2\sqrt{2\pi}p(x)$*. By a straightforward calculation, we have*

$$g(x)g''(x) - 2(g'(x))^2 \leq \left(-\frac{4}{3}\mu^2 + \frac{4}{3}\right) e^{-x^2+\mu^2}.$$

Then $p(x)p''(x) - 2(p'(x))^2 < 0$ *if* $|\mu| \in (0,1)$*. Then* $p(x)$ *is unimodal if* $|\mu| \leq 1$ *and bimodal if* $|\mu| > 1$*. The reader can refer to Li and Flury (1995) and Kieffer (1983) for the details. Li et al. (2022) gave a more general theorem that will be presented in Theorem 3.24 of this monograph. Apply their theorem to this example. The MSE-RPs are unique if* $|\mu| \leq \frac{1}{\sqrt{2}}$*.*

Example 3.7 *Generalized gamma double-sided distribution. Its density is of the form*

$$p(x) = \frac{1}{2\Gamma(a)}(\lambda|x|)^{a-1}e^{-\lambda|x|}, \tag{3.8}$$

with the variance $\sigma^2 = a(a+1)/\lambda^2$*. Kabal (1984) mentioned the sufficient condition (log-concavity) for the uniqueness of MSE-RPs of this distribution to be*

$$\frac{\partial^2 \log p(x)}{\partial x^2} < 0. \tag{3.9}$$

When $a = 1$*, it is the Laplacian distribution; for modeling the measured distribution of speech signals, the general gamma density with* $a = \frac{1}{2}$ *has the density (3.8)*

$$p(x) = \frac{1}{2\sqrt{\pi}} \frac{1}{\sqrt{\lambda|x|}} e^{-\lambda|x|}.$$

The log-concavity test for this density gives

$$\frac{\partial^2 \log p(x)}{\partial x^2} = -\frac{2(a-1)}{x^2}, x \neq 0.$$

For $a \geq 1$*, a unique stationary point exists*

For $k = 2$*, we can obtain two solutions:*

$$\xi_1 = -\tfrac{1}{\sqrt{3}}, \xi_2 = \tfrac{1}{\sqrt{3}} \text{ with MSE} = 2/3.$$
$$\xi_1 = -0.266, \xi_2 = 1.509 \text{ with MSE} = 0.599.$$

Furthermore, the uniqueness of MSE-RPs may not be guaranteed for the distributions that do not have the log-concavity property. Yamamoto and Shinozaki (2000) gave a sufficient condition of uniqueness of two MSE-RPs for univariate symmetric distributions, which are not necessarily unimodal. Primarily, a class of location mixtures, including the normal ones, is shown to have two unique MSE-RPs.

Theorem 3.11 *Let the probability density $p(x)$ of X be symmetric around the origin. If the following condition holds*

$$\frac{c}{2F(c)-1} > \frac{\int_c^{\infty}(x-c)p(x)dx}{1-F(c)} \quad \text{for } \forall c \in \{c|0.5 < F(c) < 1\}. \tag{3.10}$$

Then X has two MSE-RPs that are uniquely and symmetrically around the origin.

Section 1.1.2 introduces the mixture distributions in (1.4). A special case is given as follows:

Definition 3.3 *Let F be a distribution symmetric around the origin with pdf $p(x)$. Let F_μ is two-component location mixtures with equal proportions*

$$p_\mu(x) = p(x+\mu)/2 + p(x-\mu)/2, \tag{3.11}$$

where $p_\mu(x)$ is the pdf of F_μ and $\mu > 0$. We call this type of mixtures as mirrored mixtures.

Corollary 3.2 *Let F be a continuous distribution. If $c/(2F(c)-1)$ is non-decreasing for $c > 0$ and F satisfies condition (3.10); then the mixture distribution of the form (3.11) also satisfies condition (3.10) for arbitrary $\mu > 0$.*

Example 3.8 *Mirrored mixtures of symmetric log-concave distributions always have two MSE-RPs uniquely. The typical examples of symmetric log-concave distributions are normal distributions $N(0,\sigma^2)$ and triangular distributions $F(x) : a - |x|/a; for \ -a < x < a, a > 0$. Though Laplace distributions*

$$\frac{1}{2\lambda}\left\{\exp(-|x|/\lambda)\right\}, x \in \mathbb{R}, \lambda > 0,$$

are not strictly log-concave, we can easily show that the condition (3.10) holds for them, and therefore for their mirrored mixtures.

Example 3.9 *Scale mixtures of normal distributions. The cdf is*

$$F_\sigma(x) = \Phi(x)/2 + \Phi(x/\sigma)/2, \sigma > 1.$$

If $\sigma > \pi - 1 + \sqrt{\pi^2 - 2\pi}$, Flury (1990) showed that two MSE-RPs are not unique. Tarper (1994) also treated this situation and derived a sufficient condition on σ for the log-concave density function. We can confirm analytically that the inequality (3.10) holds, and consequently, the two MSE-RPs are unique for $\sigma < 2.21$. This range is slightly wider than that obtained by Tarpey (1994). Furthermore, we investigate both sides of the inequality (3.10) numerically and see that it holds for $\sigma \leq \pi - 1 + \sqrt{\pi^2 - 2\pi}$. And the two RPs are unique if $\sigma \leq \pi - 1 + \sqrt{\pi^2 - 2\pi}$.

Trushkin (1982) proved that for mean square error criterion in the cases of Gaussian, Laplacian, Rayleigh, and χ^2 distributions defined on a finite or infinite interval, a locally optimal k-level quantizer for any k is unique. Uniqueness and symmetry of MSE-RPs and self-consistent points have been widely discussed; see Tarpey (1994); Zoppè (1995); Graf and Luschgy (2000); Gu and Mathew (2001).

3.4 Generation of MSE-RPs of Univariate Distributions

The generation of MSE-RPs is significant for applications. Let us review the definition of MSE-RPs in Definition 2.1. Suppose that a random variable $X \sim F(x)$ has a density function $p(x)$ with finite mean μ and variance σ^2. A set of points of $\mathcal{Y}^k = \{\xi_1, \xi_2, \cdots \xi_k\}$ minimizes the mean square error $(MSE)_\sigma$

$$\text{MSE}(\mathcal{Y}) = \int_{-\infty}^{+\infty} \min_i (x - \xi_i)^2 p(x) dx \tag{3.12}$$

are called MSE-RPs of $F(x)$ or X. Section 2.3 defines MSE-RPs of multivariate continuous distributions. In this section, we use RPs short for MSE-RPs. To find a set of RPs is an optimal problem including some difficulties:

1. the objective function is multivariate on simplex $\{\xi_1 < \xi_2 < \ldots < \xi_k\}$ for the univariate distribution or on Voronoi regions in both univariate and multivariate distributions;
2. the objective function might not be differentiable in the whole domain;
3. the minimum of the objective function is not unique, and the objective function may multiply on the domain.

The classical optimization methods can not solve this kind of problem.

There are three main different approaches for the generation of RPs:
(a) theoretic approach or combining the theoretical approach and computational calculation;
(b) to employ the k-means methods to find approximate RPs that can be applied to all univariate and multivariate distributions; and
(c) to solve a system of nonlinear equations.

The approach (a) can be used for very few distributions, like the uniform distribution in a finite interval; see Example 3.1 and Section 2.4.3. Section 3.3 introduces the generation of RPs for the exponential distribution by combining the theoretical approach and computational calculation.

The approach (b) can be applied to continuous univariate and multivariate distributions. It seems to us that Pollard (1982) is the first one to propose this approach. Along this line Lloyd (1982) proposed two trial-and-error methods. This approach is easy to implement, but it needs to choose a good quality initial and a large number of training samples.

Proposed approach (c) was by Max (1960) and Fang and He (1982). The latter gave more details for the numerical algorithm of the normal distribution. From the theory of optimization, the necessary condition for the existence of RPs is the existence of the following system of equations by taking derivatives of MSE($\mathcal{Y}$) with respective to $\xi_1, \ldots, \xi_k$, separately

$$\begin{cases} \int_{-\infty}^{(\xi_1+\xi_2)/2} xp(x)dx = \xi_1 p_1, \\ \int_{(\xi_{i-1}+\xi_i)/2}^{(\xi_i+\xi_{i+1})/2} xp(x)dx = \xi_i p_i, i = 2, \ldots, k-1, \\ \int_{(\xi_{k-1}+\xi_k)/2}^{\infty} xp(x)dx = \xi_k p_k. \end{cases} \tag{3.13}$$

This system of nonlinear equations needs some efficient algorithms to solve. Fang and He (1982) proposed an algorithm and gave some theoretic adjustments. This algorithm has been applied to the normal, t-, gamma, and Weibull distributions, the Pearson family of distributions, and some mixture distributions. The details of the algorithm will be introduced in Section 3.7.

Fang and Wang (1991) suggested the use of a sequential number-theoretic method for optimization (SNTO) to solve the system of equations (3.13), where SNTO is introduced in Section 1.13.

Trushkin (1982) and Chakraborty et al. (2021) gave comprehensive reviews. When the L_1-distance error (instead of mean square error) is employed, the RPs of a univariate distribution have been studied by Graf and Luschgy (2000). Yu (2022) proved that when the density function of a univariate bounded continuous random variable fluctuates slightly, its L_1-RPs are unique. He also proved the uniqueness of L_1-RPs of some special bounded distributions which are not strongly unimodal, especially the truncated distributions of many distributions.

3.5 MSE-RPs for Some Proper Distributions

This section discusses MSE-RPs of three kinds of univariate distributions: the uniform distribution in an interval, exponential and related distributions, and Laplace distribution. The generation of MSE-RPs for many other distributions will be discussed in several sections later in this chapter. Numerical results of the most useful univariate distributions can be calculated using BNU-HKBU United International College (UIC) web software. This web can help users calculate the MSE-RPs with related probabilities online for the distributions listed in Table 3.2. https://fst.uic.edu.cn/isci/Representative/representative_real_time.htm

TABLE 3.2
MSE-RPs of the distributions in the UIC web

Normal	Log Normal	Generalized normal	Log-Skew-Normal
Beta-Normal	Skew-Normal	Alpha-Skew-Normal	Log-Skew-Normal
Inverse Gaussian	Arcsine-Gaussian	Mixtrue of two normal	
Beta	Beta Prime	Gamma	Chi-square
Exponential	Laplace	Beta Laplace	Asymmetric Laplace
Student's t	Skew t	F	Pareto
Logistic	Double Logistic	Double Gaussian	Double Exponential
Weibull	Gumbel	Generalized Extreme Value	

3.5.1 Uniform distribution on an interval

This subsection considers the RPs for the uniform distribution $U(0,1)$. The uniform distribution $U(0,1)$ has played an important role in statistical simulation. We can quickly obtain RPs for $U(0,1)$, and then we can have RPs for $U(m,M)$. The result is rewritten in the following theorem.

Theorem 3.12 *Let $X \sim U(0,1)$, then its k MSE-RPs are given by*

$$\xi_i = \frac{2i-1}{2k}, \quad i = 1, ..., k. \tag{3.14}$$

The proof of the theorem can refer to Subsection 3.13.4. It is interesting to know that MSE-RPs of $U(0,1)$ are the same as QMC-RPs or NT-net in (0,1), see Fang and Wang (1994).

Example 3.10 *Let $X \sim U(\mu-\theta, \mu+\theta)$. Then the k RPs are*

$$\frac{2j-1}{k}\theta + \mu - \theta, j = 1, 2..., k. \tag{3.15}$$

The RPs of $U(m, M)$ are

$$m + (M-m)\frac{2j-1}{2k}, j = 1, \ldots, k. \tag{3.16}$$

3.5.2 Exponential and related distributions

In this section, we focus on the exponential distribution. The cdf and pdf of the exponential distribution, denoted by $EP(\lambda)$, are

$$F(x) = \begin{cases} 0, & x \leq 0, \\ 1 - e^{-\lambda x}, & x > 0, \end{cases} \quad p(x) = \begin{cases} 0, & x \leq 0, \\ \lambda e^{-\lambda x}, & x > 0, \end{cases} \tag{3.17}$$

respectively. For $X \sim EP(\lambda)$ it is known that

$$E(X) = \frac{1}{\lambda}, \mathrm{Var}(X) = \frac{1}{\lambda^2},$$

the skewness $Sk(X) = 2$, and the kurtosis $Ku(X) = 6$. Furthermore,

$$E(X|X > a) = a + \frac{1}{\lambda}, \mathrm{Var}(X|X > a) = \frac{1}{\lambda^2}.$$

The famous memoryless property of $EP(\lambda)$ is

$$P(X > x + y | X > y) = P(X > x), \text{ for } x > 0, y > 0.$$

Denote $\xi_1^{(k)} < \xi_2^{(k)} < \ldots < \xi_k^{(k)}$ be k RPs of $EP(\lambda)$. Graf and Luschgy (2007) pointed out

$$\mathrm{MSE}(\xi_1^{(k)}, \ldots, \xi_k^{(k)}) = (\xi_1^{(k)})^2. \tag{3.18}$$

It means that $\mathrm{MSE}(\xi_1^{(k)}, ..., \xi_k^{(k)})$ only depends on the location of the first representative point $\xi_1^{(k)}$ for all k. Xu et al. (2021) gives a comprehensive study on RPs of the exponential distribution.

Theorem 3.13 *Let $X \sim EP(\lambda)$ and Y_{MSE} be its MSE approximate random variable, we have*

$$E(Y_{MSE}) = E(X) = \frac{1}{\lambda}, \ \mathrm{Var}(Y_{MSE}) = \frac{1}{\lambda^2} - (\xi_1^{(k)})^2; \tag{3.19}$$

$$\xi_k^{(k)} - \xi_{k-1}^{(k)} = \frac{2}{\lambda} \ \textit{for} \ k \geq 2; \tag{3.20}$$

The proof of the theorem is given in Subsection 3.13.5. Furthermore, the gap $\{\xi_{j+1}^{(k)} - \xi_j^{(k)}\}, j = 1, \ldots, k-1$ is increasing.

Theorem 3.14 *Let $X \sim EP(\lambda)$ and Y_{MSE} be its MSE approximate random variable. Then the sequence*

$$\{\xi_{j+1}^{(k)} - \xi_j^{(k)}\}_{j=1}^{k-1}, k \geq 3,$$

is an increasing sequence.

The proof of the theorem is long and is given in Subsection 3.13.5. Based on Theorem 3.14, we obtain the relationships RPs gaps between the number of RPs k and $k+1$.

Theorem 3.15 *Let $X \sim EP(\lambda)$. We have*

$$\xi_{i+1}^{(k+1)} - \xi_i^{(k+1)} = \xi_i^{(k)} - \xi_{i-1}^{(k)}, \quad i = 2, ..., k-1; \tag{3.21}$$

$$\xi_2^{(k+1)} - \xi_1^{(k+1)} = 2\xi_1^{(k)}, k \geq 1. \tag{3.22}$$

The proof of the theorem can refer to Subsection 3.13.5. Based on those theoretical results, Xu et al. (2021) proposed the algorithm for generating MSE-RPs of the exponential distribution. The idea of the algorithm is to use the iteration method from $k = 1$ to $k \geq 2$.

1. $k = 1$: $\xi_1^{(1)} = E(X)$ with MSE $= \sigma^2$ by Theorem 3.1.
2. From Theorem 3.15, we can obtain $\xi_1^{(2)}$ by using the bisection methods to solve equation below

$$e^{\xi_1^{(2)}}(1 - \xi_1^{(2)}) = 2e^{-1},$$

   and get $\xi_1^{(2)} = 0.593624$.
3. $\xi_2^{(2)} = 2 + \xi_1^{(2)} = 2.593624$ can be calculated by using Theorem 3.15.
4. By using Theorem 3.15 again, $\xi_1^{(3)}$ is the solution of

$$e^{\xi_1^{(3)}}(1 - \xi_1^{(3)}) = (\xi_1^{(2)} + 1)e^{-\xi_1^{(2)}},$$

   and we obtain $\xi_1^{(3)} = 0.423954$.
5. By use of Theorem 3.15 agian, $\xi_2^{(3)} = 2\xi_1^{(2)} + \xi_1^{(3)} = 1.611202$ and $\xi_3^{(3)} = \xi_2^{(3)} + \xi_2^{(2)} - \xi_1^{(2)} = 3.611202$.

In a similar way we can find RPs of $EP(1)$ for any $k \geq 1$, and the new algorithm can be expressed as follows:

Suppose we have found the RPs $\xi_j^{(k)}, j = 1, ..., k$ of $EP(1)$, the following three steps can help us find $\xi_j^{(k+1)}, j = 1, ..., k+1$.

Algorithm 3.1 (Algorithm of generation of MSE-RPs of $EP(1)$)

Step 1. *By using the bisection methods to solve $\xi_1^{(k+1)}$ from the equation*

$$e^{\xi_1^{(k+1)}}(1 - \xi_1^{(k+1)}) = (\xi_1^{(k)} + 1)e^{-\xi_1^{(k)}}.$$

Step 2. *Compute* $\xi_2^{(k+1)}$ *by using Theorem 3.15:*

$$\xi_2^{(k+1)} = 2\xi_1^{(k)} + \xi_1^{(k+1)}.$$

Step 3. *Determine* $\xi_j^{(k+1)}, j = 3, ..., k+1$ *by using Theorem 3.15:*

$$\xi_{i+1}^{(k+1)} = \xi_i^{(k+1)} + \xi_i^{(k)} - \xi_{i-1}^{(k)}, \quad i = 2, \ldots, k.$$

The MSE-RPs and related probabilities of the standard exponential distribution ($n \leq 15$) is given in Tables 3.3 and 3.4.

TABLE 3.3
MSE-RPs of $EP(1)$

k	MSE	ξ_1	ξ_2	ξ_3	ξ_4	ξ_5	ξ_6	ξ_7	ξ_8
1	1.000000	1.000000							
2	0.352390	0.593624	2.593624						
3	0.179737	0.423954	1.611202	3.611202					
4	0.108952	0.330079	1.177986	2.365234	4.365234				
5	0.073090	0.270351	0.930508	1.778415	2.965664	4.965664			
6	0.052427	0.228969	0.769671	1.429828	2.277735	3.464984	5.464984		
7	0.039439	0.198593	0.656531	1.197233	1.85739	2.705297	3.892546	5.892546	
8	0.030745	0.175343	0.572529	1.030467	1.571169	2.231326	3.079233	4.266482	6.266482
9	0.024640	0.156971	0.507656	0.904843	1.362781	1.903482	2.563640	3.411547	4.598796
10	0.020189	0.142087	0.456029	0.806715	1.203901	1.661839	2.202541	2.862698	3.710605
11	0.016844	0.129783	0.413958	0.727900	1.078585	1.475772	1.933710	2.474411	3.134569
12	0.014266	0.119442	0.379009	0.663183	0.977125	1.327811	1.724997	2.182935	2.723637
13	0.012238	0.110628	0.349512	0.609078	0.893253	1.207195	1.557880	1.955067	2.413005
14	0.010614	0.103026	0.324281	0.563165	0.822732	1.106906	1.420848	1.771534	2.168720
15	0.009293	0.096401	0.302452	0.523708	0.762592	1.022159	1.306333	1.620275	1.970961
k		ξ_9	ξ_{10}	ξ_{11}	ξ_{12}	ξ_{13}	ξ_{14}	ξ_{15}	
9		6.598796							
10		4.897854	6.897854						
11		3.982476	5.169725	7.169725					
12		3.383794	4.231701	5.418950	7.418950				
13		2.953707	3.613864	4.461771	5.649020	7.649020			
14		2.626658	3.167360	3.827517	4.675424	5.862673	7.862673		
15		2.368147	2.826085	3.366787	4.026944	4.874851	6.062100	8.062100	

3.5.3 Laplace distribution

The Laplace distribution, which is also called the double exponential distribution, is the distribution of differences between two independent variates with identical exponential distributions.

Definition 3.4 *A random variable has a Laplace*(μ, b) *distribution if its pdf is*

$$p(x|\mu, b) = \frac{1}{2b}e^{-\frac{|x-\mu|}{b}}, \tag{3.23}$$

where μ *is the location parameter and* $b > 0$ *is the scale parameter. Its cdf is given by*

$$F(x) = \begin{cases} \frac{1}{2}e^{-\frac{\mu-x}{b}}, & x \leq \mu, \\ 1 - \frac{1}{2}e^{-\frac{x-\mu}{b}}, & x \geq \mu. \end{cases} \tag{3.24}$$

TABLE 3.4
The corresponding prob. of RPs of $EP(1)$

k	ξ_1	ξ_2	ξ_3	ξ_4	ξ_5	ξ_6	ξ_7	ξ_8
1	1.000000							
2	0.796812	0.203188						
3	0.638531	0.288023	0.073446					
4	0.529534	0.300407	0.135505	0.034554				
5	0.451424	0.290490	0.164796	0.074335	0.018955			
6	0.393057	0.273989	0.176311	0.100022	0.045117	0.011505		
7	0.347903	0.256311	0.178667	0.114972	0.065224	0.029421	0.007502	
8	0.311979	0.239365	0.176347	0.122927	0.079103	0.044875	0.020242	0.005162
9	0.282738	0.223771	0.171687	0.126487	0.088171	0.056738	0.032187	0.014519
10	0.258484	0.209655	0.165930	0.127309	0.093792	0.065380	0.042072	0.023868
11	0.238047	0.196952	0.159747	0.126431	0.097003	0.071465	0.049817	0.032057
12	0.220596	0.185535	0.153506	0.124507	0.098541	0.075605	0.055700	0.038827
13	0.205522	0.175259	0.147404	0.121957	0.098918	0.078288	0.060066	0.044253
14	0.192372	0.165985	0.141544	0.119047	0.098496	0.079889	0.063228	0.048511
15	0.180800	0.157591	0.135975	0.115953	0.097524	0.080688	0.065445	0.051796
k	ξ_9	ξ_{10}	ξ_{11}	ξ_{12}	ξ_{13}	ξ_{14}	ξ_{15}	
9	0.003702							
10	0.010766	0.002745						
11	0.018186	0.008203	0.002092					
12	0.024985	0.014174	0.006394	0.001630				
13	0.030847	0.019850	0.011261	0.005080	0.001295			
14	0.035740	0.024913	0.016032	0.009095	0.004102	0.001046		
15	0.039740	0.029278	0.020409	0.013133	0.007450	0.003361	0.000857	

Laplace distribution has been widely applied in many fields, such as finance, the processing of geometric engineering data, voice and image data. The Laplace distribution is used to model symmetric data with long tails. The Laplace distribution used to model the errors, motivate the use of L_1-regression, in which parameter estimates are based on minimizing the sum of absolute values of the residuals rather than on least squares.

Chakraborty et al. (2021) mentioned the symmetry of MSE-RPs of a Laplace distribution. Due to the uniqueness and symmetry of MSE-RPs of a Laplace distribution, the k MSE-RPs of Laplace distribution can be obtained from the $k/2$ MSE-RPs of exponential distribution when k is even. In this situation, the optimal MSE of the Laplace distribution is the same as that of the exponential distribution. Chakraborty et al. (2021) mentioned the results, but they lacked details of proof. Xu et al. (2021) provided a detailed approach as follows.

Denote $\xi_1^{(k)} < \xi_2^{(k)} < \ldots < \xi_k^{(k)}$ be the k RPs of $Laplace(\mu, b)$ or $\xi_1 < \ldots < \xi_k$ for shout. Graf and Luschgy (2007) pointed out

$$E(\min_{j=1,\ldots,k} |X - \xi_j|^r) = \begin{cases} \xi^r_{k/2+1}, & \text{if } k \text{ is even,} \\ rb^r \int_0^{\xi_{(k+3)/2}/2b} x^{r-1} e^{-x} dx, & \text{if } k \text{ is odd,} \end{cases}$$

where

$$\{\xi_1, ..., \xi_k\} \in \underset{y_j \in \mathbb{R}, \quad j=1,\ldots,k}{\arg\min} \left\{ E(\min_{j=1,\ldots,k} |X - y_j|^r) \right\},$$

and $r \geq 1$.

Theorem 3.16 *Let the random variable X follow Laplace(μ, b) distribution and $\xi_1, \xi_2, ..., \xi_k$ be the n RPs of X with $\xi_1 < \xi_2 < ... < \xi_k$, $k \geq 2$, then*

$$\xi_k - \xi_{k-1} = 2b.$$

Namely, the final gap between successive MSE representative points is a constant b when Laplace(μ, b) is given.

The proof is given in Subsection 3.13.6.

Theorem 3.17 *Let random variable X follow Laplace(μ, b) and $\xi_1^{(k)}, \xi_2^{(k)}, ..., \xi_k^{(k)}$ be the k MSE-RPs of X with $\xi_1^{(k)} < \xi_2^{(k)} < ... < \xi_k^{(k)}$, $k \geq 2$, then*
(a)$\{\xi_{j+1}^{(k)} - \xi_j^{(k)}\}_{j=\lfloor (k+1)/2 \rfloor}^{k-1}$ for $k \geq 4$ is an increasing sequence.
(b) then $\xi_{i+1}^{(k+1)} - \xi_i^{(k+1)} = \xi_i^{(k)} - \xi_{i-1}^{(k)}, \quad i = \lfloor (k+1)/2 \rfloor + 1, ..., k$.

The proof is given in Subsection 3.13.6. Moreover, when the number of RPs increases from $2k$ to $2k+1$, we have the following results.

Theorem 3.18 *Let univariate random variable X follow Laplace(μ, b) distribution and $\xi_1^{(n)}, \xi_2^{(n)}, ..., \xi_n^{(n)}$ be the n MSE-RPs of X with $\xi_1^{(n)} < \xi_2^{(n)} < ... < \xi_n^{(n)}$, $n \geq 2$, then*

$$\xi_{k+2}^{(2k+1)} - \xi_{k+1}^{(2k+1)} = \xi_{k+2}^{(2k+1)} - \mu = 2\xi_{k+1}^{(2k)}.$$

If $\mu = 0$, then $\xi_{k+1}^{(2k+1)} = 0$, implying $\xi_{k+2}^{(2k+1)} = 2\xi_{k+1}^{(2k)}$.

The proof is given in Subsection 3.13.6.

Theorem 3.19 *Let $X \sim$ Laplace(μ, b). Its MSE-RPs have the following relations*

$$\xi_i^{(n+1)} < \xi_i^{(n)} < \xi_{i+1}^{(n+1)}, \quad i = 1, ..., n.$$

The proof is given in Subsection 3.13.6.

The idea of the existing algorithm for generating RPs of the double-exponential variates were formed by the difference of two independent exponential variates obtained by the inverse transform method in Tarpey (1997). Based on the above theorems, we use the following algorithm to generate MSE-RPs of Laplace distributions.

Algorithm 3.2 (Generation of MSE-RPs of a Laplace distribution)

Step 1. *If the number of MSE-RPs is $2k$, the positive entries can be obtained from the k MSE-RPs of the exponential distribution, then due to the symmetry of RPs of the Laplace distribution, all the n RPs can be determined.*

Step 2. *If the number of MSE-RPs is $2k+1$, by Theorem 3.17(b)*

$$\xi_{i+1}^{(2k+1)} = \xi_i^{(2k+1)} + \xi_i^{(2k)} - \xi_{i-1}^{(2k)}, i = k+1, \ldots, 2k.$$

Then from the symmetry property,

$$\xi_{k+1}^{(2k+1)} = 0, \quad \xi_i^{(2k+1)} = -\xi_{2k-i+2}^{(2k+1)}, \ i = 1, \ldots, k.$$

The gamma distributions are often used to model speech signal statistics. Kabal (1984) stated that the Laplace distribution occupies a unique place amongst the family of general gamma distributions. The pdf defines the generalized gamma double-sided distribution.

$$f(x) = \frac{1}{2\Gamma(a)}\lambda(\lambda|x|)^{a-1}e^{-\lambda|x|}$$

with variance $\sigma^2 = a(a+1)/\lambda^2$. When $a = 1$ it is the Laplacian distribution.

Therefore, the log-concavity seems necessary and sufficient for uniqueness, as in other cases, the gamma distributions are not log-concave.

3.6 K-Means Algorithm

The k-means clustering algorithm proposed by McQueen (1967) is an important algorithm in cluster analysis, machine learning, and computational geometry. See Hartigan (1975) and Hartigan and Wong (1979) for reference. The k-means algorithm has also played a useful role in the construction of MSE-RPs.

3.6.1 Traditional k-means algorithm

In the traditional k-means formulation, there is a set of n observations in $\mathbb{R}^p$ and the user needs to cluster those observations into k groups under a loss function ϕ. Solving this problem exactly is NP-hard, even with just two clusters.

The k-means algorithm begins with k arbitrary centers. Each point is then assigned to the nearest center, and each center is recomputed as the center of mass of all points assigned to it. These steps (assignment and center calculation) are repeated until the process stabilizes. One can check that the total error ϕ is monotonically decreasing, which ensures that no clustering is repeated during the procedure of the algorithm. The mean square error (MSE) has been popularly used as an error ϕ and is defined in Algorithm 3.3.

The process will permanently terminate since there are at most k^n possible clusterings. The convergency of the k-means refers to Diday (1974). In practice, very few iterations are usually required, making the k-means algorithm much faster than most competitors.

Algorithm 3.3 (k-means algorithm)

Given k, the k-means algorithm for a set of n data points $\boldsymbol{x}_1, \ldots, \boldsymbol{x}_n$ *in* $\mathbb{R}^p$ *is implemented in the following steps:*

Step 1: Let $t = 0$. *Choose k initial seed points* $\boldsymbol{y}_1^{(t)}, \ldots, \boldsymbol{y}_k^{(t)}$ *as the centers and input n data points into the nearest center one by one and obtain clusters* $S_j^{(t)}, j = 1, \ldots, k$. *Calculate the initial mean square error*

$$\mathrm{MSE}_t = \sum_{i=1}^{n}(\min_{1\le j\le k}\{(\boldsymbol{x}_i - \boldsymbol{y}_j^{(t)})^T(\boldsymbol{x}_i - \boldsymbol{y}_j^{(t)})\}. \tag{3.25}$$

Step 2: Compute the mean of $S_j^{(t)}, j = 1, \ldots, k$ *as the new seed points, denote as* $\boldsymbol{y}_1^{(t+1)}, \ldots, \boldsymbol{y}_k^{(t+1)}$ *and calculate the mean square error*

$$\mathrm{MSE}_{t+1} = \sum_{i=1}^{n}(\min_{1\le j\le k}\{(\boldsymbol{x}_i - \boldsymbol{y}_j^{(t+1)})^T(\boldsymbol{x}_i - \boldsymbol{y}_j^{(t+1)})\}. \tag{3.26}$$

Step 3: Assign each object to the cluster with the nearest seed point to form new clusters;

Step 4: Stop when the new clusters are the same as the previous ones; otherwise, let $t := t + 1$ *and go back to Step 2,*

In Step 4 of Algorithm 3.3, the convergence condition can be modified so that the new clusters are very close to the previous ones. For example, the L_2-norm between the new and previous clusters is very small. The k-means algorithm always converges to a local minimum based on a set of initial seed points and a set of data.

3.6.2 The k-means algorithm for finding MSE-RPs

Lloyd (1982) first applied the k-means algorithm for generating MSE-RPs of univariate distributions. Linde et al. (1980) applied the k-means algorithm to estimate the k MSE-RPs for a distribution $\boldsymbol{X} \sim F(\boldsymbol{x})$ in $\mathbb{R}^p$. Denote their method as the Linde-Buzo-Gray (LBG) algorithm. The detailed procedure is shown in Algorithm 3.4.

Algorithm 3.4 (LBG algorithm for generating MSE-RPs)

Step 1: Generating a training iid sequence $\boldsymbol{x}_1, \dots, \boldsymbol{x}_N$ *from the given distribution* $F(\boldsymbol{x})$ *by the Monte Carlo method, where* N *is much larger than* k*. Set* $t = 0$.

Step 2: Give an initial set of output vectors $\mathcal{Y}^{(t)} = \{\boldsymbol{y}_1^{(t)}, \dots, \boldsymbol{y}_k^{(t)}\}$ *and find the associated Voronoi partition* $\mathsf{S}^{(t)} = \{\mathsf{S}_1^{(t)}, \dots, \mathsf{S}_k^{(t)}\}$ *by assigning each* $\boldsymbol{x}_i$ *to the nearest region of the partition* S_t*, where*

$$\mathsf{S}_i^{(t)} = \{\boldsymbol{x} : ||\boldsymbol{x} - \boldsymbol{y}_i^{(t)}|| < ||\boldsymbol{x} - \boldsymbol{y}_j^{(t)}||, j \neq i\}, i = 1, \dots, k.$$

In the above formula, the norm can be chosen as L_p*-norm or, in particulate form,* L_2*-norm in most cases.*

Step 3: Calculate the sample mean, denoted by $\boldsymbol{y}_j^{(t+1)}$*, of* $\boldsymbol{x}_i$ *which falling in the region* $\mathsf{S}_j^{(t)}, j = 1, \dots, k$*. Let* $\mathcal{Y}^{(t+1)} = \{\boldsymbol{y}_1^{(t+1)}, \dots, \boldsymbol{y}_k^{(t+1)}\}$*. If* $\mathcal{Y}^{t+1} = \mathcal{Y}^t$*, deliver* $\mathcal{Y}^t$ *as approximated MSE-RPs, denoted as* $\boldsymbol{\xi} = \{\boldsymbol{\xi}_1, \dots, \boldsymbol{\xi}_k\}$ *with probability* $p_i = n_i/N$*, where* n_i *is the number of training sample in* $\mathsf{S}_i^{(t)}$*, otherwise to next step. The output is the discrete random vector* $\boldsymbol{Y}_{MSE}$ *with support points* $\boldsymbol{\xi} = \{\boldsymbol{\xi}_1, \dots, \boldsymbol{\xi}_k\}$.

Step 4: Let $t := t + 1$ *and go to step 2).*

The LBG algorithm suggests using a random sample of size k as the initial set of output vectors. Fang et al. (1994b) pointed out two problems of the LBG algorithm:

(i) the approach gives the local optimum, and the results depend on the initial points;

(ii) the training samples are generated by the Monte Carlo method, which is slow in the sense of the convergence rate $O_p(n^{-1/2})$.

Fang et al. (1994b) proposed the NTLBG algorithm for finding MSE-RPs of $\boldsymbol{X}$ based on the LBG algorithm and QMC method. The NTLBG algorithm will be introduced in Subsection 6.4.1.

3.6.3 Nonparametric and parametric k-means algorithms

The k-means algorithm in clustering analysis classifies the data. It does not request prior information, so it uses a nonparametric approach. To generate MSE-RPs of a given distribution F, one can create a set of random samples from F using the Monte Carlo methods.

This set of samples is regarded as a data set in the k-means algorithm. The process can be applied to any continuous distribution. Therefore, this way is the nonparametric approach.

For a univariate distribution $F(x)$ with pdf $p(x)$, one can use another process called the parametric k-means algorithm.

Algorithm 3.5 (The parametric k-means algorithm)

Step 1: For given pdf $p(x)$, the number of RPs: k, and $t = 0$, input a set of initial points $y_1^{(t)} < y_2^{(t)} < \cdots < y_k^{(t)}$. Determine a partition of $\mathbb{R}$ as

$$I_i^{(t)} = (a_{i-1}^{(t)}, a_i^{(t)}], i = 1, \ldots, k-1, I_k^{(t)} = (a_{k-1}^{(t)}, a_k^{(t)}),$$

where

$$a_0^{(t)} = -\infty, a_i^{(t)} = (y_{i-1}^{(t)} + y_i^{(t)})/2, i = 1, \ldots, k-1, a_k^{(t)} = \infty.$$

Step 2: Calculate probabilities

$$p_j^{(t)} = \int_{I_j^{(t)}} p(x)dx, j = 1, \ldots, k;$$

and the condition mean

$$y_j^{(t+1)} = \frac{\int_{I_j^{(t)}} xp(x)dx}{\int_{I_j^{(t)}} p(x)dx} = \frac{\int_{I_j^{(t)}} xp(x)dx}{p_j^{(t)}}.$$

Step 3: If two sets of $\{y_j^{(t)}\}$ and $\{y_j^{(t+1)}\}$ are identical, the process stops and deliver $\{y_j^{(t)}\}$ as MSE-RPs of the distribution with probabilities $\{p_j^{(t)}\}$; otherwise, let $t := t+1$ and go back to Step 1).

Stampfer and Stadlober (2002) called this algorithm the self-consistency algorithm, as the output set of RPs is self-consistent (not necessarily MSE-RPs).

Remarks

The initial points $y_1^{(t)} < y_2^{(t)} < \cdots < y_k^{(t)}, t = 0$ are very important. We might use the QMC-RPs of the distribution as initial points. A more detailed discussion can refer to Section 3.10. If the underlying distribution $F(x; \boldsymbol{\theta})$ with some unknown parameters $\boldsymbol{\theta}$ is to be estimated, denote an estimate by $\hat{\boldsymbol{\theta}}$, and then generate random samples as training samples. Section 5.1 will give more discussion on this issue.

Lloyd (1982) used the idea of parametric k-means method to propose the algorithm, namely *Trial-and-error algorithm*, for finding RPs of continuous univariate distributions, and gave three examples: the uniform $U(-\sqrt{3}, \sqrt{3})$, the normal $N(0, 1)$ and Laplace distributions. The asymptotic properties and the detailed proofs were also shown. Tarpey (2007) gave a detailed discussion on the parametric and nonparametric k-means algorithm. Zoppè (1995) reviewed some of the theory of MSE-RPs and presented a method of determining MSE-RPs of univariate continuous distributions.

3.6.4 k-means++

Arthur and Vassilvitskii (2007) developed k-means algorithm into k-means++. Let $\phi_{OPT,k}$ denote the optimal potential for a k-clustering on a given data set. The two authors proved that k-means++ is O(1)-competitive in the case where $\frac{\phi_{OPT,k}}{\phi_{OPT,k-1}} \leq e^2$. The question is that if this condition does not hold, the data is not well suited for clustering with the given value for k. Let ϕ_{OPT} denote the corresponding potential of ϕ. Arthur and Vassilvitskii (2007) showed that the k-means++ is $O(\log k)$-competitive. They indicated: "If $\mathcal{C}$ is constructed with k-means++, then the corresponding potential function ϕ satisfies $E[\phi] \leq 8(\ln k + 2)\phi_{OPT}$".

Arthur and Vassilvitskii (2007) proposed initializing k-means by choosing random starting centers with precise probabilities. Specifically, they decided on a point as a center with probability proportional to p's contribution to the overall potential. Letting ϕ denote the potential after choosing centers in this way, we show the following.

The k-means algorithm begins with an arbitrary set of cluster centers. We propose a specific way of choosing these centers. At any given time, let $D(\boldsymbol{x})$ denote the shortest distance from a data point $\boldsymbol{x}$ to the closest center we have already chosen. The following algorithm is called k-means++.

Algorithm 3.6 (The k-means++ algorithm)

Step 1: Set $j = 1$. *Choose an initial centroid* $\mathbf{w}_{0j}$ *randomly from the training set* $\mathcal{Y} = \{\mathbf{y}_1, \ldots, \mathbf{y}_N\}$, *i.e.,* $\mathbf{w}_{0j} = \mathbf{y}_i$ *for some* i. *In this step, each* $\mathbf{y}_i$ *has the same probability* $p_{ij} = 1/N$.

Step 2: Denote $D(\mathbf{y})$ *as the shortest distance from* $\mathbf{y}$ *to the closest centroid that has been selected. Then, the next centroid is selected by* $\mathbf{w}_{0,j+1} = \mathbf{y}_i$ *with corresponding probability computed by*

$$p_{i,j+1} = \frac{D(\mathbf{y}_i)^2}{\sum_{i=1}^{N} D(\mathbf{y}_i)^2}.$$

Step 3: If $j = k$, *then all the* k *initial centroids have been selected, and stop the procedure. Otherwise, let* $j = j + 1$ *return to Step 2.*

As the probability p_{ij} of each $\mathbf{y}_i$ in the training set becomes higher with a larger $D(\mathbf{y})^2$, so the k-means++ seeding method is trying to choose an initial centroid as far away from the selected centroids as possible each time. Yang et al. (2021) applied k-means++ for generation of MSE-RPs of multivariate distributions.

3.6.5 Spherical k-means clustering

Spherical k-means clustering addresses current data analysis challenges, especially in large collections of text documents. Solving spherical k-means clustering problems corresponds to finding optimal group memberships employing the cosine dissimilarity measure. More details refer to Hornik et al. (2012).

As the set of quasi-random numbers $\mathcal{Q}$ are on the unit sphere, a spherical k-means algorithm is applied to find the RPs of $\boldsymbol{w}$. The details of the spherical k-means algorithm refer to Dhillon et al. (2002). Let $\mathcal{S} = \{S_1, \cdots, S_N\}$ be the partition of $\boldsymbol{w}$ with initial points $\mathcal{Y}_0^{(RPs)} = \{\boldsymbol{y}_{01}^{(3)}, \cdots, \boldsymbol{y}_{0N}^{(3)}\}$ generated on unit sphere by NTM and $t = 0$, the new center of each partition is

$$\boldsymbol{y}_{t+1,j} = \frac{\sum_{\boldsymbol{q}_i \in S_j} \boldsymbol{q}_i}{\left\| \sum_{\boldsymbol{q}_i \in S_j} \boldsymbol{q}_i \right\|} \tag{3.27}$$

for $i = 1, \cdots, N_1$ and $j = 1, \cdots, N$. Let $d(\boldsymbol{q}_i, \boldsymbol{y}_j)$ be the distance between $\boldsymbol{q}_i$ and $\boldsymbol{y}_j$, so $\boldsymbol{q}_i \in S_j$ if $d(\boldsymbol{q}_i, \boldsymbol{y}_j) < d(\boldsymbol{q}_i, \boldsymbol{y}_h)$, $j \neq h$ and $j, h = 1, \cdots, N$. As the points $\boldsymbol{q}_i$ are on unit sphere, a distance suggested by Dhillon and Modha (2001) is applied, that is

$$d(\boldsymbol{a}, \boldsymbol{b}) = 1 - \cos\alpha = 1 - \frac{\boldsymbol{a}^T\boldsymbol{b}}{||\boldsymbol{a}|| \cdot ||\boldsymbol{b}||}, \tag{3.28}$$

where α is the angle between $\boldsymbol{a}$ and $\boldsymbol{b}$. The other steps of the spherical k-means algorithm are the same as the k-means algorithm.

3.6.6 Pulse-code modulation algorithm

The pulse-code modulation (PCM) algorithm is based on a problem from the transactions in the information industry. When handling a given ensemble of signals, the quantum values should be closer together in the voltage regions where signal amplitudes tend to drop, see Lloyd (1982). By dividing the voltage range of the signal into a finite number of subsets, one can transmit only the information indicating which subset a sample falls within rather than the exact value. It is defined that the noise signal is the difference between the receiver-output signal and the original signal, whereas the noise power is the average square of the noise signal. Given the signal's number of quanta and specific statistical properties, this algorithm determines which subsets and quanta will minimize the noise power. For a set of samples $[x_1, x_2, \ldots, x_n] \in \mathbb{R}^d$, where $\mathbb{R}^d$ is the data space of d dimensions, the algorithm tries to find a set of k cluster centers (centroids) $Y = [y_1, y_2, \ldots, y_k] \in \mathbb{R}^d$ that is a solution to the minimization problem:

$$E = \sum_{i=1}^{k}\sum_{j=1}^{n} d\left(y_i, x_{ij}\right)$$

Here is the pseudocode describing the algorithm:

1. Choose the number of clusters, metric (Squared Euclidian distance: $d = \sum_i^k (y_i - x_i)^2$) and the method to pick initial centroids.
2. Assign/Reassign all samples to closest cluster and update the partitions.
3. Recalculate centroids, update the new centroids to be the center of mass of the partition.
4. Repeat Steps 2 and 3 until the centroids no longer change.

This algorithm is well suited to analyses of large samples since other incremental k-means algorithms (McQueen's algorithm, Hartigan's algorithm) require storing the two nearest-cluster computations for each case, which is computationally expensive when processing large data sets.

3.6.7 Examples

A comparison will be made among the nonparametric k-means, parametric k-means and k-means ++ algorithms in terms of the mean squared error (MSE) between the estimated RPs and the actual RPs:

$$\text{MSE} = E\left[\sum_{j=1}^{k}\left|\left|\hat{\boldsymbol{\xi}_j} - \boldsymbol{\xi}_j\right|\right|^2\right]$$

where $\hat{\boldsymbol{\xi}_j}$ and $\boldsymbol{\xi}_j, j = 1, \ldots, k$, are the estimated k MSE-RPs and the true k MSE-RPs of the underlying distribution, respectively. To compute the MSE, the expectation is estimated

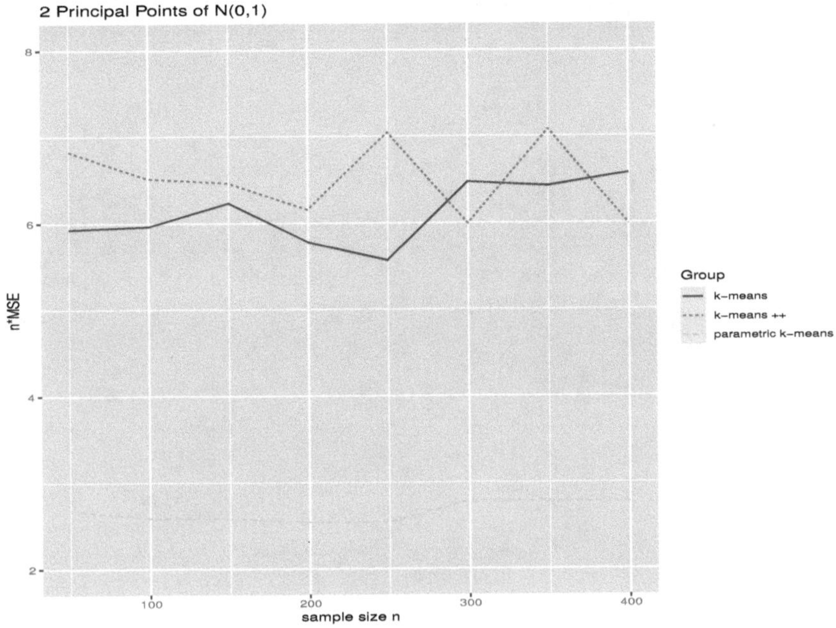

FIGURE 3.1
Plot of $nMSE$ versus sample size for estimating $k = 2$ MSE-RPs of the $N(0,1)$ distribution.

by averaging $\sum_{j=1}^{k} \left\| \hat{\boldsymbol{\xi}}_j - \boldsymbol{\xi}_j \right\|^2$, 500 simulated data sets in the examples. The parametric k-means algorithm was implemented using $n_s = 100{,}000$ for each data set.

Using the conventional non-parametric k-means algorithm and the parametric k-means algorithm as illustrations, $k = 2, 5$ MSE-RPs were estimated from a standard normal distribution $N(0,1)$. The parametric k-means algorithm was implemented by

1. estimating the sample mean $\bar{x}$ and standard deviation s as the MLE estimators for normal distributions from the simulated data set,
2. simulating 100,000 random samples from $N\left(\bar{x}, s^2\right)$,
3. using this larger simulated data set to run the k-means algorithm.

Figures 3.1 and 3.2 plot the estimated MSEs for the nonparametric and parametric k-means algorithms for sample sizes ranging from 10 to 400. The parametric k-means algorithm has a better performance.

3.7 Fang-He Algorithm

Max (1960) and Fang and He (1982) independently proposed an algorithm based on system equations for generating MSE-RPs of the standard normal distribution $N(0,1)$. Fang and He (1982) gave a more detailed discussion on theory and algorithm. The MSE-RPs of $N(\mu, \sigma^2)$ can be easily obtained from the MSE-RPs of $N(0,1)$.

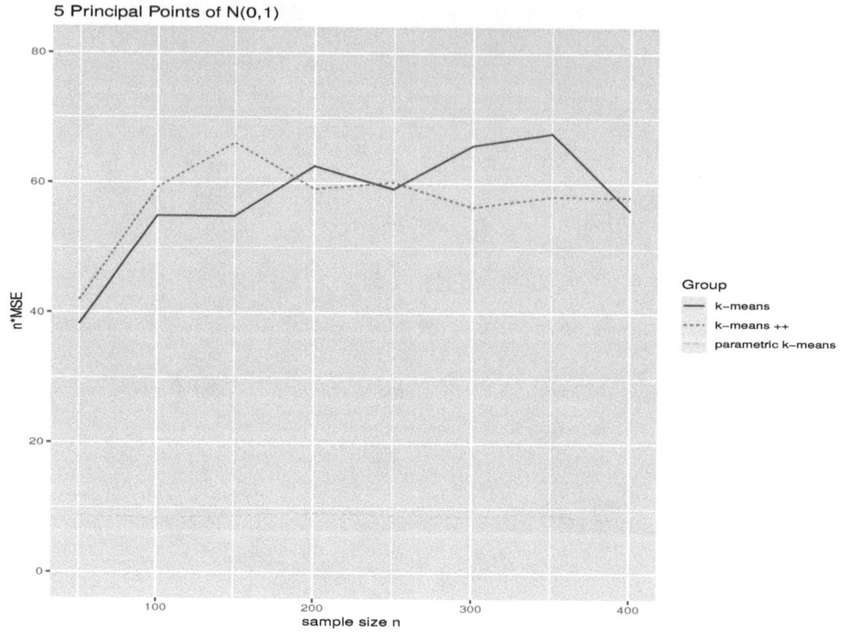

FIGURE 3.2
Plot of $nMSE$ versus sample size for estimating $k = 5$ MSE-RPs of the $N(0,1)$ distribution.

3.7.1 Method and theory

Fang and He (1982) proposed a numerical algorithm for finding the solution of (3.13) and gave some theoretic support for the algorithm. Its Chinese version can be found in Fang and He (1984). Suppose $X \sim N(0,1)$ with pdf $\varphi(x)$ and cdf $\Phi(x)$. As the pdf of $N(0,1)$ is symmetric about 0 and log-concave, its MSE-RPs are unique for a given k. Therefore, we have $\xi_i = \xi_{k-i+1}, i = 1, \ldots, k$. Take derivatives of MSE($\mathcal{Y}$) with respect to each of $\xi_1, \ldots, \xi_k$, we have a system of equations (3.13). When $k = 2m$ is even or $k = 2m+1$ is odd, the number of equations can be reduced into $m = k/2$ or $m = (k-1)/2$, respectively. For even k, by using $\xi_i = \xi_{m+i}, i = 1\ldots, m$ we have

$$\begin{cases} \varphi(0) - \varphi((\xi_1+\xi_2)/2) = \xi_1[\Phi((\xi_1+\xi_2)/2) - \Phi(0)], \\ \varphi((\xi_{i-1}+\xi_i)/2) - \varphi((\xi_i+\xi_{i+1})/2) = \xi_i[\Phi((\xi_i+\xi_{i+1})/2) - \Phi((\xi_{i-1}+\xi_i)/2)], \\ \qquad i = 2, \ldots, m-1, \\ \varphi((\xi_{m-1}+\xi_m)/2) = \xi_m[1 - \Phi((\xi_{m-1}+\xi_m)/2))]. \end{cases} \tag{3.29}$$

There are three types of equations in (3.29). Fang and He (1982) gave a detailed study of these equations for the solution's existence and uniqueness. To introduce their theory, we need to know the Mills' ratio.

Definition 3.5 *The function*

$$M(x,y) = \frac{\Phi(y) - \Phi(x)}{\varphi(x) - \varphi(y)} \tag{3.30}$$

is called as generalized Mills' ration. When $y = \infty$, it becomes the Mills' ration $M(x) = \frac{1-\Phi(x)}{\varphi(x)}$.

Now the system of equations (3.29) can be expressed as

$$\xi_i M((\xi_{i-1}+\xi_i)/2, (\xi_i+\xi_{i+1})/2) = 1, i = 1, \ldots, m, \tag{3.31}$$

when $k = 2m$, $\xi_0 = -\xi_1$; otherwise $\xi_0 = 0$. The following properties were given in Fang and He (1982).

Lemma 3.1

$$\begin{aligned} M'_x(x,y) &= \frac{\partial}{\partial x}M(x,y) = (1-e^{-\frac{1}{2}(y^2-x^2)})^{-1}(xM(x,y)-1), \\ M'_y(x,y) &= \frac{\partial}{\partial y}M(x,y) = (e^{-\frac{1}{2}(y^2-x^2)}-1)^{-1}(1-yM(x,y)), \\ M(x) &\approx \frac{1}{x} - \frac{1}{x^3} + \frac{1\cdot 3}{x^5} - \frac{1\cdot 3\cdot 5}{x^7} + \cdots + (-1)^n\frac{(2n-1)!!}{x^{2n+1}}. \end{aligned}$$

Lemma 3.2

$$\begin{aligned} M'_x(x,y) &< 0, \quad \textit{for} \quad 0 \le x < y \le \infty, \\ M'_y(x,y) &< 0, \quad \textit{for} \quad 0 \le x < y \le \infty, \\ \varphi(x)(x+2M(x)) &< 1, \quad \textit{for} \quad x > 0. \end{aligned}$$

Denote the solution for the standard normal MSE-RPs by $x_{k1} < \ldots < x_{km}$ for $k = 2m$ or $k = 2m+1$. The following theorems give the necessary and sufficient conditions for different equations.

Theorem 3.20 *For any given $x_i > 0$ the equation*

$$x_i M((x_{i-1}+x_i)/2, (x_i+x_{i+1})/2) = 1, i = 1, \ldots, m,$$

has a unique solution $x_{i+1} = h_{i+1}(x_i) = g_{i+1}(x_1)$, if $x_1 < x_{2i,1}$.

For given $k = 2m, x_1 > 0$, applying this theorem we have a unique solution $x_2 = g_2(x_1)$ if $x_1 < x_{21}$, where $g_2(x_1)$ is a strictly increasing function of x_1. From the solution x_2, applying this theorem we have a unique solution $x_3 = h_3(x_2) = g_3(x_1)$ if $x_1 < x_{41}$, where the functions h_3 and g_3 are strictly increasing functions. From the last equation and the solution x_{m-1} can find a unique solution $x_m = h_m(x_{m-1}$ if $x_1 < x_{m1}$, where the functions h_m strictly increase.

Fang and He (1982, 1984) gave detailed proofs for the above theorems; the exciting reader can refer to their paper.Many authors used a similar approach to prove the above theorems under different populations. For example, Section 3.8.1 applies Fang-He's approach to prove Theorem 3.20 under the gamma distribution, and we will copy their proof for the interested reader.

Let us introduce the process of finding RPs for $N(0,1)$. It is clear $x_{11} = E(X) = 0$. For $k = 2$ there is only one equation

$$\varphi(0) = x_1(1-\Phi(0)) = \frac{1}{2}x_1.$$

We obtain $x_{21} = 2\varphi(0) = \sqrt{2/\pi} \approx 0.797885$. For $k = 3$, $x_{30} = 0$, the equation (3.63) reduces to

$$\varphi(\frac{1}{2}x) = x(1-\Phi(\frac{1}{2}x)).$$

By some optimization algorithm, for example, the bisection or Newton algorithm, we can find the solution $x_{31} = 1.224014$.

For $k = 4$, there are two equations

$$\varphi(0) - \varphi(\frac{1}{2}(x_1 + x_2)) = x_1[\Phi(\frac{1}{2}(x_1 + x_2)) - \frac{1}{2}],$$
$$\varphi(\frac{1}{2}(x_1 + x_2)) = x_2[1 - \Phi(\frac{1}{2}(x_1 + x_2))].$$

The solution of the equations can be obtained by $x_{41} = 0.45271$ and $x_{42} = 1.510437$, respectively. Going on the similar process until to obtain k RPs.

3.7.2 Fang-He algorithm

Now, we return to our notation. Let $\xi_1^{(k)} < \xi_2^{(k)} < \cdots < \xi_k^{(k)}$ be the RPs with size k. It is easy to know $\xi_1^{(1)} = 0$ and $\xi_1^{(2)} = -\sqrt{\frac{2}{\pi}}$, $\xi_2^{(2)} = \sqrt{\frac{2}{\pi}}$. Then we can find $\xi_i^{(3)}, i = 1, 2, 3$ and $\xi_i^{(k)}, i = 1, \ldots, k; k = 4, 5, \ldots$.

(a) Give an initial $\xi_1 = \xi_1^{(1)}$, find $\xi_2^{(1)}$ by solving the first equation of (3.29);

(b) Give $\xi_1^{(1)}, \xi_2^{(1)}$, from the second equation of (3.29) we get $\xi_3^{(1)}$;

(c) By the similar way we can get $\xi_4^{(1)}, \ldots, \xi_k^{(1)}$.

(d) But from the last equation of (3.29) we can find another solution $\xi_k^{(1*)}$ from given $\xi_{k-1}^{(1)}$.

(e) If $|\xi_k^{(1)} - \xi_k^{(1*)}| < \varepsilon$, the process stop, otherwise to shift $\xi_1^{(1)}$ into $\xi_1^{(2)}$ and repeat the above steps N times until $|\xi_k^{(N)} - \xi_k^{(N*)}| < \varepsilon$.

Let $\xi_1^{(k)} < \xi_2^{(k)} < \cdots < \xi_k^{(k)}$ be the RPs with size k. For a given $k > 2$, they suggested some way to reduce the domain of RPs and the initial $\xi_1^{(k)}$.

By using the property

$$\xi_1^{(k+1)} < \xi_1^{(k)} < \xi_2^{(k+1)} < \xi_2^{(k)} < \cdots < \xi_k^{(k)} < \xi_{k+1}^{(k+1)}, \tag{3.32}$$

when the RPs with size k are obtained, we can suggest a suitable interval for initial point $\xi_1^{(k+1)}$. By this iteration, we can find RPs with size $k + 1$. Note that the property of (3.32) is obvious by intuitive judgment without theoretical proof.

3.7.3 SNTO for RPs

The above algorithm splits a system of equations into several equations, each with only one unknown solution. Fang and Wang (1990) proposed a sequential algorithm (SNTO, refer to Section 1.13) for solving a system of nonlinear equations by which they can solve the system equations in (3.13) with a better accuracy.

Each equation can be expressed as a solution of $G_1(\xi_1, \ldots, \xi_k) = 0$ where $G_1(\xi_1, \ldots, \xi_k) \geq 0$. Obtaining a solution of the system is an optimization problem to minimum

$$T(\xi_1, \ldots, \xi_k) = \sum_{j=1}^{k} G_j(\xi_1, \ldots, \xi_k)$$

on the domain $D = \{a < \xi_1 < \ldots < \xi_k < B\}$ for some $A < B$. Fang and Wang (1991) gave a detailed discussion on generation of the MSE-RPs of the normal distribution $N(0, 1)$. Fang et al. (2014) provided more accurate MSE-RPs of $N(0, 1)$ for $k \leq 36$ by the use of Fang-He algorithm.

3.8 Applications of Fang-He Algorithm to More Univariate Distributions

The Fang-He algorithm can be applied not only for the normal distribution but also for many other distributions. Most univariate continuous distributions can apply the k-means method to obtain their MSE-RPs. However, the k-means method strongly depends on the training samples and the initial points. We have compared the MSE-RPs generated by the k-means method and the Fang-He algorithm for many different univariate distributions, and find that the Fang-He algorithm can obtain more accurate MSE-RPs.

Many authors applied the Fang-He algorithm to find MSE-RPs and proved the uniqueness of the solution in the three types of equations. Let us consider the MSE-RPs of the gamma distribution.

3.8.1 Gamma distribution

Definition 3.6 *A random variable X has a gamma distribution with shape parameter α and rate parameter β if the pdf*

$$p(x) = \frac{\beta^\alpha}{\Gamma(\alpha)} x^{\alpha-1} e^{-\beta x}, \qquad \textit{for } x > 0,\ \alpha, \beta > 0, \tag{3.33}$$

where $\Gamma(\cdot)$ is the gamma function, it is denoted by $X \sim Ga(\alpha, \beta)$.

The mean, variance, skewness, and kurtosis of X are

$$E(X) = \frac{\alpha}{\beta}, \quad \mathrm{Var}(X) = \frac{\alpha}{\beta^2}, \quad \mathrm{Sk}(X) = \frac{2}{\sqrt{\alpha}} \text{ and } \mathrm{Ku}(X) = \frac{6}{\alpha}, \tag{3.34}$$

respectively. The distribution $Ga(\alpha, 1)$ is called as the standard gamma distribution; the distribution $Ga(1, \beta)$ reduces to the exponential distribution and the distribution $Ga(n/2, 1/2)$ is just the χ^2-distribution with degrees of freedom n.

Let $\mathcal{Z}_{MSE} = \{z_1, \cdots, z_k\}$ be a set of MSE-RPs for $Ga(\alpha, \beta)$ and sort $\mathcal{Z} = \{z_1, z_2, \cdots z_k\}$ as $(0 < z_1 < z_2 < \cdots < z_k < \infty)$. To minimize MSE($\mathcal{Z}$), by taking the partial derivative of MSE($\mathcal{Z}$) we have

$$\begin{cases} \int_0^{\frac{1}{2}(z_1+z_2)} (z_1 - x) p(x) \mathrm{d}x = 0 \\ \int_{\frac{1}{2}(z_1+z_2)}^{\frac{1}{2}(z_2+z_3)} (z_2 - x) p(x) \mathrm{d}x = 0 \\ \cdots\cdots\cdots\cdots \\ \int_{\frac{1}{2}(z_{k-1}+z_k)}^{\infty} (z_k - x) p(x) \mathrm{d}x = 0, \end{cases} \tag{3.35}$$

where $p(x)$ is the pdf of gamma distribution in (3.33). Ke et al. (2023) gave a systematic study on MSE-RPs of the gamma distribution. The results in this section are cited from their paper.

When $k = 1$, it is known that $z_1 = E(X) = \frac{\alpha}{\beta}$.

When $k \geq 2$, to show the existence of gamma MSE-RPs is equivalent to showing that system of equations (3.35) has a solution. After several transformations (3.35) becomes

$$\begin{cases} (z_1-\mu)[F(\frac{z_1+z_2}{2})-F(0)] = -\frac{1}{2\beta}(z_1+z_2)p(\frac{z_1+z_2}{2}) \\ (z_2-\mu)[F(\frac{z_2+z_3}{2})-F(\frac{z_1+z_2}{2})] = \frac{1}{2\beta}(z_1+z_2)p(\frac{z_1+z_2}{2})-\frac{1}{2\beta}(z_2+z_3)p(\frac{z_2+z_3}{2}) \\ \cdots\cdots\cdots\cdots \\ (z_k-\mu)[1-F(\frac{z_{k-1}+z_k}{2})] = \frac{1}{2\beta}(z_{k-1}+z_k)p(\frac{z_{k-1}+z_k}{2}), \end{cases} \tag{3.36}$$

where $F(x)$ is the cdf of X with $F(0)=0$. The following theorem and its proof follow Fang and He (1982).

Theorem 3.21

1. *For any $z_1 > 0$, equation*

$$(z_1-\mu)F(\frac{z_1+z_2}{2}) = -\frac{1}{2\beta}(z_1+z_2)p(\frac{z_1+z_2}{2}) \tag{3.37}$$

exists a unique solution z_2 if and only if $z_1 < \mu = E(X) = \frac{\alpha}{\beta}$.

2. *For given $z_{i-1} > 0,\ i = 2, \cdots, k-1$, equation*

$$\begin{aligned} &(z_i-\mu)[F(\frac{z_i+z_{i+1}}{2})-F(\frac{z_{i-1}+z_i}{2})] \\ &= \frac{1}{2\beta}(z_{i-1}+z_i)f(\frac{z_{i-1}+z_i}{2})-\frac{1}{2\beta}(z_i+z_{i+1})p(\frac{z_i+z_{i+1}}{2}) \end{aligned} \tag{3.38}$$

exists a unique solution z_{i+1} when $z_{i-1} < z_{i,i-1}$, where $z_{i,i-1}$ is the $(i-1)$th representative point in the set of gamma MSE-RPs which has $k = i$.

3. *For any $z_{k-1} > 0$, equation*

$$(z_k-\mu)[1-F(\frac{z_{k-1}+z_k}{2})] = \frac{1}{2\beta}(z_{k-1}+z_k)p(\frac{z_{k-1}+z_k}{2}) \tag{3.39}$$

exists a unique solution z_k.

The proof of the theorem can refer to Subsection 3.13.7.

The uniqueness of gamma MSE-RPs under log-concavity condition
According to the above discussion, the existence of gamma MSE-RPs is proved by solving equations. Theorem 3.9 indicates that if the density $p(x)$ is log-concavity, its MSE-RPs are unique.

Theorem 3.22 *Suppose $X \sim Ga(\alpha, \beta)$. For any $k \in \mathbb{N}^+$, the set of gamma MSE-RPs is unique if $\alpha \geq 1$.*

Proof If the pdf of $Ga(\alpha, \beta)$ is log-concavity, then the assertion follows. From (3.33) we have

$$\begin{aligned} \ln p(x;\alpha,\beta) &= (\alpha-1)\ln x + \alpha\ln\beta - bx - \ln\Gamma(\alpha), \\ (-\ln p(x;\alpha,\beta))' &= (1-\alpha)x^{-1}+\beta, \\ (-\ln p(x;\alpha,\beta))'' &= (\alpha-1)x^{-2} \geq 0. \end{aligned}$$

The above inequality holds when $\alpha \geq 1$.

The weighted Lindley distribution

There are many distributions related to the gamma distribution. The Lindley distribution was originally introduced by Lindley (1958) in the context of Bayesian statistics as a counterexample of fiducial statistics. Its pdf is given by

$$p(x|\theta) = \frac{\theta^2}{1+\theta}(1+x)e^{-\theta x}, x > 0, \theta > 0.$$

Many authors think that the Lindley distribution could be a better lifetime model than the exponential distribution for real data sets. Ghitany et al. (2011) introduced the two-parameter Weighted Lindley (shortly WL) distribution as follows,

$$p(x|\theta, m) = \frac{\theta^{m+1}}{(m+\theta)\Gamma(m)} x^{m-1}(1+x)e^{-\theta x}, x > 0, \theta > 0, m > 0,$$

which is denoted as $X \sim WL(\theta, m)$. Its cdf is given by

$$F(x|\theta, m) = 1 - \frac{(m+\theta)\Gamma(m, \theta x) + (\theta x)^m e^{-\theta x}}{(m+\theta)\Gamma(m)}, x > 0, \theta > 0, m > 0, \tag{3.40}$$

where

$$\Gamma(a, b) = \int_b^\infty x^{a-1}e^{-x}dx, a > 0, b \geq 0.$$

Wang and Wang (2017) discussed estimation of parameters of θ and m for $WL(\theta, m)$. Finding MSE-RPs of the Lindley distribution and applying MSE-RPs to accurate data is good practice for the students.

3.8.2 MSE-RPs of the class of Pearson distributions

In the period between 1894 and 1895, Karl Pearson provided flexible descriptions of the nonnormal distributions encountered in his biometric research. This is a famous class of univariate distributions, called *the class of Pearson distributions*. It includes many well-known distributions. The standard indices of skewness and kurtosis can distinguish the types. Fei (1990) studied the class of Pearson distributions where the pdf of X has the form of

$$p(x) = c\exp\left\{-\int \frac{a_0 + a_1 x}{b_0 + b_1 x + b_2 x^2} dx\right\}, \tag{3.41}$$

where c is the normalized constant and parameters a_0, a_1, b_0, b_1, and b_2 satisfy the differential equation

$$\frac{dp(x)}{dx} = \frac{a_0 + a_1 x}{b_0 + b_1 x + b_2 x^2} p(x).$$

The class of Pearson distributions includes many useful distributions. For example, the type I distribution is the beta distribution; the type II distribution curve is a symmetrical U-shaped curve; the type III distribution curve is the shifted gamma distribution; the type V is the shift inverse gamma distribution, the type VI is an inverse beta distribution, the type VII is the t-distribution; the type VIII is the power function distribution; the type X is the exponential distribution, and the type XI is the normal distribution. Fei (1990) gave some sufficient conditions for the uniqueness of the solution.

Theorem 3.23 *Let $p(x)$ be the density function of a Pearson distribution. If $p(x)$ satisfies one of the following conditions with $D = b_1^2 - 4b_0b_2$,*

1. $-\frac{1}{3} < b_2 < 0, D \geq 2, b_1 + \sqrt{D} \leq 0$;
2. $-\frac{1}{3} < b_2, D < 0$;
3. $b_2 = 0, b_1 < 0, b_0 \leq 0$;
4. $b_1 = b_2 = 0, b_0 < 0$.

There is a unique solution to the equation.

Fei (1990) gave a detailed proof and showed that the exponential, χ^2-, and F distributions satisfy one of the conditions in Theorem 3.23 and gave the MSE-RPs for the standard exponential distribution, the χ^2-distribution with 4 degrees of freedom, and F-distribution with degrees of freedom (10,10) for $k \leq 10$.

Fu (1985) applied the Fang-He algorithm to the gamma distribution $\Gamma(\frac{3}{2}, 1)$ and obtained the MSE-RPs for $k \leq 20$. Fu (1993) also obtained MSE-RPs for Weibull distribution for $k \leq 20$. Zhou and Wang (2016) gave the MSE-RPs of the t-distribution with 10 degrees of freedom for $k \leq 35$. See Table 1.4 for the definition of the above distributions. Fei (1992) proposed an algorithm for generating MSE-RPs by the Newton optimization algorithm. Chakraborty et al. (2021) proposed a high-precision algorithm to find the solutions of MSE-RPs in univariate distributions due to its accuracy and generality.

3.9 Mixtures of Univariate Distributions

A finite mixture of distributions allows for great flexibility in capturing a variety of density shapes. Research into mixture models has a long history. The most cited early publication is Pearson (1894), as he used a two-component normal mixture model for biometric data set. This model was motivated by the apparent skewness in the ratio of forehead width to body length of 1000 crabs from Naples. It was suspected that the population contained two unidentified species. The Gaussian mixture family has attracted considerable attention because it can approximate continuous distribution using a finite Gaussian mixture model. Mixtures of normal distributions have numerous applications across various disciplines, including physics, engineering, economics, biology, and finance.

Definition 3.7 *Given a finite set of probability density distributions $f(x; \boldsymbol{\theta}_1), \ldots, f(x; \boldsymbol{\theta}_m)$ and weights $w_1, \ldots, w_m$, such that $w_i \geq 0$ and $\sum w_i = 1$, the mixture distribution is defined as*

$$p(x) = \sum_{i=1}^{m} w_i f(x; \boldsymbol{\theta}_i).$$

The cdf of the mixture distribution $p(x)$ is defined as

$$P(x) = \sum_{i=1}^{m} w_i F(x; \boldsymbol{\theta}_i),$$

where $F(x; \boldsymbol{\theta}_i)$ is the cdf of $f(x; \boldsymbol{\theta}_i)$.

A general linear combination of probability density functions is not necessarily a probability density, since it may be harmful or it may integrate into something other than one.

However, a convex combination of probability density functions preserve both of these properties (non-negativity and integrating to one), and thus, mixture densities are themselves probability density functions.

3.9.1 Mixtures of two-component normal distributions

In this section, we focus on a mixture of two normal distributions.

$$\begin{aligned} p(x) &= \alpha\varphi(x;\mu_1,\sigma_1^2) + (1-\alpha)\varphi(x;\mu_2,\sigma_2^2) \qquad (3.42) \\ &= \alpha\frac{1}{\sqrt{2\pi}\sigma_1}\exp\left\{-\frac{(x-\mu_1)^2}{2\sigma_1^2}\right\} + (1-\alpha)\frac{1}{\sqrt{2\pi}\sigma_2}\exp\left\{-\frac{(x-\mu_2)^2}{2\sigma_2^2}\right\}. \end{aligned}$$

We denote $X \sim \text{MixN}(\alpha,\mu_1,\sigma_1^2,\mu_2,\sigma_2^2)$.

- **A scale mixture**: $\text{MixN.S}\left(\alpha,\mu,\sigma_1^2,\sigma_2^2\right)$, when $\mu_1=\mu_2$, $\sigma_1^2\neq\sigma_2^2$.
- **A location mixture**: $\text{MixN.L}\left(\alpha,\mu_1,\mu_2,\sigma^2\right)$, when $\mu_1\neq\mu_2$, $\sigma_1^2=\sigma_2^2$.

  Mirrored mixtures are a special case of location mixtures; the corresponding density function is

  $$p(x) = \frac{0.5}{\sigma\sqrt{2\pi}}e^{-\frac{(x-\mu)^2}{2\sigma^2}} + \frac{0.5}{\sigma\sqrt{2\pi}}e^{-\frac{(x+\mu)^2}{2\sigma^2}}, \qquad (3.43)$$

  denote it as $\text{MixN.M}\left(\mu,\sigma^2\right)$.
- **General mixture**; $\text{MixN}\left(\alpha,\mu_1,\sigma_1^2,\mu_2,\sigma_2^2\right)$, when $\mu_1\neq\mu_2$, $\sigma_1^2\neq\sigma_2^2$.

Some basic statistics for $\text{MixN}\left(\alpha,\mu_1,\sigma_1^2,\mu_2,\sigma_2^2\right)$ are:

$$\begin{aligned} E(X) &= \int_R xp(x)dx = \alpha\mu_1 + (1-\alpha)\mu_2; \\ E(X^2) &= \int_R x^2p(x)dx = \alpha(\mu_1^2+\sigma_1^2) + (1-\alpha)(\mu_2^2+\sigma_2^2); \\ Var(X) &= E((X-E(X))^2) = \alpha\sigma_1^2 + (1-\alpha)\sigma_2^2 + \alpha(1-\alpha)(\mu_1-\mu_2)^2; \\ E((X-E(X))^3) &= 3\alpha(1-\alpha)(\mu_1-\mu_2)(\sigma_1^2-\sigma_2^2) - \alpha(1-\alpha)(\alpha-(1-\alpha))(\mu_1-\mu_2)^3; \\ E((X-E(X))^4) &= 3\alpha\sigma_1^4 + 6\alpha\sigma_2^4\sigma_1^4(\mu_1-\mu_2)^4 + \alpha(1-\alpha)^4(\mu_1-\mu_2)^4 + 3(1-\alpha)\sigma_2^4 \\ &\quad + 6(1-\alpha)\sigma_1^4\sigma_2^2(\mu_1-\mu_2)^4 + \alpha^4(1-\alpha)(\mu_1-\mu_2)^4. \end{aligned}$$

To obtain the skewness and kurtosis of X by re-parametric, denote

$$w = \frac{1-\alpha}{\alpha}, \quad v = \frac{\sigma_2^2}{\sigma_1^2}, \quad \Delta = \mu_1-\mu_2,$$

then, we have

$$\begin{aligned} Sk &= \frac{E((X-E(X))^3)}{(E((X-E(X))^2))^{\frac{3}{2}}} = \frac{3w(w+1)(1-v)\,\Delta\,\sigma_1^2 + w(1-w)\,\Delta^3\,\alpha^3}{[(w+1)(wv+1)\sigma_1^2 + w\,\Delta^2]^{\frac{3}{2}}}, \\ Ku &= \frac{E((X-E(X))^4)}{[E((X-E(X))^2)]^2} - 3 \\ &= \frac{3w(w+1)^2(v-1)^2\sigma_1^4 - 6w(w^2-1)(v-1)\,\Delta^2\,\sigma_1^2 + w(w^2-4w+1)\,\Delta^4}{[(1+w)(wv+1)\sigma_1^2 + w\,\Delta^2]^2}. \end{aligned}$$

For example, let $X \sim MixN(0.6,-8,3^2,2;4^2)$, we have $E(X) = -4$, $\text{Var}(X) = 35.8$, $Sk(X) = 0.4594$ and $Ku(X) = 2.3609$.

3.9.2 Measure of disparity

Suppose the overlap in density between two mixture components is large. In that case, the estimation problem under MixN will be difficult due to the challenge of identifying the correct group for each observation. Overlapping Coefficient (OVL) is an index that measures the disparity between two distributions proposed by Inman and Bradley Jr (1989). The explicit formula of OVL for the case of normal distributions is,

$$OVL = \begin{cases} 2\Phi\left(-\frac{|\mu_1-\mu_2|}{2\sigma}\right), & \text{if } \sigma_1 = \sigma_2 = \sigma, \\ 1+\Phi\left(\frac{\tau_1-\mu_1}{\sigma_1}\right)+\Phi\left(\frac{\tau_2-\mu_2}{\sigma_2}\right)-\Phi\left(\frac{\tau_1-\mu_2}{\sigma_2}\right)-\Phi\left(\frac{\tau_2-\mu_1}{\sigma_1}\right), & \text{if } \sigma_1 \neq \sigma_2, \end{cases}$$

Where Φ denotes the cumulative distribution function of the standard normal distribution, τ_1 and τ_2 are given by

$$\tau_1 = \frac{\mu_1\sigma_2^2 - \mu_2\sigma_1^2 - \sigma_1\sigma_2\left[(\mu_1-\mu_2)^2 + (\sigma_2^2-\sigma_1^2)\ln\left(\frac{\sigma_2^2}{\sigma_1^2}\right)\right]^{1/2}}{\sigma_2^2-\sigma_1^2},$$

and

$$\tau_2 = \frac{\mu_1\sigma_2^2 - \mu_2\sigma_1^2 + \sigma_1\sigma_2\left[(\mu_1-\mu_2)^2 + (\sigma_2^2-\sigma_1^2)\ln\left(\frac{\sigma_2^2}{\sigma_1^2}\right)\right]^{1/2}}{\sigma_2^2-\sigma_1^2}.$$

Then, the disparity index (DI) is defined as:

$$DI = 1 - OVL.$$

The standard components with a significant value of DI are expected to be separated more efficiently so that the model parameters of MixN are considered to be estimated with more accuracy. We focus on the mixture of two normal components with medium to high disparities (DI≥ 0.4).

Table 5.3 displays the parameter settings for ten mixtures of two normal distributions by different mixture types and different levels of DI.

- Mirror mixtures with two model parameters (μ, σ^2) are shown in A1 and A2.
- Location mixtures with four model parameters $(\alpha, \mu_1, \mu_2, \sigma^2)$ are given in B1 and B2. A1 and B1 represent mixture distributions with a moderate level of disparity (DI=0.55), and A2 and B2 have well-separated mixture components (DI=0.95).
- Six distributions for location-scale mixtures (C1-C6) are given with DI from 0.4 to 0.9. C2 and C6 represent location-scale mixtures with large variances among distribution group C.

Densities of the six underlying distributions in group C are shown in Figure 5.2.

3.9.3 Uniqueness of MSE-RPs

Theorem 3.8 gives the sufficient condition (log-concavity) for the uniqueness of MSE-RPs of the univariate distribution is

$$\frac{\partial^2 \log p(x)}{\partial x^2} < 0.$$

The density function $p(x)$ is weakly unimodal if

$$p \cdot p'' - 2(p') < 0.$$

TABLE 3.5
Parameters of underlying distributions

Type	Distribution	α	μ_1	σ_1^2	μ_2	σ_2^2	DI
MixN. M	A1	0.50	-0.76	1	0.76	1	0.55
MixN. M	A2	0.50	-2.00	1	2.00	1	0.95
MixN. L	B1	0.41	1.00	1	2.51	1	0.55
MixN. L	B2	0.41	1.00	1	5.00	1	0.95
MixN	C1	0.70	2.70	2	1.50	1	0.40
MixN	C2	0.70	2.20	9	1.50	1	0.50
MixN	C3	0.70	3.50	2	1.50	1	0.60
MixN	C4	0.70	4.25	3	1.50	1	0.70
MixN	C5	0.70	4.60	2	1.50	1	0.80
MixN	C6	0.70	8.00	10	1.50	1	0.90

Although the normal distributions are log-concave, a convex combination of two normal distributions cannot easily satisfy this property. More conditions on the five parameters of a MixN are required. The following theorems are from Li et al. (2022).

Theorem 3.24 *For any location mixture of two normal distributions with pdf*

$$p(x) = \alpha \frac{1}{\sqrt{2\pi}\sigma} e^{-\frac{(x-\mu)^2}{2\sigma^2}} + (1-\alpha) \frac{1}{\sqrt{2\pi}\sigma} e^{-\frac{(x+\mu)^2}{2\sigma^2}},$$

the set of MSE-RPs is unique if $\frac{|\mu|}{\sigma} \leq \frac{\sqrt{2}}{2}$.

The proof of the theorem is given in Subsection 3.13.8. A more general result is given in the following theorem.

Theorem 3.25 *(Location mixture of two normal densities) Suppose* $X \sim \mathrm{MixN.L}(\alpha, \mu_1, \mu_2, \sigma^2)$. *For any* $n \in \mathbb{N}^+$, *the set of* n *MSE-RPs of* X *is unique if, for all* $\alpha \in (0, 1)$,

$$|\mu_1 - \mu_2| \leq \sqrt{2}\sigma.$$

Corollary 3.3 *For any mixture of two normal distributions in Theorem 3.24 it is unimodal when* $\frac{|\mu|}{\sigma} \leq 1$.

The proofs of Theorem 3.24 and Corollary 3.3 can be found in Subsection 3.13.8. It is strict for a mixture of two normal distributions to satisfy the log-concavity condition of uniqueness. As a result, if the strong unimodal condition is satisfied, there will be severe overlap between the two mixture components.

3.9.4 Algorithm for generation of MSE-RPs for MixN

Let $X \sim \mathrm{MixN}(\alpha, \mu_1, \sigma_1^2, \mu_2, \sigma_2^2)$ with density $p(x)$ and cdf $F(x)$. Given a set of k MSE-RPs of X, denoted as $-\infty < \xi_1^{(k)} < \ldots < \xi_k^{(k)} < +\infty$. Let $a_j^{(k)} = (\xi_j^{(k)} + \xi_{j+1}^{(k)})/2, j = 1, \ldots, k-1)$, the objective function can be expressed as

$$\begin{aligned} \mathrm{MSE}_X\left(\xi_1^{(k)}, \ldots, \xi_k^{(k)}\right) &= \int_{-\infty}^{\infty} \left(\min_{i=1,\ldots,k} \left(x - \xi_i^{(k)}\right)^2\right) p(x) dx \\ &= \sum_{i=1}^{k} \int_{I_i} \left(x - \xi_i^{(k)}\right)^2 p(x) dx, \end{aligned} \tag{3.44}$$

where $I_1 = (-\infty, a_1^{(k)}]$, $I_j = (a_{j-1}^{(k)}, a_j^{(k)}], j = 2, \dots, k-1$, and $I_k = (a_{k-1}^{(k)}, \infty)$.

The k-means and Fang-He algorithms can be applied to find MSE-RPs of a MixN distribution. For the Fang-He algorithm, the objective function (3.44) can be minimized by taking the first-order partial derivatives concerning $\left(\xi_1^{(k)}, \dots, \xi_k^{(k)}\right)$ correspondingly. Then, the optimal solutions to (3.44) can be obtained by solving a system of k non-linear equations, for $i = 1, \dots, k$,

$$\alpha \int_{I_i} \left(x - \xi_i^{(k)}\right) \varphi(x; \mu_1, \sigma_1^2) dx + (1-\alpha) \int_{I_i} \left(x - \xi_i^{(k)}\right) \varphi(x; \mu_2, \sigma_2^2) dx = 0. \tag{3.45}$$

For a normal distribution $\varphi(x; \mu, \sigma^2)$, we have $\sigma^2 \varphi'(x; \mu, \sigma^2) = (\mu - x)\varphi(x; \mu, \sigma^2)$, which implies $\int x\varphi(x; \mu, \sigma^2) dx = \mu \int \varphi(x; \mu, \sigma^2) dx - \sigma^2 \int \varphi'(x; \mu, \sigma^2) dx$. Then, the system of equations (3.45) can be expressed as

$$\begin{aligned}
&\alpha \left[\mu_1 \, \Phi(x; \mu_1, \sigma_1^2)\big|_{I_i} - \sigma_1^2 \, \varphi(x; \mu_1, \sigma_1^2)\big|_{I_i}\right] \\
&+ (1-\alpha) \left[\mu_2 \, \Phi(x; \mu_2, \sigma_2^2)\big|_{I_i} - \sigma_2^2 \, \varphi(x; \mu_2, \sigma_2^2)\big|_{I_i}\right] \\
&- \xi_i^{(k)} \, F(x)\big|_{I_i} = 0, i = 1, \dots, k,
\end{aligned} \tag{3.46}$$

where F is the cdf of X.

Li et al. (2022) proved the convergence of the three types of equations.

Type I equation is

$$\begin{aligned}
&\alpha \left[\mu_1 \Phi\left(a_1^{(k)}; \mu_1, \sigma_1^2\right) - \sigma_1^2 \varphi\left(a_1^{(k)}; \mu_1, \sigma_1^2\right)\right] \\
&+ (1-\alpha) \left[\mu_2 \Phi\left(a_1^{(k)}; \mu_2, \sigma_2^2\right) - \sigma_2^2 \varphi\left(a_1^{(k)}; \mu_2, \sigma_2^2\right)\right] \\
&- \xi_1^{(k)} F\left(a_1^{(k)}\right) = 0.
\end{aligned} \tag{3.47}$$

Type II equation is

$$\begin{aligned}
&\alpha \left\{\mu_1 \left[1 - \Phi\left(a_{k-1}^{(k)}; \mu_1, \sigma_1^2\right)\right] + \sigma_1^2 \varphi\left(a_{k-1}^{(k)}; \mu_1, \sigma_1^2\right)\right\} \\
&+ (1-\alpha) \left\{\mu_2 \left[1 - \Phi\left(a_{k-1}^{(k)}; \mu_2, \sigma_2^2\right)\right] + \sigma_2^2 \varphi\left(a_{k-1}^{(k)}; \mu_2, \sigma_2^2\right)\right\} \\
&- \xi_k^{(k)} \left[1 - F\left(a_{k-1}^{(k)}\right)\right] = 0.
\end{aligned} \tag{3.48}$$

Type III equation is

$$\begin{aligned}
&\alpha\mu_1 \left[\Phi\left(a_i^{(k)}; \mu_1, \sigma_1^2\right) - \Phi\left(a_{i-1}^{(k)}; \mu_1, \sigma_1^2\right)\right] \\
&+ \alpha\sigma_1^2 \left[\varphi\left(a_{i-1}^{(k)}; \mu_1, \sigma_1^2\right) - \varphi\left(a_i^{(k)}; \mu_1, \sigma_1^2\right)\right] \\
&+ (1-\alpha)\mu_2 \left[\Phi\left(a_i^{(k)}; \mu_2, \sigma_2^2\right) - \Phi\left(a_{i-1}^{(k)}; \mu_2, \sigma_2^2\right)\right] \\
&+ (1-\alpha)\sigma_2^2 \left[\varphi\left(a_{i-1}^{(k)}; \mu_2, \sigma_2^2\right) - \varphi\left(a_i^{(k)}; \mu_2, \sigma_2^2\right)\right] \\
&- \xi_i^{(k)} \left[F\left(a_i^{(k)}\right) - F\left(a_{i-1}^{(k)}\right)\right] = 0.
\end{aligned} \tag{3.49}$$

The following discusses the uniqueness of Types I–III equations discussed by Li et al. (2022), respectively.

Theorem 3.26 *Given* $k \geq 2$, *denote* $E(X)$ *as the population mean of* $X \sim \text{MixN}(\alpha, \mu_1, \sigma_1^2, \mu_2, \sigma_2^2)$.
(i) Type I equation (3.47) *has a unique solution denoted by* $\xi_2^{(k)} = h_1(\xi_1^{(k)})$, *if and only if,*

$$-\infty < \xi_1^{(k)} < E(X). \tag{3.50}$$

(ii) $\xi_2^{(k)} = h_1(\xi_1^{(k)})$ *is a strictly increasing function with respect to* $\xi_1^{(k)}$ *if*

$$\frac{\xi_2^{(k)} - \xi_1^{(k)}}{4} \cdot p\left(\frac{\xi_1^{(k)} + \xi_2^{(k)}}{2}\right) - F\left(\frac{\xi_1^{(k)} + \xi_2^{(k)}}{2}\right) < 0, \tag{3.51}$$

for $\xi_1^{(k)} \in (-\infty, E(X))$.

Theorem 3.26 shows that there is a unique and non-decreasing solution of $\xi_2^{(k)}$ within a specific searching range of $\xi_1^{(k)}$. Part (i) of Theorem 3.26 proves that Type I equation has a unique solution of $\xi_2^{(k)}$ if we set the initial searching range of $\xi_1^{(k)}$ with an upper bound of $E(X)$. It is feasible for finding the solution of $\xi_2^{(k)}$ given $\xi_1^{(k)}$ in (3.50) for $k \geq 2$. Based on Part (ii), the solution of $\xi_2^{(k)}$ is strictly increasing with respect to $\xi_1^{(k)}$ under the condition (3.51). In practice, the condition (3.51) is easy to meet and convenient to verify.

Due to the complexity of the density of MixN, it is difficult to prove that the inequality (3.51) holds over the entire searching range of $\xi_1^{(k)} \in (-\infty, E(X))$. Remarks 3.1–3.2 give the sufficient conditions for (3.51) to be true.

Remark 3.1 *When* $\mu_1 = \mu_2$, $\xi_2^{(k)} = h_1(\xi_1^{(k)})$ *is a strictly increasing function with respect to* $\xi_1^{(k)}$ *if*

$$E(X) - \sqrt{2}\max\{\sigma_1, \sigma_2\} < \xi_1^{(k)} < E(X).$$

Remark 3.2 *When* $\mu_2 < \mu_1 < \mu_2 + \sigma_1/(1-\alpha)$, $\xi_2^{(k)} = h_1(\xi_1^{(k)})$ *is a strictly increasing function with respect to* $\xi_1^{(k)}$ *if*

$$\mu_2 < \xi_1^{(k)} < E(X).$$

Proofs of Theorem 3.26 and Remarks 3.1–3.2 can be found in Subsection 3.13.8

Theorem 3.27 proves that the last equation in the non-linear system (3.46) has unique and non-decreasing solution of $\xi_k^{(k)}$ given $\xi_{k-1}^{(k)}$.

Theorem 3.27 *Given* $k \geq 2$ *and* $\xi_1^{(k)} \in (-\infty, E(X))$, *Type II equation* (3.48) *has a unique solution of* $\xi_k^{(k)} = h^*(\xi_{k-1}^{(k)})$ *and* $h^*(\xi_{k-1}^{(k)})$ *is a strictly increasing function with respect to* $\xi_{k-1}^{(k)}$.

Subsection 3.13.8 presents a proof of Theorem 3.27.

For any given $\xi_1^{(2)}$ satisfies the conditions in Theorem 3.26, we can obtain $\xi_2^{(2)} = h_1(\xi_1^{(2)})$ from the first equation in (3.46) and $\xi_2^{(2)*} = h^*(\xi_1^{(2)})$ from the second equation in (3.46), respectively. Following the procedure of the Fang-He algorithm, the iteration will stop until $|\xi_2^{(2)} - \xi_2^{(2)*}| < \epsilon$.

For example, let $X \sim \text{MixN}(0.7, 4.6, 2, 1.5, 1)$, we get a set of 2 MSE-RPs $\xi_1^{(2)} = -0.552945$ and $\xi_2^{(2)} = 1.989754$ with probabilities $p_1^{(2)} = 0.696014$ and $p_2^{(2)} = 0.303986$, respectively. When $k \geq 3$, we want to obtain $\xi_{i+1}^{(k)} = h_i(\xi_1^{(k)})$ from Type III equation in (3.46) based on $\xi_1^{(k)}$ and $\xi_2^{(k)} = h_1(\xi_1^{(k)})$ for $i = 2, \ldots, k$. Theorem 3.28 proves that, when $k = 3$, Type III equation has the unique solution of $\xi_3^{(3)}$ given a $\xi_1^{(3)}$ in the range of $(-\infty, \xi_1^{(2)})$, where $\xi_1^{(2)}$ is the first point of the set of 2 MSE-RPs.

Theorem 3.28 *When $k = 3$, the Type III equations (3.49) in (3.46) have a unique solution of $\xi_3^{(3)} = h_2(\xi_1^{(3)})$, if and only if,*

$$\xi_1^{(3)} < \xi_1^{(2)}.$$

The proof will be given in Subsection 3.13.8.

According to Theorem 3.28, for given $\xi_1^{(k)}$, where $k > 3$, we obtain $\xi_2^{(k)} = h_1(\xi_1^{(k)})$, $\xi_3^{(k)} = h_2(\xi_1^{(k)}), \ldots, \xi_k^{(k)} = h_{k-1}(\xi_1^{(k)})$ from the 1st, 2nd, $\ldots, (k-1)^{th}$ equation of (3.46), in turn. Then, the equation (3.49) has a unique solution of $\xi_j^{(k)} = h_{j-1}(\xi_1^{(k)})$ if and only if $\xi_1^{(k)} < \xi_1^{(j-1)}$ for $j = 3, \ldots, k$.

As we discussed above, given an initial value of $\xi_1^{(k)}$, we can calculate the values of $\xi_{(i+1)}^{(k)} = h_i(\xi_1^{(k)})$, where $i = 2, \ldots, k-1$, and the value of $\xi_{(k)}^{(k)*} = h^*(\xi_1^{(k)})$. Based on Theorems 3.26, 3.27, and 3.28, the non-linear equation (3.46) has a unique solution. Given $\xi_1^{(k)}$, when $\xi_k^{(k)*}$ is significantly smaller than $\xi_k^{(k)}$, we can reduce the initial value of $\xi_1^{(k)}$ to narrow down the difference between $\xi_k^{(k)}$ and $\xi_k^{(k)*}$. When $\xi_k^{(k)*}$ is significantly larger than $\xi_k^{(k)}$, we can increase the value $\xi_1^{(k)}$ to shrink the difference between the two approximations of the last point. A more detailed discussion can refer to Li et al. (2022).

3.9.5 Mixture of two Weibull distributions

As a specialized variant of the mixture Weibull distributions, the bimodal mixture Weibull distributions serve as an ideal model for heterogeneous data sets within reliability studies and survival analysis. A principal challenge in dealing with the mixture of Weibull distributions lies in the estimation of parameters. Inspired by the exemplary efficacy of the quasi-Monte Carlo method in quantile estimation, Yan et al. (2024) introduced an innovative approach. This approach employs the Harrell-Davis and three Sfakianakis and Verginis quantile estimators (see Section 1.7) to enhance the representativeness of the data, thereby improving the accuracy of parameter estimation. Given the difficulty in deriving analytical expressions for the parameters of the mixture Weibull distributions and their propensity for convergence to local maxima, this research adopts the sequential number-theoretic optimization (SNTO) algorithm for the numerical resolution of parameter estimation. The initial optimization area for SNTO is determined via the Weibull Probability Plot (WPP). Simulation studies have demonstrated that the proposed method significantly enhances estimation precision and reduces dependence on the "quality" of the sample. Furthermore, this methodology has been applied to two real data sets, showcasing the effectiveness of the proposed approach.

3.10 Generation of RPs of a Function $g(X)$ Based on RPs of X

There are many relationships between the useful distributions. For example, the uniform, exponential, normal, log-normal, the χ^2-, the t-, gamma distributions, and any continuous distributions have the following relationships:

1. The uniform distribution $X \sim U(0,1)$ and $Z = F^{-1}(X) \sim F(x)$ where F is a cdf of Z.
2. The uniform distribution $X \sim U(0,1)$ and the standard exponential distribution $Z \sim EP(1)$, where $Z = -\ln(X)$.

3. The normal distribution $X \sim N(\mu, \sigma^2)$ and $Z \sim LN(\mu, \sigma^2)$ the log-normal distribution, where $X = \ln(Z)$.
4. The normal distribution $X \sim N(0,1)$ and $Z = X^2 \sim \chi^2(1)$, the χ^2-distribution with degrees of freedom 1.
5. The t-distribution is proportional to a ratio of the normal distribution and χ distribution.
6. The Gamma distribution $Z \sim Ga(\alpha, \beta)$ is the distribution of $X_1 + X_2$, where $X_1 \sim Ga(\alpha_1, \beta)$ and $X_2 \sim Ga(\alpha_2, \beta)$ are independent with $\alpha = \alpha_1 + \alpha_2$.

Let two random variables X and Z, where $X \sim F(x)$ and $Z = g(X)$ where g is a continuous function on $\mathcal{D}_X$, the domain of X. Can we extend this relation to RPs between X and Z? Let's consider three kinds of RPs.

MC-RPs: If $x_1, \ldots, x_k$ are random samples from $F(x)$, then $g(x_1), \ldots, g(x_k)$ are MC-RPs of Z.

QMC-RPs: If $\{q_1, \ldots, q_k\}$ is a set of QMC-RPs of X, then $\{g(q_1), \ldots, g(q_k)\}$ is an approximate set of QMC-RPs of Z.

MSE-RPs: If $\{\xi_1, \ldots, \xi_k\}$ is a set of MSE-RPs of X, then $\{g(\xi_1), \ldots, g(\xi_k)\}$ is an approximate set of MSE-RPs of Z in the sense of consistency.

Let $\{\xi_1, \ldots, \xi_k\}$ be a set of MSE-RPs of X with size k, in general, the set of $\{g(\xi_1), \ldots, g(\xi_k)\}$ is not a set of MSE-RPs of $g(X)$ in general. Cheng and Lei (2022) discussed relationships between the two sets: MSE-RPs of Z and MSE-RPs of X under some nonlinear regression models, but they did not show any applications. Li et al. (2024c) pointed out that although $\{g(\xi_1), \ldots, g(\xi_k)\}$ is not a set of MSE-RPs of $g(X)$, it converges to the set of MSE-RPs of $Z = g(X)$ as well as the related approximation distributions via the following fact (Hogg et al. (2005)).

Lemma 3.3 *Let $Y^{(k)}$ be a discrete random variable with distribution in the first two rows below, and g is a continuous function on $\mathbb{R}$. Then the random variable $W^{(k)} = g(Y^{(k)})$ has the distribution in the last two rows below:*

$Y^{(k)}$	ξ_1	ξ_2	$\cdots$	ξ_k
p_Y	p_1	p_2	$\cdots$	p_k
$W^{(k)} = g(Y^{(k)})$	$g(\xi_1)$	$g(\xi_2)$	$\cdots$	$g(\xi_k)$
p_W	p_1	p_2	$\cdots$	p_k

(3.52)

For a discrete random variable $Y^{(k)}$ the distribution of $W^{(k)} = g(Y^{(k)})$ is very easy to obtain, especially the probabilities $p_Y = p_W$. It is a great advantage in discrete distributions. Note that all Y_{MC}, Y_{QMC}, and Y_{MSE} of X are discrete random variables. The random variables of $g(Y_{MC}), g(Y_{QMC})$, and $g(Y_{MSE})$ have a simple structure in (3.52). Based on Lemma 3.3 and MSE-RPs of X, Li et al. (2024c) proposed the following algorithm for generating approximation MSE-RPs of Z (App-RPs of Z for short):

Algorithm 3.7 (The App-RPs algorithm)

Step 1: Find the MSE-RPs of X: $\xi_1, \ldots, \xi_k$ with $P(Y = \xi_j) = p_j, j = 1, \ldots, k$ and form the random variable $Y^{(k)}$.

Step 2: Calculate $g(\xi_1), g(\xi_2), \cdots, g(\xi_k)$ as the support points of a random variable $W^{(k)}$ with corresponding probabilities $P(W^{(k)} = g(\xi_j)) = p_j, j = 1, \ldots, k$.

Step 3: Deliver the distribution of $W^{(k)}$ *as an approximation distribution to* F_Z *and points* $\{g(\xi_1), g(\xi_2), \cdots, g(\xi_k)\}$ *as App-RPs of* Z *with size* k.

Lemma 3.4 *Let* $Z \sim G$ *and* $Z^{(k)} \sim G_{mse}$. *If* $E(|Z|^m) < \infty$, *we have*

$$E(|Z^{(k)}|^m) \leq E(|Z|^m) < \infty.$$

Lemma 3.4 directly results from Pages (2018, Equation 5.10). It shows that the mth moments of $Z^{(k)}$ are controlled by mth absolute moment of Z. By this Lemma, we can obtain the following theorem.

Theorem 3.29 *Let random variable* $X \sim F(x)$, *where* $F(x)$ *has a density and finite second moments. Let* $\xi_1, \ldots, \xi_k$ *be MSE-RPs of* X *with* $P(Y_{mse}^{(k)} = \xi_j) = p_j, j = 1, \ldots, k$. *Let* $Z = g(X)$ *have finite second moments where function* $g(x)$ *is continuous in* $\mathbb{R}$. *Define a discrete random variable* $W^{(k)}$ *on the support points* $\{g(\xi_1), g(\xi_2), \cdots, g(\xi_k)\}$ *with*

$$P(W^{(k)} = g(\xi_j)) = p_j, j = 1, \ldots, k. \tag{3.53}$$

Then

1. $W^{(k)} \xrightarrow{D} Z = g(X)$ *with* MSE$(W^{(k)}) \to 0$ *as* $k \to \infty$.
2. *If* $E\left(|g(X)|^{r+\varepsilon}\right) < \infty$ *with some* $r \geq 3, \varepsilon > 0$, *then,*

$$E\left(W^{(k)}\right)^r \to E\left(Z^r\right) \text{ as } k \to \infty.$$

Especially we have

$$\mathrm{Var}\left(W^{(k)}\right) \to \mathrm{Var}(Z) \text{ as } k \to \infty.$$

Proof Corollary 3.1 shows that $Y^{(k)} \to X$ in mean square and in distribution. By Theorem 4.1 we have the first conclusion in the next chapter. The second assertion follows Theorem 4.2 in the next chapter.

Therefore, the set of $\{g(\xi_1), g(\xi_2), \cdots, g(\xi_k)\}$ is regarded as App-RPs of $Z = g(X)$ in the sense of consistency. However, there is a potential to improve the quality of App-RPs of Z. Choosing $X \sim U(0,1)$ and $Z = F^{-1}(X) \sim F(x)$, both QMC-RPs and MSE-RPs of $U(0,1)$ are $\{\frac{2j-1}{2k}, j = 1, \ldots, k\}$ with equal probability $1/k$ at each RP. In this case, a discrete random variable $W^{(k)} \sim F_{QMC}^{(k)}$ with the support points $\{F^{-1}(\frac{2j-1}{2k}), j = 1, \ldots, k\}$ each having probability $1/k$. These support points are QMC-RPs of Z (Fang et al. (2014)). By Theorem 3.29 we have

Corollary 3.4 $W^{(k)} \xrightarrow{D} Z = F^{-1}(X)$, *i.e.,* $F_{QMC}^{(k)} \to F$ *in distribution as* $k \to \infty$.

Example 3.11 *Let* $X \sim N(0,1)$ *and* $Z = X^2 \sim \chi^2(1)$. *Let MSE-RPs of* $N(0,1)$ *be* $\xi_1, \ldots, \xi_k$ *with* $P(Y = \xi_j) = p_j, j = 1, \ldots, k$. *Now the set of* $\xi_1^2, \ldots, \xi_k^2$ *with* $P(Y = \xi_j) = p_j, j = 1, \ldots, k$. *Cheng and Lei (2022) gave a study by linear regression between the true RPs of* $\chi^2(1)$ *and* $\xi_1^2, \ldots, \xi_k^2$.

Example 3.12 *Let* $X \sim N(0,1)$ *and* $Z \sim RN(0,1)$, *the log-normal distribution, where* $X = \ln(Z)$. *Now* $Z = e^X$. *Let RPs of* $N(0,1)$ *be* $\xi_1, \ldots, \xi_k$ *with* $P(Y = \xi_j) = p_j, j = 1, \ldots, k$. *Now the set of* $\{e^{\xi_1}, \ldots, e^{\xi_k}\}$ *with* $P(Y = \xi_j) = p_j, j = 1, \ldots, k$ *be App-RPs of* $RN(0,1)$.

TABLE 3.6
Comparisons of discrete approximations to $N(0,1)$: $Var(\hat{Z}) - Var(Z)$

k	F_{MSE}^{Normal}	$F^{-1}(U_{qmc})$	$F^{-1}(U_{fm})$	$F^{-1}(U_{fm2})$	$F^{-1}(U_{fm4})$	$g(Y_{mse}^{\chi^2})$
10	-0.0229	-0.1202	-0.0654	-0.1008	-0.0827	0
20	-0.0062	-0.0614	-0.0421	-0.0551	-0.0478	0
30	-0.0028	-0.0414	-0.0121	-0.0382	-0.0341	0
50	-0.0010	-0.0251	-0.0144	-0.0238	-0.0210	0
100	-0.0003	-0.0127	-0.0102	-0.0123	-0.0119	0

3.10.1 A new App-RPs for $N(0,1)$

For convenience, we call $X \sim F$ as *initial distribution* and the distribution $Z = g(X) \sim G$ as *target distribution* in this subsection.

Denote
MSE-RPs of X by $\xi_1^F, \ldots, \xi_k^F$,
MSE-RPs of Z by $\xi_1^G, \ldots, \xi_k^G$ and
App-RPs of Z by $g(\xi_i^F), i = 1, \ldots, k$.
Their corresponding discrete approximation random variables are denoted by Y_{MSE}^F, Y_{MSE}^G and Y_{F-G}, respectively. For example, For the target distribution $Z \sim N(0,1)$ and the initial distribution $X \sim \chi^2(1)$ we use $\xi_1^{\chi^2(1)}, \ldots, \xi_k^{\chi^2(1)}$ as MSE-RPs of $\chi^2(1)$, and $\xi_1^{N(0,1)}, \ldots, \xi_k^{N(0,1)}$ as MSE-RPs of $N(0,1)$. The transformation from X to Z is given by

$$Z = g(X) \stackrel{d}{=} \begin{cases} \sqrt{X} & \text{with } p = 0.5 \\ -\sqrt{X} & \text{with } p = 0.5 \\ 0 & \text{if } X = 0. \end{cases} \tag{3.54}$$

The App-RPs form a discrete variable $Y_{\chi^2(1)-N(0,1)}$. Li et al. (2024c) considered the log-normal $LN(0,1)$, $\chi^2(1)$ and uniform $U(0,1)$ distributions as initial and $N(0,1)$ as the target distribution. For the uniform distribution $U(0,1)$, they consider three discrete approximations:

(1) U_{QMC}, U_{MSE}: the uniform distributed on $\{\frac{2j-1}{2k}, j = 1, \ldots, k\}$;

(2) U_{FM}: It will be introduced in Section 4.4 that incorporate moment conditions into the mean square error criterion;

(3) U_{FM2}: It relaxes to match the first two moments only in the FM criterion introduced in Section 4.4.

They find that the discrete approximation $Y_{\chi^2(1)-N(0,1)}$ has the best performance in the estimation of the first four moments of $N(0,1)$ and find that the transformed discrete approximation from the MSE-RPs of the $\chi^2(1)$ distribution has the exact mean, variance, and the skewness of the target distribution.

Theorem 3.30 *Under the above notation when the initial distribution $X \sim \chi^2(1)$ for the target distribution $N(0,1)$ the discrete approximation by $Y_{\chi^2(1)-N(0,1)}$ has the same mean, variance and the skewness with $N(0,1)$. The kurtosis of $Y_{\chi^2(1)-N(0,1)}$ equals to* $\mathrm{Var}(Y_{MSE}^{\chi^2(1)})$.

The proof is given in Subsection 3.13.10. More detailed discussion can refer to Li et al. (2024c).

The variance of the distribution of $g(Y_{mse}^{\chi^2})$ exactly equals the variance of $N(0,1)$, which is exciting since even MSE-RPs derived from $N(0,1)$ exhibit variance bias. Some simulation studies are given in Tables 3.6 and 3.7. We observed that

1. The transformed discrete approximation generated from the MSE-RPs of the $\chi^2(1)$ distribution has the exact mean, variance, and skewness as $N(0,1)$.

TABLE 3.7
Comparisons of discrete approximations to $N(0,1)$: $Ku(\hat{Z}) - Ku(Z)$

k	F_{MSE}^{Normal}	$F^{-1}(U_{qmc})$	$F^{-1}(U_{fm})$	$F^{-1}(U_{fm2})$	$F^{-1}(U_{fm4})$	$g(Y_{mse}^{\chi^2})$
10	-0.2023	-0.7512	-0.5897	-0.7368	-0.6726	**-0.1411**
20	-0.0627	-0.4943	-0.4245	-0.4852	-0.4545	**-0.0385**
30	-0.0302	-0.3813	-0.2014	-0.3751	-0.3564	**-0.0176**
50	-0.0117	-0.2707	-0.1957	-0.2672	-0.2513	**-0.0065**
100	-0.0031	-0.1658	-0.1429	-0.1644	-0.1624	**-0.0017**

2. U_{fm} is susceptible to overfitting.
3. $F^{-1}(U_{fm2})$ and $F^{-1}(U_{fm4})$ exhibit superior performance compared to the traditional QMC-RPs ($F^{-1}(U_{qmc})$).

3.11 A Generalized Inverse Transformation Method

The inverse transformation method (ITM) has been applied in statistical simulation for decades where the initial distribution is $X \sim U(0,1)$, and the target distribution is any continuous distribution $Z \sim F$ (Theorem 2.1). There are some questions related to ITM:

(a) Can we use the approximation distributions U_{QMC} and U_{MSE} to replace $U(0,1)$ for improving ITM performance by the MC method in a statistical simulation?

(b) Can we find an alternative approximation distribution to $U(0,1)$ that is better than U_{QMC} and U_{MSE}?

Li et al. (2024c) gave the following study on these questions.

(A) A new initial distribution for $U(0,1)$
One can reversely use $X \sim F$ as the initial distribution and the uniform distribution $Z \sim U(0,1)$ as the target distribution. Li et al. (2024c) chose several initial distributions
the normal distribution $N(0,1)$,
the exponential distribution $EX(1)$,
the log-normal distribution $LN(0,1)$,
the $\chi^2(1)$-distribution, and
the $Y_{\chi^2(1)-N(0,1)}$ that is constructed from $\chi^2(1)$-distribution to $N(0,1)$ discussed in the previous subsection.

The following algorithm carries out the above consideration:

Algorithm 3.8 (Generalized inverse transformation algorithm)
Step 1: Generate a set of MSE-RPs of the initial distribution with size k, $\xi_1 < \cdots < \xi_k$ *and related probabilities* $p_i, i = 1, \ldots, k$ *for each initial distribution.*

Step 2: Find the function h *for each initial distribution of* X *to the target distribution* $Z \sim U(0,1)$.

Step 3: Deliver a discrete distribution with the support points, $\{h(\xi_i), i = 1, \ldots, k\}$ *(see Table 3.8, where the function* $\text{erf}(x) = \frac{2}{\sqrt{\pi}} \int_0^x e^{-t^2} dt$, $\gamma(s,t)$ *is the lower incomplete gamma function).*

TABLE 3.8
Notation for generalized inverse transformation to $U(0,1)$

Notation	Initial Distribution	h function
TY_{mse}^{Normal}	Initial $N(0,1)$ MSE-RPs	$h(x) = \frac{1}{2}\left[1 + \text{erf}\left(\frac{x}{\sqrt{2}}\right)\right]$
TY_{mse}^{EP}	Initial $Exp(1)$ MSE-RPs	$h(x) = 1 - e^{-x}$
TY_{mse}^{LN}	Initial $LN(0,1)$ MSE-RPs	$h(x) = \frac{1}{2}\left[1 + \text{erf}\left(\frac{\ln x}{\sqrt{2}}\right)\right]$
$TY_{mse}^{\chi^2}$	Initial $\chi^2(1)$ MSE-RPs	$h(x) = \frac{1}{\Gamma(1/2)}\gamma\left(\frac{1}{2}, \frac{x}{2}\right)$
$Tg(Y_{mse}^{\chi^2})$	Initial $g(Y_{mse}^{\chi^2})$	$h(x) = \frac{1}{2}\left[1 + \text{erf}\left(\frac{x}{\sqrt{2}}\right)\right]$

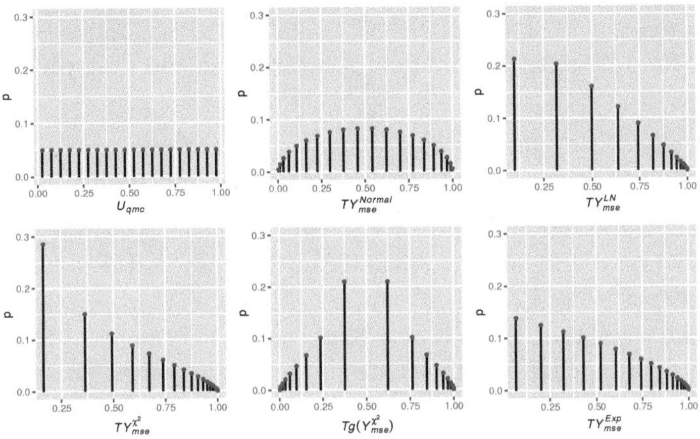

FIGURE 3.3
Different discrete approximation distributions to $U(0,1)$ based on 20 support points.

The new-ITM approach provides more efficient and reliable alternatives to obtain discrete approximations to target distribution $U(0,1)$, name themes as U_{qmc}, TY_{mse}^{normal}, TY_{mse}^{LN}, $TY_{mse}^{\chi^2}$, $Tg(Y_{mse}^{\chi^2})$, and TY_{MSE}^{EP}, respectively. Figure 3.3 shows plots of six approximation distributions to $U(0,1)$, and we have some observations:

- The transformed log-normal U^{LN}, transformed chi-squared U^{χ^2} and transformed exponential U^{EP} exhibit asymmetry and are unsuitable candidates.
- The uniform U_{qmc}, transformed normal U^{Normal}, and double-transformed chi-squared $Tg(Y_{mse}^{\chi^2})$ are symmetric around 0.5. These distributions have the same expectation, variance, and skewness as the $U(0,1)$ distribution.
- Both U^{Normal} and $Tg(Y_{mse}^{\chi^2})$ exhibit more weights on support points in the middle of $(0,1)$, with U^{Normal} showing a more uniformly scattered support point pattern.
- Surprisingly, the kurtosis of U^{Normal} is closer to the kurtosis of $U(0,1)$ compared to U_{qmc}, which are directly drawn from $U(0,1)$. From now on, we use U_{new} to replace U^{Normal}.

We found that U_{new} matches the first four moments of $U(0,1)$ to be the best among others. Li et al. (2024c) provided the RPs of U_{new} and related probabilities.

(B). Performance of the new-ITM

Li et al. (2024c) examined the ITM from U_{QMC} and the new-ITM from U_{new} for estimation of the first four moments for the eight target distributions: log-normal, chi-squared, exponential, exponentially modified Gaussian (exGau), Student-t, arcsine, gamma, and the mixture of two-component normal distributions. They observed

- The accuracy of the new-ITM distribution of U_{new} is very close to the target ones in terms of mean, variance, skewness, and kurtosis.
- The heavy-tailed distributions may require larger k points to form a valid approximation using the new U_{new}.
- It is interesting that the distribution of U_{new} is not uniform on the support points. A non-uniform discrete distribution is better than the uniform discrete approximation by U_{QMC}.

Li et al. (2024c) gave comparisons of new-ITM via U_{qmc} and U_{new} in the bias of the mean, variance, skewness, and kurtosis and the IG concerning the eight target distributions. The Summary of our findings are

1. The new-ITM via U_{new} significantly improves the estimation compared to the new-ITM via U_{qmc}.
2. Our simulation results consistently show that the new-ITM via U_{new} has a smaller bias in the first four moments and higher IG values for scenarios (2)–(5) and (7)–(8).
3. In scenario (1), although the IGs of new-ITM via U_{new} are slightly lower, it provides better estimates for the moments.
4. It is worth mentioning that Jiang et al. (2015) proved that the new-ITM via U_{qmc} has identical moments to the target distribution when the target distribution is the arc-sine distribution. We observed the same results in scenario (6).

Li et al. (2024c) provided a comparison between the non-parametric k-means algorithm and the inverse transformation via U_{qmc} for estimating the distribution $MixN(0.70, 8, 10, 1.5, 1)$ (mixture of two normal distributions of $N(8, 10)$ and $N(1.5, 1)$ with $\alpha = 0.70$, notation can refer to Subsection 3.9.1). In their study, the non-parametric k-means method requires generating a large training sample of size $N = 100{,}000$. Their results show that the New-ITM via U_{new} yields better estimation in the first four moments than the non-parametric k-means approach. The non-parametric k-means method relies heavily on the training sample's quality.

3.12 Discrete Approximation to the Distribution Constructed by Convolution

In many case studies, the distribution of interest is the convolution of two random variables. Let random variable $Z = X+Y$, where the joint density of (X, Y) is $f_{X+Y}(z)$ (or $(f_X * f_Y)(z)$ in the literature) with marginal densities f_X and f_Y. Variable Z is called as the convolution of X and Y and has the density function

$$f_{X+Y}(z) = \int_{-\infty}^{+\infty} f(z-y, y)dy = \int_{-\infty}^{+\infty} f(x, z-x)dx.$$

When X and Y are independent, the about formulas become

$$f_{X+Y}(z) = \int_{-\infty}^{+\infty} f_X(z-y) f_Y(y) dy = \int_{-\infty}^{+\infty} f_X(x) f_Y(z-x) dx.$$

In many cases, the density of Z does not have an analytic formula, and finding its MSE-RPs requires more computational work. Can we use a seminar approach to consider two approximate distributions ξ_X and ξ_Y to X and Y, respectively? Their convolution $\xi_Z = \xi_X + \xi_Y$ approximates to Z.

As ξ_X and ξ_Y are discrete variables, their convolution ξ_Z can be easily obtained.

Lemma 3.5 *When ξ_X and ξ_Y are independent and both have discrete distributions*

$$\begin{array}{c|cccc} \xi_X & x_1 & x_2 & \cdots & x_u \\ \hline p & p_1 & p_2 & \cdots & p_u \end{array} \tag{3.55}$$

and

$$\begin{array}{c|cccc} \xi_Y & y_1 & y_2 & \cdots & y_v \\ \hline p & q_1 & q_2 & \cdots & q_v \end{array} \tag{3.56}$$

Then $\xi_Z = \xi_X + \xi_Y$ has the discrete distribution with support points $z_{ij} = x_i + y_j$ with probabilities $p_{ij} = p_i q_j, i = 1, \ldots, u; j = 1 \ldots, v$. When some z_{ij} has duplicated value, the corresponding probabilities p_{ij} should be aggregated together. We have

$$E(\xi_Z) = E(\xi_X) + E(\xi_Y), \;\; \mathrm{Var}(\xi_Z) = \mathrm{Var}(\xi_X) + \mathrm{Var}(\xi_Y).$$

For clarifying notation, we denote the MSE-RPs of Z by the support points $\{w_1, \ldots, w_k, k \le uv\}$ with probabilities $p_{r_1}, \ldots, p_{r_k}$.

Theorem 3.31 *Let $Z = X+Y$, where $X \sim F_1$ and $Y \sim F_2$, and X and Y are independent. Let $\xi_X \sim F_{MSE}^X$ and $\xi_Y \sim F_{MSE}^Y$ be the discrete approximation formed by the respective MSE-RPs of X and Y and their corresponding probabilities. Let ξ_Z be formed by ξ_X and ξ_Y as in Lamma 3.5. We have*

$$E(Z) = E(\xi_X) + E(\xi_Y) = E(\xi_Z), \tag{3.57}$$

and

$$\begin{aligned} \mathrm{Var}(Z) &= \mathrm{Var}(\xi_X) + \mathrm{Var}(\xi_Y) + \mathrm{MSE}(\xi_X) + \mathrm{MSE}(\xi_Y) && (3.58) \\ &= \mathrm{Var}(\xi_Z) + \mathrm{MSE}(\xi_X) + \mathrm{MSE}(\xi_Y). && (3.59) \end{aligned}$$

Proof *The assertions follow from Theorem 3.5 and Lemma 3.5.*

This approach offers three primary advantages:

1. it is straightforward to implement as it does not require the density of Z and its derivation;
2. the proposed discrete approximation ensures that the mean of the approximation aligns with the mean of the actual distribution of Z;
3. the variance of Z is given by the sum of the variance of the proposed discrete approximation and the mean squared errors of MSE-RPs obtained from X and Y.

Example 3.13 *An exponentially modified Gaussian distribution (EMG, also known as ex-Gaussian distribution) describes the sum of independent normal and exponential random variables. Let $X \sim N(\mu, \sigma^2)$ and $Y \sim EP(\lambda)$ be independent. Let an exGaussian random variable $Z = X + Y$. The density of Z is*

$$f_Z(z; \mu, \sigma, \lambda) = \frac{\lambda}{2} e^{\left(\frac{\lambda}{2}(2\mu + \lambda\sigma^2 - 2z)\right)} erfc\left(\frac{\mu + \lambda\sigma^2 - z}{\sqrt{2}\sigma}\right),$$

where erfc is the complementary error function defined as

$$erfc(z) = \frac{2}{\sqrt{\pi}} \int_x^\infty e^{-t^2} dt.$$

From the above approach, we employ MSE-RPs of X and MSE-RPs of Y to generate approximate MSE-RPs of Z.

In a simulation study, Li et al. (2024c) performed the following process to evaluate the performance of the proposed discrete approximation of Z (called CONV-MSE):

1. *To generate k_1 MSE-RPs from $N(\mu, \sigma^2)$ and form ξ_X and k_2 MSE-RPs from $EP(\lambda)$ and form ξ_Y*

2. *Using the relationship $\xi_Z = \xi_X + \xi_Y$ and Lemma 3.5, we construct a discrete approximation ξ_Z of Z. This approximation is labeled as CONV-MSE.*

3. *To generate discrete approximations of Z using the inverse transformation of U_{qmc} (ITM-QMC) and the inverse transformation of U_{fm4} (ITM-FM4) methods.*

4. *To assess the performance of CONV-MSE and compare it with the direct MSE-RPs obtained from the distribution of Z (Direct-MSE).*

By a simulation Li et al. (2024c) found that the CONV-MSE method provides an effective means of approximating the distribution of Z based on the MSE-RPs of its constituent random variables with accurate estimates of the mean and reasonable approximations of variance, skewness, and kurtosis.

More general case is $Z = X_1 + X_2 + \cdots + X_d$, where $X_i \sim F_i, i = 1 \ldots, d$ are independent each having density $f_i, i = 1, \ldots, d$, respectively.

Example 3.14 *Let random variables $X_i \sim Gamma(\alpha_i, \lambda), i = 1 \ldots, d$ be independent and $Z = X_1 + X_2 + \cdots + X_d \sim Gamma(\sum_{i=1}^d \alpha_i, \lambda)$.*

3.13 Proofs of Theorems in Chapter 3

3.13.1 Proof of Theorem 3.5

Proof *Denote the MSE-RPs of X by $\xi_1 < \xi_2 < \ldots < \xi_k$. Then*

$$\text{MSE}(Y_{MSE}) = \sum_{i=1}^k \int_{a_i}^{a_{i+1}} (x - \xi_i)^2 p(x) dx, \tag{3.60}$$

where $a_1 = -\infty, a_i = (\xi_i + \xi_{i-1})/2,\ i = 2, \cdots, k$ and $a_{k+1} = \infty$.

Take derivatives of $\text{MSE}(\mathcal{Y})$ *with respective to each of* $\xi_1, \ldots, \xi_k$ *and we find*

$$\begin{cases} \int_{-\infty}^{(\xi_1+\xi_2)/2} xp(x)dx = \xi_1 p_1, \\ \int_{(\xi_{i-1}+\xi_i)/2}^{(\xi_i+\xi_{i+1})/2} xp(x)dx = \xi_i p_i, i = 2, \ldots, k-1, \\ \int_{(\xi_{k-1}+\xi_k)/2}^{\infty} xp(x)dx = \xi_k p_k. \end{cases} \tag{3.61}$$

Sum both sides, we find that

$$E(X) = \xi_1 p_1 + \sum_{i=2}^{k-1} \xi_i p_i + \xi_k p_k = E(Y_{MSE})$$

that implies conclusion (A). Similarly, we can prove (B) by

$$\begin{aligned} E(XY_{MSE}^{(t-1)}) = & \int_{-\infty}^{(\xi_1+\xi_2)/2} x\xi_1^{(t-1)} p(x)dx + \sum_{i=2}^{k-1} \int_{(\xi_{i-1}+\xi_i)/2}^{(\xi_i+\xi_{i+1})/2} x\xi_i^{(t-1)} p(x)dx \\ & + \int_{(\xi_{k-1}+\xi_k)/2}^{\infty} x\xi_k^{(t-1)} p(x)dx \\ & = \sum_{i=1}^{k} \xi_i^t p_i = E(Y_{MSE}^t) \end{aligned}$$

by the use of (3.61), then the assertion (B) follows. When $t = 2$, *we have* $E(XY_{MSE}) = E(Y_{MSE}^2)$ *that is needed for proving the assertion (C).*

$$\begin{aligned} \text{MSE}(Y_{MSE}) &= E(X - Y_{MSE})^2 = E(X^2) + E(Y_{MSE}^2) - 2E(XY_{MSE}) \\ &= E(X^2) + E(Y_{MSE}^2) - 2E(Y_{MSE}^2) \\ &= E(X^2) - E(Y_{MSE}^2) \\ &= [E(X^2) - E^2(X)] - [E(Y_{MSE}^2) - (E(Y_{MSE})^2] \\ &= \text{Var}(X) - \text{Var}(Y_{MSE}). \end{aligned}$$

The conclusion (C) is true. For proving (D), assume that the domain of X *is* $\mathbb{R}$. *Denote* $\mathcal{Y}^{(k)} = \{\xi_1, \ldots, \xi_k\}$ *where* $\xi_1 < \xi_2 < \ldots < \xi_k$. *If we choose* $\mathcal{Y}_+ = \{\xi_1, \ldots, \xi_k, \xi_+\}$ *where* $\xi_+ = \xi_k + 2\delta, \delta > 0$, *and show that* $\text{MSE}(\mathcal{Y}^{(k)}) \geq \text{MSE}(\mathcal{Y}_+)$, *the proof is complete. In fact*

$$\begin{aligned} \text{MSE}(\mathcal{Y}^{(k)}) - \text{MSE}(\mathcal{Y}_+) &= \int_{\xi_k+\delta}^{\infty} [(x - \xi_k)^2 - (x - \xi_+)^2] p(x)dx \\ &= 2\int_{\xi_k+\delta}^{\infty} (x - \xi_k - \delta) p(x)dx \geq 0. \end{aligned}$$

In a similar way, we can prove that the support of X *is finite.*

3.13.2 Proof of Theorem 3.7

Proof *According to Proposition 2 (iii) in Liu and Pagès (2020), the assertion* $m \leq \xi_1 < \ldots < \xi_k \leq M$ *is obvious. Xu et al. (2023) gave a different way to prove that* $\xi_1 \neq m$ *and* $\xi_n^{(k)} \neq M$ *for all* $k \geq 2$. *Suppose there exists* $k \geq 2$ *such that* $\xi_1 = m$. *After taking the derivative, we obtain*

$$\int_m^{(\xi_1+\xi_2)/2} (x - \xi_1) f(x)\mathrm{d}x = 0.$$

Putting $\xi_1 = m$, *it yields*

$$\int_m^{(m+\xi_2)/2} (x-m)f(x)\mathrm{d}x = 0. \tag{3.62}$$

Notice that

$$m = \frac{m+\xi_1}{2} < \frac{m+\xi_2}{2} < \frac{m+\xi_n}{2} \le \frac{m+M}{2} < M.$$

Define

$$g_1(y) = \int_m^y (x-m)f(x)\mathrm{d}x, \; m < y < M.$$

Since the derivative of g_1 *with respect to* y *is* $g_1'(y) = (y-m)f(y) > 0$, *it follows that* $g_1(y)$ *is strictly increasing on* (m, M), *hence,* $g_1(y) > g_1(m) = 0$ *for all* $y \in (m, M)$. *This means that* $g_1(y) = 0$ *is impossible when* $y \in (m, M)$.

In fact, (3.62) is equivalent to $g_1(\xi_1) = 0$ *where* $\xi_1 = (m+a_2)/2$ *and* $\xi_1 \in (m, M)$, *which makes the contradiction. Thus,* $\xi_1 \neq m$ *for all* $k \geq 2$, *hence* $a_1 > m$.

Similarly, suppose $a_k = M$. *After taking the derivative, we obtain*

$$\int_{(\xi_{k-1}+\xi_k)/2}^{M} (x-\xi_k)f(x)\mathrm{d}x = 0.$$

Putting $\xi_k = M$, *it yields*

$$\int_{(\xi_{k-1}+M)/2}^{M} (x-M)f(x)\mathrm{d}x = 0. \tag{3.63}$$

Notice that

$$m < \frac{m+M}{2} \le \frac{\xi_1+M}{2} < \frac{\xi_{k-1}+M}{2} < \frac{\xi_k+M}{2} = M.$$

Define

$$g_2(z) = \int_z^M (x-M)f(x)\mathrm{d}x, \; m < z < M.$$

Since the derivative of g_2 *with respect to* z *is* $g_2'(z) = -(z-M)f(z) > 0$, *it follows that* $g_2(z)$ *is strictly increasing on* (m, M), *hence,* $g_2(z) < g_2(M) = 0$ *for all* $z \in (m, M)$. *This means that* $g_2(z) = 0$ *is impossible when* $y \in (m, M)$. *In fact, (3.63) is equivalent to* $g_2(z) = 0$ *where* $z = (\xi_{k-1} + M)/2$ *and* $z \in (m, M)$, *which makes the contradiction. Thus,* $\xi_k \neq M$ *for all* $k \geq 2$, *hence* $a_n < M$. *In conclusion,* $m < \xi_1 < \ldots < \xi_k < M$ *for all* $k \geq 2$, *which completes the proof.*

3.13.3 Proof of Theorem 3.9

Let $Q(u)$ denote the quantile function of $F(x)$. If F has a density $p(x)$, the quantile

$$Q(u) = F^{-1}(u) \Rightarrow F(Q(u)) = u.$$

Differentiating u on both sides, we have,

$$\begin{aligned} &p(Q(u))Q'(u) = 1, \\ &g(u) = -\log p(Q(u)), \\ &g'(u) = -\frac{p'(Q(u))Q'(u)}{p(Q(u))} = -p'(Q(u))\left(Q'(u)\right)^2, \end{aligned}$$

and

$$\begin{aligned}
g''(u) &= -\left[p''(Q(u))\,(Q'(u))^3 + 2Q'(u)Q''(u)p'(Q(u))\right] \\
&= -\left[p''(Q(u))\,(Q'(u))^3 - 2Q'(u)p'(Q(u))\,(Q'(u))^2\,p'(Q(u))/p(Q(u))\right] \\
&= -\left[p''(Q(u))(p(Q(u))) - 2\,(p'(Q(u)))^2\right]/(p(Q(u)))^4 \\
&= \left[2\,(p'(Q(u)))^2 - p(Q(u))p''(Q(u))\right]/(p(Q(u)))^4.
\end{aligned}$$

Thus, $g''(u) > 0$ if and only if

$$p \cdot p'' - 2(p')^2 < 0.$$

3.13.4 Proof of Theorem 3.12

As the density function of $U(0,1)$ is

$$p(x) = \begin{cases} 1 & 0 < x < 1 \\ 0 & \text{otherwise} \end{cases},$$

its RPs $\xi_1 < \ldots < \xi_k$ satisfy (refer to (3.61))

$$\begin{cases} \displaystyle\int_0^{(\xi_1+\xi_2)/2} (x-\xi_1)\mathrm{d}x = 0, \\ \displaystyle\int_{(\xi_{i-1}+\xi_i)/2}^{(\xi_i+\xi_{i+1})/2} (x-\xi_i)\mathrm{d}x = 0, \ i = 2, \ldots, k-1, \\ \displaystyle\int_{(\xi_{k-1}+\xi_k)/2}^{1} (x-\xi_k)\mathrm{d}x = 0. \end{cases} \tag{3.64}$$

From the first equation of (3.64) we have

$$\int_0^{(\xi_1+\xi_2)/2} (x-\xi_1)\mathrm{d}x = \frac{1}{2}\left[\left(\frac{\xi_2-\xi_1}{2}\right)^2 - (\xi_1)^2\right],$$

and it follows $\xi_1 = \frac{\xi_2-\xi_1}{2}$. Similarly, we can find

$$\xi_1 = \frac{1}{2}\left(\frac{\xi_1+\xi_2}{2}\right), \xi_2 = \frac{1}{2}\left(\frac{\xi_1+\xi_2}{2} + \frac{\xi_2+\xi_3}{2}\right), \cdots \xi_k = \frac{1}{2}\left(1 + \frac{\xi_{k-1}+\xi_k}{2}\right)$$

It follows

$$\xi_2 = 3\xi_1, \xi_i = 2\xi_{i-1} - \xi_{i-2}, i = 3, \ldots, k, 3\xi_k - \xi_{k-1} = 2.$$

And we can find

$$3\xi_1 = \xi_2, 5\xi_1 = \xi_3, 7\xi_1 = \xi_4, \ldots, (2k-1)\xi_1 = \xi_k.$$

From $3\xi_k - \xi_{k-1} = 2$ and the above relationships we have

$$3(2k-1)\xi_1 - (2(k-2)-1)\xi_1 - 2 = 0,$$

Which implies $\xi_1 = \frac{1}{2k}$ and $\xi_i = \frac{2i-1}{2k}$, $i = 1, ..., k$.

There is another proof for this result as follows: from the second equation of (3.64), we have

$$\begin{aligned}\int_{(\xi_{i-1}+\xi_i)/2}^{(\xi_i+\xi_{i+1})/2}(x-\xi_i)\mathrm{d}x &= \left[\frac{1}{2}(x-\xi_i)^2\right]_{\frac{\xi_{i-1}+\xi_i}{2}}^{\frac{\xi_i+\xi_{i+1}}{2}} \\ &= \frac{1}{2}\left[\left(\frac{\xi_{i+1}-\xi_i}{2}\right)^2-\left(\frac{\xi_i-\xi_{i-1}}{2}\right)^2\right],\end{aligned}$$

it follows that

$$\xi_{i+1}-\xi_i=\xi_i-\xi_{i-1}\equiv\delta,\ i=2,...,k-1.$$

From the last equation, we similarly get

$$\int_{(\xi_{k-1}+\xi_k)/2}^{1}(x-\xi_k)\mathrm{d}x=\frac{1}{2}\left[(1-\xi_k)^2-\left(\frac{\xi_k-\xi_{k-1}}{2}\right)^2\right],$$

and

$$1-\xi_k=\frac{\xi_k-\xi_{k-1}}{2}.$$

As $\xi_1=\frac{1}{2}\delta$ and $\xi_k=1-\frac{1}{2}\delta$, we have

$$\xi_1^{(k)}+\sum_{i=2}^{k}(\xi_i-\xi_{i-1})+(1-\xi_k)=1,$$

it follows that

$$\frac{\delta}{2}+(k-1)\delta+\frac{\delta}{2}=1,$$

hence, $\delta=\frac{1}{k}$. Thus, the RPs of $U(0,1)$ are list in (3.14).

3.13.5 Proofs of theorems for exponential distributions

All the proofs in this subsection are from Xu et al. (2021).

Proof for Theorem 3.13

The assertion (3.19) is known about the mean $E(Y_{MSE})$, for the variance of $E(Y_{MSE})$

$$\begin{aligned}(\xi_1^{(k)})^2 &= E(X^2)-E(Y_{MSE}^2)\\ &= E(X^2)-(E(X))^2-(E(Y_{MSE}^2)-(E(Y_{MSE}))^2)\\ &= \mathrm{Var}(X)-\mathrm{Var}(Y_{MSE}).\end{aligned}$$

Therefore,

$$\mathrm{Var}(Y_{MSE})=\mathrm{Var}(X)-(\xi_1^{(k)})^2=\frac{1}{\lambda^2}-(\xi_1^{(k)})^2.$$

For finding the gap of $\xi_k^{(k)} - \xi_{k-1}^{(k)}$ note that the last equation in (3.61) gives us the following equation

$$\int_{(\xi_{k-1}^{(k)}+\xi_n^{(k)})/2}^{\infty} (x - \xi_n^{(k)})p(x)dx = 0. \tag{3.65}$$

Then,

$$\begin{aligned}
&\int_{(\xi_{k-1}^{(k)}+\xi_k^{(k)})/2}^{\infty} (x - \xi_k^{(k)})p(x)dx \\
&= \int_{(\xi_{k-1}^{(k)}+\xi_k^{(k)})/2}^{\infty} x\lambda e^{-\lambda x}dx - \xi_k^{(k)} \int_{(\xi_{k-1}^{(k)}+\xi_k^{(k)})/2}^{\infty} \lambda e^{-\lambda x}dx \\
&= -xe^{-\lambda x}|_{(\xi_{k-1}^{(k)}+\xi_k^{(k)})/2}^{\infty} + \frac{1}{\lambda}\int_{(\xi_{k-1}^{(k)}+\xi_k^{(k)})/2}^{\infty} \lambda e^{-\lambda x}dx - \xi_k^{(k)} \int_{(\xi_{k-1}^{(k)}+\xi_k^{(k)})/2}^{\infty} \lambda e^{-\lambda x}dx \\
&= \frac{1}{2}(\xi_{k-1}^{(k)} + \xi_k^{(k)})e^{-\frac{\lambda}{2}(\xi_{k-1}^{(k)}+\xi_k^{(k)})} + (\frac{1}{\lambda} - \xi_k^{(k)}) \int_{(\xi_{k-1}^{(k)}+\xi_k^{(k)})/2}^{\infty} \lambda e^{-\lambda x}dx \\
&= \frac{1}{2}(\xi_{k-1}^{(k)} + \xi_k^{(k)})e^{-\frac{\lambda}{2}(\xi_{k-1}^{(k)}+\xi_k^{(k)})} + (\frac{1}{\lambda} - \xi_k^{(k)})[-e^{-\lambda x}]_{(\xi_{k-1}^{(k)}+\xi_k^{(k)})/2}^{\infty} \\
&= \frac{1}{2}(\xi_{k-1}^{(k)} + \xi_k^{(k)})e^{-\frac{\lambda}{2}(\xi_{k-1}^{(k)}+\xi_k^{(k)})} + (\frac{1}{\lambda} - \xi_k^{(k)})e^{-\frac{\lambda}{2}(\xi_{k-1}^{(k)}+\xi_k^{(k)})} \\
&= (\frac{\xi_{k-1}^{(k)} + \xi_k^{(k)}}{2} + \frac{1}{\lambda} - \xi_k^{(k)})e^{-\frac{\lambda}{2}(\xi_{k-1}^{(k)}+\xi_k^{(k)})} \\
&= \frac{1}{2}(\xi_{k-1}^{(k)} - \xi_k^{(k)} + \frac{2}{\lambda})e^{-\frac{\lambda}{2}(\xi_{k-1}^{(k)}+\xi_k^{(k)})}.
\end{aligned}$$

Since $e^{-\frac{\lambda}{2}(\xi_{k-1}^{(k)}+\xi_k^{(k)})} > 0$, it follows that $\xi_{k-1}^{(k)} - \xi_k^{(k)} + \frac{2}{\lambda} = 0$, which completes the proof.

Remark 3.3 *Let* $y = x - \xi_k^{(k)}$ *in (3.65), then we can obtain*

$$\int_{(\xi_{k-1}^{(k)}-\xi_k^{(k)})/2}^{\infty} yp(y + \xi_k^{(k)})dy = 0. \tag{3.66}$$

From the memoryless property of the exponential distribution, we have

$$p(y + \xi_k^{(k)}) = \lambda e^{-\lambda(y+\xi_k^{(k)})} = e^{-\lambda\xi_k^{(k)}}\lambda e^{-\lambda y} = e^{-\lambda\xi_k^{(k)}}p(y).$$

Thus, equation (3.66) becomes

$$e^{-\lambda\xi_k^{(k)}} \int_{(\xi_{k-1}^{(k)}-\xi_k^{(k)})/2}^{\infty} yp(y)dy = 0;$$

and it follows that

$$\int_{(\xi_{k-1}^{(k)}-\xi_k^{(k)})/2}^{\infty} yp(y)dy = 0. \tag{3.67}$$

By using a similar way to analyze (3.67), proof for the assertion (3.20) is completed.

Proof of Theorem 3.14

For proving Theorem 3.14, we need the following Lemma.

Lemma 3.6 *If* $0 < \theta_1 \le \frac{1}{\lambda}$ *and* $\theta_2 > 0$ *satisfy the following equation*

$$e^{\lambda\theta_2}\left(\frac{1}{\lambda} - \theta_2\right) = \left(\theta_1 + \frac{1}{\lambda}\right)e^{-\lambda\theta_1}, \tag{3.68}$$

then θ_1 *and* θ_2 *have to satisfy the relation:* $0 < \theta_2 < \theta_1 \le \frac{1}{\lambda}$*. Further, if* θ_1 *is given, then* θ_2 *must be unique, namely, there is a unique solution in the following equation*

$$e^{\lambda x}\left(\frac{1}{\lambda} - x\right) = \left(\theta_1 + \frac{1}{\lambda}\right)e^{-\lambda\theta_1}, \quad \text{when } 0 < x < \theta_1.$$

Proof *Let*

$$f(x) = \left(x + \frac{1}{\lambda}\right)e^{-\lambda x}, x > 0,$$

then equation (3.68) becomes $f(\theta_1) = f(-\theta_2)$*. Define*

$$h(x) = f(x) - f(-x) = \left(x + \frac{1}{\lambda}\right)e^{-\lambda x} - \left(-x + \frac{1}{\lambda}\right)e^{\lambda x}, x > 0.$$

First, we show the monotonicity of $h(x)$*. The derivative of* $h(x)$ *is*

$$\begin{aligned} h'(x) &= e^{-\lambda x} + (x + \frac{1}{\lambda})(-\lambda)e^{-\lambda x} - (-e^{\lambda x} + (-x + \frac{1}{\lambda})(\lambda)e^{\lambda x}) \\ &= -\lambda x e^{-\lambda x} + \lambda x e^{\lambda x} = \lambda x e^{-\lambda x}(-1 + e^{2\lambda x}). \end{aligned}$$

It follows that $h(x)$ *is a strict increasing function on* $(0, +\infty)$ *as* $h'(x) > 0$ *when* $x > 0$*. Thus,* $h(x) > h(0) = 0$*, namely,* $f(x) > f(-x)$ *when* $x > 0$*. As a consequence,* $f(\theta_1) < f(-\theta_1)$ *when* $0 < \theta_1 \le \frac{1}{\lambda}$*. Under the condition* $f(-\theta_2) = f(\theta_1)$*, it is clear that* $f(-\theta_2) = f(\theta_1) > f(-\theta_1)$*. Now, since*

$$f'(x) = e^{-\lambda x} + \left(x + \frac{1}{\lambda}\right)(-\lambda)e^{-\lambda x} = -\lambda x e^{-\lambda x} > 0$$

when $x < 0$*, it follows that* $f(x)$ *is a strict increasing function on* $(-\infty, 0)$*. Thus,* $f(-\theta_2) > f(-\theta_1) \Leftrightarrow -\theta_2 > -\theta_1$*, namely* $\theta_1 > \theta_2$*.*
Next, we will show the uniqueness of the solution of $g(x) = 0$ *on* $(0, \theta_1)$*, where* $0 < \theta_1 \le \frac{1}{\lambda}$ *and*

$$g(x) = e^{\lambda x}\left(\frac{1}{\lambda} - x\right) - \left(\theta_1 + \frac{1}{\lambda}\right)e^{-\lambda\theta_1}.$$

Since $g'(x) = \lambda e^{\lambda x}(\frac{1}{\lambda} - x) + e^{\lambda x}(-1) = -\lambda x e^{\lambda x} < 0$ *when* $0 < x < \theta_1 \le \frac{1}{\lambda}$*, it follows that* $g(x)$ *is a strict decreasing function on* $(0, \theta_1)$*. Then, the sign of* $g(0)$ *can be determined as follows*

$$\begin{aligned} g(0) &= \frac{1}{\lambda} - \left(\theta_1 + \frac{1}{\lambda}\right)e^{-\lambda\theta_1} \\ &= \frac{1}{\lambda}e^{-\lambda\theta_1}(e^{\lambda\theta_1} - (\lambda\theta_1 + 1)) \\ &> 0 \quad (\textit{since } e^x > x + 1 \textit{ for } x \in \mathbb{R}^+). \end{aligned}$$

Because $g(x)$ is a strict decreasing function on $(0, \theta_1)$ and $g(0) > 0$, if we could prove $g(\theta_1) < 0$, then $g(x)$ must have a unique solution on $(0, \theta_1)$ based on intermediate value theorem. The function $g(\theta_1)$ can be expressed as

$$\begin{aligned} g(\theta_1) &= e^{\lambda\theta_1}\left(\frac{1}{\lambda} - \theta_1\right) - \left(\theta_1 + \frac{1}{\lambda}\right)e^{-\lambda\theta_1} \\ &= \frac{1}{\lambda}e^{-\lambda\theta_1}((1-\lambda\theta_1)e^{2\lambda\theta_1} - (\lambda\theta_1 + 1)). \end{aligned}$$

In order to prove $g(\theta_1) < 0$, it is equivalent to prove the following inequality:

$$(1-\lambda\theta_1)e^{2\lambda\theta_1} - (\lambda\theta_1 + 1) < 0, \quad 0 < \theta_1 \leq \frac{1}{\lambda}. \tag{3.69}$$

Let $k(x) = (1-x)e^{2x} - (x+1), 0 < x \leq 1$. It is easy to calculate that

$$\begin{aligned} k'(x) &= -e^{2x} + 2(1-x)e^{2x} - 1 = e^{2x} - 2xe^{2x} - 1, \\ k''(x) &= 2e^{2x} - 2(e^{2x} + 2xe^{2x}) = -4xe^{2x} < 0. \end{aligned}$$

Since $k''(x) < 0$ when $0 < x \leq 1$, it follows that $k'(x)$ is a strict decreasing function on $(0, 1]$ and we can obtain $k'(x) < k'(0) = 0$ when $0 < x \leq 1$. It leads that $k(x)$ is a strict decreasing function on $(0, 1]$, so $k(x) < k(0) = 0$ when $0 < x \leq 1$ and equation (3.69) holds, i.e., $g(\theta_1) < 0$. Based on the intermediate value theorem, the proof is complete.

Corollary 3.5 *Let univariate random variable $X \sim EP(\lambda)$. Its MSE-RP $\xi_1^{(k+1)}$ can be obtained by solving*

$$e^{\lambda\xi_1^{(k+1)}}\left(\frac{1}{\lambda} - \xi_1^{(k+1)}\right) = \left(\xi_1^{(n)} + \frac{1}{\lambda}\right)e^{-\lambda\xi_1^{(k)}},$$

for given $\xi_1^{(k)}$. Further, there is a unique solution $\xi_1^{(k+1)}$ satisfying the above equation.

Now, we prove Theorem 3.14. From equation (3.61), we can obtain

$$\left(\frac{1}{\lambda} + \frac{\xi_{i+1}^{(k)} - \xi_i^{(k)}}{2}\right)e^{-\lambda\frac{\xi_{i+1}^{(k)} - \xi_i^{(k)}}{2}} = \left(\frac{1}{\lambda} - \frac{\xi_i^{(k)} - \xi_{i-1}^{(k)}}{2}\right)e^{\lambda\frac{\xi_i^{(k)} - \xi_{i-1}^{(k)}}{2}} \tag{3.70}$$

for $i = 2, ..., n-1$. From Theorem 3.13, we know the final gap $(\xi_n^{(k)} - \xi_{n-1}^{(k)})$ is constant $\frac{2}{\lambda}$. By the mathematical induction, if we set $\theta_1 = (\xi_{i+1}^{(k)} - \xi_i^{(k)})/2$ and $\theta_2 = (\xi_i^{(k)} - \xi_{i-1}^{(k)})/2$ in Lemma 3.6 and the condition is satisfied for each $i = 2, ..., n-1$, then we can obtain

$$\frac{\xi_{i+1}^{(k)} - \xi_i^{(k)}}{2} > \frac{\xi_i^{(k)} - \xi_{i-1}^{(k)}}{2}, \quad i = 2, ..., k-1,$$

which completes the proof.

Proof of Theorem 3.15

According to equations (3.61) and Theorem 3.14, we can obtain the following two systems of equations. The first one is for k RPs and the second is for $k+1$ case.

$$\begin{cases} \left(\frac{1}{\lambda} + \frac{\xi_{j+1}^{(k+1)} - \xi_j^{(k+1)}}{2}\right)e^{-\lambda\frac{\xi_{j+1}^{(k+1)} - \xi_j^{(k+1)}}{2}} \\ \quad = \left(\frac{1}{\lambda} - \frac{\xi_j^{(k+1)} - \xi_{j-1}^{(k+1)}}{2}\right)e^{\lambda\frac{\xi_j^{(k+1)} - \xi_{j-1}^{(k+1)}}{2}}, j = 3, ..., k-1 \\ \xi_{k+1}^{(k+1)} - \xi_k^{(k+1)} = \frac{2}{\lambda}. \end{cases}$$

The proof for the first equation is completed by using the Theorem 3.14 and Lemma 3.6 directly.

To prove the second assertion of the theorem, set $\delta = (\xi_1^{(k+1)} + \xi_2^{(k+1)})/2$ and $\rho = 1$ in Theorem 3.5 (C), then $X + \delta$ has has the density function

$$q(x) = \frac{\lambda e^{-\lambda x}}{e^{-\lambda\delta}} I_{(x \geq \delta)},$$

and its RPs are

$$\beta_i^{(k)} = \xi_i^{(k)} + \frac{\xi_1^{(k+1)} + a_2^{(k+1)}}{2}. \tag{3.71}$$

Consider the $k+1$ RPs of $EP(\lambda)$ and denote them by $\xi_1^{(k+1)} < \ldots < \xi_{n+1}^{(k+1)}$. And those points must satisfy similar with equations (3.61)

$$\begin{cases} \int_0^{(\xi_1^{(k+1)}+\xi_2^{(k+1)})/2} (x - \xi_1^{(k+1)}) p(x) dx = 0, \\ \int_{(\xi_{i-1}^{(k+1)}+\xi_i^{(k+1)})/2}^{(\xi_i^{(k+1)}+\xi_{i+1}^{(k+1)})/2} (x - \xi_i^{(k+1)}) p(x) dx = 0, \quad i = 2, ..., n, \\ \int_{(\xi_k^{(k+1)}+\xi_{k+1}^{(k+1)})/2}^{\infty} (x - \xi_{n+1}^{(k+1)}) p(x) dx = 0, \end{cases}$$

where $p(x) = \lambda e^{-\lambda x}, x \geq 0$. With some simple calculation, we can obtain the following k equations by deleting the first original equation

$$\begin{cases} \left(\frac{\xi_{i-1}^{(k+1)} - \xi_i^{(k+1)}}{2} + \frac{1}{\lambda} \right) e^{-\frac{\lambda}{2}(\xi_{i-1}^{(k+1)}+\xi_i^{(k+1)})} \\ \quad = \left(\frac{\xi_{i+1}^{(k+1)} - \xi_i^{(k+1)}}{2} + \frac{1}{\lambda} \right) e^{-\frac{\lambda}{2}(\xi_{i+1}^{(k+1)}+\xi_i^{(k+1)})}, i = 2, ..., k \\ \xi_{k+1}^{(k+1)} - \xi_k^{(k+1)} = \frac{2}{\lambda} \end{cases} \tag{3.72}$$

For $q(x)$, as we know that $\beta_1^{(k)}, ..., \beta_k^{(k)}$ are the n MSE-RPs of $q(x)$ where $a = (\xi_1^{(k+1)} + \xi_2^{(k+1)})/2$, then we can obtain the following n equations

$$\begin{cases} \left(\frac{\xi_1^{(k+1)} - (2\beta_1^{(k)} - \xi_2^{(k+1)})}{2} + \frac{1}{\lambda} \right) e^{-\frac{\lambda}{2}(\xi_1^{(k+1)}+\xi_2^{(k+1)})} \\ \quad = \left(\frac{\beta_2^{(k)} - \beta_1^{(k)}}{2} + \frac{1}{\lambda} \right) e^{-\frac{\lambda}{2}(\beta_1^{(k)}+\beta_2^{(k)})} \left(\frac{\beta_{i-1}^{(k)} - \beta_i^{(k)}}{2} + \frac{1}{\lambda} \right) e^{-\frac{\lambda}{2}(\beta_{i-1}^{(k)}+\beta_i^{(k)})} \\ \quad = \left(\frac{\beta_{i+1}^{(k)} - \beta_i^{(k)}}{2} + \frac{1}{\lambda} \right) e^{-\frac{\lambda}{2}(\beta_i^{(k)}+\beta_{i+1}^{(k)})}, \quad i = 2, ..., k-1 \\ \beta_k^{(k)} - \beta_{k-1}^{(k)} = \frac{2}{\lambda} \end{cases} \tag{3.73}$$

Based on the sufficient condition (3.9), the solutions of equation (3.73) must be unique. Thus, compared equations (3.72) with (3.73), we can obtain

$$\beta_i^{(k)} = \xi_{i+1}^{(k+1)}, \quad i = 1, ..., n.$$

With equation (3.71), we conclude

$$\xi_i^{(k)} + \frac{\xi_1^{(k+1)} + \xi_2^{(k+1)}}{2} = \xi_{i+1}^{(k+1)}, \quad i = 1, ..., k.$$

Particularly,

$$\xi_2^{(k+1)} = \xi_1^{(k)} + \frac{\xi_1^{(k+1)} + \xi_2^{(k+1)}}{2}.$$

3.13.6 Proofs of theorems for Laplace distributions

All the proofs in this subsection are from Xu et al. (2021).

Proof for Theorem 3.16

Without loss of generality, we assume $\mu = 0$, namely $p(x) = \frac{1}{2b}e^{-\frac{|x|}{b}}$. From the following equation

$$\int_{\frac{1}{2}(\xi_{k-1}+\xi_k)}^{\infty} (x - \xi_k)p(x)dx = 0$$

we have

$$\begin{aligned}
&\int_{\frac{1}{2}(\xi_{k-1}+\xi_k)}^{\infty} (x - \xi_k)p(x)dx \\
&= \int_{\frac{1}{2}(\xi_{k-1}+\xi_k)}^{\infty} x\frac{1}{2b}e^{-\frac{|x|}{b}}dx - \xi_k \int_{\frac{1}{2}(\xi_{k-1}+\xi_k)}^{\infty} \frac{1}{2b}e^{-\frac{|x|}{b}}dx \\
&= -\frac{1}{2}xe^{-\frac{x}{b}}|_{\frac{1}{2}(\xi_{k-1}+\xi_k)}^{\infty} + \frac{b}{2}\int_{\frac{1}{2}(\xi_{k-1}+\xi_k)}^{\infty} \frac{1}{b}e^{-\frac{x}{b}}dx - \frac{\xi_k}{2}\int_{\frac{1}{2}(\xi_{k-1}+\xi_k)}^{\infty} \frac{1}{b}e^{-\frac{x}{b}}dx \\
&= \frac{1}{2}(\frac{\xi_{k-1}+\xi_k}{2})e^{-\frac{1}{2b}(\xi_{k-1}+\xi_k)} + \frac{b-\xi_k}{2}\int_{\frac{1}{2}(\xi_{k-1}+\xi_k)}^{\infty} \frac{1}{b}e^{-\frac{x}{b}}dx \\
&= \frac{1}{2}(\frac{\xi_{k-1}+\xi_k}{2})e^{-\frac{1}{2b}(\xi_{k-1}+\xi_k)} + \frac{b-\xi_k}{2}[-e^{-\frac{x}{b}}]_{\frac{1}{2}(\xi_{k-1}+\xi_k)}^{\infty} \\
&= \frac{1}{2}(\frac{\xi_{k-1}+\xi_k}{2})e^{-\frac{1}{2b}(\xi_{k-1}+\xi_k)} + \frac{b-\xi_k}{2}e^{-\frac{1}{2b}(\xi_{k-1}+\xi_k)} \\
&= \frac{1}{2}(\frac{\xi_{k-1}+\xi_k}{2} + b - \xi_k)e^{-\frac{1}{2b}(\xi_{k-1}+\xi_k)} \\
&= \frac{1}{4}(\xi_{k-1} - \xi_k + 2b)e^{-\frac{1}{2b}(\xi_{k-1}+\xi_k)}.
\end{aligned}$$

Since $e^{-\frac{1}{2b}(\xi_{k-1}+\xi_k)} > 0$, it follows that $\xi_{k-1} - \xi_k + 2b = 0$.

Proof of Theorem 3.17

We prove part (a) first. Without loss of generality, set $\mu = 0$, namely $p(x) = \frac{1}{2b}e^{-\frac{|x|}{b}}$.

Case 1: The number of MSE-RPs is $2k+1$

Based on (2.33), we can obtain

$$\begin{cases} \int_{(\xi_{i+1}^{(2k+1)}+\xi_{i+2}^{(2k+1)})/2}^{(\xi_{i+2}^{(2k+1)}+\xi_{i+3}^{(2k+1)})/2} (x - \xi_{i+2}^{(2k+1)})p(x)dx = 0, \quad i = k, ..., 2k-2 \\ \int_{(\xi_{2k}^{(2k+1)}+\xi_{2k+1}^{(2k+1)})/2}^{\infty} (x - \xi_{2k+1}^{(2k+1)})p(x)dx = 0. \end{cases} \tag{3.74}$$

Since the MSE PRs are symmetric around 0, it follows that $(\xi_{k+1}^{(2k+1)}+\xi_{k+2}^{(2k+1)})/2>0$. Thus, (3.74) can be simplified into

$$\begin{cases} \int_{(\xi_{i+1}^{(2k+1)}+\xi_{i+2}^{(2k+1)})/2}^{(\xi_{i+2}^{(2k+1)}+\xi_{i+3}^{(2k+1)})/2} (x-\xi_{i+2}^{(2k+1)})e^{-\lambda x}dx=0, \quad i=k,...,2k-2 \\ \int_{(\xi_{2k}^{(2k+1)}+\xi_{2k+1}^{(2k+1)})/2}^{\infty} (x-\xi_{2k+1}^{(2k+1)})e^{-\lambda x}dx=0, \end{cases} \tag{3.75}$$

where $\lambda=\frac{1}{b}$. According to the proving processing of Theorem 3.14, the proof is completed in case 1.

Case 2: The number of MSE-RPs is $2k$

According to the proving processing of Theorem 3 in Xu et al. (2021), we only need to add

$$\xi_{k+1}^{(2k)}-\xi_k^{(2k)}<\xi_{k+2}^{(2k)}-\xi_{k+1}^{(2k)}$$

From the symmetry of MSE-RPs of Laplace distribution, we know $\xi_k^{(2k)}=-\xi_{k+1}^{(2k)}$. Thus, we need to prove

$$2\xi_{k+1}^{(2k)}<\xi_{k+2}^{(2k)}-\xi_{k+1}^{(2k)}. \tag{3.76}$$

Based on (2.33), we can obtain

$$\int_0^{(\xi_{k+1}^{(2k)}+\xi_{k+2}^{(2k)})/2} (x-\xi_{k+1}^{(2k)})p(x)dx=0. \tag{3.77}$$

The above equation can be simplified into

$$e^{\lambda\xi_{k+1}^{(2k)}}(\frac{1}{\lambda}-\xi_{k+1}^{(2k)})=(\frac{\xi_{k+2}^{(2k)}-\xi_{k+1}^{(2k)}}{2}+\frac{1}{\lambda})e^{-\frac{\lambda}{2}(\xi_{k+2}^{(2k)}-\xi_{k+1}^{(2k)})}, \tag{3.78}$$

where $\lambda=\frac{1}{b}$. Applying Lemma 3.6, (3.76) can be directly obtained.

Moreover, according to the processing of proving Theorem 3.15, it is easy to prove the part (b). The proof is finished.

Proof of Theorem 3.18

Set

$$p(x)=\frac{1}{2b}e^{-\frac{|x|}{b}}, \quad x\in\mathbb{R}$$

When $n=2k+1$

$$\int_{(\xi_{k+1}^{(2k+1)}+\xi_{k+2}^{(2k+1)})/2}^{(\xi_{k+2}^{(2k+1)}+\xi_{k+3}^{(2k+1)})/2} (x-\xi_{k+2}^{(2k+1)})p(x)dx=0$$

$$\int_{(\xi_{k+1}^{(2k+1)}+\xi_{k+2}^{(2k+1)})/2}^{(\xi_{k+2}^{(2k+1)}+\xi_{k+3}^{(2k+1)})/2} (x-\xi_{k+2}^{(2k+1)})\frac{1}{2b}e^{-\frac{|x|}{b}}dx=0$$

Because of $(\xi_{k+1}^{(2k+1)}+\xi_{k+2}^{(2k+1)})/2=\xi_{k+2}^{(2k+1)}/2>0$,

$$\int_{\xi_{k+2}^{(2k+1)}/2}^{(\xi_{k+2}^{(2k+1)}+\xi_{k+3}^{(2k+1)})/2} (x-\xi_{k+2}^{(2k+1)})\frac{1}{2b}e^{-\frac{x}{b}}dx=0$$

$$\int_{\xi_{k+2}^{(2k+1)}/2}^{(\xi_{k+2}^{(2k+1)}+\xi_{k+3}^{(2k+1)})/2}(x-\xi_{k+2}^{(2k+1)})e^{-\frac{x}{b}}dx=0$$

Let $y = x - \xi_{k+2}^{(2k+1)}$,

$$\int_{-\xi_{k+2}^{(2k+1)}/2}^{(\xi_{k+3}^{(2k+1)}-\xi_{k+2}^{(2k+1)})/2} ye^{-\frac{y+\xi_{k+2}^{(2k+1)}}{b}}dy=0$$

Thus,

$$\int_{-\xi_{k+2}^{(2k+1)}/2}^{(\xi_{k+3}^{(2k+1)}-\xi_{k+2}^{(2k+1)})/2} ye^{-\frac{y}{b}}dy=0 \tag{3.79}$$

When $n = 2k$

$$\int_{(\xi_k^{(2k)}+\xi_{k+1}^{(2k)})/2}^{(\xi_{k+1}^{(2k)}+\xi_{k+2}^{(2k)})/2}(x-\xi_{k+1}^{(2k)})p(x)dx=0$$

Because of $\xi_k^{(2k)} = -\xi_{k+1}^{(2k)}$,

$$\int_0^{(\xi_{k+1}^{(2k)}+\xi_{k+2}^{(2k)})/2}(x-\xi_{k+1}^{(2k)})e^{-\frac{x}{b}}dx=0$$

Let $y = x - \xi_{k+1}^{(2k)}$,

$$\int_{-\xi_{k+1}^{(2k)}}^{(\xi_{k+2}^{(2k)}-\xi_{k+1}^{(2k)})/2} ye^{-\frac{y+\xi_{k+1}^{(2k)}}{b}}dy=0$$

Thus,

$$\int_{-\xi_{k+1}^{(2k)}}^{(\xi_{k+2}^{(2k)}-\xi_{k+1}^{(2k)})/2} ye^{-\frac{y}{b}}dy=0 \tag{3.80}$$

Compared (3.79) with (3.80) based on Lemma 3.6, we can obtain the following equation.

$$\xi_{k+2}^{(2k+1)} = 2\xi_{k+1}^{(2k)}$$

By the transformation property of RPs, the proof is completed.

Proof of Theorem 3.19

I. $2k \to 2k+1$

When $n = 2k$, we need to prove

$$\xi_i^{(2k+1)} < \xi_i^{(2k)} < \xi_{i+1}^{(2k+1)}, \quad i = 1, ..., 2k.$$

Due to the symmetry of MSE-RPs of Laplace distribution, we only need to consider the positive entry.

Thus, we only need to prove

$$\xi_i^{(2k+1)} < \xi_i^{(2k)} < \xi_{i+1}^{(2k+1)}, \quad i = k+1, ..., 2k.$$

Obviously, $\xi_{k+1}^{(2k+1)} = 0 < \xi_{k+1}^{(2k)} < 2\xi_{k+1}^{(2k)} = \xi_{k+2}^{(2k+1)}$ based on the previous theorem. Thus, we only need to prove

$$\xi_i^{(2k+1)} < \xi_i^{(2k)} < \xi_{i+1}^{(2k+1)}, \quad i = k+2, ..., 2k. \tag{3.81}$$

First, we consider

$$\xi_i^{(2k)} < \xi_{i+1}^{(2k+1)}, \quad i = k+2, ..., 2k. \tag{3.82}$$

We have

$$\xi_{i-1}^{(2k)} < \xi_i^{(2k+1)}, \tag{3.83}$$

and

$$\xi_{i+1}^{(2k+1)} - \xi_i^{(2k+1)} = \xi_i^{(2k)} - \xi_{i-1}^{(2k)} \tag{3.84}$$

if we apply (3.83), (3.84) for $i = k+2, ..., 2k$ step by step, then we can deduce that

$$\xi_i^{(2k)} < \xi_{i+1}^{(2k+1)}, \quad i = k+2, ..., 2k. \tag{3.85}$$

Second, we consider

$$\xi_i^{(2k+1)} < \xi_i^{(2k)}, \quad i = k+2, ..., 2k.$$

From Case 2 in Theorem 3 and Theorem 5, we have

$$\xi_{k+2}^{(2k+1)} = 2\xi_{k+1}^{(2k)} < \xi_{k+2}^{(2k)}$$

We have

$$\begin{aligned}
\xi_i^{(2k+1)} &= \xi_{i-1}^{(2k+1)} + (\xi_i^{(2k+1)} - \xi_{i-1}^{(2k+1)}) \\
&< \xi_{i-1}^{(2k+1)} + (\xi_{i+1}^{(2k+1)} - \xi_i^{(2k+1)}) \\
&< \xi_{i-1}^{(2k)} + (\xi_i^{(2k)} - \xi_{i-1}^{(2k)}) \\
&= \xi_i^{(2k)}
\end{aligned}$$

for $i = k+3, ..., 2k$ step by step bu the use of Mathematical Induction ($\xi_{i-1}^{(2k+1)} < \xi_{i-1}^{(2k)}$).

II. $2k+1 \rightarrow 2k+2$
When $n = 2k+1$, we need to prove

$$\xi_i^{(2k+2)} < \xi_i^{(2k+1)} < \xi_{i+1}^{(2k+2)}, \quad i = 1, ..., 2k+1.$$

Due to the symmetry of MSE-RPs of Laplace distribution, we only need to consider the positive entry.
Thus, we only need to prove

$$\xi_i^{(2k+2)} < \xi_i^{(2k+1)} < \xi_{i+1}^{(2k+2)}, \quad i = k+2, ..., 2k+1.$$

First, we consider

$$\xi_i^{(2k+1)} < \xi_{i+1}^{(2k+2)}, \quad i = k+2, ..., 2k+1.$$

Based on Theorem 4,

$$\xi_{k+2}^{(2k+1)} - \xi_{k+1}^{(2k+1)} = \xi_{k+2}^{(2k+1)} - 0 = \xi_{k+3}^{(2k+2)} - \xi_{k+2}^{(2k+2)}$$

Since $\xi_{k+2}^{(2k+2)} > 0$, it follows that

$$\xi_{k+2}^{(2k+1)} < \xi_{k+3}^{(2k+2)}$$

And if we apply

$$\xi_{i-1}^{(2k+1)} < \xi_{i}^{(2k+2)}$$

and

$$\xi_{i+1}^{(2k+2)} - \xi_{i}^{(2k+2)} = \xi_{i}^{(2k+1)} - \xi_{i-1}^{(2k+1)}$$

for $i = k+3, ..., 2k+1$ step by step, then we can deduce

$$\xi_{i}^{(2k+1)} < \xi_{i+1}^{(2k+2)}, \quad i = k+3, ..., 2k+1.$$

Second, we consider

$$\xi_{i}^{(2k+2)} < \xi_{i}^{(2k+1)}, \quad i = k+2, ..., 2k+1.$$

It is noticeable that $\xi_{k+2}^{(2k+2)}$ and $\xi_{k+1}^{(2k)}$ are the first positive MSE-RPs of Laplace distribution, as well as the first MSE-RPs of exponential distribution.
From the result in Xu et al. (2021), we have

$$\xi_{k+2}^{(2k+2)} < \xi_{k+1}^{(2k)} < 2\xi_{k+1}^{(2k)} = \xi_{k+2}^{(2k+1)}$$

Next,

$$\begin{aligned} \xi_{i}^{(2k+2)} &= \xi_{i-1}^{(2k+2)} + (\xi_{i}^{(2k+2)} - \xi_{i-1}^{(2k+2)}) \\ &< \xi_{i-1}^{(2k+2)} + (\xi_{i+1}^{(2k+2)} - \xi_{i}^{(2k+2)}) \\ &< \xi_{i-1}^{(2k+1)} + (\xi_{i}^{(2k+1)} - \xi_{i-1}^{(2k+1)}) \\ &= \xi_{i}^{(2k+1)} \end{aligned}$$

for $i = k+3, ..., 2k+1$ step by step (Mathematical Induction) ($\xi_{i-1}^{(2k+2)} < \xi_{i-1}^{(2k+1)}$).

3.13.7 Proof of theorem for the gamma distribution

All the proofs in this subsection are from Ke et al. (2023).

Proof of Theorem 3.21

Consider gamma distribution with $\beta = 1$ (i.e., $\mu = \alpha$) in the proof without loss of any generality.

Proof of conclusion 1.

Let

$$H(z_1, z_2) = (z_1 - \alpha)F(\frac{z_1 + z_2}{2}) + \frac{1}{2}(z_1 + z_2)p(\frac{z_1 + z_2}{2}),$$

there is

$$H(z_1, z_1) = (z_1 - \alpha)F(z_1) + z_1 p(z_1).$$

Because $\lim_{z_1 \to 0} H(z_1, z_1) = 0$ and

$$\begin{aligned} H'_{z_1}(z_1, z_1) &= F(z_1) + (z_1 - \alpha)p(z_1) + p(z_1) + (\alpha - 1)p(z_1) - z_1 p(z_1) \\ &= F(z_1) > 0, \end{aligned}$$

Hence $H(z_1, z_1) > 0$.

$$\begin{aligned} H'_{z_2}(z_1, z_2) &= \frac{1}{2}(z_1 - \mu)p(\frac{z_1 + z_2}{2}) + \frac{1}{2}p(\frac{z_1 + z_2}{2}) \\ &\quad + \frac{1}{2}(\mu - 1)p(\frac{z_1 + z_2}{2}) - \frac{1}{4}(z_1 + z_2)p(\frac{z_1 + z_2}{2}) \\ &= \frac{1}{4}(z_1 - z_2)p(\frac{z_1 + z_2}{2}) < 0, (z_1 < z_2). \end{aligned}$$

In addition,

$$H(z_1, \infty) = \lim_{z_2 \to \infty} H(z_1, z_2) = z_1 - \alpha,$$

we have

$$H(z_1, \infty) < 0 \Leftrightarrow z_1 < \alpha. \tag{3.86}$$

Combine (3.86) with the condition $H(z_1, z_2)$ is continuous for $z_2 \in [z_1, \infty)$ and $H(z_1, z_1) > 0$, assertion 1 is proved.

Proof of conclusion 3.

When $b = 1$, equation (3.39) becomes

$$(z_k - \alpha)[1 - F(\frac{z_{k-1} + z_k}{2})] = \frac{1}{2}(z_{k-1} + z_k)p(\frac{z_{k-1} + z_k}{2}). \tag{3.87}$$

Let

$$U(z_{k-1}, z_k) = (z_k - \alpha)[1 - F(\frac{z_{k-1} + z_k}{2})] - \frac{1}{2}(z_{k-1} + z_k)p(\frac{z_{k-1} + z_k}{2}),$$

therefore

$$U(z_{k-1}, z_{k-1}) = (z_{k-1} - \alpha)(1 - F(z_{k-1})) - z_{k-1}p(z_{k-1}).$$

Since $U(0,0) = -\alpha < 0$, $\lim_{z_{k-1} \to \infty} U(z_{k-1}, z_{k-1}) = 0$ and

$$\begin{aligned} U'_{z_{k-1}}(z_{k-1}, z_{k-1}) &= 1 - F(z_{k-1}) - (z_{k-1} - \alpha)p(z_{k-1}) - p(z_{k-1}) \\ &\quad - (\alpha - 1)p(z_{k-1}) + z_{k-1}p(z_{k-1}) \\ &= 1 - F(z_{k-1}) > 0, \end{aligned}$$

we have

$$U(z_{k-1}, z_{k-1}) < 0 \quad (z_{k-1} > 0). \tag{3.88}$$

Next, from the property of gamma distribution that

$$p(x; \alpha, 1) = F(x; \alpha - 1, 1) - F(x; \alpha, 1),$$

$$\begin{aligned}
U(z_{k-1}, z_k) &= (z_k - \alpha)[1 - F(\frac{z_{k-1}+z_k}{2}; \alpha, 1)] - \frac{1}{2}(z_{k-1}+z_k)[F(\frac{z_{k-1}+z_k}{2}; \alpha-1, 1) \\
&\quad - F(\frac{z_{k-1}+z_k}{2}; \alpha, 1)] \\
&= (z_k - \alpha) + F(\frac{z_{k-1}+z_k}{2}; \alpha, 1)(-z_k + \alpha + \frac{1}{2}z_{k-1} + \frac{1}{2}z_k) \\
&\quad + F(\frac{z_{k-1}+z_k}{2}; \alpha-1, 1)(-\frac{1}{2}z_{k-1} - \frac{1}{2}z_k) \\
&\geq (z_k - \alpha) + F(\frac{z_{k-1}+z_k}{2}; \alpha, 1)(\alpha - z_k).
\end{aligned}$$

Next, we show that for $z_k \in [z_{k-1}, \infty)$, $U(z_{k-1}, z_k)$ is firstly monotone increasing and then monotone decreasing. Derive $U(z_{k-1}, z_k)$ by z_k, get

$$\begin{aligned}
U'_{z_k}(z_{k-1}, z_k) &= (1 - F(\frac{z_{k-1}+z_k}{2})) - \frac{1}{2}(z_k - \alpha)p(\frac{z_{k-1}+z_k}{2}) \\
&\quad - \frac{1}{2}f(\frac{z_{k-1}+z_k}{2}) - \frac{1}{2}(\alpha-1)p(\frac{z_{k-1}+z_k}{2}) + \frac{1}{4}(z_{k-1}+z_k)p(\frac{z_{k-1}+z_k}{2}) \\
&= 1 - F(\frac{z_{k-1}+z_k}{2}) - \frac{1}{4}(z_k - z_{k-1})p(\frac{z_{k-1}+z_k}{2}).
\end{aligned}$$

Let

$$V(z_{k-1}, z_k) = U'_{z_k}(z_{k-1}, z_k), \quad z = \frac{1}{2}(z_{k-1}+z_k),$$

we have

$$\begin{aligned}
V'_{z_k}(z_{k-1}, z_k) &= -\frac{1}{2}p(\frac{z_{k-1}+z_k}{2}) - \frac{1}{4}p(\frac{z_{k-1}+z_k}{2}) \\
&\quad - \frac{1}{4}(z_k - z_{k-1})[\frac{2}{z_{k-1}+z_k} \cdot \frac{\alpha-1}{2} - \frac{1}{2}]p(\frac{z_{k-1}+z_k}{2}) \\
&= [-\frac{3}{4} + \frac{1}{8}(z_k - z_{k-1}) - \frac{(\alpha-1)(z_k - z_{k-1})}{4(z_{k-1}+z_k)}]p(\frac{z_{k-1}+z_k}{2}) \\
&= [z^2 - (\alpha + 2 + z_{k-1})z + (\alpha-1)z_{k-1}]p(z)/z. \qquad (3.89)
\end{aligned}$$

Note that in equation (3.89) $z^2 > 0$, as long as

$$V(z_{k-1}, z_{k-1}) = 1 - F(z_{k-1}) > 0, \text{ and } \lim_{z_k \to \infty} H(z_{k-1}, z_k) = 0,$$

there must exist $C_0 > z_{k-1}$ that satisfies

$$\text{when } z_k < C_0,\ V'_{z_k}(z_{k-1}, z_k) < 0;$$
$$\text{when } z_k > C_0,\ V'_{z_k}(z_{k-1}, z_k) > 0.$$

Therefore, there exist C^* $(z_{k-1} < C^* < C_0)$ satisfies

$$\text{when } z_k < C^*,\ U'_{z_k}(z_{k-1}, z_k) = V(z_{k-1}, z_k) < 0;$$
$$\text{when } z_k > C^*,\ U'_{z_k}(z_{k-1}, z_k) = V(z_{k-1}, z_k) > 0,$$

which means that $U(z_{k-1}, z_k)$ is firstly monotone increasing and then monotone decreasing. In addition, we have

$$\lim_{z_k \to \infty} U(z_{k-1}, z_k) = 0 \qquad (3.90)$$

and $GUz_{k-1}, z_{k-1}) < 0$. Thus the function $U(z_{k-1}, z_k)$ must cross the x-axis and the solution z_k exists. One more step, as we know

$$\frac{\mathrm{d}z_k}{\mathrm{d}z_{k-1}} = -U'_{z_{k-1}}(z_{k-1}, z_k)/U'_{z_k}(z_{k-1}, z_k)$$

and $U'_{z_k}(z_{k-1}, z_k) > 0$ in the neighborhood domain, furthermore

$$G'_{z_{k-1}}(z_{k-1}, z_i) = -\frac{1}{4}(z_k - z_{k-1})p(\frac{z_{k-1}+z_k}{2}) < 0,$$

we find z_k is a monotone increasing function of z_{k-1}.
Proof of conclusion 2.
To prove point 2 is complicated. Here, we give the proof of a particular case when $X \sim G(1,1)$. Let

$$\begin{aligned} G(z_{i-1}, z_i, z_{i+1}) &= (z_i - \alpha)[F(\frac{z_i + z_{i+1}}{2}) - F(\frac{z_{i-1}+z_i}{2})] \\ &\quad + \frac{1}{2}(z_i + z_{i+1})p(\frac{z_i + z_{i+1}}{2}) - \frac{1}{2}(z_{i-1}+z_i)p(\frac{z_{i-1}+z_i}{2}), \end{aligned}$$

thus:

$$\begin{aligned} G(z_{i-1}, z_i, z_i) = (z_i - \alpha)[F(z_i) - F(\frac{z_{i-1}+z_i}{2})] \\ + z_i p(z_i) - \frac{z_{i-1}+z_i}{2}p(\frac{z_{i-1}+z_i}{2}). \end{aligned}$$

Since

$$p(x;\alpha,1) = F(x;\alpha-1,1) - F(x;\alpha,1), \tag{3.91}$$

Derive $G(z_{i-1}, z_i, z_i)$ by z_i we have

$$G'_{z_i}(z_{i-1}, z_i, z_i) = F(z_i) - F(\frac{z_{i-1}+z_i}{2}) - \frac{z_i - z_{i-1}}{4}p(\frac{z_{i-1}+z_i}{2}). \tag{3.92}$$

Let

$$H(z_{i-1}, z_i) = G'_{z_i}(z_{i-1}, z_i, z_i),$$

we have

$$H'_{z_i}(z_{i-1}, z_i) = p(z_i) + [-\frac{3}{4} + \frac{1}{8}(z_i - z_{i-1}) - (\alpha - 1)\frac{z_i - z_{i-1}}{4(z_{i-1}+z_i)}]p(\frac{z_{i-1}+z_i}{2}).$$

For $X \sim G(1,1)$, the above equation can be simplify to

$$H'_{z_i}(z_{i-1}, z_i) = p(z_i) + [-\frac{3}{4} + \frac{1}{8}(z_i - z_{i-1})]p(\frac{z_{i-1}+z_i}{2}), \tag{3.93}$$

set $x = \frac{z_i - z_{i-1}}{2}$, we have

$$\begin{aligned} & p(z_i) + [-\tfrac{3}{4} + \tfrac{1}{8}(z_i - z_{i-1})]p(\tfrac{z_{i-1}+z_i}{2}) = 0 \\ \Rightarrow \quad & \frac{e^{-z_i}}{e^{-\frac{z_{i-1}+z_i}{2}}} = \tfrac{3}{4} - \tfrac{1}{8}(z_i - z_{i-1}) \\ \Rightarrow \quad & e^{-\frac{z_{i-1}-z_i}{2}} = \tfrac{3}{4} - \tfrac{1}{4}\tfrac{z_{i-1}-z_i}{2} \\ \Rightarrow \quad & e^{-x} = \tfrac{3}{4} - \tfrac{1}{4}x. \end{aligned}$$

Therefore, $H'_{z_i}(z_{i-1}, z_i)$ crosses the x-axis twice for $z_i > z_{i-1}$.

As $H'_{z_i}(z_{i-1}, z_{i-1}) = \frac{1}{4}f(z_{i-1}) > 0$ and $\lim_{z_i\to\infty} \frac{\partial H(z_{i-1},z_i)}{\partial z_i} = 0$, combining with the facts $H(z_{i-1}, z_{i-1}) = 0$ and $\lim_{z_i\to\infty}(H(z_{i-1}, z_i)) = 0$, we know that for $z_i \in [z_{i-1}, \infty)$, $G(z_{i-1}, z_i, z_i)$ is first monotone increasing and then monotone decreasing. In addition that $G(z_{i-1}, z_{i-1}, z_{i-1}) = 0$ and $\lim_{z_i\to\infty} G(z_{i-1}, z_i, z_i) = 0$, we conclude $G(z_{i-1}, z_i, z_i) > 0$ when $z_{i-1} > 0$. Next, consider

$$\begin{aligned} G'_{z_{i+1}}(z_{i-1}, z_i, z_{i+1}) &= \frac{z_i - \alpha}{2} p(\frac{z_i + z_{i+1}}{2}) + \frac{1}{2} p(\frac{z_i + z_{i+1}}{2}) \\ &\quad + \frac{\alpha - 1}{2} p(\frac{z_i + z_{i+1}}{2}) - \frac{1}{4}(z_i + z_{i+1}) p(\frac{z_i + z_{i+1}}{2}) \\ &= \frac{z_i - z_{i+1}}{4} p(\frac{z_i + z_{i+1}}{2}) < 0, \end{aligned}$$

therefore the solution z_{i+1} exists if

$$G(z_{i-1}, z_i, \infty) < 0.$$

We find that

$$G(z_{i-1}, z_i, \infty) = (z_i - \alpha)\left[1 - F(\frac{z_{i-1} + z_i}{2})\right] - \frac{z_{i-1} + z_i}{2} p(\frac{z_{i-1} + z_i}{2})$$

is exactly (3.88). From the analysis in the proof of assertion 2, we conclude that the solution z_{i+1} exists when $z_{i-1} < z_{i,i-1}$.

3.13.8 Proofs of theorems for the normal distribution

All the proofs in this subsection are from Li et al. (2022).

Proof of Theorem 3.24

Proof *Without loss of generality, set $\sigma = 1$.*

$$p(x)' = \frac{\alpha}{\sqrt{2\pi}}[-(x-\mu)]e^{-\frac{(x-\mu)^2}{2}} + \frac{1-\alpha}{\sqrt{2\pi}}[-(x+\mu)]e^{-\frac{(x+\mu)^2}{2}},$$

$$p(x)'' = \frac{\alpha}{\sqrt{2\pi}}[(x-\mu)^2 - 1]e^{-\frac{(x-\mu)^2}{2}} + \frac{(1-\alpha)}{\sqrt{2\pi}}[(x+\mu)^2 - 1]e^{-\frac{(x+\mu)^2}{2}},$$

$$p(x)p(x)'' = \frac{\alpha^2}{2\pi}[(x-\mu)^2 - 1]e^{-(x-\mu)^2} + \frac{(1-\alpha)\alpha}{\pi}(x^2 + \mu^2 - 1)e^{-(x^2+\mu^2)},$$
$$+ \frac{(1-\alpha)^2}{2\pi}[(x+\mu)^2 - 1]e^{x+\mu)^2},$$

$$\begin{aligned} (p(x)')^2 &= \frac{\alpha^2}{2\pi}(x - \mu^2)^2 e^{-(x-\mu)^2} + \frac{(1-\alpha)^2}{2\pi}(x+\mu)^2 e^{-(x+\mu)^2} \\ &\quad + \frac{\alpha(1-\alpha)}{\pi}(x^2 - \mu^2)e^{-(x^2+\mu^2)}. \end{aligned}$$

$$\begin{aligned} &p(x)p(x)'' - 2(p(x)')^2 \leq -[\frac{\alpha^2}{2\pi}[(x-\mu)^2]e^{-(x-\mu)^2} + \frac{(1-\alpha)^2}{2\pi}[(x+\mu)^2]e^{-(x+\mu)^2} \\ &+ (\frac{\alpha(1-\alpha)}{\pi}x^2 - \frac{3\alpha(1-\alpha)}{\pi}\mu^2 + \frac{\alpha(1-\alpha)}{\pi})e^{-(x^2+\mu^2)}] \\ &\leq -[2\sqrt{\frac{\alpha^2}{2\pi}(x-\mu)^2 \frac{(1-\alpha)^2}{2\pi}(x+\mu)^2 e^{-(x^2+\mu^2)}} \\ &+ (\frac{\alpha(1-\alpha)}{\pi}x^2 - \frac{3\alpha(1-\alpha)}{\pi}\mu^2 + \frac{\alpha(1-\alpha)}{\pi})e^{-(x^2+\mu^2)}] \\ &= -[\frac{\alpha(1-\alpha)}{\pi} \mid \mu^2 - x^2 \mid + \frac{\alpha(1-\alpha)}{\pi}x^2 - \frac{3\alpha(1-\alpha)}{\pi}\mu^2 + \frac{\alpha(1-\alpha)}{\pi}]e^{-(x^2+\mu^2)} \\ &= -\frac{\alpha(1-\alpha)}{\pi}(-2\mu^2 + 1)e^{-(x^2+\mu^2)}. \end{aligned}$$

Then

$$p(x)p(x)'' - (p(x)')^2 \leq 0 \Rightarrow -2\mu^2 + 1 \geq 0 \Rightarrow \mu^2 \leq \frac{1}{2} \Rightarrow |\mu| \leq \frac{\sqrt{2}}{2}.$$

When $\sigma \geq 1$ *we can also get* $\frac{|\mu|}{\sigma} \leq \frac{\sqrt{2}}{2}$.

Proof of Theorem 3.25

Proof *The random variable* X *has the pdf*

$$p(x) = \alpha\varphi(x;\mu_1,\sigma^2) + (1-\alpha)\varphi(x;\mu_2,\sigma^2).$$

According to the second-order concavity test, $p(x)$ *is log-concave if*

$$p(x)p''(x) - (p'(x))^2 < 0, \tag{3.94}$$

for all x *in* $\mathbb{R}$. *The first-order derivative of* $p(x)$ *is*

$$p'(x) = \alpha\left(-\frac{(x-\mu_1)}{\sigma_1^2}\right)\varphi_1(x) + (1-\alpha)\left(-\frac{(x-\mu_2)}{\sigma_2^2}\right)\varphi_2(x).$$

The second-order derivative of $p(x)$ *is*

$$p''(x) = \left(\frac{(x-\mu_1)^2}{\sigma_1^4} - \frac{1}{\sigma_1^2}\right)\alpha\varphi_1(x) + \left(\frac{(x-\mu_2)^2}{\sigma_2^4} - \frac{1}{\sigma_2^2}\right)(1-\alpha)\varphi_2(x).$$

Accordingly, we have

$$\begin{aligned} p(x)p''(x) &= \alpha^2\varphi_1^2(x)\left(\frac{(x-\mu_1)^2}{\sigma_1^4} - \frac{1}{\sigma_1^2}\right) + (1-\alpha)^2\varphi_2^2(x)\left(\frac{(x-\mu_2)^2}{\sigma_2^4} - \frac{1}{\sigma_2^2}\right) \\ &\quad + \alpha(1-\alpha)\varphi_1(x)\varphi_2(x)\left(\frac{(x-\mu_1)^2}{\sigma_1^4} + \frac{(x-\mu_2)^2}{\sigma_2^4} - \frac{1}{\sigma_1^2} - \frac{1}{\sigma_2^2}\right), \end{aligned}$$

and

$$\begin{aligned} (p'(x))^2 &= \alpha^2\varphi_1^2(x)\left(\frac{(x-\mu_1)^2}{\sigma_1^4}\right) + (1-\alpha)^2\varphi_2^2(x)\left(\frac{(x-\mu_2)^2}{\sigma_2^4}\right) \\ &\quad + \alpha(1-\alpha)\varphi_1(x)\varphi_2(x)\left(\frac{2(x-\mu_1)(x-\mu_2)}{\sigma_1^2\sigma_2^2}\right). \end{aligned} \tag{3.95}$$

By (3.94)–(3.95), *we have*

$$-\frac{\alpha^2\varphi_1^2}{\sigma_1^2} - \frac{(1-\alpha)^2\varphi_2^2}{\sigma_2^2} - \alpha(1-\alpha)\varphi_1(x)\varphi_2(x)\left[\frac{1}{\sigma_1^2} + \frac{1}{\sigma_2^2} - \left(\frac{(x-\mu_1)}{\sigma_1^2} - \frac{(x-\mu_2)}{\sigma_2^2}\right)^2\right] < 0. \tag{3.96}$$

If $\frac{1}{\sigma_1^2} + \frac{1}{\sigma_2^2} - \left(\frac{(x-\mu_1)}{\sigma_1^2} - \frac{(x-\mu_2)}{\sigma_2^2}\right)^2 \geq 0$ *holds for all* $x \in \mathbb{R}$, *then* (3.96) *holds. As* $\sigma_1^2 = \sigma_2^2 = \sigma^2$, $\frac{2}{\sigma^2} - \left(\frac{(\mu_2-\mu_1)}{\sigma^2}\right)^2 \geq 0$ *if* $|\mu_1 - \mu_2| \leq \sqrt{2}\sigma$ *given any* $\alpha \in (0,1)$.

(For Corollary 3.3) Without loss of generality, again set $\sigma = 1$.

$$\begin{aligned} &p(x)p(x)'' - (p(x)')^2 \\ &= -\frac{1}{2\pi}[\alpha^2 e^{-(x-\mu)^2} + (1-\alpha)^2 e^{-(x+\mu)^2} - 2\alpha(1-\alpha)(2\mu^2-1)]e^{-(x^2+\mu^2)} \\ &\leq -\frac{1}{2\pi}[2\alpha(1-\alpha)e^{-(x^2+\mu^2)} - 2\alpha(1-\alpha)(2\mu^2-1)]e^{-(x^2+\mu^2)} \\ &= -\frac{2}{\pi}\alpha(1-\alpha)(1-\mu^2)e^{-(x^2+\mu^2)}. \end{aligned}$$

Then

$$p(x)p(x)'' - (p(x)')^2 \leq 0 \Rightarrow (1-\mu^2) \geq 0 \Rightarrow \mu^2 \leq 1 \Rightarrow |\mu| \leq 1.$$

When $\sigma > 1$, *we can also get* $\frac{|\mu|}{\sigma} \leq 1$.

Proof of Theorem 3.26

For simplicity, we denote $\phi_1(x)$ and $\Phi_1(x)$ as the pdf and the cdf of $N(\mu_1, \sigma_1^2)$, respectively, and denote $\phi_2(x)$ and $\Phi_2(x)$ as the pdf and the cdf of $N(\mu_2, \sigma_2^2)$, respectively.

Proof *Part (i). Denote the left-hand side of* (3.47) *as a function of two variables, i.e.,*

$$\begin{aligned} G_{u,v}(u,v) &= \alpha\left[\mu_1\Phi_1\left(\frac{u+v}{2}\right) - \sigma_1^2\varphi_1\left(\frac{u+v}{2}\right)\right] \\ &+ (1-\alpha)\left[\mu_2\Phi_2\left(\frac{u+v}{2}\right) - \sigma_2^2\varphi_2\left(\frac{u+v}{2}\right)\right] - uF\left(\frac{u+v}{2}\right), \end{aligned}$$

where $u \in \mathbb{R}$ *and* $v \in (u, +\infty)$ *represent* $\xi_1^{(k)}$ *and* $\xi_2^{(k)}$, *respectively, for shorts; and* φ_1 *and* φ_2 *denote the first and second normal densities in* $p(x)$, *and* Φ_1 *and* Φ_2 *are their cdfs, respectively. Define an intermediate function of* u *by setting* $v = u$ *in* $G_{u,v}(u,v)$ *as*

$$G_u(u) = \alpha\left[\mu_1\Phi_1(u) - \sigma_1^2\varphi_1(u)\right] + (1-\alpha)\left[\mu_2\Phi_2(u) - \sigma_2^2\varphi_2(u)\right] - uF(u). \tag{3.97}$$

From (3.97), *it is easy to verify that* $G_u(u) \to 0$ *as* $u \to -\infty$ *and* $G_u'(u) = -F(u) < 0$ *for* $\forall u \in \mathbb{R}$. *Hence,* $G_u(u) < 0$ *for* $u \in \mathbb{R}$, *which implies*

$$\lim_{v \to u} G_{u,v}(u,v) < 0. \tag{3.98}$$

The first-order partial derivative of $G(u,v)$ *with respect to* v *is positive as* $-\infty < u < v < \infty$,

$$\frac{\partial G_{u,v}(u,v)}{\partial v} = \frac{v-u}{4} \cdot p\left(\frac{u+v}{2}\right) > 0. \tag{3.99}$$

By (3.98)–(3.99), *the function* $G_{u,v}(u,v) = 0$ *has a unique solution of* v, *if and only if,*

$$\lim_{v \to +\infty} G_{u,v}(u,v) = E(X) - u > 0,$$

where $E(X)$ *is the population mean of* $X \sim \text{MixN}(\alpha, \mu_1, \sigma_1^2, \mu_2, \sigma_2^2)$. *Therefore, when* $a_1^{(k)} < E(X)$, *the equation* (3.47) *has a unique solution of* $a_2^{(k)}$.

Part (ii). Given $v = h_1(u)$. *According to the implicit function theorem, we have*

$$\frac{\partial G_{u,v}}{\partial u}(u, h_1(u)) + \frac{\partial G_{u,v}}{\partial v}(u, h_1(u))h_1'(u) = 0,$$

which implies

$$h_1'(u) = -\frac{\frac{\partial G_{u,v}}{\partial u}(u,v)}{\frac{\partial G_{u,v}}{\partial v}(u,v)}. \tag{3.100}$$

Given (3.99), *if* (3.101)< 0 *for all* $u \in (-\infty, E(X))$, *so that* (3.100)> 0, *then* $v = h_1(u)$ *is a strictly increasing function with respect to* u.

$$\frac{\partial G_{u,v}}{\partial u}(u,v) = \frac{v-u}{4} \cdot p\left(\frac{u+v}{2}\right) - F\left(\frac{u+v}{2}\right). \tag{3.101}$$

Next, the strictly decreasing conditions of (3.101) *are discussed. Denote* $g_{u,v}(u,v) = \frac{\partial G_{u,v}}{\partial u}(u,v)$. *Given* $u \in (-\infty, E(x))$, *we have* $\lim_{v \to u} g_{u,v}(u,v) < 0$ *and the partial derivative of* $g_{u,v}(u,v)$ *with respect to* v *is*

$$\begin{aligned} \frac{\partial g_{u,v}(u,v)}{\partial v} &= \frac{v-u}{8} f'(\frac{u+v}{2}) - \frac{1}{4} f(\frac{u+v}{2}) \\ &= -\frac{\alpha}{4}\phi_1(\frac{u+v}{2}) \left[\frac{(v-u)(u+v-2\mu_1)}{4\sigma_1^2} + 1\right] \\ &- \frac{(1-\alpha)}{4}\phi_2(\frac{u+v}{2}) \left[\frac{(v-u)(u+v-2\mu_2)}{4\sigma_2^2} + 1\right]. \end{aligned} \tag{3.102}$$

$g_{u,v}(u,v)$ *is strictly decreasing, i.e.,* $\frac{\partial g_{u,v}(u,v)}{\partial v}$ *in* (3.102) *is less than 0, if both* (3.103) *and* (3.104) *hold.*

$$\frac{(v-u)(u+v-2\mu_1)}{4\sigma_1^2} + 1 > 0, \tag{3.103}$$

$$\frac{(v-u)(u+v-2\mu_2)}{4\sigma_2^2} + 1 > 0 \tag{3.104}$$

(For Remark 3.1) Assume that $\mu_1 = \mu_2 = E(X)$. *If* $v \geq E(X)$, *the equations* (3.103) *and* (3.104) *hold. If* $c < u < v < E(X)$, *where* $c > -\infty$ *is a lower bound of the range of* u. *Let* $\sigma^2 = \max\{\sigma_1^2, \sigma_2^2\}$, *if* $(v-u)(u+v-2E(X)) - 4\sigma^2 > 0$, *which implies*

$$u > v - \frac{4\sigma^2}{2E(X)-(u+v)}, \tag{3.105}$$

then the equations (3.103) *and* (3.104) *hold. Since* $u+v > 2u$ *and* $v < E(X)$, *we have*

$$E(X) - \frac{2\sigma^2}{E(X)-u} > v - \frac{2\sigma^2}{E(X)-u} > v - \frac{4\sigma^2}{2E(X)-(u+v)}. \tag{3.106}$$

Then, if $u > E(X) - \frac{2\sigma^2}{E(X)-u}$, *which implies* $u > E(X) - \sqrt{2}\sigma$, *then* (3.105) *holds. Therefore, when* $\mu_1 = \mu_2$, $a_2^{(n)} = h_1(a_1^{(n)})$ *is a strictly increasing function with respect to* $a_1^{(n)}$ *for* $a_1^{(n)} \in [E(X) - \sqrt{2}\sigma, E(X))$.

(For Remark 3.2) Assume that $\mu_2 < \mu_1$, *so that* $\mu_2 < E(X) < \mu_1$. *Suppose* $c < u < E(X)$, *where* $c > -\infty$ *is a lower bound for the range of* u. *If* $c \geq \mu_2$ *so that* $u+v-2\mu_2 > 2(c-\mu_2) \geq 0$, *then* (3.104) *holds. When* $u+v-2\mu_1 \geq 0$, (3.103) *holds, we have* $v > \mu_1 > E(X) > u > c \geq \mu_2$ *because* $v \geq 2\mu_1 - u > 2\mu_1 - E(X)$ *and* $\mu_1 > E(X)$. *Therefore, when* $u+v-2\mu_1 \geq 0$, $g(u) = v$ *is strictly increasing with respect to* u *if* $u \in (c, E(X)) \subseteq (\mu_2, E(X))$, *where* $c \geq \mu_2$. *When* $u+v-2\mu_1 < 0$, *we have* $2\mu_1 - u > v > u$, *and* (3.103) *holds if*

$$0 < v - u < \frac{4\sigma_1^2}{2\mu_1 - (u+v)} \tag{3.107}$$

Since $u+v > 2u$, *we have* $\frac{2\sigma_1^2}{\mu_1 - u} < \frac{4\sigma_1^2}{2\mu_1-(u+v)}$. *Then, if* $v - u < \frac{2\sigma_1^2}{\mu_1-u}$ *so that*

$$u > v - \frac{2\sigma_1^2}{\mu_1 - u}, \tag{3.108}$$

the inequality (3.107) *holds. Since* $2\mu_1 - u > v$, *we have* $2\mu_1 - u - \frac{2\sigma_1^2}{\mu_1-u} > v - \frac{2\sigma_1^2}{\mu_1-u}$. *Then, if* $u > \mu_1 - \sigma_1$ *and* $\mu_2 < \mu_1 < \mu_2 + \sigma_1/(1-\alpha)$, *we have*

$$E(X) > u > \mu_1 - \frac{\sigma_1^2}{\mu_1 - u}, \tag{3.109}$$

which implies the inequality (3.108) *holds. When* $\mu_2 < \mu_1 < \mu_2 + \sigma_1/(1-\alpha)$, $\mu_1 - \sigma_1 < \mu_2$. *Therefore,* $g(u) = v$ *is strictly increasing with respect to* u *if* $\mu_2 < u < E(X)$ *given* $\mu_2 < \mu_1 < \mu_2 + \sigma_1/(1-\alpha)$. *Therefore, for a general location-scale mixture of two normal distributions with* $\mu_2 < \mu_1 < \mu_2 + \sigma_1/(1-\alpha)$, *for* $a_1^{(n)} \in (\mu_2, E(X))$, $a_2^{(n)} = h_1(a_1^{(n)})$ *is a strictly increasing function.*

Proof of Theorem 3.27

Proof *Denote the left-hand side of* (3.48) *as a function of two variables, i.e.,*

$$\begin{aligned} D_{u,v}(u,v) = &\alpha \left\{\mu_1 \left[1 - \Phi_1\left(\frac{u+v}{2}\right)\right] + \sigma_1^2 \varphi_1\left(\frac{u+v}{2}\right)\right\} \\ &+ (1-\alpha)\left\{\mu_2 \left[1 - \Phi_2\left(\frac{u+v}{2}\right)\right] + \sigma_2^2 \varphi_2\left(\frac{u+v}{2}\right)\right\} \\ &- v\left[1 - F\left(\frac{u+v}{2}\right)\right], \end{aligned} \tag{3.110}$$

where $-\infty < u < v < +\infty$. *In* (3.110), *the variables* u *and* v *represent* $\xi_{k-1}^{(k)}$ *and* $\xi_k^{(k)}$, *respectively. Define a intermediate function of* u *by letting* $v = u$ *as*

$$\begin{aligned} D_u(u) = &\alpha\left\{\mu_1\left[1-\Phi_1(u)\right] + \sigma_1^2\varphi_1(u)\right\} + (1-\alpha)\left\{\mu_2\left[1-\Phi_2(u)\right] + \sigma_2^2\varphi_2(u)\right\} \\ &- u\left[1 - F(u)\right]. \end{aligned} \tag{3.111}$$

From (3.111), *we have* $D_u(u) \to +\infty$ *as* $u \to -\infty$, $D_u(u) \to 0$ *as* $u \to +\infty$, *and* $D_u'(u) = F(u) - 1 < 0$ *for* $u \in \mathbb{R}$. *Hence, we have, for* $u \in \mathbb{R}$,

$$\begin{aligned} &\lim_{v \to u} D_{u,v}(u,v) > 0, \\ &\lim_{v \to +\infty} D_{u,v}(u,v) = 0. \end{aligned} \tag{3.112}$$

$D_{u,v}(u,v) = 0$ *has only one solution of* v *if* $D_{u,v}(u,v)$ *first decreases from a positive value to a negative value and then increases and approaches 0 as* v *increases.*

Next, we prove that the first-order partial derivative of $D_{u,v}(u,v)$ *concerning* v *is firstly negative and then be positive as* v *increases, where* $v \in (u, +\infty)$ *and* $u \in \mathbb{R}$. *The first-order partial derivative of* $D_{u,v}(u,v)$ *with respective to* v *is*

$$\frac{\partial D_{u,v}(u,v)}{\partial v} = \left(\frac{v-u}{4}\right) \cdot p\left(\frac{u+v}{2}\right) + F\left(\frac{u+v}{2}\right) - 1. \tag{3.113}$$

For $u \in \mathbb{R}$, *we have*

$$\begin{aligned} &\lim_{v \to u} \frac{\partial D_{u,v}(u,v)}{\partial v} = F(u) - 1 < 0, \\ &\lim_{v \to +\infty} \frac{\partial D_{u,v}(u,v)}{\partial v} = 0. \end{aligned} \tag{3.114}$$

The second-order partial derivative of $D_{u,v}(u,v)$ *with respect to* v *is*

$$\begin{aligned} \frac{\partial^2 D_{u,v}(u,v)}{\partial v^2} &= \frac{3}{4} p\left(\frac{u+v}{2}\right) + \frac{v-u}{8} p'\left(\frac{u+v}{2}\right) \\ &= \frac{1}{4}\left[\alpha\varphi_1\left(\frac{u+v}{2}\right) M_1(u,v) + (1-\alpha)\varphi_2\left(\frac{u+v}{2}\right) M_2(u,v)\right], \end{aligned} \tag{3.115}$$

where

$$M_1(u,v) = 3 - \frac{(v-u)(u+v-2\mu_1)}{4\sigma_1^2},$$
$$M_2(u,v) = 3 - \frac{(v-u)(u+v-2\mu_2)}{4\sigma_2^2}.$$

Both of $M_1(u,v)$ *and* $M_2(u,v)$ *are quadratic function with respect to* v. *For* $i = 1,2$, $\lim_{v\to u} M_i(u,v) = 3$, $\lim_{v\to+\infty} M_i(u,v) = -\infty$. *Both of* $M_1(u,v)$ *and* $M_2(u,v)$ *firstly increase from* 3 *then decrease from a positive number to* $-\infty$ *with respect to* v.

$$\begin{aligned} \lim_{v\to u} \frac{\partial^2 M_{u,v}(u,v)}{\partial v^2} &= \frac{3}{4} p(u) > 0, \\ \lim_{v\to+\infty} \frac{\partial^2 M_{u,v}(u,v)}{\partial v^2} &= 0. \end{aligned} \tag{3.116}$$

Based on (3.116), *the second-order partial derivative* (3.115) *is firstly positive and then be negative as* v *increases. And when* v *approaches* $+\infty$, (3.113) *approaches to 0 from a negative value as* φ_1 *and* φ_2 *converge 0 faster then* M_1 *and* M_2.

From (3.114)–(3.116), *the first-order partial derivative* (3.113) *increases from a negative number to a positive number and then decreases and approaches 0 concerning* v. *Therefore, we prove that* $D_{u,v}(u,v) = 0$ *has a unique solution of* v *given* $u \in \mathbb{R}$.

When v *in the neighborhood of* $(u, h^*(u))$, *we have*

$$\begin{cases} \frac{\partial D_{u,v}(u,v)}{\partial v} < 0, \\ \frac{\partial D_{u,v}(u,v)}{\partial u} = \frac{v-u}{4} \cdot p\left(\frac{u+v}{2}\right) > 0. \end{cases}$$

According to the implicit function theorem, we have $h^{*\prime}(u) > 0$. *Therefore,* $v = h^*(u)$ *is a strictly increasing function concerning* u.

3.13.9 Proof of Theorem 3.28

Proof *Denote the left-hand side of* (3.49) *as a function of three variables, i.e.,*

$$\begin{aligned} &H_{u,v,z}(u,v,z) = \\ &\alpha\left\{\mu_1\left[\Phi_1\left(\frac{v+z}{2}\right) - \Phi_1\left(\frac{u+v}{2}\right)\right] + \sigma_1^2\left[\varphi_1\left(\frac{u+v}{2}\right) - \varphi_1\left(\frac{v+z}{2}\right)\right]\right\} \\ &+ (1-\alpha)\left\{\mu_2\left[\Phi_2\left(\frac{v+z}{2}\right) - \Phi_2\left(\frac{u+v}{2}\right)\right] + \sigma_2^2\left[\varphi_2\left(\frac{u+v}{2}\right) - \varphi_2\left(\frac{v+z}{2}\right)\right]\right\} \\ &- v\left[F\left(\frac{v+z}{2}\right) - F\left(\frac{u+v}{2}\right)\right], \end{aligned} \tag{3.117}$$

where $-\infty < u < v < z < +\infty$. *Variables* u, v, z *represent* $\xi_{i-1}^{(k)}$, $\xi_i^{(k)}$ *and* $\xi_{i+1}^{(k)}$ *in the* i^{th} *equation* $(i > 1)$ *of* (3.46), *respectively.*

Similarly to the proofs of Theorems 3.26 & 3.27. Denote a intermediate function $H_{u,v}(u,v)$ *by letting* $z = v$ *in* (3.117), *we have*

$$\frac{\partial H_{u,v}(u,v)}{\partial u} = \frac{v-u}{4} \cdot f\left(\frac{u+v}{2}\right) > 0, \tag{3.118}$$

$$\lim_{u\to v} H_{u,v}(u,v) = 0, \tag{3.119}$$

$$\lim_{u\to-\infty} H_{u,v}(u,v) < 0. \tag{3.120}$$

When $u \to -\infty$, *we have* $\lim_{u\to-\infty} H_{u,v}(u,v) = R(v)$, *where*

$$R(v) = \alpha\left(\mu_1\Phi_1(v) - \sigma_1^2\varphi_1(v)\right) + (1-\alpha)\left(\mu_2\Phi_2(v) - \sigma_2^2\varphi_2(v)\right) - vF(v).$$

Since $R'(v) = -F(v) < 0$ *and* $R(v) \to 0$ *as* $v \to u$, *where* $u \to -\infty$. *Therefore,* (3.120) *holds. By* (3.118)–(3.120), $H_{u,v,z}(u,v,z)$ *strictly increases from a negative number and approaches zero as* u *increases from* $-\infty$ *to* v.

Hence,

$$\lim_{z\to v} H_{u,v,z}(u,v,z) < 0. \tag{3.121}$$

The first-order partial derivative of $H_{u,v,z}(u,v,z)$ *with respect to* z *is positive, i.e.,*

$$\frac{\partial H_{u,v,z}}{\partial z} = \frac{z-v}{4}\cdot p\left(\frac{v+z}{2}\right) > 0, \quad \text{as } z > v. \tag{3.122}$$

Based on (3.121)–(3.122), $H_{u,v,z}(u,v,z)$ *has unique solution of* z *if and only if*

$$\lim_{z\to+\infty} H_{u,v,z}(u,v,z) = D_{u,v}(u,v) > 0, \tag{3.123}$$

where $D_{u,v}(u,v)$ *is the Type II function defined in* (3.110). *From the proof of Theorem 3.27 and the discussion of a set of* $k=2$ *MSE-RPs,* $D_{u,v}(u,v)$ *is positive if and only if* $v < h^*(u)$ *and* $h^*(u)$ *is a strictly increasing function concerning* u. *Given a set of* $k=3$ *MSE-RPs,* $H(\xi_1^{(3)}, \xi_2^{(3)}, \xi_3^{(3)}) > 0$ *when* $\xi_3^{(3)} \to \infty$, *if and only if,* $\xi_1^{(3)} < \xi_1^{(2)}$.

3.13.10 Proof of Theorem 3.30

The expectation of $Y_{\chi^2(1)-N(0,1)}$ is

$$E\left(Y_{\chi^2(1)-N(0,1)}\right) = \sum_{i=1}^{2k}\beta_i q_i = \sum_{i=1}^{k} -\sqrt{\xi_i}(p_i/2) + \sum_{i=1}^{k}\sqrt{\xi_i}(p_i/2) = 0,$$

where $\xi_i, i = 1,\dots,k$ are MSE-RPs of the $\chi^2(1)$ distribution and $\beta_j, j = 1\dots,2k$ be App-RPs. As $E\left(Y_{\chi^2(1)-N(0,1)}\right) = E(Z) = 0$, the variance of $Y_{\chi^2(1)-N(0,1)}$ is

$$\begin{aligned} Var\left(Y_{\chi^2(1)-N(0,1)}\right) &= E\left(Y^2_{\chi^2(1)-N(0,1)}\right) = \sum_{i=1}^{2k}\beta_i^2 q_i = 2\sum_{i=1}^{k}\xi_i(p_i/2) \\ &= \sum_{i=1}^{k}\xi_i p_i = E(Y^{\chi^2(1)}) = E(X) = 1. \end{aligned}$$

Due to $Var\left(Y_{\chi^2(1)-N(0,1)}\right) = 1$, the skewness of $Y_{\chi^2(1)-N(0,1)}$ is

$$\begin{aligned} Sk\left(Y_{\chi^2(1)-N(0,1)})\right) &= E\left(Y^3_{\chi^2(1)-N(0,1)}\right) \\ &= \sum_{i=1}^{2k}\beta_i^3 q_i = \sum_{i=1}^{k} -\xi_i^{3/2}p_i/2 + \sum_{i=1}^{k}\xi_i^{3/2}p_i/2 = 0. \end{aligned}$$

Hence, $Sk\left(Y_{\chi^2(1)-N(0,1)}\right) = Sk(Z) = 0$. Finally, the kurtosis of $Y_{\chi^2(1)-N(0,1)}$ is

$$Ku\left(Y_{\chi^2(1)-N(0,1)}\right) = \sum_{i=1}^{k}\xi_i^2 p_i = \mathrm{Var}(Y^{\chi^2(1)}).$$

4

Statistical Simulation via Distributions formed by RPs

This chapter aims to provide an overview of statistical simulation based on approximate distributions by RPs, mainly constructed by MSE-RPs. Statistical simulation depends on the population distribution and statistics of interest. This chapter examines several distribution types, such as the gamma distribution, a mixture of normal distributions, an extreme value distribution, and the beta distribution family. Several measures for the closeness of two distributions are introduced, and numerical comparisons are given.

4.1 Statistical Simulation

Statistical simulation has played an important role in statistical research and applications (see Section 1.8). The traditional statistical simulation samples are from the underlying distribution by some computational software. For simplicity, assume that the population distribution is univariate $F(x, \boldsymbol{\theta})$. The parameters $\boldsymbol{\theta}$ may be known or unknown. In this chapter, we consider the case $\boldsymbol{\theta}$ as known; there is a comprehensive study on the generation of random samples from the population $F(x, \boldsymbol{\theta})$ (Ross (2013)). In the next chapter, we study the case of $\boldsymbol{\theta}$ as unknown, and then we have to estimate $\boldsymbol{\theta}$ by some well-known methods, such as the maximum likelihood estimation (MLE) or the moment estimation (ME) methods.

When the parameter $\boldsymbol{\theta}$ is known, we use F instead of $F(x, \boldsymbol{\theta})$. Let τ be the parameter of interest, and one wants to estimate it by a random sample $X_1, \ldots, X_n$. Let $T_n = T_n(X_1, \ldots, X_n; F)$ be a statistical estimate of τ. The choice of T_n should satisfy criteria such as unbiasedness, minimum variance, and consistency in different criteria (Section 1.3). If T_n is unbiased, one wishes that its variance is as small as possible. Suppose that there are two unbiased statistics $T_n^{(1)}$ and $T_n^{(2)}$, and their variances are different. We prefer to use one with a smaller variance. Example 1.1 gives a clear explanation.

Let $Y \sim \tilde{F}$ where $\tilde{F}$ be an approximation distribution to F, such as F_{MC}, F_{QMC} and F_{MSE}. Let $Y_1, \ldots, Y_n$ be a sample from $\tilde{F}$ and let $T_n(Y_1, \ldots, Y_n; \tilde{F})$ be the statistic under the population $\tilde{F}$. One wish that $T_n(Y_1, \ldots, Y_n; \tilde{F})$ has a better performance than that of $T_n(X_1, \ldots, X_n; F)$ in a certain sense: unbiased mean, smaller variance, and consistency. At first, we have to examine the mean, variance, and other statistics like median and skewness, kurtosis, quantiles, and density estimates. Section 2.2 shows that the discrete approximation F_{QMC} and F_{MSE} have a better estimation for density function and lower order moments under the normal or mixture of two normal distributions. But it needs some theoretical adjustment.

DOI: 10.1201/9781003589389-4

4.2 Statistics Inference under $\tilde{F}$ When the Population F is Known

There are many criteria for evaluating statistical simulation methods. The consistency of the statistic is a traditional criterion. The concept of consistency in different senses has been introduced in Section 1.3.3. When an estimator T_n for an unknown parameter τ, we say T_n is a weakly consistent estimator of τ as $n \to \infty$ if $T_n \xrightarrow{P} \tau$ or $T_n \xrightarrow{D} \tau$, and say T_n is a strongly consistent estimator of τ as $n \to \infty$ if $T_n \xrightarrow{a.s.} \tau$ (see Definition 1.4). The Law of large numbers in Definition 1.1 gives way to checking consistency. Moreover, we can consider the convergence of the distribution of T_n to the population F.

4.2.1 Consistency of statistical simulation by MSE-RPs

The probability distribution of a statistic provides a theoretical basis en route to statistical inference. Let $\{X_1, \ldots, X_n\}$ be a random sample of size n from a population distribution F. Let Z_n be a random-sample-based estimator of the parameter τ. Denote a random variable of interest as

$$T(X_1, \ldots, X_n; F) = n^{1/2}(Z_n - \tau)$$

depending upon the underlying distribution F. The corresponding sampling distribution is defined as

$$H_n(x) = P(T(X_1, \ldots, X_n; F) \leq x). \tag{4.1}$$

This motivates us to study the sampling distribution approximation by resampling from $F_{MSE,k}$ with a fixed set of k MSE-RPs. If the distribution F is known, then the distribution of $F_{MSE,k}$ is completely known. A new plug-in estimate of $T(X_1, \ldots, X_n; F)$ is defined as

$$T(Y_1, \ldots, Y_n; F_{MSE,k}) = n^{1/2}\left(\tilde{Z}_n - \tilde{\tau}_k\right),$$

where $\tilde{\tau}_k$ is a plug-in estimate of τ under $F_{MSE,k}$ and $\tilde{Z}_n$ the counterpart of Z_n but based on the i.i.d. samples $Y_1, \ldots, Y_n$ from $F_{MSE,k}$. The sampling distribution of $T(Y_1, \ldots, Y_n; F_{MSE,k})$ is defined as

$$\tilde{H}_n^{(k)}(y) = P(T(Y_1, \ldots, Y_n; F_{MSE,k}) \leq y), \tag{4.2}$$

which is used to estimate the sampling distribution $H_n(x)$ defined in (4.1). Denote $Y_k = Y_{MSE}^{(k)} \sim F_{MSE,k}$. From Theorem 3.5 and Corollary 3.1 we have

$$E(Y_k) = E(X), and \lim_{k\to\infty} E[(X - Y_k)^2] = \lim_{k\to\infty} [\mathrm{Var}(X) - \mathrm{Var}(Y_k)] = 0, \tag{4.3}$$

which implies Y_k converges to X in mean square as $k \to \infty$. Hence, $Y_k \xrightarrow{D} X$. Flury (1993) proved that all MSE-RPs are self-consistent, that is,

$$\xi_i^{(k)} = E(X \mid Y_k = \xi_i^{(k)}) = E(X \mid X \in I_i^{(k)}), \quad i = 1, \ldots, k. \tag{4.4}$$

The following two theorems given by Xu et al. (2024) establish the theoretic foundation for the estimation under the approximation distribution by MSE-RPs.

Theorem 4.1 *Let $X \sim F$ and $Y_{MSE}^{(k)} \sim F_{MSE,k}$.*

(i) If $g : \mathcal{X} \to \mathbb{R}$ is continuous, then $g(Y_{MSE}^{(k)}) \to g(X)$ in distribution.

(ii) If g is continuous and bounded, then $\lim_{k\to\infty} E\left[g(Y_{MSE}^{(k)})\right] = E[g(X)]$.

The proof of Theorem 4.1 is presented in Subsection 4.8.1. Part (i) of Theorem 4.1 shows that continuous functions preserve the convergence of MSE-RPs. For example, it can be applied to estimate the confidence interval of a parameter from F. Part (ii) shows that MSE-RPs can be used to estimate the expectation of $g(X)$ consistently for any continuous and bounded function g, such as the average waist-hip ratio.

Moments are important attributes of a distribution that can characterize the shape of population density. The first four moments measure the population's mean, variance, skewness and kurtosis, respectively. From (4.3), we know that $E(Y_k) = E(X)$ and $\mathrm{Var}(Y_k) \to \mathrm{Var}(X)$ when $k \to \infty$. Equation 5.10 of Pagès (2018) proved that $E[g(Y_k)]$ is always bounded by $E[g(X)]$ if g is a convex function.

Theorem 4.2 provides the generalized conditions for the convergence of Y_k at the rth moment, where $r > 2$.

Theorem 4.2 *Let $X \sim F$ and $Y_{MSE}^{(k)} \sim F_{MSE,k}$. If there exist some $r \in \mathbb{N}^+$ and $r \geq 3$ such that $E(|X|^{r+\varepsilon}) < \infty$ with some $\varepsilon > 0$, then*

(i) $Y_{MSE}^{(k)}$ converges to X in the rth moment, that is,

$$\lim_{k\to\infty} E([Y_{MSE}^{(k)}]^r) = E(X^r);$$

(ii) $Y_{MSE}^{(k)}$ converges to X in the rth absolute central moment, that is,

$$\lim_{k\to\infty} E(|Y_{MSE}^{(k)} - \mu|^r) = E(|X - \mu|^r),$$

where $\mu = E(Y_{MSE}^{(k)}) = E(X)$.

The proof of Theorem 4.2 will be presented in Section 4.8.2. According to Jensen's inequality and the self-consistent property of representative points, the inequality (4.5) below was given by Pagès (2018).

Lemma 4.1 *Let $X \sim F$ and $Y_{MSE}^{(k)} \sim F_{MSE,k}$. The denote $L^1(P)$ consists of all complex-valued measurable functions on the sample space $\mathcal{X}$ of X that satisfy $\int_{\mathcal{X}} |g(x)|dP(x) < \infty$. Then, for every convex function $g : \mathbb{R} \to \mathbb{R}$ such that $g(X) \in L^1(P)$, we have*

$$E[g(Y_{MSE}^{(k)})] \leq E[g(X)], \quad \textit{for all } k \in \mathbb{N}^+. \tag{4.5}$$

Generally, convergence in the distribution cannot imply convergence at the moment. Additional conditions are required. One famous condition is uniform integrability. Sufficient conditions for uniform integrability are in Lemma 4.2 of Billingsley (1995).

Lemma 4.2 *Let r be a positive integer. If*

$$Y_{MSE}^{(k)} \xrightarrow{D} X, \textit{ and } \sup_{k\in\mathbb{N}^+} E\left(|Y_{MSE}^{(k)}|^{r+\varepsilon}\right) < \infty,$$

where $\varepsilon > 0$, then $E\left(|X|^r\right) < \infty$ and $E([Y_{MSE}^{(k)}]^r) \to E(X^r)$ as $k \to \infty$.

The generalized convergence conditions of Theorem 4.2 play an essential role in establishing the consistency of the distribution of the mean standardized sample and the distribution of the variance of the sample under $F_{MSE,k}$. Many useful statistics can be simulated from $F_{MSE,k}$. The following subsections show that statistical simulation under $F_{MSE,k}$ or some other approximation distributions like F_{QMC} have a good performance for many statistics.

4.2.2 Moments estimation

The first four moments, the mean, variance, skewness, and kurtosis of the population $X \sim F(x)$, are most useful in practice. Let Y be an approximate discrete random variable of X with distribution $F_Y(y)$. The first four moments can be calculated by (2.17). An important question is how close the first four moments are between X and Y. For the standard normal $X \sim N(0,1)$ it is known that $E(X) = 0,\ \sigma^2 = 1$ and $Sk(Z) = Ku(Z) = 0$. Fang et al. (2014) gave the numerical comparisons of these statistics among MC, QMC, and MSE methods. Fang et al. (2020) added RMC in the comparison. Table 2.2 in Chapter 2 compares the four methods for $k = 5, 15, 21, 30, 35$. It is clearly shown that.
(1) MC has the worst estimation, and MSE gives the best results in most cases;
(2) RMC improves MC significantly;
(3) QMC gives a better estimation than RMC except for kurtosis.

Many authors gave similar results for different distributions like t- (Zhou and Wang (2016), Gamma (Ke et al. (2023)), skew-normal (Yang et al. (2021)) distributions.

4.3 An Extraordinary Property of the Arcsine Distribution

In the previous sections, the MSE-RPs performed better in estimation than QMC-RPs and MC-RPs in most comparisons. However, it is not true for estimation under other distributions, including the arcsine distribution. A random variable X follows the standard arcsine distribution (arcsine distribution for short) if its pdf is

$$f(x) = \frac{1}{\pi\sqrt{x(1-x)}},\quad 0 \le x \le 1. \tag{4.6}$$

Its corresponding cdf is

$$F(x) = \frac{2}{\pi} arcsin(\sqrt{x}),\ 0 \le x \le 1. \tag{4.7}$$

(see Figure 4.3). The arcsine distributions occur naturally in statistical communications theory and have played an important role in the study of Brownian motion and prime numbers. The arcsine distribution is useful in the study of the fluctuations of random walks (Feller (1967)) and the *Arcsine Law* in Brownian motion which was first proved by Lévy (1940, 1948). Besides, the arcsine distribution has been widely used and accepted in physical and financial stochastic models, such as economic default time and occupation times for the pricing of options is modeled by arcsine distribution (see e.g. Guo et al. (2014) and Pechtl (1999)).

4.3.1 QMC-RPs of the arcsine distribution

The approximate discrete random variable Y_{QMC} constructed by QMC has been introduced in Subsection 2.1.3. It takes $P(Y_{QMC} = \xi_j) = \frac{1}{k}, j = 1, \ldots, k$, where

$$\xi_j = F^{-1}(q_j),\ q_j = \frac{2j-1}{2k},\quad j = 1, \ldots, k, \tag{4.8}$$

see (2.7). It is easy to know that the inverse for arcsine distribution, $F^{-1}(x) = \sin^2(\frac{\pi x}{2})$ and the QMC-RPs of X are

$$\xi_j = \sin^2(\frac{2j-1}{2k} \cdot \frac{\pi}{2}),\ j = 1, ..., k. \tag{4.9}$$

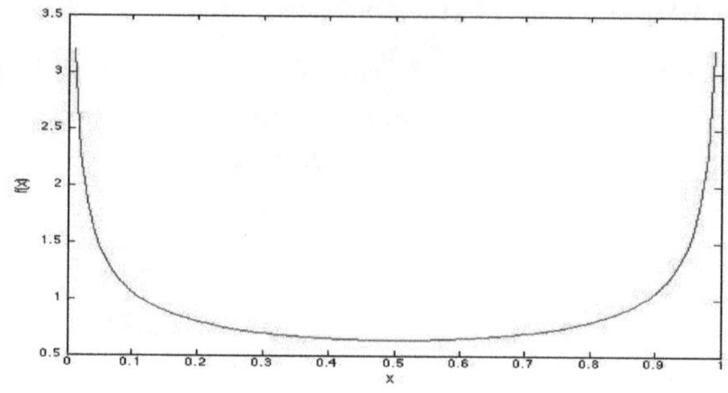

FIGURE 4.1
Probability density function of the arcsine distribution.

The following theorem from Jiang et al. (2015) derives a surprising approximation property when using QMC-RPs to approximate the arcsine distribution.

Theorem 4.3 *Let random variable X follow the standard arcsine distribution and let Y_{QMC} be defined by (4.9) where the support points are $\xi_j = \sin^2(\frac{2j-1}{2k} \cdot \frac{\pi}{2})$, $j = 1, 2, ..., k$, each occupying the same probability $\frac{1}{k}$. Then X and Y_{QMC} share all the same moments of order m $(m < k)$ as follows*

$$\mu_m = E(X^m) = E(Y_{QMC}^m) = \prod_{j=1}^{m} \frac{2j-1}{2j}, \quad m = 1, 2, ..., k-1. \tag{4.10}$$

The proof will be given in Subsection 4.8.3. According to Theorem 4.3, we can conclude that QMC-RPs provide the best discrete approximation for arcsine distribution in the sense of matching moments.

4.3.2 MSE-RPs of the arcsine distribution

Jiang et al. (2015) applied the Fang-He algorithm for generating MSE-RPs of the arcsine distribution. To find a set of MSE-RPs which minimizes the loss function $L(\xi_1, \ldots, \xi_m)$, we set

$$\frac{\partial L(\xi_1, \ldots, \xi_m)}{\partial \xi_i} = 0, \ i = 1, \ldots, m,$$

and then obtain the following system of equations

$$\begin{aligned} &(2\xi_i - 1)\left(arcsin\left(\sqrt{b_{i+1}}\right) - arcsin\left(\sqrt{b_i}\right)\right) \\ = \ &\left(b_i - b_i^2\right)^{\frac{1}{2}} + \left(b_{i+1} - b_{i+1}^2\right)^{\frac{1}{2}}, i = 1, \ldots, m. \end{aligned} \tag{4.11}$$

In the system of equations (4.11), when $k = 2m$ is even,

$$b_i = \begin{cases} 0, & i = 1, \\ \frac{1}{2}(\xi_{i-1} + \xi_i), & 2 \le i \le m, \\ \frac{1}{2}, & i = m+1. \end{cases}$$

when $k = 2m + 1$ is odd,

$$b_i = \begin{cases} 0, & i = 1, \\ \frac{1}{2}(\xi_{i-1} + \xi_i), & 2 \le i \le m, \\ \frac{1}{2}(\xi_m + \frac{1}{2}), & i = m+1. \end{cases}$$

After determining the solution $\{\xi_1^*, \ldots, \xi_m^*\}$ of system of equations (4.11), a set of MSE-RPs is of $\xi^* = \{\xi_1^*, \ldots, \xi_m^*, 1-\xi_1^*, \ldots, 1-\xi_m^*\}$ when k is even and of $\xi^* = \{\xi_1^*, \ldots, \xi_m^*, \frac{1}{2}, 1-\xi_1^*, \ldots, 1-\xi_m^*\}$ when k is odd.

In addition, the associated probabilities of MSE-RPs are calculated as

$$\begin{aligned} p_i &= \int_{b_i}^{b_{i+1}} f(x)dx = \int_{b_i}^{b_{i+1}} \frac{1}{\pi\sqrt{x(1-x)}}dx \\ &= \frac{2}{\pi}\left(arcsin\left(\sqrt{b_{i+1}}\right) - arcsin\left(\sqrt{b_i}\right)\right). \end{aligned} \tag{4.12}$$

The MSE-RPs for $n \leq 31$ and related probabilities p_i can be found in Jiang et al. (2015). Moreover, the algorithm for solving the system of equations (4.11) is shown in Algorithm 4.1.

Algorithm 4.1 (Algorithm for solving the system of equations (4.11))

Step 1 Choose an appropriate initial value for $\xi_1 > 0$. Set $LP = \xi_1$ and $RP = \frac{1}{2}$ such that ξ_2 must lie in the interval (LP, RP).

Step 2 Fix ξ_1, solve the first equation and obtain the value of ξ_2. And then fix ξ_1 and ξ_2, solve the second equation to get ξ_3.

Step 3 In the same manner, fix ξ_{i-1} and ξ_i, solve the ith equation and obtain the solution of ξ_{i+1}, for i from 2 to $m-1$. Thus ξ_m is found by fixing ξ_{m-2} and ξ_{m-1} and solving the second last equation.

Step 4 Fix ξ_{m-1}, solve the last equation and find the solution of $\xi_m^{'}$.

Step 5 Let ϵ be the threshold of convergence (assume $\epsilon = 10^{-6}$), if

[a] $|\xi_m - \xi_m^{'}| < \epsilon$, then the $\{\xi_1, \ldots, \xi_m\}$ are the solution for a system of equations (18).

[b] $\xi_m < \xi_m^{'} - \epsilon$, then the initial value ξ_1 is too small, modify $LP = \xi_1$, $\xi_1 = \frac{1}{2}(LP + RP)$, go to Step 2.

[c] $\xi_m > \xi_m^{'} + \epsilon$, then the initial value ξ_1 is too large: modify $RP = \xi_1$, $\xi_1 = \frac{1}{2}(LP + RP)$, go to Step 2.

After determining the solution $\{\xi_1^*, \ldots, \xi_m^*\}$ of the system of equations (4.11), the MSE-RPs are the set of $\xi^* = \{\xi_1^*, \ldots, \xi_m^*, 1-\xi_1^*, \ldots, 1-\xi_m^*\}$ when k is even and the set of $\xi^* = \{\xi_1^*, \ldots, \xi_m^*, \frac{1}{2}, 1-\xi_1^*, \ldots, 1-\xi_m^*\}$ when k is odd.

4.3.3 Simulation based on RPs generated by QMC and MSE

The L_p distance between two pdf functions is defined in (2.3). The L_p-distance between the original distribution $F(x)$ and its discrete approximation, $F_\xi(x)$, that is,

$$D_p(F, F_\xi) = \left[\int_{-\infty}^{+\infty} |F_\xi(x) - F(x)|^p \, dx\right]^{\frac{1}{p}}. \tag{4.13}$$

It has been shown that it is an instrumental measurement for evaluating discrete approximation.

To compare the performance of discrete approximation by QMC-RPs and MSE-RPs, we compute the value of $D_2(F, F_{Y_{QMC}})$ and $D_2(F, F_{Y_{MSE}})$ for different k RPs . The results are shown in Table 4.1. The distribution function of Y_{QMC} is much closer to the arcsine distribution function. Jiang et al. (2015) also show that $F_{Y_{QMC}}$ has a better performance in simulation than other discrete approximation distributions, including F_{MSE}.

TABLE 4.1
L_2-distance between the approximate distribution and the arcsine distribution

k	5	10	15	20	25	30	31
$D_2(F, F_{Y_{QMC}})$	0.05807	0.02891	0.01926	0.01444	0.01155	0.00962	0.00931
$D_2(F, F_{Y_{MSE}})$	0.10376	0.08621	0.07541	0.06778	0.06204	0.05751	0.05672

4.4 FM Criterion

The F-discrepancy is based on the difference between the empirical distribution function of the discrete approximation variable ξ and the population cdf of X. In contrast, the MSE criterion is based on the L2.cdf between X and ξ (see (2.3)). Both criteria do not consider the difference of the moments between ξ and X. It is known that the Maclaurin's series of the moment generating function of X, $E(e^{tX})$, can be expanded by the moments, then the moments of a random variable are important issues. Therefore, it is reasonable to require that the first m moments of the n-point discrete distribution ξ can be equal to or close to that of X. Inspired by the result in Theorem 4.3, Zhou and Fang (2019) proposed a new criterion for RPs, named as *FM criterion*, which combines the requirement of the distribution and the moments of ξ.

For a random vector X with finite first m moments, consider the following optimization problem for finding a discrete random vector ξ constructed by a set of k RPs,

$$\begin{aligned} \min \quad & O_F = \int_{-\infty}^{\infty} |F_\xi(x) - F(x)|^2 dx \\ \text{s.t.} \quad & E(\xi^r) = E(X^r), r = 1, \ldots, m, \end{aligned} \tag{4.14}$$

where $F_\xi(x)$ and $F(x)$ are respectively the distribution of ξ and X. A set of points $\xi = \{\xi_1, \ldots, \xi_k\}$ which has smallest O_F-value in (4.14) under the constraint is called as the *FM-representation points*, denoted by FM-RPs. Zhou and Fang (2019) consider $m = k - 1$. Li et al. (2024c) suggested to consider $m = 2$ or $m = 4$ that can have robust performance in statistical simulation and used FM2 and FM4 for these criteria, respectively.

4.4.1 FM-RPs of $U(0, 1)$

Zhou and Fang (2019) considered the FM-RPs of $X \sim U(0, 1)$, then the objective function of the FM criterion is

$$\begin{aligned} O_F &= \int_{-\infty}^{\infty} |F(x) - F(x)|^2 dx = \int_0^1 |F_\xi(x) - x|^2 dx \\ &= \frac{1}{3} + \frac{1}{n} \sum_{i=1}^{k} x_i^2 - \frac{1}{k^2} \sum_{i=1}^{k} (2i - 1) x_i, \end{aligned}$$

and $E(X^r) = \int_0^1 x^r dx = 1/(r + 1)$. Deleting the constant $1/3$ and multiplying k^2, the FM-RPs of $U(0, 1)$ are the solution of the following optimization problem:

$$\begin{aligned} \min \quad & k \sum_{i=1}^{k} x_i^2 - \sum_{i=1}^{k} (2i - 1) x_i \\ s.t. \quad & (x_1, \cdots, x_k) \in [0, 1]^k, \; \frac{1}{k} \sum_{i=1}^{k} x_i^r = \frac{1}{r + 1}, \quad r = 1, \cdots, k - 1, \end{aligned} \tag{4.15}$$

TABLE 4.2
The n-point FM-RPs of the uniform distribution

k	FM-RPs	ϵ
2	0.2500 0.7500	0
3	0.1464 0.5000 0.8536	2.19×10^{-13}
4	0.1127 0.3709 0.6291 0.8873	1.22×10^{-13}
5	0.0838 0.3127 0.5000 0.6873 0.9162	2.91×10^{-14}
6	0.0708 0.2561 0.4205 0.5795 0.7439 0.9292	9.77×10^{-15}
7	0.0581 0.2352 0.3380 0.5000 0.6620 0.7648 0.9419	5.55×10^{-17}
8	0.0514 0.2001 0.3070 0.4302 0.5698 0.6930 0.7999 0.9486	1.39×10^{-16}
9	0.0442 0.1995 0.2356 0.4160 0.5000 0.5840 0.7644 0.8005 0.9558	8.33×10^{-17}
10	0.0401 0.1708 0.2263 0.3570 0.4615 0.5385 0.6430 0.7737 0.8292 0.9599	1.11×10^{-16}

where $(x_1, \cdots, x_k) \in [0,1]^k$ means that each $x_i \in [0,1]$.

$$\Phi_{uniform} = \left\{ \boldsymbol{x} : \boldsymbol{x} = (x_1, \cdots, x_k) \in [0,1]^k, \sum_{i=1}^{k} x_i^r = \frac{k}{r+1}, r = 1, \cdots, k-1 \right\}. \tag{4.16}$$

For solving the optimization problem (4.15), Zhou and Fang (2019) prove the constraint in (4.15) is meaningful, that is, the set $\Phi_{uniform}$ in (4.16) is nonempty.

Theorem 4.4 *The FM-RPs of the uniform distribution $U(0,1)$ exists, i.e., the set $\Phi_{uniform}$ in (4.16) is nonempty.*

The proof will show in Subsection 4.8.4. For solving the optimization problem (4.15), we can obtain the k FM-RPs of the uniform distribution $U(0,1)$, as shown in Table 4.2 for $k = 2, \ldots, 10$. When $k = 2$, the FM-RPs are the same as the MSE-RPs, and differ when $n > 2$. Moreover, the FM-RPs are symmetric about the center point 0.5 for each k, and the point 0.5 is one of the FM-RPs when k is even. The err-value $\epsilon = \sum_{i=1}^{k-1} \left| \frac{1}{k} \sum_{j=1}^{k} x_j^i - \frac{1}{i+1} \right|$ as the L_1-norm of the difference vector between the sample and population moments. Table 4.2 shows that the err-values for $n = 2, \cdots, 10$ are very small.

In general, the FM-RPs do not have the linearity property. For example, if $\{x_1, \cdots, x_k\}$ are the n FM-RPs of the uniform distribution $U(0,1)$, $\{a + (b-a)x_1, \cdots, a + (b-a)x_k\}$ may not be the FM-RPs of the uniform distribution $U(a,b)$.

Zhou and Fang (2019) also gave the FM-RPs of the power distribution. However, they showed that many distributions, including the normal distribution, do not exist in the corresponding FM-RPs.

Zhou and Fang (2019) gave some numerical comparisons among FM-RPs and MSE-RPs, QMC-RPs, and MC-RPs of the uniform distribution and power distribution. They compared the mean, variance, and kurtosis of each type of representative point and the corresponding statistics of the distribution, then obtain the biases for each case. Their simulation shows the advantages of FM-RPs.

Li et al. (2024c) found that the FM-criterion can be overly restrictive as it requires matching the first $k-1$ order moments. To reduce this limitation, they relaxed the criterion only to require matching the first 2 or 4 moments, leading to the definition of FM2-RPs and FM4-RPs. Some discussion is given in Subsection 3.10.1.

4.5 Different Distances for Approximation Distributions

How to measure an approximation distribution $G(x) = \hat{F}(x)$ to the underlying distribution $F(x)$? Many criteria or distances exist between the objective distribution $F(x)$ and the approximation distribution $G(x)$.

4.5.1 Some popularly used distances

Some popularly used distances are listed as follows.

1. Kolmogorov-Smirnov (KS) distance,

$$D^*(F, G) = \sup_{x \in \mathbb{R}} |F(x) - G(x)|, \tag{4.17}$$

which is the supremum absolute difference between the two cdfs. It is called the *star discrepancy* in number-theoretic methods. It is also denoted as $\rho_\infty(F, G)$.

2. Weighted L_p-distance

$$D_p(F, G) = \left[\int |F(x) - G(x)|^p w(x) dx \right]^{1/p}, \tag{4.18}$$

where $w(x) \geq 0$. The following are some special cases Barbiero and Hitaj (2022):

◇ When $w(x) \equiv 1, p = 2$, it becomes the second-order squared Cramér distance, which is closely related to the so-called *energy distance*. The energy distance was defined by Székely and Rizzo (2013) and Rizzo and Székely (2016) and readers can refer to Subsection 4.5.2 for the details.

◇ When $w(x)$ is the pdf of F and $p = 2$, it becomes the so-called Cramér-von Mises distance, which can also be conveniently written as

$$\int_0^1 [u - G(F^{-1}(u))]^2 du, \tag{4.19}$$

provided that $F(x)$ is invertible.

◇ When $w(x) = f(x)[F(x)(1 - F(x))]^{-1}$, it becomes the Anderson-Darling distance.

3. Mallows–Wasserstein metric: For two distributions, F and G in $\mathcal{F}_2 = \left\{ F \in \mathcal{F} : \int_{-\infty}^{\infty} x^2 dF(x) < \infty \right\}$, the Mallows–Wasserstein distance is

$$\rho_2(F, G) = \inf_{\mathcal{T}_{U,V}} \left(E|U - V|^2 \right)^{1/2}, \tag{4.20}$$

where $\mathcal{T}_{U,V}$ is the collection of all possible joint distributions of the pairs (U, V) whose marginal distributions are F and G, respectively.

4. When $F(\boldsymbol{x})$ is the uniform distribution on the unit cube C^p, Subsection 1.11.2 introduces many discrepancies (similar to distance) between the uniform distribution $U(C^p)$ and the distribution constructed by a set of points in $C^p = [0, 1]^p$.

4.5.2 Support points based on energy distance

Mak and Joseph (2018) proposed a new method for compacting a continuous probability distribution F into a set of representative points (RPs) for F, which is called the *support points*. The key idea of support points is that it optimizes the energy distance, which was defined by Székely and Rizzo (2013).

Definition 4.1 *Let F and G be two distribution functions on the domain $\mathcal{X} \subset \mathbb{R}^p$. Let $\boldsymbol{X} \overset{iid}{=} \boldsymbol{X}^* \sim F$ and $\boldsymbol{Y} \overset{iid}{=} \boldsymbol{Y}^* \sim G$. The energy distance between F and G is defined as*

$$D(F,G) = 2E\|\boldsymbol{X} - \boldsymbol{Y}\| - E\|\boldsymbol{X} - \boldsymbol{X}^*\| - E\|\boldsymbol{Y} - \boldsymbol{Y}^*\|, \tag{4.21}$$

where the norm is L_2-norm. When $G = F_n$ is the empirical distribution of F based on a random sample $\boldsymbol{x}_i, i = 1,\dots,n$, then the energy distance between F and F_n is

$$D(F,F_n) = \frac{2}{n}\sum_{i=1}^{n} E\|\boldsymbol{x}_i - \boldsymbol{X}\| - \frac{1}{n^2}\sum_{i=1}^{n}\sum_{j=1}^{n} \|\boldsymbol{x}_i - \boldsymbol{x}_j\| - E\|\boldsymbol{X} - \boldsymbol{X}^*\|. \tag{4.22}$$

Székely and Rizzo (2013) pointed out that $D(F,G) \geq 0$, with equality holding if and only if $F = G$.

Definition 4.2 *Let $\boldsymbol{X} \sim F$. For a fixed-point set size n, the support points of F based on the energy distance is defined as*

$$\begin{aligned} \{\boldsymbol{\xi}_i\}_{i=1}^n &\in \operatorname*{argmin}_{\boldsymbol{x}_1,\dots,\boldsymbol{x}_n} D(F,F_n) \\ &= \operatorname*{argmin}_{\boldsymbol{x}_1,\dots,\boldsymbol{x}_n} \left\{ \frac{2}{n}\sum_{i=1}^{n} E\|\boldsymbol{x}_i - \boldsymbol{X}\| - \frac{1}{n^2}\sum_{i=1}^{n}\sum_{j=1}^{n} \|\boldsymbol{x}_i - \boldsymbol{x}_j\| \right\}. \end{aligned} \tag{4.23}$$

Mak and Joseph (2018) pointed out that the energy distance has two appealing features. First, its distance-based structure allows us to exploit the duality between powers of the Euclidean distance and its Fourier transform for theoretical analysis. Using this duality, we show that support points converge in distribution to F and enjoy an improved error rate for Monte Carlo when integrating a large class of functions. Second, the minimization of the energy distance can be formulated as a difference-of-convex program, which can be used to generate representative point sets efficiently.

4.5.3 Representative points based on power exponential kernel discrepancy

Most existing criteria can only ensure the properties are represented in the whole variable space. Xiong et al. (2022) proposed a new kernel discrepancy, named power exponential kernel discrepancy (PEKD), to measure the representativeness of the point set concerning the general multivariate distribution. Unlike the commonly used criteria, PEKD can improve the projection properties of the point set, which is essential in high-dimensional circumstances. Some theoretical results are presented to understand the new discrepancy better and guide the hyperparameter setting. An efficient algorithm for searching rep-points under the PEKD criterion is presented, and its convergence has also been proven. Examples illustrate its potential applications in numerical integration, uncertainty propagation, and reduction of Markov Chain Monte Carlo chains.

4.5.4 KL divergence

The Kullback-Leibler (KL) divergence or relative entropy of two probability distributions with probability density functions $p(\boldsymbol{x})$ and $q(\boldsymbol{x})$ on $\mathcal{X} \subset \mathbb{R}^p$ is defined as the expected logarithmic ratio of densities concerning $p(\boldsymbol{x})$, that is,

$$D_{KL}(p(\boldsymbol{x})|q(\boldsymbol{x})) = \int_{\mathcal{X}} \log\left(\frac{p(\boldsymbol{x})}{q(\boldsymbol{x})}\right) p(\boldsymbol{x})d\boldsymbol{x} = E_{p(\boldsymbol{x})}\left[\log\left(\frac{p(\boldsymbol{x})}{q(\boldsymbol{x})}\right)\right]. \tag{4.24}$$

The KL divergence is nonnegative and is equal to 0 if and only if the two distributions are identical. In general, the KL divergence is not symmetric:

$$D_{KL}(p(\boldsymbol{x})|q(\boldsymbol{x})) \neq D_{KL}(q(\boldsymbol{x})|p(\boldsymbol{x})).$$

The symmetrized KL-divergence is defined as

$$D_{SKL}(p(\boldsymbol{x}), q(\boldsymbol{x})) = D_{KL}(p(\boldsymbol{x})|q(\boldsymbol{x})) + D_{KL}(q(\boldsymbol{x})|p(\boldsymbol{x})). \tag{4.25}$$

The KL divergence or symmetrized KL-divergence is regarded as a measure of discrepancy between two probability distributions. The following are some basic properties from Kullback (1959).

1. The symmetrized KL-divergence has a lower bound

$$D_{SKL}(p(\boldsymbol{x}), q(\boldsymbol{x})) \geq \max\{D_{KL}(p(\boldsymbol{x})|q(\boldsymbol{x})), D_{KL}(q(\boldsymbol{x})|p(\boldsymbol{x}))\} \geq 0.$$

2. For the two normal densities $\varphi_1(x) = \varphi(x, \mu_1, \sigma_1^2)$ and $\varphi_2(x) = \varphi(x, \mu_2, \sigma_2^2)$, we have

$$D_{KL}(\varphi_1(x)|\varphi_2(x)) = \frac{1}{2}\left[\frac{(\mu_1 - \mu_2)^2}{\sigma_2^2} + \frac{\sigma_1^2}{\sigma_2^2} + \log\left(\frac{\sigma_2^2}{\sigma_1^2}\right) - 1\right]. \tag{4.26}$$

3. The symmetrized KL-divergence of $\varphi_1(x)$ and $\varphi_2(x)$ is

$$D_{SKL}(\varphi_1(x), \varphi_2(x)) = \frac{1}{2}\left[(\mu_1 - \mu_2)^2\left(\frac{1}{\sigma_1^2} + \frac{1}{\sigma_2^2}\right) + \frac{(\sigma_1^2 - \sigma_2^2)^2}{2\sigma_1^2\sigma_2^2}\right]. \tag{4.27}$$

4. When $\sigma_1 = \sigma_2$ and $\mu_2 = \mu_1 + c\sigma_1$ of the above two normal densities, we have

$$D_{KL}(\varphi_1|\varphi_2) = \frac{1}{2}c^2; \; D_{SKL}(\varphi_1, \varphi_2) = c^2.$$

5. When $\mu_1 = \mu_2$ and $\sigma_2 = (1+v)\sigma_1$ of the above two normal densities, we have

$$D_{KL}(\varphi_1|\varphi_2) = \frac{1}{2(1+v)^2} + \log(1+v) - \frac{1}{2},$$
$$D_{SKL}(\varphi_1, \varphi_2) = \frac{v^2(v+2)^2}{2(v+1)^2}.$$

6. When φ_1 is the pdf of $N(0,1)$, the KL-divergence and symmetrized KL-divergence of $\varphi_1(x)$ and $\varphi_2(x)$ are

$$D_{KL}(\varphi(\mu_1, \sigma_1^2)) = D_{KL}(\varphi_1|\varphi_2) = -\log(\sigma_1) + \frac{\mu_1^2 + \sigma_1^2}{2} - \frac{1}{2};$$
$$D_{SKL}(\varphi(\mu_1, \sigma_1^2)) = D_{SKL}(\varphi_1, \varphi_2) = \frac{(\sigma_1^2 - 1)^2}{\sigma_1^2} + \frac{\mu_1^2}{2}\left(1 + \frac{1}{\sigma_1^2}\right).$$

When φ_1 is the pdf of $N(0,1)$, the KL-divergence and symmetrized KL-divergence of $\varphi_1(x)$ in the above formulas are zero.

When $\sigma_1 > 1$, the quantity of $\frac{1}{2}\sigma_1^2$ increases and $-\log(\sigma_1)$ decreases;

When $\sigma_1 < 1$, the quantity of $\frac{1}{2}\sigma_1^2$ decreases and $-\log(\sigma_1)$ increases.

Rover and Friede (2017) used the KL divergence and symmetrized KL-divergence to define a discrete approximation to a continuous distribution $F(x)$ with density $p(x)$. Let a random variables $X \sim F(x)$ and a discrete approximation Y to X. The joint distribution of (X, Y) is $p(x, y) = p(x)p(y|x)$. Consider the joint distribution of (X, Y) with density $p(x, y)$ and define another probability distribution with density $q(x, y)$ that has the same marginal density (mixing distribution), $q(x) = p(x)$, and whose conditional probability density is given by

$$q(y|x) = p(y|z_i), \text{ for } x \in I_i,$$

where $z_1 < z_2 < \cdots < z_k$ and each interval I_i defined in (2.14). The density $q(x, y)$ is approximate to $p(x, y)$. Similarly, $q(x, y) = q(x)q(y|x)$ and $q(y) = \int q(y|x)q(x)dx$. The resulting marginal distribution of Y, $q(y)$, is a discrete sum of the conditional densities

$$q(y) = \sum_{i=1}^{k} p_i p(y|z_i), \tag{4.28}$$

where $z_1 < z_2 < \cdots < z_k$. For each interval I_i, let

$$d_i = \max_{x \in I_i}\{D_{SKL}(p(y|x), p(y|z_i)\} = \max_{x \in I_i}\{D_{SKL}(p(y|x), q(y|x)\}. \tag{4.29}$$

The maximum (symmetrized) divergence between distributions $p(y|x)$ corresponds to points within I_i and the corresponding ith support point. The chain rule for symmetrized KL divergence is given by

$$D_{SKL}(p(x, y), q(x|y)) = D_{SKL}(p(x), q(x)) + E_{p(x)}[D_{SKL}(p(y|x), q(y|x))].$$

As $D_{SKL}(p(x), q(x)) = 0$, we have

$$D_{SKL}(p(x, y), q(x|y)) = E_{p(x)}[D_{SKL}(p(y|x), q(y|x))].$$

By the chain rule again, we have

$$\begin{aligned} D_{SKL}(p(y), q(y)) &= D_{SKL}(p(x, y), q(x|y)) - E_{p(x)}[D_{SKL}(p(y|x), q(y|x))] \\ &\leq D_{SKL}(p(x, y)|q(x|y) = E_{p(x)}[D_{SKL}(p(y|x), q(y|x))] \\ &\leq \sum_{i=1}^{k} \boldsymbol{p}_i d_i \leq \max_{1 \leq i \leq k} d_i. \end{aligned}$$

We want to find a set of $z_i = \xi_i, i = 1, \ldots, k$, to reach the upper bound $\max_{1 \leq i \leq k} d_i$. It can be called as *SKL-RPs*.

Rover and Friede (2017) proposed a sequential DIRECT algorithm for finding SKL-RPs for t-distribution and convolution of two distributions.

4.6 Consistency of Statistics by Resampling

The bootstrap method is mentioned in Subsection 1.8.2. This section offers a concise review of the bootstrap method's theoretical underpinnings and highlights recent developments in resampling. In particular, we introduce the recent development by Xu et al. (2024).

4.6.1 Consistency of statistics by bootstrap

Let $\{X_1, \ldots, X_n\}$ be a random sample from a population distribution F and let Z_n be a random-sample-based estimator of the parameter θ of F. Denote a random variable of interest as $T(X_1, \ldots, X_n; F) = n^{1/2}(Z_n - \theta)$ depending upon the underlying distribution F. The corresponding sampling distribution is defined as

$$H_n(x) = P(T(X_1, \ldots, X_n; F) \leq x). \tag{4.30}$$

The bootstrap allows estimation of the sampling distribution of a statistic by random sampling from the empirical distribution F_n, where $X_1, \ldots, X_n$ as its support points and each having the same probability $1/n$. The bootstrap method is to approximate the distribution of $T(X_1, \cdots, X_n; F)$ under F by that of $T(X_1^*, \cdots, X_n^*; F_n)$ where $\{X_1^*, \ldots, X_n^*\}$ denotes a random sample of size n from F_n.

Many theoretical analyses contribute to finding the asymptotic justification of Efron's proposal. Singh (1981) established the consistency of the bootstrap approximation to the distribution of sample mean defined in (4.30), where $Z_n = \bar{X} = n^{-1}\sum_{i=1}^{n} X_i$. He also proved that the bootstrap sampling distribution of standardized sample mean is asymptotically more accurate than the approximation by the limiting normal distribution when F is nonlattice. Giné and Zinn (1989) and Hall (1990) further proved that the condition $E(X^2) < \infty$ is sufficient and necessary for the strong consistency of the bootstrap sampling distribution of $n^{1/2}\left[\bar{X} - E(X)\right]$. Bickel and Freedman (1981) proved that the bootstrap distributional approximation is asymptotically valid for a large group of statistics, including t-statistics, U-statistics, and von Mises functionals. Shi (1986) weakened the conditions required in Bickel and Freedman (1981) for the U-statistics. Hall (1992) contributed to understanding the bootstrap theory given Edgeworth expansion. Shao and Tu (1995) provided a comprehensive review of the theoretical properties of the bootstrap.

Most existing analyses for the bootstrap theory are restricted to the consistency of resampling from the empirical distribution. Bickel and Freedman (1981) showed that the consistency results of the bootstrap distribution of the sample mean they hold not only for resampling from the empirical distribution but also from other approximating discrete distributions. However, they did not provide an example of another type of approximating distribution. Motivated by the plug-in principle of the bootstrap method, Fang et al. (2014) proposed two more ways to construct approximating distribution, i.e., F_{QMC} and F_{MSE}, Fang et al. (2020) added one more F_{RMC} (see Section 2.2). They found that statistical estimation by resampling from F_{QMC} and F_{MSE} performs better in most cases. However, they did not provide some theoretical foundation. Recently, Xu et al. (2024) proposed some theoretic approach to resampling from F_{MSE}.

4.6.2 New resampling method via representative points

Let us consider distributions of sample mean and sample variance under $F_{mse,k}$. To prove the new resampling method works well, we need to justify $\tilde{H}_n^{(k)}$ defined in (4.2) converges to H_n defined in (4.30) and show $\tilde{H}_n^{(k)}$ is closed to H_n. There are several ways to define a distance between H_n and $\tilde{H}_n^{(k)}$. In this section, we use the Kolmogorov distance and the Mallows–Wasserstein distance defined in Section 4.5 to measure the distance between H_n and $\tilde{H}_n^{(k)}$. If one of the metrics converges to 0, $\tilde{H}_n^{(k)}$ is considered to be consistent to H_n.

Mallows–Wasserstein distance has many nice properties (Panaretos and Zemel, 2019). Some useful results for ρ_2 were also proved by Bickel and Freedman (1981). In particular, denote G_n as the sampling distribution of $n^{1/2}[\bar{X} - E_G(X)]$, if G_n, $G \in \mathcal{F}_2$, then $\rho_2(G_n, G) \to 0$, if and only if, $G_n \to G$ in distribution and $\lim_{n\to\infty}\int_{-\infty}^{\infty} x^r \mathrm{d}G_n(x) =$

$\int_{-\infty}^{\infty} x^r \mathrm{d}G(x)$, $r = 1, 2$. Hence, the convergence under the Mallows–Wasserstein metric (4.20) is stronger than convergence in distribution. In Xu et al. (2024), F and $F_{mse,k}$ are assumed in $\mathcal{F}_2$. Lemma 3.4 in Graf and Luschgy (2007) gave the square relationship between the Mallows–Wasserstein distance (4.20) and the mean squared function, that is,

$$\rho_2(F_{mse,k}, F) = (\mathrm{MSE}_X(\mathcal{A}))^{1/2},$$

which is equivalent to $\rho_2(F_{mse,k}, F) = E^{1/2}[(X - Y_k)^2]$.

Let the sample mean $\bar{X} = n^{-1}\sum_{i=1}^n X_i$ be an estimator of $\mu = E(X)$, where $X_1, \ldots, X_n$ are i.i.d. from F. Let $Y_1, \ldots, Y_n$ be i.i.d. from $F_{mse,k}$. Denote $\bar{Y} = n^{-1}\sum_{i=1}^n Y_i$ and $\tilde{\mu}_k = E(Y_k)$.

Theorem 4.5 *The distribution of $n^{1/2}(\bar{Y} - \tilde{\mu}_k)$ converges to the distribution of $n^{1/2}(\bar{X} - \mu)$, that is,*

(i) $\rho_2(\tilde{H}_n^{(k)}, H_n) \to 0$ as $k \to \infty$;

(ii) $\rho_\infty(\tilde{H}_n^{(k)}, H_n) \to 0$ as $n \to \infty$ and $k \to \infty$.

The two measures $\rho_2(\cdot,\cdot)$ and $\rho_\infty(\cdot,\cdot)$ can refer to Section 4.5.

Corollary 4.1 *Given a set of k-RPs,*

$$\rho_2(\tilde{H}_n^{(k)}, H_n) \le E^{1/2}[(X - Y_k)^2]$$

holds for any $n \in \mathbb{N}^+$.

Corollary 4.2 *Given a finite number of representative points k,*

$$\lim_{n\to\infty} \rho_\infty\left(\tilde{H}_n^{(k)}, H_n\right) \le \rho_\infty\left(\Phi\left(\frac{x}{\sigma}\right), \Phi\left(\frac{x}{\tilde{\sigma}_k}\right)\right),$$

where σ^2 is the variance of F, $\tilde{\sigma}_k^2$ is the variance of $F_{mse,k}$, and Φ is the cdf of $N(0,1)$.

Theorem 4.5 shows that the sampling distribution H_n of $n^{1/2}(\bar{X} - \mu)$ can be estimated by the sampling distribution $\tilde{H}_n^{(k)}$ of $n^{1/2}(\bar{Y} - \tilde{\mu}_k)$. Its proof can refer to Subsection 4.8.5. Under the Mallows-Wasserstein metric, no matter how many samples n are used to calculate the statistic $\bar{X}$ and $\bar{Y}$, the distance between the two sampling distributions is always bounded by the square root of the mean squared error of k representative points from the underlying distribution F. Given a fixed number of points k, the RPs have the minimal mean squared error from F. Therefore, the sampling distribution of the mean under $F_{mse,k}$ is considered close to the sampling distribution under F. Under the Kolmogorov metric, when the sample size n and the number of RPs k increase, the distance between the two sampling distributions converges to 0. If we choose $n = k$, it is similar to the bootstrap method, using the same number of points as the sample size to form an approximating distribution of F. The following theorems are from Xu et al. (2024).

Theorem 4.6 *Suppose $E\left(|X|^{3+\varepsilon}\right) < \infty$ with $\varepsilon > 0$. Denote $E(|X - \mu|^3) = \alpha < \infty$. For any $n \in \mathbb{N}^+$,*

$$\limsup_{k\to\infty}\left\{\left(\alpha^{-1}\sigma^3 n^{1/2}\right) \sup_{-\infty<x<\infty} \left|\mathrm{pr}(\sqrt{n}(\bar{X} - \mu)/\sigma \le t) - \mathrm{pr}(\sqrt{n}(\bar{Y} - \tilde{\mu}_k)/\tilde{\sigma}_k \le t)\right|\right\} \le 2C,$$

where C is the positive constant that appears in the Berry–Esséen bound. When $k, n \to \infty$,

$$\rho_\infty\left(P\left(\frac{n^{1/2}(\bar{X} - \mu)}{\sigma} \le t\right), P\left(\frac{n^{1/2}(\bar{Y} - \tilde{\mu})}{\tilde{\sigma}_k} \le t\right)\right) \to 0. \tag{4.31}$$

The proof of Theorem 4.6 is given in Subsection 4.8.6. Theorem 4.6 establishes the consistency of the distribution of the standardized sample mean under the Kolmogorov metric. The constant C is a positive constant independent of sample size n and the distributions of F and $F_{mse,k}$. Next, we consider consistency in the distribution approximation of sample variance. Let $X_1, \ldots, X_n$ be i.i.d. from F and $S^2 = (n-1)^{-1}\sum_{i=1}^n (X_i - \bar{X})^2$ is an unbiased estimator of $\sigma^2 = \text{var}(X)$. Let $Y_1, \ldots, Y_n$ be i.i.d. from $F_{mse,k}$ and $\tilde{S}^2 = (n-1)^{-1}\sum_{i=1}^n \left(Y_i - \bar{Y}\right)^2$.

Theorem 4.7 *Suppose F is asymmetric and/or unimodal with $E\left(|X|^{4+\epsilon}\right) < \infty$ with some $\epsilon > 0$. The distribution of $n^{1/2}(\tilde{S}^2 - \tilde{\sigma}_k^2)$ converges to the distribution of $n^{1/2}(S^2 - \sigma^2)$, i.e., $\rho_\infty(\tilde{H}_n^{(k)}, H_n) \to 0$ as $n \to \infty$ and $k \to \infty$. Given a fixed number of representative points k, if $\tilde{\mu}_{4,k} > \tilde{\sigma}_k^4$, then*

$$\lim_{n\to\infty} \rho_\infty(\tilde{H}_n^{(k)}, H_n) \le \rho_\infty\left(\Phi\left[\frac{x}{(\mu_4 - \sigma^4)^{1/2}}\right], \Phi\left[\frac{x}{(\tilde{\mu}_{4,k} - \tilde{\sigma}_k^4)^{1/2}}\right]\right), \tag{4.32}$$

where μ_4 is the fourth central moment of F and $\tilde{\mu}_{4,k}$ is the fourth central moment of $F_{mse,k}$.

Subsection 4.8.7 gives a proof of Theorem 4.7. Theorem 4.7 shows the distribution H_n of $n^{1/2}(S^2 - \sigma^2)$ can be estimated by the distribution $\tilde{H}_n^{(k)}$ of $n^{1/2}(\tilde{S}^2 - \tilde{\sigma}_k^2)$, where $\tilde{\sigma}_k^2 = \text{var}(Y_k)$. According to Theorem 4.2, $E\left(|X|^{4+\varepsilon}\right) < \infty$ guarantees the fourth central moment convergence of the representative points. Given a fixed number of representative points, the upper bound in (4.32) is measurable.

4.7 Statistical Simulation via Resampling

Fang et al. (2014) proposed a new approach to statistical simulation via resampling from the discrete approximation distributions to $N(0,1)$ formed by different kinds of RPs, i.e., Y_{MC}, Y_{QMC}, Y_{MSE}. Note that Y_{MC} is the empirical distribution of a random sample. Xu et al. (2023) gave numerical comparisons among Y_{MC}, Y_{QMC}, Y_{MSE} for the Weibull, Beta prime, Gamma, F-, lognormal, and inverse Gaussian distributions. They consider estimating the first four moments: mean, variance, skewness, and kurtosis.

Let Y_{MSE} be a discrete random variable in Definition 2.1, and its four statistics (mean, variance, skewness, and kurtosis) are given by

$$\mathbb{E}(Y_{MSE}) = \sum_{j=1}^k \xi_j^{(k)} p_j^{(k)} \equiv \mu = E(X),$$

$$\text{Var}(Y_{MSE}) = \sum_{j=1}^k (\xi_j^{(k)} - \mu)^2 p_j^{(k)} \equiv \sigma_{Y_{MSE}}^2,$$

$$\text{Sk}(Y_{MSE}) = \frac{1}{\sigma_{Y_{MSE}}^3} \sum_{j=1}^k (\xi_j^{(k)} - \mu)^3 p_j^{(k)},$$

$$\text{Ku}(Y_{MSE}) = \frac{1}{\sigma_{Y_{MSE}}^4} \sum_{j=1}^k (\xi_j^{(k)} - \mu)^4 p_j^{(k)} - 3.$$

4.7.1 Resampling under the Weibull distribution

A Weibull distribution has the following CDF,

$$F(x|m,\sigma) = \begin{cases} 1 - \exp\{-(x/\sigma)^m\}, & x \geq 0, \\ 0, & x < 0. \end{cases}$$

The pdf is

$$f(x;m,\sigma) = \frac{m}{\sigma}\left(\frac{x}{\sigma}\right)^{m-1} \exp\left\{-\left(\frac{x}{\sigma}\right)^m\right\}, x \geq 0.$$

The mean, variance, skewness, and kurtosis of the Weibull distribution are

$$\begin{aligned}
\mathbb{E}(X) &= \sigma\Gamma(1+\frac{1}{m}) \equiv \mu, \\
\text{Var}(X) &= \sigma^2[\Gamma(1+\frac{2}{m}) - (\Gamma(1+\frac{1}{m}))^2] \equiv \sigma_{sd}^2, \\
\text{Sk}(X) &= \frac{\mathbb{E}((X-\mu)^3)}{\sigma_{sd}^3} = \frac{\sigma^3\Gamma(1+\frac{3}{m}) - 3\mu\sigma_{sd}^2 - \mu^3}{\sigma_{sd}^3}, \\
\text{Ku}(X) &= \frac{\mathbb{E}((X-\mu)^4)}{\sigma_{sd}^4} - 3 = \frac{\sigma^4\Gamma(1+\frac{4}{m}) - 4\text{Sk}(X)\sigma_{sd}^3\mu - 6\mu^2\sigma_{sd}^2 - \mu^4}{\sigma_{sd}^4} - 3.
\end{aligned}$$

respectively, where $\Gamma(\cdot)$ is the Gamma function.

Section 2.4 shows that the limiting behavior of $\Delta_k^{(k)}$ is different for the different parameter m in the Weibull distribution. Recall in Table 2.4 that as $k \to \infty$, $\Delta_k^{(k)} \to \infty$ when $0 < m < 1$ and $\Delta_k^{(k)} \to 0$ when $m > 1$. We first discuss the effect of m on simulation in the Weibull distribution. The MSE-RPs of the Weibull distribution can be obtained from the numerical method mentioned in Chakraborty et al. (2021). Consider three different shapes parameters $m = 0.5, 1, 2$ to exemplify the simulation results, provided the scale parameters $\sigma = 1$. Notice that the Weibull distribution degenerates into the exponential distribution when $m = 1$.

From the pdf plots of three different cases in Figure 4.2, we observe that the black curve ($m = 2$) has the fastest convergence rate for the tail in these three cases while the red curve ($m = 0.5$) has the slowest convergence rate for the tail in these three cases. Table 4.3 provides MSE-RPs with $k = 5$ and corresponding probabilities. Based on the results in Table 4.3, there are some interesting observations.

(1) The range $\xi_5^{(5)} - \xi_1^{(5)}$ in the case $m = 0.5$ is much larger than the range in other cases.

(2) In the case $m = 0.5$, the first MSE-RP represents almost 80% of the population, while the last RP represents less than 0.2% of the population.

(3) The gap $\Delta_5^{(5)}$ in the case $m = 0.5$ is much larger than the gap $\Delta_5^{(5)}$ in the case $m = 2$. The gap $\Delta_5^{(5)}$ in the case $m = 1$ is 2, which is consistent with Theorem 2 in Xu et al. (2023).

Biases of mean, variance, skewness, and kurtosis between the Weibull distributions and their Y_{MSE} are listed in Table 4.4, where the number of MSE-RPs $k = 10, 15, 20, 30$ and $m = 0.5, 1, 2$. The estimation biases are defined as the difference between the estimated value and the true value, namely $\mathbb{E}(Y_{MSE}) - \mathbb{E}(X)$, $\text{Var}(Y_{MSE}) - \text{Var}(X)$, $\text{Sk}(Y_{MSE}) - \text{Sk}(X)$, and $\text{Ku}(Y_{MSE}) - \text{Ku}(X)$.

From the results in Table 4.4, we raise the following observations:

(1) The estimation bias of the mean is 0, which is consistent with Theorem 3.5(i). The estimation bias of variance is less than 0, which is consistent with Theorem 3.5(ii).

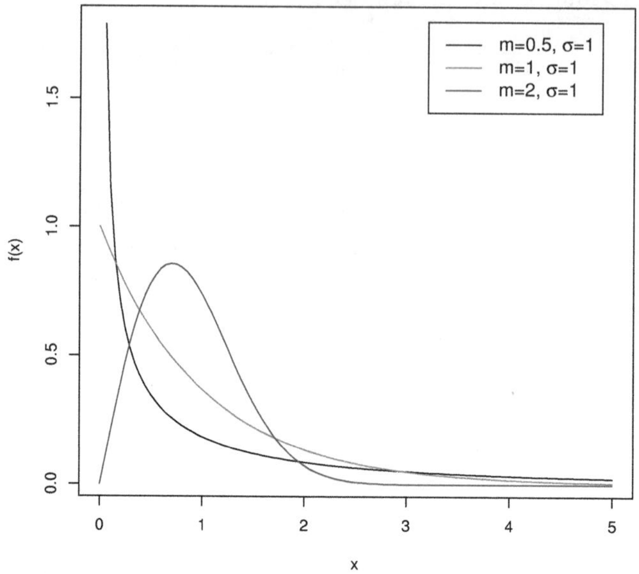

FIGURE 4.2
The pdf of Weibull distribution for $m = 0.5, 1, 2$ with fixed $\sigma = 1$.

TABLE 4.3
MSE-RPs and corresponding probabilities of the Weibull distribution with $k = 5$

MSE-RPs	ξ_1	ξ_2	ξ_3	ξ_4	ξ_5	$\Delta_5^{(5)}$
$m = 0.5$	0.5296	4.4248	11.6122	24.8865	54.0096	29.1231
prob.	0.7928	0.1483	0.0450	0.0121	0.0019	
$m = 1$	0.2704	0.9305	1.7784	2.9657	4.9657	2.0000
prob.	0.4514	0.2905	0.1648	0.0743	0.0190	
$m = 2$	0.3218	0.6681	1.0070	1.3909	1.9190	0.5281
prob.	0.2173	0.2869	0.2583	0.1729	0.0646	

(2) With the increase in the number of MSE-RPs k, the accuracy of estimation becomes better for three m cases.

(3) When the number of MSE-RPs k is fixed, the smaller m In the Weibull distribution, the estimation of variance, skewness, and kurtosis is less accurate. It indicates that the approximation to the original distribution is slightly worse regarding moment estimation when m is small.

(4) All values in the estimation of skewness and kurtosis are negative, and it is conjectured that the true value is always larger than the estimated value, namely $\text{Sk}(X) > \text{Sk}(Y_{MSE})$ and $\text{Ku}(X) > \text{Ku}(Y_{MSE})$ for all $k \in \mathbb{N}^+$.

The estimation biases by resampling for the Weibull distribution are given in Table 4.5, where the number of MSE-RPs $k = 15$, $m = 0.5, 1, 2$ and $N = 1000, 2000, 5000, 10000$. The

TABLE 4.4
Bias of mean, variance, skewness and kurtosis estimated by Y_{MSE}

	Category	True value	$k=10$	$k=15$	$k=20$	$k=30$
	$m=0.5$	2.0000	0.0000	0.0000	0.0000	0.0000
Mean	$m=1$	1.0000	0.0000	0.0000	0.0000	0.0000
	$m=2$	0.8862	0.0000	0.0000	0.0000	0.0000
	$m=0.5$	20.0000	-0.5271	-0.2504	-0.1448	-0.0663
Variance	$m=1$	1.0000	-0.0202	-0.0093	-0.0053	-0.0024
	$m=2$	0.2146	-0.0041	-0.0019	-0.0011	-0.0005
	$m=0.5$	6.6188	-0.2895	-0.1613	-0.1012	-0.0514
Skewness	$m=1$	2.0000	-0.0494	-0.0242	-0.0143	-0.0067
	$m=2$	0.6311	-0.0161	-0.0078	-0.0046	-0.0022
	$m=0.5$	84.7200	-15.7698	-9.7197	-6.5615	-3.6651
Kurtosis	$m=1$	6.0000	-0.7550	-0.3935	-0.2416	-0.1180
	$m=2$	0.2451	-0.1835	-0.0916	-0.0549	-0.0261

TABLE 4.5
Estimation biases by resampling for the Weibull distribution with $k=15$

	Category	1000	2000	5000	10000
	$m=0.5$	-0.0508	0.0048	-0.0293	0.0108
Mean	$m=1$	0.0031	-0.0021	-0.0020	-0.0005
	$m=2$	0.0098	0.0018	0.0015	-0.0010
	$m=0.5$	-2.3841	-1.0398	-2.1991	-1.5503
Variance	$m=1$	-0.0496	-0.0943	-0.0829	-0.0741
	$m=2$	-0.0126	-0.0150	-0.0150	-0.0157
	$m=0.5$	-4.5853	-4.5587	-4.5731	-4.5620
Skewness	$m=1$	-0.7866	-0.8269	-0.8091	-0.8054
	$m=2$	-0.1885	-0.2003	-0.1911	-0.1873
	$m=0.5$	-81.2084	-81.1122	-81.1756	-81.1162
Kurtosis	$m=1$	-4.9815	-5.1238	-5.0790	-5.0733
	$m=2$	-0.5861	-0.6158	-0.5922	-0.5945

estimation biases by resampling are the difference between the corresponding $\bar{T}$ and the true value.

Based on Table 4.5, we find that the estimation bias of the mean by resampling in the case $m=0.5$ is slightly larger than in other cases. Further, the simulation results show that the estimation biases of variance, skewness, and kurtosis by resampling in the case $m=0.5$ are relatively more significant than those values in the instances $m=1$ and $m=2$.

4.7.2 Resampling under other distributions

Next, we compare simulation performance for the following distributions that can refer to Table 1.4.

• Case I. ($x \in \mathbb{R}^+$) The beta prime distribution ($\alpha=1$ and $\beta=5$), gamma distribution ($\alpha=2$ and $\beta=1/\sqrt{2}$), F-distribution ($d_1=10$ and $d_2=10$), lognormal distribution ($\mu=1$ and $\sigma^2=1$), and inverse Gaussian distribution ($\mu=1$ and $\lambda=1$).

• Case II. ($x \in \mathbb{R}$) The normal distribution ($\mu=0$ and $\sigma^2=1$), logistic distribution ($\mu=0$ and $\sigma=\sqrt{3}/\pi$), and Student's t-distribution ($\nu=6$).

TABLE 4.6
Estimation biases by resampling for $k = 15$, Case I

	Category	1000	2000	5000	10000
Mean	Beta prime distribution	0.0037	0.0021	0.0014	0.0019
	Gamma distribution	-0.0062	-0.0040	0.0009	0.0017
	F-distribution	-0.0059	0.0047	0.0032	0.0017
	Lognormal distribution	-0.0317	-0.0263	0.0160	-0.0073
	Inverse Gaussian distribution	-0.0183	-0.0106	-0.0073	0.0031
Variance	Beta prime distribution	-0.0118	-0.0146	-0.0114	-0.0094
	Gamma distribution	-0.0963	-0.0705	-0.0754	-0.0719
	F-distribution	-0.0808	-0.0217	-0.0829	-0.0795
	Lognormal distribution	-6.4960	-5.7334	-1.9009	-3.3368
	Inverse Gaussian distribution	-0.1549	-0.1199	-0.1122	-0.0589
Skewness	Beta prime distribution	-3.1119	-3.1187	-3.1026	-3.1028
	Gamma distribution	-0.5746	-0.5353	-0.5443	-0.5226
	F-distribution	-2.3725	-2.3835	-2.4149	-2.4041
	Lognormal distribution	-4.5662	-4.5982	-4.5794	-4.5770
	Inverse Gaussian distribution	-1.5830	-1.5614	-1.5668	-1.5398
Kurtosis	Beta prime distribution	-68.9705	-68.9979	-68.9271	-68.9390
	Gamma distribution	-2.8004	-2.6829	-2.7130	-2.6589
	F-distribution	-43.8878	-43.9430	-44.0103	-43.9952
	Lognormal distribution	-108.6470	-108.7789	-108.7329	-108.7059
	Inverse Gaussian distribution	-13.4422	-13.3471	-13.3750	-13.2845

TABLE 4.7
Estimation biases by resampling for $k = 15$, Case II

	Category	1000	2000	5000	10000
Mean	Normal distribution	-0.0028	-0.0047	0.0065	-0.0000
	Logistic distribution	0.0087	-0.0053	-0.0026	-0.0010
	Student's t-distribution	-0.0213	0.0085	-0.0022	-0.0001
Variance	Normal distribution	-0.0667	-0.0805	-0.0734	-0.0791
	Logistic distribution	-0.0869	-0.0829	-0.0747	-0.0717
	Student's t-distribution	-0.1071	-0.1418	-0.1183	-0.1191
Skewness	Normal distribution	-0.0262	-0.0052	-0.0005	-0.0028
	Logistic distribution	0.0186	0.0082	0.0018	-0.0048
	Student's t-distribution	-0.0167	0.0042	-0.0110	-0.0083
Kurtosis	Normal distribution	-0.4407	-0.3973	-0.3999	-0.3848
	Logistic distribution	-1.1935	-1.1651	-1.2345	-1.2133
	Student's t-distribution	-2.8461	-2.8643	-2.8612	-2.8763

• Case III. ($x \in [a, b]$) The uniform distribution ($a = 0$ and $b = 1$), and beta distribution ($\alpha = 2$ and $\beta = 2$).

By the numerical methods in Fang and He (1982) or Chakraborty et al. (2021), we obtain the MSE-RPs of those distributions. We discuss simulation performance separately. Some simulation results in the subsection are from Xu et al. (2023), and some are new.

The estimation biases by resampling for those distributions are illustrated in Tables 4.6-4.8, respectively, where $k = 15$ and $N = 1000, 2000, 5000, 10000$. Based on the results in Tables 4.6, 4.7, and 4.8, we can conclude that

(1) The estimation biases of skewness and kurtosis by resampling for the beta prime distribution, F-distribution, lognormal distribution, and Student's t-distribution are slightly larger than others.

(2) For distribution whose $\Delta_k^{(k)} \to \infty$ as $k \to \infty$ like the beta prime distribution, F-distribution, lognormal distribution, and Student's t-distribution, the simulation results are unstable. The estimations perform slightly better for distribution whose $\Delta_k^{(k)}$ tends to a non-negative real number as $k \to \infty$.

TABLE 4.8
Estimation biases by resampling for $k = 15$, Case III

	Category	1000	2000	5000	10000
Mean	Uniform distribution	-0.0003	0.0013	0.0010	-0.0002
	Beta distribution	0.0028	0.0013	-0.0006	0.0000
Variance	Uniform distribution	-0.0065	-0.0054	-0.0064	-0.0057
	Beta distribution	-0.0033	-0.0036	-0.0035	-0.0035
Skewness	Uniform distribution	0.0023	-0.0115	-0.0051	-0.0048
	Beta distribution	-0.0210	0.0001	0.0043	-0.0014
Kurtosis	Uniform distribution	0.1664	0.1364	0.1475	0.1407
	Beta distribution	0.0571	0.0265	0.0352	0.0239

(3) It seems hard for MSE-RPs to capture the characteristics of skewness and kurtosis, especially when the distributions in Case I.

For the univariate continuous random variable X with $\mathbb{E}(X^2) < +\infty$, the existence of MSE-RPs can be guaranteed. However, the existence of skewness and kurtosis is not restricted because they may fail.

4.7.3 Mixture of two normal distributions

Section 3.9 discusses the two normal distributions' mixtures. In this subsection, we give more studies on density estimation and real case studies.

Consider a data transmission problem. Suppose the input signal X follows the distribution MixN $(\alpha, \mu_1, \sigma_1^2, \mu_2, \sigma_2^2)$ with the cdf $F(x)$ and the density function $p(x)$, and the output signal is a set of k points. Three types of points (MC-RPs, QMC-RPs, and MSE-RPs) from a MixN signal are compared for kernel density estimation. Consider two mixtures of normal distributions C1–C2 as follows:

C1: MixN$(0.6, 9, 9, 2, 1)$ with mean 6.2 and variance 17.56
C2: MixN$(0.6, -5, 16, 9, 9)$ with mean 0.6 and variance 60.24

For the underlying distributions, the density comparisons are given in Figure 4.3 and the corresponding L_2-distance between the estimated kernel density and the population density $p(x)$ are summarised in Table 4.9. We can observe that the MSE method offers better performance in density fitting than the QMC method. According to Table 4.9, kernel density estimations based on $k = 25$ MSE points are more accurate than one based on $k = 25$ QMC points.

TABLE 4.9
The L_2-distance between the kernel estimate $\hat{p}_h(x)$ and the underlying mixture density $p(x)$ $(k = 25)$

Distribution	MC	QMC	MSE
C1	0.1273	0.0420	0.0351
C2	0.0701	0.0165	0.0131

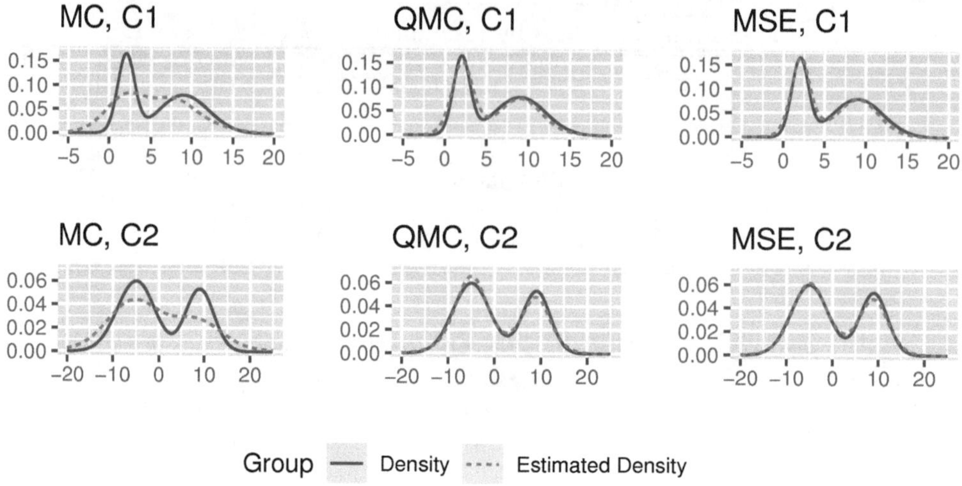

FIGURE 4.3
Results of kernel density estimation comparisons ($k = 25$).

Case studies

Mixtures of two-component normal distributions have applications across many disciplines. Consider the analysis of the cell nucleus data from the diagnostic database of the University of Wisconsin Clinical Sciences Center, see Wolberg et al. (1995). Li et al. (2022) fitted the perimeter of the cell nucleus by using MixN models based on the revised penalized maximum likelihood estimation and the application of MSE-RPs in density estimations. The MixN model presented below

$$\text{MixN}(0.5806, 75.4189, 89.4452, 117.1346, 667.5068) \tag{4.33}$$

is taken. They generated 38 MSE-RPs from the MixN model (4.33) using the Fang-He algorithm and the k-means algorithm. Considering the large variance in the model (4.33), five million training samples are used for the k-means algorithm. Table 4.10 presents the comparisons of the two algorithms. MSE-RPs generated by the k-means algorithm have a total error of 0.0572, which does not satisfy the non-linear system (3.46). Using the Fang-He algorithm to generate more accurate numerical approximations of the 50 MSE-RPs with a smaller mean squared error.

TABLE 4.10
Comparisons of the Fang-He algorithm and the k-means algorithm in terms of the total error on the non-linear system and the mean squared error (MSE)

Algorithm	Total Error	MSE
Fang-He	0.0000	1.0165
k-means	0.0572	3.1044

Li et al. (2022) generated 38 QMC points and 38 MC points from the model (4.33). Table 4.11 summarizes the estimates of the mean, variance, skewness, and kurtosis of the distribution (4.33) based on the MC, QMC, and MSE methods. We can see that the MSE method estimates the moments of the model more accurately, especially the MSE-RPs generated by the Fang-He algorithm, which have better results. Comparisons of kernel density estimates based on MC, QMC, and MSE methods are presented in Figure 4.4. The corresponding L_2-distance between the kernel estimates and the density of the model (4.33) are 0.0752 for MC, 0.0498 for QMC and 0.0164 for MSE.

TABLE 4.11
The estimations of the first four moments of the fitted MixN model are based on 38 sampling points from the model

Method	Mean	Variance	Skewness	Kurtosis
Model	92.9145	755.6292	0.9853	0.2471
MSE(Fang-He)	92.9145	754.5623	0.9850	0.2338
MSE(k-means)	92.9013	751.1684	0.9683	0.1248
QMC	92.8736	736.6777	0.9446	-0.0306
MC	100.0654	866.3720	0.6641	-0.7553

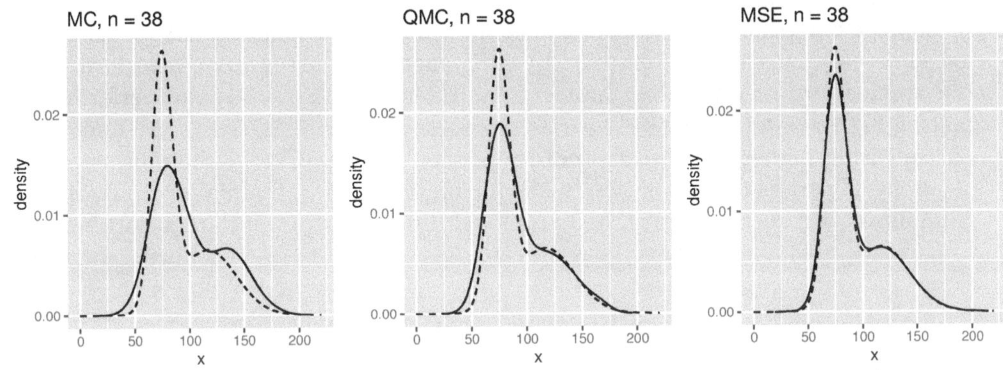

FIGURE 4.4
Results of kernel density estimation comparisons. The dashed lines represent the distribution's density (4.33).

4.7.4 Generalized alpha skew-t distribution

Zhou et al. (2024a) studied the Generalized Alpha Skew-t (GAST) distribution, which has been studied by Altun et al. (2018).

Definition 4.3 *(GAST distribution). A random variable X is said to follow the GAST distribution, denoted as $X \sim GAST(\alpha, s, \nu)$, if it has the following pdf*

$$p(x; \alpha, s, \nu) = \frac{(1-\alpha x)^2 + 1}{c(\alpha, s, \nu)} t(x; \nu) T\left(\sqrt{\frac{1+\nu}{x^2+\nu}} sx; \nu+1\right), \qquad \nu > 2, \quad x \in \mathcal{R},$$

where

$$c(\alpha, s, \nu) = 1 - \alpha \left[\frac{s}{\sqrt{1+s^2}} \left(\frac{\nu}{\pi}\right)^{1/2} \Gamma\left(\frac{\nu-1}{2}\right) / \Gamma\left(\frac{\nu}{2}\right)\right] + \frac{\alpha^2}{2}\left(\frac{\nu}{\nu-2}\right).$$

The GAST distribution involves several popular useful distributions:

- If $\alpha = 0$, the GAST distribution reduces to the skew-t (ST) distribution.
- If $s = 0$, the GAST distribution reduces to the alpha-skew-t (AST) distribution.
- If $\alpha = 0$ and $s = 0$, the GAST distribution reduces to the Student's-t distribution.
- If $\nu \to \infty$, the GAST distribution reduces to the alpha-skew-normal (ASN) distribution.
- If $\nu \to \infty$, $\alpha = 0$, the GAST distribution reduces to the skew-normal (SN) distribution.
- If $\nu \to \infty$, $\alpha = 0$ and $s = 0$ the GAST distribution reduces to the normal distribution.

Zhou et al. (2024a) studied some properties of the GAST distribution, including the calculation of the moments, and the relationship between the number of peaks and the GAST parameters. By the experiment design, they considered 16 different groups of the GAST distribution according to the three parameters α, s, ν. They studied the unimodal and bimodal property and pointed out that (1) The skew-t distribution is always unimodal if $\nu > 2$; (2) giving the sufficient and necessary condition for the bimodal of the AST distribution. They gave the generation RPs algorithm based on the stochastic representation of the GAST distribution given by Altun et al. (2018). They applied the QMC-MLE method proposed by Li and Fang (2022) for the maximum likelihood estimation of parameters α, s, ν to improve the traditional MLE. For both unimodal and bimodal GAST distributions they found reveal that the RPs obtained by MSE method consistently outperform the others in the applications of estimating moments and densities. However, the performance on estimating higher-order moments, such as skewness and kurtosis, shows the limitations of RPs on capturing higher-order statistical properties. the QMC-MLE method can improve the accuracy of parameter estimation, while for bimodal cases, the plain MLE is more effective. Combined with such property, we can model different types of data accordingly.

4.8 Proofs of Theorems in Chapter 4

4.8.1 Proof of Theorem 4.1

Proof *According to the properties of representative points, $Y_{MSE}^{(k)}$ converges X in distribution as k approaches infinity. Then, part (i) follows the continuous mapping theorem. Part (ii) follows the Portmanteau theorem (see Dudley (1989), Theorem 11.1.1).*

4.8.2 Proof of Theorem 4.2

The proof is from Xu et al. (2024).

Proof *Part (i). Given $E\left(|X|^{r+\varepsilon}\right) < \infty$ for some $r \geq 3$ and $r \in \mathbb{N}^+$, where $\varepsilon > 0$. Let $g_1(x) = |x|^{r+\varepsilon}$. Since $g_1(x)$ is convex for any $r \geq 3$ and satisfies $\int_{\mathcal{X}} |x|^{r+\varepsilon} dF(x) < \infty$, according to Lemma 4.1, we have*

$$\sup_{k \in \mathbb{N}^+} E\left(|Y_{MSE}^{(k)}|^{r+\varepsilon}\right) \leq E\left(|X|^{r+\varepsilon}\right) < \infty. \tag{4.34}$$

Since $Y_{MSE}^{(k)} \to X$ in distribution and $\sup_{k \in \mathbb{N}^+} E\left(|Y_{MSE}^{(k)}|^{k+\varepsilon}\right) < \infty$, according to Lemma 4.2, we have $E\left(|X|^k\right) < \infty$ and $E([Y_{MSE}^{(k)}]^k) \to E(X^k)$ as $k \to \infty$.

Part (ii). Under the same condition, let a continuous function $g_2(x) = |x - \mu|$, where $\mu = E(X)$. It is also the expectation of $F_{MSE,k}$ for any $k \in \mathbb{N}^+$. By the continuous mapping theorem, $Y_{MSE}^{(k)} \to X$ in distribution implies $|Y_{MSE}^{(k)} - \mu| \to |X - \mu|$ in the same mode. Using the Minkowski inequality, we have

$$\left[E(|Y_{MSE}^{(k)} - \mu|^p)\right]^{\frac{1}{p}} \le \left[E(|Y_{MSE}^{(k)}|^p)\right]^{\frac{1}{p}} + [E(|\mu|^p)]^{\frac{1}{p}} = \left[E(|Y_{MSE}^{(k)}|^p)\right]^{\frac{1}{p}} + |\mu|, \quad (4.35)$$

for $1 \le p < \infty$. By (4.34)-(4.35),

$$\sup_{k\in\mathbb{N}^+} E\left(|Y_{MSE}^{(k)} - \mu|^{r+\varepsilon}\right) \le \sup_{k\in\mathbb{N}^+} \left\{\left[E(|Y_{MSE}^{(k)}|^{r+\varepsilon})\right]^{\frac{1}{r+\varepsilon}} + |\mu|\right\}^{r+\varepsilon} < \infty \quad (4.36)$$

holds when $E\left(|X|^{r+\varepsilon}\right) < \infty$. Hence, $E(|Y_{MSE}^{(k)} - \mu|^r) \to E(|X - \mu|^r)$ as $k \to \infty$.

4.8.3 Proof of Theorem 4.3

The proof is from Jiang et al. (2015).

Proof *The mth moment of the arcsine distribution is easy to find*

$$\begin{aligned} E(X^m) &= \int_0^1 x^m \frac{1}{\pi\sqrt{x(1-x)}} dx = \int_0^1 \frac{2u^{2m}}{\pi\sqrt{1-u^2}} du \\ &= \int_0^{\frac{\pi}{2}} \frac{2}{\pi} \sin^{2m}(\theta) d\theta = \prod_{j=1}^{m} \frac{2j-1}{2j}. \end{aligned}$$

The mth moment of the approximate distribution is given by

$$E(\xi^m) = \frac{1}{k}\sum_{j=1}^{k} \sin^{2m}(\frac{\pi}{2}\frac{2j-1}{2k}) = \sum_{j=1}^{k} \sin^{2m}(\theta_j),$$

where $\theta_j = \frac{\pi}{2}\frac{2j-1}{2k}$. To prove the theorem, we need to show

$$\frac{1}{k}\sum_{j=1}^{k} \sin^{2m}(\theta_j) = \prod_{j=1}^{m} \frac{2j-1}{2j}. \quad (4.37)$$

$$\begin{aligned} E(\xi^m) &= \frac{1}{k}\sum_{j=1}^{k} \sin^{2m}(\theta_j) = \frac{1}{k}\sum_{j=1}^{k} [\frac{e^{i\theta_j} - e^{-i\theta_j}}{2i}]^{2m} \\ &= (-1)^m(2^{-2m})\frac{1}{k}\sum_{j=1}^{k}\sum_{l=0}^{2m} \binom{2m}{l}(-1)^l(e^{i\theta_j})^{2l-2m} \\ &= (-1)^m(2^{-2m})\frac{1}{k}\sum_{l=0}^{2m} \binom{2m}{l}(-1)^l \sum_{j=1}^{k} e^{i(\frac{l-m}{k}\cdot\frac{\pi}{2})(2j-1)}. \end{aligned} \quad (4.38)$$

Note that on the right-hand side of the above equation, we have

$$\sum_{j=1}^{k} \left(e^{i(\frac{l-m}{k}\cdot\frac{\pi}{2})}\right)^{(2j-1)} = \begin{cases} k, & \text{if } l = m, \\ \frac{y(1-y^{2k})}{1-y^2}, & \text{if } l \ne m, \end{cases} \quad (4.39)$$

where $y = e^{i(\frac{l-m}{n}\cdot\frac{\pi}{2})}$. *Moreover,* $1 - y^{2n} = 1 - e^{i(l-m)\pi} = 2$ *when the value of* $(l-m)$ *is odd, otherwise* $1 - y^{2n} = 0$.

Let $\frac{l-m}{n}\cdot\frac{\pi}{2} = \alpha\pi$, *we have* $y = e^{i\alpha\pi}$ *and*

$$\frac{y}{1-y^2} = \frac{e^{i\alpha\pi}}{1-e^{i2\alpha\pi}} = \frac{2i(\sin\alpha\pi)}{[1-\cos(2\alpha\pi)]^2 + [\sin(2\alpha\pi)]^2}. \tag{4.40}$$

From (4.40), for $l \neq m$ *and* $k > m$ *we have that*

$$\text{real part of} \sum_{j=1}^{k}\left(e^{i(\frac{l-m}{n}\cdot\frac{\pi}{2})}\right)^{(2j-1)} = 0 \text{ and} \tag{4.41}$$

$$\text{imaginary part of} \sum_{j=1}^{k}\left(e^{i(\frac{l-m}{k}\cdot\frac{\pi}{2})}\right)^{(2j-1)}$$
$$= -\text{imaginary part of} \sum_{j=1}^{k}\left(e^{i(\frac{l-k}{n}\cdot\frac{\pi}{2})}\right)^{(2j-1)} \tag{4.42}$$

Therefore, from (4.39)-(4.42), we have

$$\begin{aligned}
&\sum_{l=0}^{2m}\binom{2m}{l}(-1)^k\sum_{j=1}^{k}\left(e^{i(\frac{l-m}{n}\cdot\frac{\pi}{2})}\right)^{(2j-1)} \\
= \;& (-1)^m\binom{2m}{m}\cdot k + \sum_{l=0}^{m-1}\binom{2m}{l}(-1)^k\sum_{j=1}^{k}(e^{i(\frac{l-m}{k})\cdot\frac{\pi}{2}})^{(2j-1)} \\
& + \sum_{l=m+1}^{2m}\binom{2m}{l}(-1)^k\sum_{j=1}^{k}(e^{i(\frac{l-m}{k})\cdot\frac{\pi}{2}})^{(2j-1)} \\
= \;& (-1)^m\binom{2m}{m}k. \qquad (4.43)
\end{aligned}$$

Substituting (4.43) into (4.38), we obtain

$$\begin{aligned}
E(\xi^m) &= \frac{1}{k}(-1)^m(2^{-2m})(-1)^m\binom{2m}{m}k = \binom{2m}{m}(2^{-2m}) \\
&= \frac{1\times 3\times 5\times\cdots\times(2m-1)}{2^{2m}\times m!} \\
&= \prod_{j=1}^{m}\frac{2j-1}{2j} = E(X^m).
\end{aligned}$$

4.8.4 Proof of Theorem 4.4

The proof is from Zhou and Fang (2019).

Proof of Theorem 4.4. For each k, denote

$$\Phi_k = \left\{\boldsymbol{x} : \boldsymbol{x} \in [0,1]^n,\ \sum_{i=1}^{n} x_i^k = \frac{n}{k+1}\right\}. \tag{4.44}$$

Then, Φ_k can be considered as a part of a high-dimensional spheroid under geometric meaning, and $\Phi_{uniform} = \Phi_1 \bigcap \Phi_2 \bigcap \cdots \bigcap \Phi_{n-1}$. Denote $\mathbf{0} = (0, \cdots, 0)$ be the origin. Consider the minimum and maximum squared L_2-distance between $\mathbf{0}$ and Φ_k, denoted as $d_{k,min}$ and $d_{k,max}$, respectively. Obviously, $d_{k,min} = \min_{\boldsymbol{x}\in\Phi_k} ||\boldsymbol{x}||^2$ and $d_{k,max} = \max_{\boldsymbol{x}\in\Phi_k} ||\boldsymbol{x}||^2$, where $||\cdot||$ is the L_2-norm. When $k = 2$, $d_{k,min} = d_{k,max} = n/3$. When $k \neq 2$, for obtaining $d_{k,min}$, denote the Lagrange multiple

$$L = \sum_{i=1}^{n} x_i^2 + \lambda \left(\sum_{i=1}^{n} x_i^k - \frac{n}{k+1} \right).$$

Let $\partial L/\partial x_i = 0$ and $\partial L/\partial \lambda = 0$, we have

$$\begin{cases} x_i(2 + k\lambda x_i^{k-2}) = 0, i = 1, \cdots, n, \\ \sum_{i=1}^{n} x_i^k = \frac{n}{k+1}. \end{cases} \tag{4.45}$$

When $k = 1$, $x_1 = \cdots = x_n = 1/2$ and $d_{k,min} = n/4$. When $k \geq 3$, from (4.45), $x_i = 0$ or $x_i = \left(-\frac{2}{k\lambda}\right)^{1/(k-2)}$, $i = 1, \cdots, n$. Suppose there exist m $x_i \neq 0$ among $\{x_1, \cdots, x_n\}$, and denote $x_{i_1} = \cdots = x_{i_m} = \left(-\frac{2}{k\lambda}\right)^{1/(k-2)}$, then $x_{i_{m+1}} = \cdots = x_{i_n} = 0$, where $\{i_1, \cdots, i_n\}$ is a permutation of $\{1, \cdots, n\}$. According to the condition $\sum_{i=1}^{n} x_i^k = \frac{n}{k+1}$, we have

$$x_{i_1} = \cdots = x_{i_m} = \left(\frac{n}{m(k+1)} \right)^{1/k}, x_{i_{m+1}} = \cdots = x_{i_n} = 0.$$

Moreover, since $x_i \in [0, 1]$, $m \geq \left\lceil \frac{n}{k+1} \right\rceil$ and $\sum_{i=1}^{n} x_i^2 = m^{1-2/k} \left(\frac{n}{k+1}\right)^{2/k}$, which is an increase function of m, then $d_{k,min} = m_k^{1-2/k} \left(\frac{n}{k+1}\right)^{2/k}$ with $m_k = \left\lceil \frac{n}{k+1} \right\rceil$, and $d_{k,max} = n \left(\frac{1}{k+1}\right)^{2/k}$, i.e., for any $\boldsymbol{x} \in \Phi_k$ and $k \geq 3$,

$$||\boldsymbol{x}||^2 \in \left[m_k^{1-\frac{2}{k}} \left(\frac{n}{k+1} \right)^{\frac{2}{k}}, n \left(\frac{1}{k+1} \right)^{\frac{2}{k}} \right] \doteq [a_k, b_k]. \tag{4.46}$$

Moreover, when $k = 1$, to find the maximum distance $d_{1,max}$ is equivalent to the optimization problem whose objective is $\sum_{i=1}^{n} x_i^2$ with constraint $\sum_{i=1}^{n} x_i = n/2, x_i \in [0, 1]$. Its solution is $x_{i_1} = \cdots = x_{i_r} = 1, x_{i_{r+1}} = n/2 - r, x_{i_{r+2}} = \cdots = x_{i_n} = 0$, where $r = \lfloor \frac{n}{2} \rfloor$. Then, $d_{1,max} = \lfloor \frac{n}{2} \rfloor + (\frac{n}{2} - \lfloor \frac{n}{2} \rfloor)^2$, and for any $\boldsymbol{x} \in \Phi_1$, $||\boldsymbol{x}||^2 \in [\frac{n}{4}, \lfloor \frac{n}{2} \rfloor + (\frac{n}{2} - \lfloor \frac{n}{2} \rfloor)^2] \doteq [a_1, b_1]$. It can be checked that $d_{2,min} = d_{2,max} = \frac{n}{3} \in [\xi_k, b_k]$ when $k \neq 2$, $[a_1, b_1] \bigcap [a_k, b_k] \neq \varphi$ with $k \geq 2$ and $[a_{k_1}, b_{k_1}] \subset [a_{k_2}, b_{k_2}]$ if $3 \leq k_1 < k_2 \leq n - 1$. Then, for the $n - 1$ continuous convex surfaces $\Phi_k, k = 1, \cdots, n - 1$, there exists at least one point $\boldsymbol{x} = (x_1, \cdots, x_n)$ such that $\boldsymbol{x} \in \Phi_1 \bigcap \cdots \bigcap \Phi_{n-1}$.

4.8.5 Proof of Theorem 4.5

A proof of Theorem 4.5 is from Xu et al. (2024) and needs the following lemma.

Lemma 4.3 *(Bickel and Freedman (1981)) Denote $Z_1(G), \cdots, Z_n(G)$ as independent random variables from common distribution function G. Let $G^{(n)}$ be the distribution of the statistic*

$$n^{-1/2} \sum_{j=1}^{n} \{Z_j(G) - E[Z_j(G)]\}.$$

Denote $K^{(n)}$ as the sampling distribution of the same statistics based on n random variables from distribution K. If $G, K \in \mathcal{F}_2$, then

$$\rho_2(G^{(n)}, K^{(n)}) \leq \rho_2(G, K). \tag{4.47}$$

The proof of Lemma 4.3 can be found in Lemma 3 of Mallows (1972) and Lemma 8.6 and Lemma 8.7 of Bickel and Freedman (1981).

Part (i). Denote the law of $n^{1/2}(\bar{X}-\mu)$ and $n^{1/2}(\bar{Y}-\tilde{\mu}_k)$ as H_n and $\tilde{H}_n^{(k)}$, respectively. Since $F \in \mathcal{F}_2$, so dose $F_{mse,k}$. By Lemma 4.3, for any $n, k \in \mathbb{N}^+$, we have

$$\rho_2(\tilde{H}_n^{(k)}, H_n) \leq \rho_2(F_{mse,k}, F). \tag{4.48}$$

According to the properties of representative points, $\tilde{\mu}_k = \mu$ and Y_k converge to X in the mean square. Hence, $F_{mse,k} \to F$ and

$$\lim_{k\to\infty} \int_{-\infty}^{\infty} y^2 dF_{mse,k}(y) = \int_{-\infty}^{\infty} x^2 dF(x),$$

which implies $\lim_{k\to\infty} \rho_2(F_{mse,k}, F) = 0$. By (4.48), it follows that $\rho_2(\tilde{H}^{(n)}, H^{(n)}) \to 0$ as $k \to \infty$ for any $n \in \mathbb{N}^+$. According to Lemma 3.4, page 33 of Graf and Luschgy (2007), the relationship between the Mallows–Wasserstein distance and the mean squared function is

$$\rho_2(F_{mse,k}, F) = (\text{MSE}_X(\mathcal{A}))^{1/2} = E^{1/2}[(X-Y_k)^2]. \tag{4.49}$$

For a fixed number of representative points k, we can calculate the upper bound of the Mallows–Wasserstein distance between $\tilde{H}_n^{(k)}$ and H_n.

Part (ii). Denote the variances of X and Y_k as σ^2 and $\tilde{\sigma}_k^2$. Let $\Phi(x)$ be the cumulative distribution function of the standard normal distribution, which is a uniformly continuous function. It leads to

$$\rho_\infty\left(\Phi\left(\frac{x}{\sigma}\right), \Phi\left(\frac{x}{\tilde{\sigma}_k}\right)\right) \to 0 \text{ as } k \to \infty. \tag{4.50}$$

Due to Lindeberg–Levy central limit theorem, we obtain

$$\rho_\infty\left(H_n(x), \Phi\left(\frac{x}{\sigma}\right)\right) \to 0 \text{ as } n \to \infty, \tag{4.51}$$

and

$$\rho_\infty\left(\tilde{H}_n^{(k)}(x), \Phi\left(\frac{x}{\tilde{\sigma}_k}\right)\right) \to 0 \text{ as } n \to \infty. \tag{4.52}$$

Using the triangle inequality, we have

$$\begin{aligned}\rho_\infty\left(\tilde{H}_n^k(x), H_n(x)\right) \leq & \rho_\infty\left(\tilde{H}_n^{(k)}(x), \Phi\left(\frac{x}{\tilde{\sigma}_k}\right)\right) + \rho_\infty\left(H_n(x), \Phi\left(\frac{x}{\sigma}\right)\right) + \\ & \rho_\infty\left(\Phi\left(\frac{x}{\sigma}\right), \Phi\left(\frac{x}{\tilde{\sigma}_k}\right)\right).\end{aligned}$$

By (4.50)–(4.52), we have

$$\rho_\infty\left(\tilde{H}_n^k(x), H_n(x)\right) \to 0 \text{ as } n \to \infty \text{ and } k \to \infty.$$

Given a set of k-RPs,

$$\rho_\infty\left(\tilde{H}_n^k(x), H_n(x)\right) \leq \rho_\infty\left(\Phi\left(\frac{x}{\sigma}\right) - \Phi\left(\frac{x}{\tilde{\sigma}_k}\right)\right)$$

holds when $n \to \infty$.

4.8.6 Proof of Theorem 4.6

The proof is from Xu et al. (2024) and needs the following lemma.

Lemma 4.4 *(The Berry–Esséen's inequality) If $Z_1, \ldots, Z_n$ are independent random variables with $E(Z) = 0$, $var(Z) = \sigma^2 > 0$, and $E(|Z|^3) = \alpha < \infty$, then*

$$\sup_{-\infty<z<\infty} \left| \text{pr}\left\{ \frac{n^{1/2}\bar{Z}}{\sigma} \leq t \right\} - \Phi(t) \right| \leq \frac{C\alpha}{\sigma^3 n^{1/2}},$$

where $\Phi(t)$ is cdf of $N(0,1)$ and C is a positive constant independent of n and the distributions of Z.

Proof *Applying triangle inequality and Berry–Esséen's inequality, we have*

$$\begin{aligned} &\sup_{-\infty<t<\infty} \left| \text{pr}\left(\frac{n^{1/2}(\bar{X}-\mu)}{\sigma} \leq t \right) - \text{pr}\left(\frac{n^{1/2}(\bar{Y}-\tilde{\mu}_k)}{\tilde{\sigma}_k} \leq t \right) \right| \\ \leq &\sup_{-\infty<t<\infty} \left| \text{pr}\left(\frac{n^{1/2}(\bar{X}-\mu)}{\sigma} \leq t \right) - \Phi(t) \right| + \sup_{-\infty<t<\infty} \left| \text{pr}\left(\frac{n^{1/2}(\bar{Y}-\tilde{\mu}_k)}{\tilde{\sigma}_k^2} \leq t \right) - \Phi(t) \right| \\ \leq &\frac{C\alpha}{\sigma^3 n^{1/2}} + \frac{C\tilde{\alpha}_k}{\tilde{\sigma}_k^3 n^{1/2}}, \end{aligned}$$

where $\alpha = E(|X-\mu|^3)$ and $\tilde{\alpha}_k = E(|Y_k - \tilde{\mu}_k|^3)$. Then,

$$\left(\frac{\sigma^3 n^{1/2}}{\alpha} \right) \sup_t \left| P\left(\frac{n^{1/2}(\bar{X}-\mu)}{\sigma} \leq t \right) - P\left(\frac{n^{1/2}(\bar{Y}-\tilde{\mu}_k)}{\tilde{\sigma}_k} \leq t \right) \right| \leq C\left(1 + \frac{\sigma^3}{\tilde{\sigma}_k^3} \frac{\tilde{\alpha}_k}{\alpha} \right).$$

According to Theorem 4.2, when $E(|X|^{3+\epsilon}) < \infty$ with $\epsilon > 0$, $\lim_{k\to\infty} \tilde{\alpha}_k = \alpha$. Hence,

$$\limsup_{k\to\infty} \left(\frac{\sigma^3}{\tilde{\sigma}_k^3} \frac{\tilde{\alpha}_k}{\alpha} \right) \leq \left(\limsup_{k\to\infty} \frac{\sigma^3}{\tilde{\sigma}_k^3} \right) \left(\limsup_{k\to\infty} \frac{\tilde{\alpha}_k}{\alpha} \right) = 1,$$

which implies the upper bound of $(2C\alpha)/(\sigma^3 n^{1/2})$ between the sampling distribution of $n^{1/2}(\bar{X}-\mu)/\sigma$ and the sampling distribution of $n^{1/2}(\bar{Y}-\tilde{\mu}_k)/\tilde{\sigma}_k$ under Kolmogorov metric, that is,

$$\lim_{k\to\infty} \sup_t \left| P\left(\frac{n^{1/2}(\bar{X}-\mu)}{\sigma} \leq t \right) - P\left(\frac{n^{1/2}(\bar{Y}-\tilde{\mu}_k)}{\tilde{\sigma}_k} \leq t \right) \right| \leq \left(\frac{2C\alpha}{\sigma^3 n^{1/2}} \right). \tag{4.53}$$

When $n \to \infty$,

$$\lim_{k\to\infty} \sup_t \left| P\left(\frac{n^{1/2}(\bar{X}-\mu)}{\sigma} \leq t \right) - P\left(\frac{n^{1/2}(\bar{Y}-\tilde{\mu}_k)}{\tilde{\sigma}_k} \leq t \right) \right| \leq 0. \tag{4.54}$$

Therefore, we establish the consistency of the resampling approximation based on the representative points to the sampling distribution of standardized sample mean, that is,

$$\sup_t \left| P\left(\frac{n^{1/2}(\bar{X}-\mu)}{\sigma} \leq t \right) - P\left(\frac{n^{1/2}(\bar{Y}-\tilde{\mu})}{\tilde{\sigma}_k} \leq t \right) \right| \to 0, \tag{4.55}$$

when $k, n \to \infty$.

4.8.7 Proof of Theorem 4.7

The proof is from Xu et al. (2024)

Proof *Applying triangle inequality and Berry–Esséen's inequality, we have*

$$\begin{aligned}
&\sup_{-\infty<t<\infty}\left|\mathrm{pr}\left(\frac{n^{1/2}(\bar{X}-\mu)}{\sigma}\le t\right)-\mathrm{pr}\left(\frac{n^{1/2}(\bar{Y}-\tilde{\mu}_k)}{\tilde{\sigma}_k}\le t\right)\right|\\
&\le \sup_{-\infty<t<\infty}\left|\mathrm{pr}\left(\frac{n^{1/2}(\bar{X}-\mu)}{\sigma}\le t\right)-\Phi(t)\right|+\sup_{-\infty<t<\infty}\left|\mathrm{pr}\left(\frac{n^{1/2}(\bar{Y}-\tilde{\mu}_k)}{\tilde{\sigma}_k^2}\le t\right)-\Phi(t)\right|\\
&\le \frac{C\alpha}{\sigma^3 n^{1/2}}+\frac{C\tilde{\alpha}_k}{\tilde{\sigma}_k^3 n^{1/2}},
\end{aligned}$$

where $\alpha = E(|X-\mu|^3)$ *and* $\tilde{\alpha}_k = E(|Y_k-\tilde{\mu}_k|^3)$*. Then,*

$$\left(\frac{\sigma^3 n^{1/2}}{\alpha}\right)\sup_t\left|\mathrm{pr}\left(\frac{n^{1/2}(\bar{X}-\mu)}{\sigma}\le t\right)-\mathrm{pr}\left(\frac{n^{1/2}(\bar{Y}-\tilde{\mu}_k)}{\tilde{\sigma}_k}\le t\right)\right|\le C\left(1+\frac{\sigma^3}{\tilde{\sigma}_k^3}\frac{\tilde{\alpha}_k}{\alpha}\right).$$

According to Theorem 2, when $E(|X|^{3+\epsilon})<\infty$ *with* $\epsilon>0$, $\lim_{k\to\infty}\tilde{\alpha}_k=\alpha$*. Hence,*

$$\limsup_{k\to\infty}\left(\frac{\sigma^3}{\tilde{\sigma}_k^3}\frac{\tilde{\alpha}_k}{\alpha}\right)\le\left(\limsup_{k\to\infty}\frac{\sigma^3}{\tilde{\sigma}_k^3}\right)\left(\limsup_{k\to\infty}\frac{\tilde{\alpha}_k}{\alpha}\right)=1,$$

which implies the upper bound of $(2C\alpha)/(\sigma^3 n^{1/2})$ *between the sampling distribution of* $n^{1/2}(\bar{X}-\mu)/\sigma$ *and the sampling distribution of* $n^{1/2}(\bar{Y}-\tilde{\mu}_k)/\tilde{\sigma}_k$ *under Kolmogorov metric, that is,*

$$\lim_{k\to\infty}\sup_t\left|P\left(\frac{n^{1/2}(\bar{X}-\mu)}{\sigma}\le t\right)-P\left(\frac{n^{1/2}(\bar{Y}-\tilde{\mu}_k)}{\tilde{\sigma}_k}\le t\right)\right|\le\left(\frac{2C\alpha}{\sigma^3 n^{1/2}}\right). \tag{4.56}$$

When $n\to\infty$,

$$\lim_{k\to\infty}\sup_t\left|P\left(\frac{n^{1/2}(\bar{X}-\mu)}{\sigma}\le t\right)-P\left(\frac{n^{1/2}(\bar{Y}-\tilde{\mu}_k)}{\tilde{\sigma}_k}\le t\right)\right|\le 0. \tag{4.57}$$

Therefore, we establish the consistency of the resampling approximation based on the representative points to the sampling distribution of standardized sample mean, that is,

$$\sup_t\left|P\left(\frac{n^{1/2}(\bar{X}-\mu)}{\sigma}\le t\right)-P\left(\frac{n^{1/2}(\bar{Y}-\tilde{\mu})}{\tilde{\sigma}_k}\le t\right)\right|\to 0, \tag{4.58}$$

when $k,n\to\infty$.

5

Estimation of MSE-RPs and Resampling

The previous chapters have studied several kinds of RPs of continuous univariate distributions. The statistical simulation is based on the approximation distributions of $F(x, \boldsymbol{\theta})$ constructed by a given set of RPs. This chapter concerns statistical simulation under the underlying distribution with some unknown parameters. Assume that the type of $F(x, \boldsymbol{\theta})$ is known but with some unknown parameters $\boldsymbol{\theta}$. By a random sample one can find the estimates of the unknown parameters, denoted by $\hat{\boldsymbol{\theta}}$, then we can generate MSE-RPs of $F(x, \hat{\boldsymbol{\theta}})$ and construction approximation distributions for $F(x, \boldsymbol{\theta})$. Statistical estimation via resampling from these approximation distributions is considered. A general theory for the consistency of statistical estimation is introduced. In this chapter, we pay more attention to estimating MSE-RPs of univariate distributions. In particular, the estimation of MSE-RPs for the location-scale distributions with unknown parameters is discussed. The moment estimation and maximum likelihood estimation and their bias correction are studied.

5.1 Estimation of MSE-RPs of Distributions

In the previous chapters, we study statistical simulation based on discrete approximation distributions of $F(x, \boldsymbol{\theta})$ constructed by RPs, where $F(x, \boldsymbol{\theta})$ is completely known. If the parameter $\boldsymbol{\theta}$ is unknown, can we still find RPs of the population distribution?

Let a random vector $\boldsymbol{X} \in \mathbb{R}^d$ with the domain $\mathcal{D}(\boldsymbol{X})$ and $\boldsymbol{X} \sim F(\boldsymbol{x}; \boldsymbol{\theta})$ with unknown parameters $\boldsymbol{\theta} \in \Omega$. We want to estimate the RPs including the QMC-RPs and the MSE-RPs denoted by $\{\boldsymbol{\xi}_1, \ldots, \boldsymbol{\xi}_k\}$ based on a set of random sample $\{\boldsymbol{x}_1, \ldots, \boldsymbol{x}_n\}$ from $F(\boldsymbol{x}; \boldsymbol{\theta})$. There are two kinds of approaches:

(a) Nonparametric estimator. Section 1.7.4 gives a way to estimate QMC-RPs by a nonparametric approach. To find MSE-RPs estimates $\{\hat{\boldsymbol{\xi}}_1, \ldots, \hat{\boldsymbol{\xi}}_k\}$ satisfying

$$\{\hat{\boldsymbol{\xi}}_1, \ldots, \hat{\boldsymbol{\xi}}_k\} = \arg \min_{\boldsymbol{y}_1 \cdots, \boldsymbol{y}_k \in \mathcal{D}(\boldsymbol{X})} \sum_{i=1}^{n} \min_{1 \le j \le k} (\boldsymbol{x}_i - \boldsymbol{y}_j)^T (\boldsymbol{x}_i - \boldsymbol{y}_j)$$

is an optimization problem. It is computationally infeasible when n and k are large, especially in high dimensions. The k-means algorithm (Section 3.6) is a fast and popular procedure for finding nonparametric estimators of MSE-RPs. However, the k-means algorithm can often give a local minimum, not a global minimum.

(b) Parametric estimator. First, one needs to find an estimate of $\boldsymbol{\theta}$, denoted by $\hat{\boldsymbol{\theta}}$, then generate the MSE-RPs of $F(\boldsymbol{x}; \hat{\boldsymbol{\theta}})$. An approximation distribution $\tilde{F}(\boldsymbol{x}; \hat{\boldsymbol{\theta}})$ constructed by estimated MSE-RPs is also an approximation distribution to $F(\boldsymbol{x}; \boldsymbol{\theta})$ if the estimate of

DOI: 10.1201/9781003589389-5

$\boldsymbol{\theta}$ is good enough. A natural estimation way is to employ the moment estimation (ME, Section 1.4) or the maximum likelihood estimation (MLE, Section 1.5). However, these two methods are not always good enough and need bias correction. Li and Fang (2022) proposed some methods for improvement of ME and MLE estimates under the mixture of two normal distributions. Their approach can apply to other distributions. In the literature, the MLE of $\boldsymbol{\theta}$ is often used since MLE has the invariance property, that is, $h(\hat{\boldsymbol{\theta}})$ is also the MLE of $h(\boldsymbol{\theta})$, where $\hat{\boldsymbol{\theta}}$ is the MLE of $\boldsymbol{\theta}$ and h is measurable/continuous. As a consequence, the MSE-RPs of $F(\boldsymbol{x};\hat{\boldsymbol{\theta}})$ is a MLE of MSE-RPs of $F(\boldsymbol{x};\boldsymbol{\theta})$, if $\hat{\boldsymbol{\theta}}$ is the MLE of $\boldsymbol{\theta}$.

If the underlying distribution F of $\boldsymbol{X} \in \mathbb{R}^d$ belongs to a location-scale family (1.6), i.e., its pdf is of the form

$$p(\boldsymbol{x};\boldsymbol{\mu},\sigma) = \frac{1}{\sigma^d} g\left(\frac{\boldsymbol{x}-\boldsymbol{\mu}}{\sigma}\right), \tag{5.1}$$

where $\boldsymbol{\mu}$ is a location parameter and $\sigma > 0$ is a scale parameter. When $d = 1$, let $Z = (X-\mu)/\sigma$, where X follows a location-scale family with pdf (5.1) and $J(z)$ be the cdf of Z. Usually, J is the standard distribution of the parametric family, e.g. the standard normal distribution $N(0,1)$ and the exponential distribution $EP(1)$. Suppose that one obtains a set of k MSE-RPs of Z with any k, denoted by $\boldsymbol{\xi} = \{\xi_1^{(k)},\dots,\xi_k^{(k)}\}$, then X has MSE-RPs,

$$\mu + \sigma\xi_j^{(k)}, j = 1,\dots,k.$$

Similarly, for the multivariate location-scale family, we have: let $\boldsymbol{\xi}_1^{(k)},\dots,\boldsymbol{\xi}_k^{(k)}$ be k MSE-RPs of $\boldsymbol{Z}$, then $\boldsymbol{X}$ has MSE-RPs

$$\boldsymbol{\mu} + \sigma\boldsymbol{\xi}_j^{(k)}, j = 1,\dots,k. \tag{5.2}$$

Suppose that the values of $\boldsymbol{\mu}$ and σ are unknown and $\hat{\boldsymbol{\mu}}$ and $\hat{\sigma}$ are the MLEs of $\boldsymbol{\mu}$ and σ, respectively. Therefore, the MLEs of the MSE-RPs of $\boldsymbol{X}$ are given by

$$\hat{\boldsymbol{\xi}_j} = \hat{\boldsymbol{\mu}} + \hat{\sigma}\boldsymbol{\xi}_j^{(k)}, j = 1,\dots,k. \tag{5.3}$$

5.2 Estimation of Two MSE-RPs for Univariate Distributions

Let random variable $X \sim F(x,\boldsymbol{\theta})$ with unknown $\boldsymbol{\theta}$ and $\{x_1,\dots,x_n\}$ be a random sample from F. For the case of $k = 1$, the estimator of the MSE-RP is just the sample mean $\bar{x}$. Tarpey (1997) consider the estimation of MSE-RPs with $k = 2$ and propose the following four methods (A) - (D) with some exciting examples.

(A) Maximum likelihood estimator

Let $\hat{\boldsymbol{\theta}}$ be the MLE of $\boldsymbol{\theta}$. We can find two MSE-RPs of $F(x,\hat{\boldsymbol{\theta}})$ that are the MLEs $\{\xi_1,\xi_2\}$ of $F(x,\boldsymbol{\theta})$ with $\xi_1 < \xi_2$. Define the mean square error of these estimators by

$$\text{MSE}(\hat{\boldsymbol{\theta}}) = E[(y_1-\xi_1)^2 + (y_1-\xi_2)^2]. \tag{5.4}$$

Let us see some examples which are from Tarpey (1997).

Example 5.1 *Consider estimation for two MSE-RPs $\xi_1 < \xi_2$ of $N(\mu,\sigma^2)$. The MLEs of μ and σ are the sample mean $\bar{x}$ and $\frac{n-1}{n}s^2$, respectively, where s^2 is the sample variance.*

Fang and He (1982) pointed out that the two MSE-RPs of $N(\mu, \sigma^2)$ *are* $\mu \pm \sigma\sqrt{2/\pi}$ *(also see Section 3.7). When* $\mu = 0, \sigma = 1$*, the MSE in (5.4) is*

$$\begin{aligned}\text{MSE}(\hat{\boldsymbol{\theta}}) &= E\left[\bar{x} - s\sqrt{\frac{2}{\pi}}\right]^2 + E\left[\bar{x} + s\sqrt{\frac{2}{\pi}}\right]^2 \\ &= \frac{2}{n} + \frac{8}{\pi}\left[1 - \frac{1}{2n} - \sqrt{\frac{2}{\pi}}\frac{\Gamma(n/2)}{\Gamma((n-1)/2)}\right],\end{aligned}$$

where we use the fact that ns^2 *follows the gamma distribution* $Ga(\frac{n-1}{2}, \frac{1}{2})$*.*

Example 5.2 *Let* $X \sim U(a, b)$*. Its two MSE-RPs are* $\xi_1 = a + (b-a)/4$ *and* $\xi_2 = b - (b-a)/4$*, respectively (see Example 3.10). The MLEs of* a *and* b *are the first and last order statistics* $x_{(1)}$ *and* $x_{(n)}$*, respectively. The MLEs of* ξ_1 *and* ξ_2 *are given by*

$$\hat{\xi}_1 = x_{(1)} + \frac{x_{(n)} - x_{(1)}}{4}, \ \hat{\xi}_2 = x_{(n)} - \frac{x_{(n)} - x_{(1)}}{4},$$

respectively. When $a = -1, b = 1$*, the corresponding* $\xi_1 = -\frac{1}{2}$ *and* $\xi_2 = \frac{1}{2}$*. The MSE of* $\hat{\xi}_1$ *and* $\hat{\xi}_2$ *is*

$$\begin{aligned}\text{MSE}(\hat{\boldsymbol{\theta}}) =& E\left[x_{(1)} + \frac{x_{(n)} - x_{(1)}}{4} + \frac{1}{2}\right]^2 + E\left[x_{(n)} - \frac{x_{(n)} - x_{(1)}}{4} - \frac{1}{2}\right]^2 \\ =& \frac{7}{(n+1)(n+2)},\end{aligned}$$

where $\boldsymbol{\theta} = (a, b)^T$*.*

Example 5.3 *The Laplace or double-exponential distribution has a density (3.23)* $p(x) = \frac{1}{2b}\exp(-\frac{|x-\mu|}{b}), x \in \mathbb{R}$*. Its two MSE-RPs are* $\mu \pm b$*, and the MLEs of two RPs are*

$$\tilde{x} \pm \frac{1}{n}\sum_{i=1}^{n}|x_i - \tilde{x}|,$$

where $\tilde{x}$ *is the sample median. For* $\mu = 0, b = 1$*, the MSE of the MLEs are*

$$\text{MSE}(\hat{\boldsymbol{\theta}}) = E[\tilde{x} - \frac{1}{n}\sum_{i=1}^{n}|x_i - \tilde{x}| + 1]^2 + E[\tilde{x} + \frac{1}{n}\sum_{i=1}^{n}|x_i - \tilde{x}| - 1]^2.$$

Example 5.4 *The Logistic distribution has the density*

$$p(x) = \frac{\exp[-(x-\mu)/\lambda]}{\lambda[1 + \exp(-(x-\mu)/\lambda)]^2}.$$

Two RPs are $\mu \pm 2\lambda\ln(2)$*. The MLEs of* (μ, λ) *do not exist in a close form.*

Example 5.5 *When the distribution* F *is symmetric about the mean* μ*, its two MSE-RPs are* $\mu \pm \xi$*, where* $\xi = E|X - \mu|$ *(see Theorm 3.3) and* ξ *can be estimated by*

$$\hat{\xi} = \frac{1}{n}\sum_{i=1}^{n}|x_i - \bar{x}|,$$

then semi-parametric estimators of ξ_1 *and* ξ_2 *are*

$$\bar{x} \pm \frac{1}{n}\sum_{i=1}^{n}|x_i - \bar{x}|.$$

(B) Quantile estimators

The quantiles of X can be estimated by their sample quantiles. When $k = 2$ and the pdf is symmetric with mean 0, the two MSE-RPs are $\pm\xi = \pm E|X|$. Let $\alpha = P(X \leq -E|X|)$. Two MSE-RPs can be estimated by α and $1-\alpha$ sample quantiles. More discussion is given by Tarpey (1997).

(C) Nonparametric estimators

Let $\tilde{\xi}_i, i = 1, 2$ be the k-means estimators of the two MSE-RPs $\pm\xi$ of a symmetric random variable X. Assume that X has the first three moments with $p(0) > 0$ and denote the variance by σ^2. Hartigan (1975) pointed out

$$n^{1/2}\begin{pmatrix} \tilde{\xi}_1 + \xi \\ \tilde{\xi}_2 - \xi \end{pmatrix} \to N_2(\mathbf{0}, \boldsymbol{U}) \text{ in distribution,}$$

where

$$\boldsymbol{U} = \frac{2(\sigma^2 - \xi^2)}{[1 - 2\xi p(0)]^2}\begin{bmatrix} [1-\xi p(0)]^2 + \xi^2 p^2(0) & 2\xi p(0)[1-\xi p(0)] \\ 2\xi p(0)[1-\xi p(0)] & [1-\xi p(0)]^2 + \xi^2 p^2(0) \end{bmatrix}.$$

The MSE of $\{\tilde{\xi}_i, i = 1, 2\}$ is approximately to

$$\frac{4(\sigma^2 - \xi^2)[1 - 2\xi p(0) + 2\xi^2 p^2(0)]}{n[1 - 2\xi p(0)]^2}.$$

Tarpey (1997) compared these four methods and pointed out that the k-means method is the worst and MLE is the best. He also applied these estimators to the head measurement example in Flury and Riedwyl (1988). Stampfer and Stadlober (2002) gave some discussion on the estimation of self-consistent and principal points for a given sample $x_1, \ldots, x_n$ from F.

5.3 RPs of General Location-Scale Families

This section considers more general cases than the location-scale family. Let $Z \sim J(z)$, where $J(z)$ is a continuous distribution with finite variance. Denote a set of k MSE-RPs of J as $\boldsymbol{\xi} = \left\{\xi_1^{(k)}, \ldots, \xi_k^{(k)}\right\}$.

Assume that a random variable $X \sim F$ has a linear relationship with Z, i.e.

$$X \stackrel{d}{=} mZ + b, \quad m, b \in \mathbb{R} \text{ and } m \neq 0. \tag{5.5}$$

Three special families satisfy the relationship defined in (5.5). They are the location-scale family ($m > 0$), the location family ($m = 1$) and the scales family ($m > 0$, $b = 0$). We call all distributions satisfying (5.5) as *general location-scale family* in this monograph. It is known that X and Z exist in the same order of moments.

The MSE-RPs of a standardized distribution J can be extended to general cases by linear transformations (Zoppè, 1995). A set of MSE-RPs of $X \sim F$ is

$$\mathcal{A} = \left\{a_i^{(k)} = m\xi_i^{(k)} + b, i = 1, \ldots, k\right\}, \tag{5.6}$$

where $\text{MSE}_X(\mathcal{A}) = m^2\,\text{MSE}_Z(\boldsymbol{\xi})$. It is also easy to prove that the corresponding probability of $a_i^{(k)}$ is equal to the same probability of $\xi_i^{(k)}$ for $i = 1, \ldots k$. Therefore, for a distribution F in the general location-scale family, we only need to estimate the MSE-RP values without their corresponding probabilities. Given a random sample of size n from F, the unknown constants m and b in (5.5) can be estimated using the maximum likelihood method, denoted as $\hat{m}_n$ and $\hat{b}_n$, respectively. The maximum likelihood estimators of the k MSE-RPs of F are defined as

$$\hat{a}_i^{(k)} = \hat{m}_n \xi_i^{(k)} + \hat{b}_n, \quad i = 1, \ldots, k. \tag{5.7}$$

Denote a discrete random variable $W_k^{mse} \sim \hat{F}_{mse,k}$ in $\left\{\hat{a}_1^{(k)}, \ldots, \hat{a}_k^{(k)}\right\}$ with probability function $P(W_k^{mse} = \hat{a}_i^{(k)}) = P(Z \in V_i^{(k)})$, for $i = 1, \ldots, k$, where $V_i^{(k)} = V(\boldsymbol{\xi}, V_i^{(k)})$ is the Voronoi region associated with $\xi_i^{(k)}$ (Definition 2.4). As the number of random samples n increases to infinity, the maximum likelihood estimators $\hat{m}_n$ and $\hat{b}_n$ converge to m and b almost surely, respectively.

The new simulation-based method estimates the sampling distribution of a statistic by random sampling of $\hat{F}_{mse,k}$. In the remainder of the discussion, we follow Xu et al. (2024) to set $k = n$ and cite their theoretical approach. Denote $P(\cdot)$ and $P^*(\cdot)$ as probabilities under F and $\hat{F}_{mse,n}$, respectively. Let a random variable $X \sim F$ with the variance σ^2 and $\mu_4 = E(X-\mu)^4$, and denote $X_1, \ldots, X_n$ as the i.i.d. samples from F with the sample mean $\bar{X}$ and the sample variance s^2. Let $W_n^{mse} \sim \hat{F}_{mse,n}$ with the variance σ^{*2} and the fourth central moment μ_4^*. Denote $W_1, \ldots, W_n$ as the i.i.d. samples of $\hat{F}_{mse,n}$ with the mean of the sample $\bar{W}$ and the variance of the sample s^{*2}. Theorem 5.1 establishes the consistency of the distribution of the sample mean under $\hat{F}_{mse,n}$.

Theorem 5.1 *Suppose $E(|X|^{3+\epsilon}) < \infty$ with some $\epsilon > 0$. The distribution of the sample mean under $\hat{F}_{mse,n}$ converges to the distribution of the sample mean under F almost surely, i.e.,*

$$\rho_\infty \left\{P\left(n^{1/2}(\bar{X} - E(X)) \leq t\right), P^*\left(n^{1/2}(\bar{W} - E(W_n^{mse})) \leq t\right)\right\} \to 0 \quad a.s.$$

where $\rho_\infty(\cdot,\cdot)$ is defined in Section 4.5.

Many statistics can be represented by the U-statistics, including the sample mean and variance. The details of U-statistics can refer to Section 1.10 of this book and Section 5 of Serfling (1980). To justify the convergence for the sample variance under $\hat{F}_{mse,n}$, we utilize the asymptotic normality property of the U-statistic. Let $X, X_1, \ldots, X_n$ be iid distributed F. The variance $\text{Var}(X) = E_F\{h(X_1, X_2)\}$, where $h(x_1, x_2) = (x_1 - x_2)^2/2$. The corresponding U-statistic is the sample variance of $X_1, \ldots, X_n$. Based on the asymptotic normality of the sample variance, we demonstrate the consistency of the distribution of the sample variance under $\hat{F}_{mse,n}$ in Theorem 5.2.

Theorem 5.2 *If $E(|X|^{6+\epsilon}) < \infty$ with some $\epsilon > 0$, then the distribution of the sample variance under $\hat{F}_{mse,n}$ converges to the distribution of the sample variance under F almost surely, in the following sense,*

$$\rho_\infty \left\{P\left(n^{1/2}(S^2 - \sigma^2) \leq t\right), P^*\left(n^{1/2}(\hat{S}^{*2} - \sigma^{*2}) \leq t\right)\right\} \to 0 \quad a.s. \tag{5.8}$$

Theorem 5.1 proves the consistency of the distribution of the standardized sample mean under $\hat{F}_{mse,n}$ while Theorem 5.2 indicates the consistency of the distribution of the sample variance under $\hat{F}_{mse,n}$. With these two guarantees, given a set of n random samples, the sampling approximation under $\hat{F}_{mse,n}$ is a valid alternative to the traditional bootstrap method in statistical inference, especially MSE-RPs have many good properties as a special mapping of the underlying population F. The proofs of these two theorems can refer to Xu et al. (2024).

5.4 Estimation of Confidence Interval

In this section, we construct confidence intervals based on the proposed resampling method in this chapter and compare them with the bootstrap interval. Confidence intervals are usually constructed by considering a pivotal quantity $T_n = T(X_1, \ldots, X_n; F)$ whose distribution is known. Suppose θ is a parameter of interest. Consider a studentized pivot $T_n = (\hat{\theta}_n - \theta)/\hat{\sigma}_n$, where $\hat{\theta}_n$ is an estimator of θ and $\hat{\sigma}_n$ is an estimator of the standard deviation of $\hat{\theta}_n$. Denote the distribution of T_n as H_n. If the inverse of H_n is unique, an exact $100(1-\alpha)\%$ confidence interval for θ is

$$\left[\hat{\theta}_n - \hat{\sigma}_n H_n^{-1}(1-\alpha/2), \hat{\theta}_n - \hat{\sigma}_n H_n^{-1}(\alpha/2)\right], \tag{5.9}$$

where $H_n^{-1}(q)$ is the q-quantile of the distribution H_n. It is often not easy to find a T_n with a known distribution H_n. In the traditional asymptotic approach, we replace the unknown H_n by its limit H. There are two main disadvantages of the traditional asymptotic approach. It requires the limit H to be derived analytically and explicitly and lacks high-order precision. The bootstrap method can be applied to obtain an easy-to-use confidence interval with higher accuracy by replacing H_n in (5.9) with its bootstrap estimator H_{boot}. Xu et al. (2024) found that the proposed resampling method tends to be more accurate for interval estimation than the bootstrap method.

If the distribution H_n of T_n is unknown, the resampling approximation of H_n under $\hat{F}_{mse,n}$ is defined by

$$H_{re}(x) = P^*\left(\frac{\hat{\theta}^*_{re,n} - \hat{\theta}_{mse,n}}{\hat{\sigma}^*_{re,n}} \leq t\right), \tag{5.10}$$

where $\hat{\theta}^*_{re,n}$ and $\hat{\sigma}^*_{re,n}$ are analogs of $\hat{\theta}_n$ and $\hat{\sigma}_n$ based on i.i.d. samples from $\hat{F}_{mse,n}$, respectively, and $\hat{\theta}_{mse,n}$ is the parameter of $\hat{F}_{mse,n}$. The resulting $100(1-\alpha)\%$ confidence interval for θ is

$$[\underline{\theta}_{\text{re}}, \bar{\theta}_{\text{re}}] = \left[\hat{\theta}_n - \hat{\sigma}_n H_{re}^{-1}(1-\alpha/2),\ \hat{\theta}_n - \hat{\sigma}_n H_{re}^{-1}(\alpha/2)\right]. \tag{5.11}$$

Xu et al. (2024) considered the following three kinds of distributions in their numerical studies:

A. Normal distribution, $X = mZ + b$, where $m \neq 0$ and $Z \sim N(0,1)$.

B. Exponential distribution, $X = mZ$, where $m \neq 0$ and $Z \sim EP(1)$, with pdf $p(z) = e^{-z}$, $z \geq 0$.

C. Weibull distribution, $X = mZ$, where $m \neq 0$ and $Z \sim We(2,1)$ with pdf $p(z) = 2ze^{-z^2}$, $z \geq 0$.

The baseline of the comparison is the bootstrap-t interval; see Section 12.5 of Efron and Tibshirani (1993). Xu et al. (2024) examined the coverage probability and the expected interval length under three distributions A-C, respectively. Their simulation consider

1. $m = 1$ and $b = 0$ and taking $N = 10000$ random samples of size $n, n = 10, 15, 25$, and 50 from each assumed population.

TABLE 5.1
Coverage probabilities and expected length, $\theta = E(X)$, $n = 15$

Distribution	Method	$\alpha = 0.80$ CP	$\alpha = 0.80$ EL	$\alpha = 0.90$ CP	$\alpha = 0.90$ EL	$\alpha = 0.95$ CP	$\alpha = 0.95$ EL
Normal	H_{re}	**0.8057**	**0.6821**	**0.9038**	**0.8929**	0.9513	**1.0888**
	H_{boot}	0.8059	0.6894	0.9049	0.9103	**0.9505**	1.1196
Exponential	H_{re}	**0.8012**	**0.7422**	**0.8999**	**0.9973**	**0.9483**	**1.2511**
	H_{boot}	0.7943	0.7608	0.8921	1.0297	0.9435	1.3150
Weibull	H_{re}	**0.8021**	**0.3192**	**0.9043**	**0.4209**	**0.9487**	**0.5158**
	H_{boot}	0.8074	0.3218	0.9096	0.4276	0.9531	0.5280

2. The sampling distribution approximations are based on $B = 5000$ resamples of size $n = k$ from $\hat{F}_{mse,n}$ and the empirical distribution F_n, respectively.
3. The coverage probability is defined as the proportion of N time the $100(1-\alpha)\%$ confidence interval covers the true parameter θ.
4. The expected interval length is the average absolute difference between the upper lower bound of the $100(1-\alpha)\%$ confidence intervals. Taking $\alpha = 0.2, 0.1, 0.05$, respectively.

Consider the confidence interval estimation of the population mean $\theta = E(X)$, where X follows the distributions A, B, and C. As these distributions have finite fourth moments, H_{re} is convergent to H_n almost surely according to Theorem 5.1. The studentized pivot is $t_n = (\bar{X} - E(X))/\hat{\sigma}_n$, where $\hat{\sigma}_n = s/n^{1/2}$ and s is the sample standard error. To get H_{re} defined in (5.10), we then calculate the sample mean $\hat{\theta}^*_{re,n}(l)$ and the sample variance $\hat{\sigma}^*_{re,n}(l)$ of the lth resample from $\hat{F}_{mse,n}$, for $l = 1, \ldots, L$.

Tables 5.1–5.2 present the converge probabilities (CP) and the expected lengths (EL) of the proposed resampling intervals and the bootstrap-t intervals based on the sample sizes of 15 and 10, respectively. Bold highlights in the tables indicate which method performs better at each level of α. The proposed method provides a more reliable interval estimation of the population mean for the given exponential and Weibull distributions are shown in distribution B and distribution C. The resampling intervals tend to have smaller expected lengths for the normal distribution, and the coverage probabilities of the two types of intervals are close. Xu et al. (2024) also compared the predicted interval lengths with the exact interval lengths under distributions A and B. As a result, the simulation interval lengths are more significant than the exact lengths in all cases. Therefore, we consider shorter expected lengths to be more accurate in our simulation comparisons.

Xu et al. (2024) also consider the interval estimation of the population variance $\sigma^2 = \text{Var}(X)$. Define the used pivot for constructing the two types of intervals as $T_n = (s^2 - \sigma^2)/\text{se}(s^2)$, where s^2 is the sample variance and $\text{se}(s^2)$ is the standard error of s^2 estimated by the bootstrap method. According to Chapter 6 of Efron and Tibshirani (1993), we calculated $\text{se}(s^2)$ based on 1000 random samples of size n from the empirical distribution F_n formed by the initial random sample generated from the underlying distribution. For constructing H_{re}, the analog $\hat{\sigma}^*_{re,n}$ of the standard error of s^2 defined in (5.10) is calculated based on 500 bootstrap replicates for each resample from $\hat{F}_{mse,n}$. For constructing H_{boot}, the same replication size of 500 is used for calculating the analog of the standard error of s^2.

TABLE 5.2
Coverage probabilities and expected length, $\theta = E(X)$, $n = 10$

Distribution	Method	$1-\alpha=0.80$ CP	$1-\alpha=0.80$ EL	$1-\alpha=0.90$ CP	$1-\alpha=0.90$ EL	$1-\alpha=0.95$ CP	$1-\alpha=0.95$ EL
Normal	H_{re}	**0.7992**	**0.8487**	0.8964	**1.1288**	0.9442	**1.3926**
	H_{boot}	0.8025	0.8749	**0.8968**	1.1850	**0.9453**	1.4987
Exponential	H_{re}	**0.7973**	**0.9582**	**0.9004**	**1.3165**	**0.9484**	**1.6901**
	H_{boot}	0.7878	1.0235	0.8921	1.4039	0.9412	1.8844
Weibull	H_{re}	**0.7959**	**0.3992**	**0.8991**	**0.5333**	**0.9515**	**0.6666**
	H_{boot}	0.8071	0.4100	0.9059	0.5574	0.9549	0.7131

5.5 The Expected Mean Squared Distance

This section introduces the measure of expected mean squared distance for the estimation of MSE-RPs of a distribution on location-scale family.

5.5.1 A property of location-scale family

Let X follow a location-scale family $F(x;\mu,\sigma)$ with pdf

$$p(x;\mu,\sigma) = \frac{1}{\sigma} g\left(\frac{x-\mu}{\sigma}\right), \quad \Omega = \{(\mu,\sigma) : \mu \in \mathbb{R}, \sigma > 0\} \tag{5.12}$$

for some function g (refer to (5.1)). If μ_0 and σ_0 represent the true values of μ and σ for a given case, then the density of the standardized variate $Z = (X-\mu_0)/\sigma_0$ is given by $g(z)$ which is independent of both parameters μ and σ. For example, the normal distribution $N(\mu,\sigma)$ belongs to the location-scale family, where $\boldsymbol{\theta} = (\mu,\sigma^2)$, $Z \sim N(0,1)$, i.e., $\mu = 0, \sigma = 1$.

Antle and Bain (1969) found a property of the maximum likelihood estimates of μ and σ for the location-scale family in the following theorem. Its proof can be found in Subsection 5.8.1.

Theorem 5.3 *Suppose the distribution of X belongs to the location-scale family with the density* (5.12)*. Let $\hat{\mu}$ and $\hat{\sigma}$ be the MLEs of μ and σ, respectively. Then*

$$\hat{\sigma}_s \equiv \frac{\hat{\sigma}}{\sigma_0}, \quad \hat{\mu}_s \equiv \frac{\hat{\mu}-\mu_0}{\sigma_0}, \quad \hat{\mu}_p \equiv \frac{\hat{\mu}-\mu_0}{\hat{\sigma}}$$

are each distributed independently of μ and σ.

Example 5.6 *Let $X \sim N(\mu,\sigma)$ with unknown μ and σ. The MLEs of μ and σ are*

$$\bar{x} = \frac{1}{n}\sum_{i=1}^{n} x_i, \; s^2 = \frac{1}{n}\sum_{i=1}^{n}(x_i - \bar{x})^2, \tag{5.13}$$

respectively. It is easy to find MSE-RPs of $N(0,1)$ from Max (1960), Fang and He (1982) or Fang et al. (2014). By (5.3) we can find the estimators of k MSE-RPs for $N(\mu,\sigma)$ for $1 \le k \le 36$ and the related probabilities.

5.5.2 Expected mean squared distance (EMSD)

Estimators of k MSE-RPs of a given distribution $F(x;\mu,\sigma)$ depend on the random sample. From the statistical consideration, we need a measure to evaluate these estimators of k MSE-RPs. Matsuura et al. (2015) proposed a measure called as *expected mean squared distance (EMSD)* and extend their results to the multivariate distributions.

Definition 5.1 *Let a continuous random variable $X \sim F(x, \boldsymbol{\theta})$ have the finite second moments with unknown parameter $\boldsymbol{\theta}$ and $\xi_1, \ldots, \xi_k$ be its MSE-RPs of $F(x)$. Let $\hat{\xi}_1, \ldots, \hat{\xi}_k$ be MSE-RPs of $F(x, \hat{\boldsymbol{\theta}})$, where $\hat{\boldsymbol{\theta}}$ is an estimate of $\boldsymbol{\theta}$ based on random sample from $F(x)$ with size n. The expected mean squared distance (EMSD) is defined as*

$$EMSD = E[MSD(\hat{\xi}_1 \ldots, \hat{\xi}_k)] = E[\min \|\boldsymbol{x} - \hat{\xi}_j)\|^2], \tag{5.14}$$

where $MSD(\hat{\xi}_1 \ldots, \hat{\xi}_k)$ is MSE of $\{\hat{\xi}_1 \ldots, \hat{\xi}_k\}$ under the distribution $F(x, \hat{\boldsymbol{\theta}})$.

The mean square distance (MSD) is the mean square error (MSE). A detailed discussion and examples not only on univariate cases but also on multivariate cases are given by Matsuura et al. (2015). Note, minimum EMSD depends on the method for estimating $\boldsymbol{\theta}$, such as moment estimation, maximum likelihood estimation, or other methods. For the normal distribution, there are more detailed results.

Theorem 5.4 *Let $X \sim N(\mu, \sigma^2)$ with unknown μ and σ^2. Let $x_1, \ldots, x_n$ be a random sample of size n from $N(\mu, \sigma^2)$. Denote the MLEs of μ and σ^2 by $\hat{\mu}$ and $\hat{\sigma}^2$ in (5.13), respectively. Let $z_1, \ldots, z_k$ be MSE-RPs of $N(0,1)$. The MLE of MSE-RPs for $N(\mu, \sigma^2)$ are given by*

$$\hat{\xi}_j = \hat{\mu} + \hat{\sigma} z_j, j = 1, \ldots, k.$$

Consider a class of estimators of the simple form

$$\hat{\xi}_j = \hat{\mu} + c\hat{\sigma} z_j, j = 1, \ldots, k.$$

Then the value of c minimizing the EMSD satisfies

$$c = \frac{\sum_{j=1}^k \int_{c(z_{j-1}+z_j)/2}^{c(z_j+z_{j+1})/2} t z_j h_{n+1}(t)dt}{\sum_{j=1}^k \int_{c(z_{j-1}+z_j)/2}^{c(z_j+z_{j+1})/2} z_j^2 h_{n+1}(t)dt}, \tag{5.15}$$

where $z_0 = -\infty, z_{k+1} = \infty$ and $h_{n+1}(t)$ is the pdf of t-distribution with $n+1$ degrees of freedom.

The proof will be given in Subsection 5.8.3. Matsuura et al. (2015) proposed an iteration algorithm to find the optimal c as follows:

1. Set an initial value of $c^{(0)}$ and set an error bound ϵ.
2. Calculate

$$c^{(1)} = \frac{\sum_{j=1}^k \int_{c^{(0)}(z_{j-1}+z_j)/2}^{c^{(0)}(z_j+z_{j+1})/2} t z_j h_{n+1}(t)dt}{\sum_{j=1}^k \int_{c^{(0)}(z_{j-1}+z_j)/2}^{c^{(0)}(z_j+z_{j+1})/2} z_j^2 h_{n+1}(t)dt}.$$

3. If $|c^{(1)} - c^{(0)}| < \epsilon$ holds, then stop the iteration. Otherwise, go back to step 2 with $c^{(0)} := c^{(1)}$.

Matsuura et al. (2015) also proposed another estimator of $\hat{\xi}_j$ of the form below,

$$\hat{\xi}_j = \hat{\mu} + a_j\hat{\sigma}, j = 1, \ldots, k. \tag{5.16}$$

Theorem 5.5 *Let $X \sim N(\mu, \sigma^2)$ with unknown μ and σ^2. Let $x_1, \ldots, x_n$ be a random sample from $N(\mu, \sigma^2)$. Denote the MLEs of μ and σ^2 by $\hat{\mu}$ and $\hat{\sigma}^2$, respectively. Then, $a_1, \ldots, a_k$ in (5.16) minimizing the EMSD are given by k MSE-RPs of the t-distribution with $n+1$ degrees of freedom.*

The proof will be given in Subsection 5.8.4.

Theorem 5.6 *Let X be a random variable with a finite second moment, and the pdf g in (5.12) is known with unknown μ and σ. Let $x_1, \ldots, x_n$ be a random sample from the distribution of X and $\{X, x_1, \ldots, x_n\}$ are independent. Denote the MLEs of μ and σ^2 by $\hat{\mu}$ and $\hat{\sigma}^2$, respectively. Let $U = \frac{X-\hat{\mu}}{\sigma}$ and $W = \frac{\hat{\sigma}}{\sigma}$. Let $\psi(u|W)$ be the pdf of $U|W$ (i.e., the pdf of the conditional distribution of U given W). Also, let $\hat{\xi}_j = \hat{\mu} + a_j\hat{\sigma}, j = 1, \ldots, k$. Then, $a_1, \ldots, a_k$ minimizing the EMSD are given by k MSE-RPs of the distribution $h(z)$, where*

$$h(z) = \frac{E[W^3\psi(zW|W)]}{E(W^2)}. \tag{5.17}$$

The proof will be given in Subsection 5.8.2.

Example 5.7 *Consider the exponential distribution with pdf*

$$p(x; \mu, \sigma) = \begin{cases} 0, & x < \mu, \\ \frac{1}{\sigma}e^{-\frac{x-\mu}{\sigma}}, & x \geq \mu. \end{cases}$$

The MLEs of μ and σ are

$$\hat{\mu} = \min\{x_1, \ldots, x_n\}, \ \hat{\sigma} = \bar{x} - \hat{\mu},$$

respectively, and they are independent. The pdf of $h(z)$ is

$$h(z) = \begin{cases} (1-z)^{-(n+2)}, & z < 0, \\ (1+\frac{z}{n})^{-(n+2)}, & z \geq 0. \end{cases}$$

Matsuura et al. (2015) gave comparisons between the EMSDs of the maximum likelihood estimator and the minimum EMSD estimator of k MSE-RPs for $n = 2, 5$ and $k = 2, 5$.

5.6 Parameter Estimation and Its Improvement

Let $X \sim F(x; \boldsymbol{\theta})$, where $\boldsymbol{\theta}$ is the unknown parameters. In a parametric approach, it is recommended to use MSE-RPs of $F(x; \hat{\boldsymbol{\theta}})$ as estimators for MSE-RPs of $F(x; \boldsymbol{\theta})$, where $\hat{\boldsymbol{\theta}}$ is obtained through moment estimation (ME) or maximum likelihood estimation (MLE) using a random sample. Sections 1.4 and 1.5 provide a brief introduction to the ME and MLE methods. It is important to note that the accuracy of the estimator $\hat{\boldsymbol{\theta}}$ depends on the quality of the random sample and the computational algorithm employed.

This section considers MLEs of five parameters of the MixN distribution $X \sim \text{MixN}(\alpha, \mu_1, \sigma_1^2, \mu_2, \sigma_2^2)$ in Section 3.9 with the pdf

$$\begin{aligned} p(x) &= \alpha\varphi(x; \mu_1, \sigma_1^2) + (1-\alpha)\varphi(x; \mu_2, \sigma_2^2) \\ &= \alpha\frac{1}{\sqrt{2\pi}\sigma_1}\exp\left\{-\frac{(x-\mu_1)^2}{2\sigma_1^2}\right\} + (1-\alpha)\frac{1}{\sqrt{2\pi}\sigma_2}\exp\left\{-\frac{(x-\mu_2)^2}{2\sigma_2^2}\right\}. \end{aligned}$$

where the parameters $\boldsymbol{\theta} = (\theta_1, \ldots, \theta_s)$ with the domain $\Omega = \{\alpha \in (0,1), \mu_1, \mu_2 \in \mathbb{R}, \sigma_1, \sigma_2 > 0\}$.

5.6.1 Challenges problems in estimation under MixN

We meet several challenges and problems.

1. **Optimization algorithm:** The maximum likelihood estimation is to find $\hat{\boldsymbol{\theta}}$ that maximizes the log-likelihood function over the parameter space Ω, that is

$$\hat{\boldsymbol{\theta}}_{MLE} = \arg\max_{\boldsymbol{\theta}\in\Omega} \ell(\boldsymbol{\theta}|\boldsymbol{x}) = \arg\max_{\boldsymbol{\theta}\in\Omega} \sum_{i=1}^{n} \ln p(x_i, \boldsymbol{\theta}),$$

   where $\boldsymbol{x} = (x_1, x_2, \cdots, x_n)$ be a set of i.i.d. samples from a distribution $F(x, \boldsymbol{\theta})$ with pdf $p(x, \boldsymbol{\theta})$. It is known that there are so many local maxima of the log-likelihood function in Ω and the Newton-Raphson method and its extensions do not work well (Fang and Wang (1994)). It is crucial to note that the estimation results should be interpreted carefully due to the possibility of estimated local maxima deviating from the actual model parameters. This is attributed to countless critical points in the parameter space Ω. Therefore, we need effective algorithms, like the SNTO algorithm (Section 1.13).

   Furthermore, the log-likelihood function under MixN is unbounded in $\Omega = \{\alpha \in (0,1), \mu_1, \mu_2 \in \mathbb{R}, \sigma_1, \sigma_2 > 0\}$, and no global maxima exists to the MLE problem. Moreover, finite mixtures are hardly to have identifiability without conditions.

2. **Data randomness:** All the statistical inferences meet the sample randomness problem. We need some method to enhance the representativeness of observed data. In this section, we apply Li-Fang's approach (Li and Fang (2022)) to establish the so-called QMC-data by using the distribution-free Harrell-Davis quantile estimators (see Subsection 1.7.4).

3. **Identifiability:** The loss of identifiability poses challenges in establishing the asymptotic consistency of maximum likelihood estimation (MLE) under the MixN distribution. This issue is less critical regarding density fitting, as equivalent solutions are considered equally good. However, the loss of identifiability becomes significant when interpreting the estimated model parameter values in practical applications. Incorporating certain assumptions before parameter estimation can be beneficial to mitigate this issue. For instance, assumptions such as $\mu_1 < \mu_2$ or $\alpha < 0.5$ can help avoid estimations that lean toward the opposite set of parameters.

Besides the use of SNTO and QMC-data, Li and Fang (2022) employed the penalized MLE, EM algorithm, and bootstrap bias-corrected methods for more accurate and reliable parameter estimation. The following study follows their approach. For a more detailed discussion, you can refer to their article.

5.6.2 Penalized MLE under MixN (PMLE)

Regularizing the likelihood function is a natural way to obtain consistency. Adding a penalty to the likelihood function is one form of regularization. For example, for the MixN distribution, the penalized log-likelihood under MixN can be defined as follows:

$$\tilde{\ell}_n(\boldsymbol{\theta}) = \ell_n(\boldsymbol{\theta}) + p_n(\boldsymbol{\theta}), \tag{5.18}$$

where $\ell_n(\boldsymbol{\theta})$ is the log-likelihood function and $p_n(\boldsymbol{\theta}) = \tilde{p}_n\left(\sigma_1^2\right) + \tilde{p}_n\left(\sigma_2^2\right)$.

By giving conditions on $\tilde{p}_n\left(\sigma^2\right)$, the penalized maximum-likelihood estimator is proved to be strongly consistent by Ciuperca et al. (2003) and Chen (2017). Ciuperca et al. (2003) also proved that the penalized MLE under finite mixture normal distributions is asymptotic efficient. The required conditions for penalty functions can be found in the above two papers.

Suggested by Chen (2017), a prior gamma distribution placed on σ^{-2} with the density function

$$f(\sigma^2) = \frac{\beta^\alpha}{\Gamma(\alpha)}\left(\sigma^{-2}\right)^{\alpha-1}\exp\left\{-\beta\sigma^{-2}\right\}, \quad \alpha > 0, \beta > 0,$$

could be used as a penalty function as follows

$$\tilde{p}_n(\sigma^2) = n^{-1}\log f(\sigma^2|\alpha = 2, \beta = 1) = -n^{-1}\left\{\sigma^{-2} + \log\sigma^2\right\}. \tag{5.19}$$

To retain scale in variance, σ should be replaced by σ/s_n in application, where s_n is the sample variance. Ridolfi and Idier (2001) applied the inverted gamma distribution on variance of normal components σ^2 as the penalty:

$$f\left(\sigma^2\right) = \frac{\beta^\alpha}{\Gamma(\alpha)}\left(\sigma^{-2}\right)^{(\alpha+1)}\exp\left\{-\beta\sigma^{-2}\right\}, \quad \alpha > 0, \beta > 0.$$

Within the Bayesian framework, the penalized likelihood function can be interpreted as a *posterior density* and the penalty function can be regarded as a *prior* in the likelihood function. The choice of α and β could be determined based on practical experience or prior information. In general, users are free to choose another kind of prior distribution as a penalty function as long as it satisfies the penalty requirements for a bounded likelihood function.

The penalty choice of inverted gamma function introduces no "structural" changes in the log-likelihood function based on complete data (Ridolfi and Idier, 2001). Li and Fang (2022) found that the SNTO algorithm can deal with any penalty function without explicit formulation. The rate of convergence could also be well controlled.

Note that there is a bias-variance trade-off between the MLE and the penalized MLE. Adding a penalty can reduce the estimate's variability and improve its concentration properties. However, it will lead to some estimation bias in the meantime. Comparisons of the MLE and the penalized MLE under MixN based on small samples are conducted in Section 5.7.3.

5.6.3 MLE-QMC: MLE via QMC-data

Many authors, for example Hosmer (1973) and Lourens et al. (2013), found that the MLE under mixture normal distributions cannot yield satisfactory results, especially when the sample size is not large enough or the mixture components are not well separated. Due to the complexities of MLE under MixN, many potential problems could occur and influence the estimation performance. Although the penalized MLE can avoid the singular estimates and the SNTO algorithm can overcome some difficulties from the computational aspect, the

TABLE 5.3
Parameters of underlying distributions

Type	Distribution	α	μ_1	σ_1^2	μ_2	σ_2^2	DI
MixN	C1	0.70	2.70	2	1.50	1	0.40
MixN	C2	0.70	2.20	9	1.50	1	0.50
MixN	C3	0.70	3.50	2	1.50	1	0.60
MixN	C4	0.70	4.25	3	1.50	1	0.70
MixN	C5	0.70	4.60	2	1.50	1	0.80
MixN	C6	0.70	8.00	10	1.50	1	0.90

estimation results could still be unsatisfactory when the collected random sample has poor quality.

By using the QMC-data, $z_i = Q((2i-1)/2n)$, defined in Subsection 1.7.4, the "plug-in" log-likelihood function is defined as:

$$\widehat{\ell}(\boldsymbol{\theta}|\boldsymbol{z}) = \sum_{i=1}^{n} \ln f(z_i|\boldsymbol{\theta}) = \sum_{i=1}^{n} \ln\left(\alpha\varphi(z_i|\mu_1,\sigma_1^2) + (1-\alpha)\varphi(z_i|\mu_2,\sigma_2^2)\right). \tag{5.20}$$

An approach to improve the estimation results is by maximizing equation (5.20) concerning the parameter vector $\boldsymbol{\theta}$ (Li and Fang, 2022). This method is called MLE-QMC. The underlying idea of MLE-QMC is that by treating the QMC-RPs as a random sample with good representativeness, the MLE based on this new sample can yield more robust and promising estimates that are less dependent on the representativeness of the original random sample.

The MLE-QMC can also be applied to the penalized MLE to maintain the estimation consistency. In subsection 5.7.3, the PMLE based on the exact QMC-RPs (assume $\boldsymbol{\theta}$ is known) is given to illustrate the efficiency of QMC-RPs in the maximum-likelihood framework. Numerical comparisons between the plain penalized MLE (PMLE) and the QMC-PMLE (via SNTO algorithm) are given in Subsection 5.7.3.

Qi et al. (2019) considered using the density estimation for revising location-biased datasets for reducing biased datasets. The kernel estimation of the density function is a nonparametric method similar to the QMC-Data structure. Wang et al. (2022a) proposed a new bias-correction numerical method based on the SNTO algorithm that can reduce the bias and variance of parameter estimators in three types of extreme value distributions. Many case studies can find in Qi et al. (2019), Li and Fang (2022) and Wang et al. (2022a).

5.7 Simulation Comparisons

Some numerical comparisons among various estimation methods mentioned in this chapter on six models are given in Table 5.3.

5.7.1 Estimation accuracy measures

There are many measures for estimation accuracy. There are three kinds of accuracy measures (a)-(c) are considered, i.e., the L_2-distances between two cdf's F_{true} and F_{est}, and pdf's p_{true} and p_{est} in Subsection 1.6.1; the Kullback-Leibler divergence (KL) introduced in Section 4.5.4; and the absolute bias index (ABI) defined below.

(a) **L2-distance**

$$L_2(F_{true}, F_{est}) = \left\{ \int_{\mathbf{R}^d} (F_{true}(\boldsymbol{x}) - F_{est}(\boldsymbol{x}))^2 d\boldsymbol{x} \right\}^{1/2},$$

$$L_2(p_{true}, p_{est}) = \left\{ \int_{\mathbf{R}^d} (p_{true}(\boldsymbol{x}) - p_{est})^2 d\boldsymbol{x} \right\}^{1/2}.$$

(b) **Kullback-Leibler divergence** (KL)

$$D_{\mathrm{KL}}(F_{true} \| F_{est}) = \int_{-\infty}^{\infty} p_{true}(\boldsymbol{x}) \ln \left(\frac{p_{true}(\boldsymbol{x})}{p_{est}(\boldsymbol{x})} \right) d\boldsymbol{x}.$$

(c) **Absolute bias index** (ABI)

$$ABI = \frac{1}{2} \left[\frac{|\mu - \hat{\mu}|}{\mu} + \frac{|\sigma - \hat{\sigma}|}{\sigma} \right],$$

where $\hat{\mu}$ and $\hat{\sigma}$ are the estimated expectation and estimated standard deviation of the MixN distribution, satisfying $\mu \neq 0$ and $\sigma^2 > 0$.

5.7.2 MLE via SNTO vs. MLE via EM

This subsection compares SNTO and EM when applied in MixN's MLE. Figure 5.1 displays the visualization of the worst density estimations, measured by the largest L_2-distance, obtained from 99 simulations using both the EM and SNTO algorithms. The estimated density with the largest distance allows us to visualize the worst estimation result among the 99 random samples. The x-axis label in the figure indicates the type of algorithm being compared. The plot shows that the SNTO algorithm can converge to more reasonable estimates of the five parameters, even when the given random sample is of poor quality. This further underscores the advantage of the SNTO algorithm in generating more reliable and accurate parameter estimates than the EM algorithm.

For further interpretation and in-depth analysis of the singularity comparisons between the EM and SNTO algorithms, readers can refer to (Li and Fang, 2022, Subsection 5.3.1.).

Maximum Likelihood Estimation (MLE) performance using the SNTO algorithm and MLE using the EM algorithm are compared based on a sample size of $n = 49$ by Li and Fang (2022). The simulation comparisons are given in Table 5.4. The log-likelihood function value based on the actual parameters is calculated for each random sample, and the average of the 99 actual log-likelihood values are listed in the "Loglike.True" column. The average maximum log-likelihood values of 99 random samples are shown in the "Loglike" column. "Iter" represents average iteration times.

The results suggest that the SNTO algorithm generates more robust estimation results, indicating improved accuracy compared to the EM algorithm. Note that the EM algorithm reaches the larger log-likelihood value on average ("Loglike") for six distributions. However, the larger log-likelihood values do not lead to more precise estimations regarding L_2-distance, KL-divergence, and ABI. Li and Fang (2022) indicated that the values of "Loglike" are always larger than the "Loglike.True" value for every distribution. Although the SNTO algorithm converges to ML estimates with smaller "Loglike", the estimates are more accurate, and the average likelihood function values are closer to the "Loglike.True". The large, more significant of the log-likelihood function does not always give us a better estimation. Refer to Li and Fang (2022, Subsection 5.3.2) for more detailed discussion.

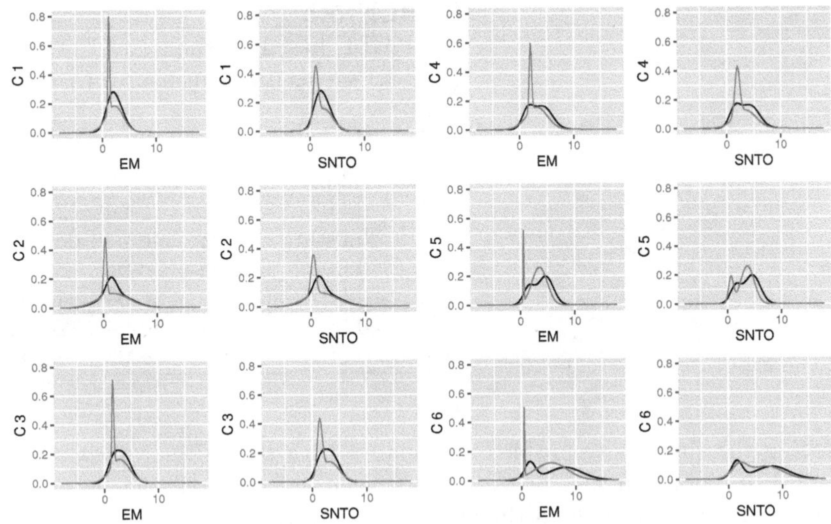

FIGURE 5.1
Density estimation comparisons, MLE via EM algorithm and SNTO algorithm ($n = 49, N = 99$). Black curves represent the densities of underlying distributions. Red curves represent the density of the ML estimates with the largest L_2-distance (L2.pdf).

TABLE 5.4
Comparisons on average accuracy measures between MLE via SNTO and EM (n=49, N=99)

Case	Method	Loglike.True	Loglike	L2.pdf	L2.cdf	KL	ABI	Iter
C1	EM	-86.0776	**-82.7980**	0.1247	0.1067	0.0727	0.0656	270
C1	SNTO	-86.0776	-83.7465	**0.1027**	**0.1023**	**0.0501**	**0.0647**	**27**
C2	EM	-114.2630	**-111.2074**	0.1056	0.1392	0.0683	0.1408	75
C2	SNTO	-114.2630	-111.9544	**0.0887**	**0.1224**	**0.0379**	**0.1363**	**27**
C3	EM	-91.8240	**-88.7017**	0.1185	0.1154	0.0717	0.0597	229
C3	SNTO	-91.8240	-89.4867	0.0982	**0.1101**	**0.0520**	**0.0592**	**27**
C4	EM	-102.2025	**-99.3789**	0.1054	0.1284	0.0722	0.0635	136
C4	SNTO	-102.2025	-99.9037	**0.0856**	**0.1185**	**0.0466**	**0.0630**	**27**
C5	EM	-100.2699	**-97.2734**	0.1028	0.1253	0.0678	0.0572	109
C5	SNTO	-100.2699	-97.8160	**0.0876**	**0.1178**	**0.0537**	**0.0557**	**27**
C6	EM	-131.6298	**-128.9529**	0.0767	0.1780	0.0579	0.0730	31
C6	SNTO	-131.6298	-129.4028	**0.0660**	**0.1544**	**0.0411**	**0.0664**	**27**

Overall, based on these evaluations, the SNTO algorithm demonstrates its effectiveness in avoiding singularities and achieving robust estimation results compared to the EM algorithm.

5.7.3 QMC-PMLE

Li and Fang (2022) discussed and compared the performance of MLE and PMLE based on the SNTO algorithm with the same initial searching region D. They concluded that Based on the sample size of $n = 50$, PMLE can yield superior estimation accuracy for the six types of distributions (C1–C6). They also found that the PMLE estimations are more robust in

TABLE 5.5
The PMLE based on the exact QMC-RPs—accuracy measures (n=49; N=99)

Case	L2.pdf	L2.cdf	KL	ABI
C1	0.0087	0.0053	8e-04	2e-04
C2	0.0054	0.0078	4e-04	0e+00
C3	0.0069	0.0058	5e-04	3e-04
C4	0.0072	0.0047	5e-04	4e-04
C5	0.0050	0.0047	3e-04	2e-04
C6	0.0070	0.0080	-1e-03	6e-04

handling the singularity problem associated with the MLE framework. This suggests that the PMLE approach can provide more accurate and reliable parameter estimates than traditional MLE methods in the experimental setup.

In this subsection, the revised PMLE based on the distribution-free Harrell-Davis quantile estimator (QMC-PMLE) is discussed. The plain PMLE ("plain") refers to the MLE by maximizing (5.18) based on (5.19). The method of the QMC-PMLE based on the distribution-free Harrell-Davis quantile estimator is defined in Subsection 5.6.3. The results of the "plain" and "QMC" methods are summarized based on the average of 100 Monte Carlo repetitions. The PMLE based on the exact QMC-RPs ("ideal") will only be performed once per distribution, and it is only for simulation and comparison purposes because parameters $\boldsymbol{\theta}$ are assumed to be known.

Estimation results of the PMLE based on the exact QMC-RPs

Table 5.5 presents the results of PMLE based on the exact QMC-RPs of size 49. The table shows that even for distributions C1 and C2, which have low DI, the PMLE based on the exact QMC-RPs can yield significant estimates.

QMC-PMLE vs. PMLE

Table 5.6 provides an example that illustrates the significant improvement achieved by QMC-PMLE compared to plain PMLE. The table demonstrates the improvement in estimation results for the six distributions. This suggests that incorporating the QMC technique into the PMLE framework can lead to notable enhancements in the accuracy and precision of parameter estimation. Figure 5.2 gives examples of improvement in density fitting of the QMC-PMLE from the estimation results based on the plain PMLE. More comparisons refer to (Li and Fang, 2022, Subsection 5.5.2).

Ke et al. (2023) gave a systematic study on the gamma distribution by different RPs. They applied the idea of QMC-data in estimation under the gamma distribution.

5.8 Proofs of Theorems in Chapter 5

5.8.1 Proof of Theorem 5.3

The proof is from Antle and Bain (1969).

TABLE 5.6
Comparison of the PMLE and the QMC-PMLE on average accuracy measures (n=49; N=99)

Case	Method	L2.pdf	L2.cdf	KL	ABI
C1	PMLE	0.1137	0.1040	0.0550	**0.0644**
C1	QMC-PMLE	**0.0965**	**0.1004**	**0.0470**	0.0648
C2	PMLE	0.0847	0.1220	0.0367	**0.1360**
C2	QMC-PMLE	**0.0729**	**0.1193**	**0.0301**	0.1368
C3	PMLE	0.1025	0.1100	0.0541	0.0586
C3	QMC-PMLE	**0.0891**	**0.1058**	**0.0460**	**0.0586**
C4	PMLE	0.0853	0.1182	0.0465	0.0631
C4	QMC-PMLE	**0.0774**	**0.1149**	**0.0407**	**0.0619**
C5	PMLE	0.0880	0.1183	0.0543	0.0562
C5	QMC-PMLE	**0.0790**	**0.1140**	**0.0447**	**0.0559**
C6	PMLE	0.0615	0.1534	0.0358	0.0664
C6	QMC-PMLE	**0.0569**	**0.1501**	**0.0310**	**0.0662**

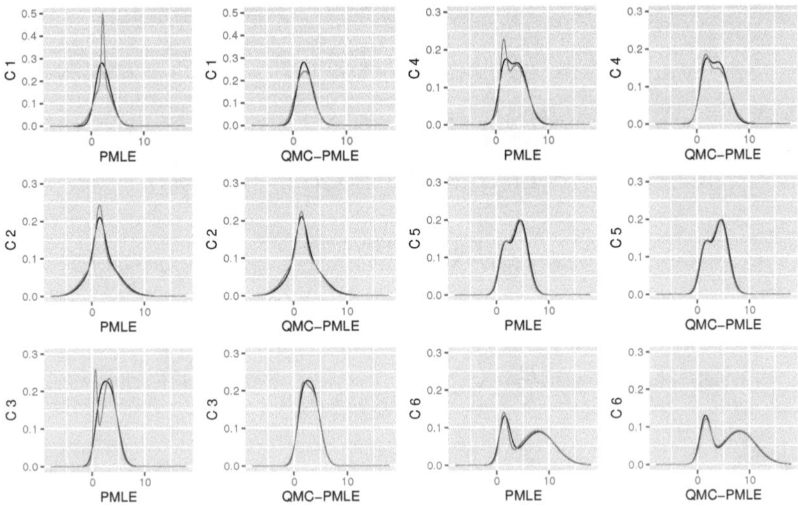

FIGURE 5.2
The PMLE versus the QMC-PMLE – example of improvements in density fitting ($n = 49$). Black curves represent the densities of underlying distributions.

Proof *Let $x_1, \ldots, x_n$ be a random sample from X. The likelihood function is*

$$L(x; \mu, \sigma) = \sigma^{-n} \prod_{i=1}^{n} g\left(\frac{x_i - \mu}{\sigma}\right).$$

A straightforward step

$$\begin{aligned} L(x; \hat{\mu}, \hat{\sigma}) &= \hat{\sigma}^{-n} \prod_{i=1}^{n} g\left(\frac{x_i - \hat{\mu}}{\hat{\sigma}}\right) = \max_{\Omega} \left\{ \sigma^{-n} \prod_{i=1}^{n} g\left(\frac{x_i - \mu}{\sigma}\right) \right\} \\ &= \max_{\Omega} \left\{ \sigma_0^{-n} (\frac{\sigma}{\sigma_0})^{-n} \prod_{i=1}^{n} g\left(\frac{(x_i - \mu_0)/\sigma_0 - (\mu - \mu_0)/\sigma_0}{\sigma/\sigma_0}\right) \right\}. \end{aligned}$$

That implies

$$\hat{\sigma}_s^{-n}\prod_{i=1}^{n} g\left(\frac{z_i-\hat{\mu}_s}{\hat{\sigma}_s}\right) = \max_{\Omega^*}\left\{(\sigma^*)^{-n}\prod_{i=1}^{n} g\left(\frac{z_i-\mu^*}{\sigma^*}\right)\right\},$$

where $\Omega^* = \Omega, \mu^* = (\mu-\mu_0)/\sigma_0, \sigma^* = \sigma/\sigma_0$ *and* $z_i = (x_i-\mu_0)/\sigma_0$. *Since* $\Omega^*, \hat{\mu}_s$ *and* $\hat{\sigma}_s$ *correspond exactly to the maximum likelihood estimators of* μ *and* σ *when the sampling is actually on a standardized variate* z. *Thus the joint density of* $\hat{\mu}_s$ *and* $\hat{\sigma}_s$ *does not depend on* μ *and* σ.

5.8.2 Proof of Theorem 5.6

The proofs for the following three theorems are from Matsuura et al. (2015)

Proof *Consider the loss function* $L(a_1,\ldots,a_k)$ *as*

$$\begin{aligned}
L(a_1,\ldots,a_k) &= E[\mathrm{MSD}(\hat{\mu}+a_1\hat{\sigma},\ldots,\hat{\mu}+a_k\hat{\sigma})]\\
&= E[\min_j(x-\hat{\mu}-a_j\hat{\sigma})^2]\\
&= \sigma^2 E[\min_j(\frac{x-\hat{\mu}}{\sigma}-a_j\frac{\hat{\sigma}}{\sigma})^2]\\
&= \sigma^2 E[\min_j(U-a_jW)^2]\\
&= \sigma^2 E_U\{E_W[\min_j(U-a_jw)^2|W=w]\}\\
&= \sigma^2 E_W\left\{\sum_{j=1}^{k}\int_{w(a_{j-1}+a_j)/2}^{w(a_j+a_{j+1})/2}(x-a_jw)^2\psi(x|w)dx\right\} \qquad \textit{let } x = zw\\
&= \sigma^2 E_W\left\{\sum_{j=1}^{k}\int_{(a_{j-1}+a_j)/2}^{(a_j+a_{j+1})/2}(z-a_j)^2 w^3\psi(zw|w)dz\right\}\\
&= \sigma^2\sum_{j=1}^{k}\int_{(a_{j-1}+a_j)/2}^{(a_j+a_{j+1})/2}(z-a_j)^2 E_W[w^3\psi(zw|w)]dz\\
&= \sigma^2 E(W^2)\sum_{j=1}^{k}\int_{(a_{j-1}+a_j)/2}^{(a_j+a_{j+1})/2}(z-a_j)^2 h(z)dz.
\end{aligned}$$

Then we need to confirm that $h(z)$ *defined in (5.17) is a density function.*

$$\begin{aligned}
\int_{-\infty}^{\infty} h(z)dz &= \int_{-\infty}^{\infty}\frac{E(W^3\psi(zw|w))}{E(W^2)}dz\\
&= \frac{1}{E(W^2)}E[W^3\int_{-\infty}^{\infty}\psi(zw|w)dz]\\
&= \frac{1}{E(W^2)}E[W^3\int_{-\infty}^{\infty}\frac{1}{w}\psi(x|w)dx]\\
&= \frac{1}{E(W^2)}E[W^2\int_{-\infty}^{\infty}\psi(x|w)dx]\\
&= \frac{1}{E(W^2)}E[W^2] = 1.
\end{aligned}$$

Hence we have

$$\begin{aligned} L(a_1,\dots,a_k) &= \sigma^2 E(W^2)\sum_{j=1}^{k}\int_{(a_{j-1}+a_j)/2}^{(a_j+a_{j+1})/2}(z-a_j)^2 h(z)dz \\ &= \sigma^2 E(W^2)E[\min_j (Z-a_j)^2], \end{aligned}$$

where Z has the pdf (5.17) and $a_1,\dots,a_j$ are MSE-RPs of Z.

5.8.3 Proof of Theorem 5.4

Proof *Define a function $L(c)$ as*

$$L(c) = E[MSD(\hat{\mu}+cz_1\hat{\sigma}z_k)], z_1 < z_2 < \cdots < z_k.$$

It follows the proof of Theorem 5.6 we have

$$L(c) = \sigma^2 E(W^2)\sum_{j=1}^{k}\int_{c(z_{j-1}+z_j)/2}^{c(z_j+z_{j+1})/2}(z-cz_j)^2 h(z)dz,$$

It is well-known that $\hat{\mu}$ and $\hat{\sigma}$, the MLEs of μ and σ of $N(\mu,\sigma^2)$, are independent, and $\hat{\mu}\sim N(\mu,\sigma^2/n)$ and $n\hat{\sigma}^2/\sigma^2 \sim \chi^2_{n-1}$. As a consequence, $U = (X-\mu)/\sigma$ and $W = \hat{\sigma}/\sigma$ are independent, and $U \sim N(0,(n+1)/n$ and $\sqrt{n}W \sim \chi_{n-1}$, respectively. Then $E(W^2) = (n-1)/n$. Since U and W are independent, $U|W \sim N(0,(n+1)/n)$ with pdf

$$\psi(u|W) = \frac{\sqrt{n}}{\sqrt{n+1}}\phi\left(\frac{\sqrt{n}}{\sqrt{n+1}}u\right).$$

The pdf of $V = \sqrt{n}W$ is

$$\kappa(v) = \frac{2^{1-(n-1)/2}}{\Gamma(\frac{n-1}{2})}v^{n-2}e^{-v^2/2}.$$

Detailed steps are

$$\begin{aligned} h(z) &= \frac{E[W^3\psi(zW|W)]}{E(W^2)} \\ &= \frac{n}{n-1}E\left[W^3\frac{\sqrt{n}}{\sqrt{n+1}}\phi\left(\frac{\sqrt{n}}{\sqrt{n+1}}zW\right)\right] \\ &= \frac{1}{(n-1)\sqrt{n+1}}E\left[(\sqrt{n}W)^3\phi\left(\frac{\sqrt{n}W}{\sqrt{n+1}}z\right)\right] \\ &= \frac{1}{(n-1)\sqrt{n+1}}\int_0^\infty v^3\phi\left(\frac{v}{n+1}z\right)\frac{2^{1-(n-1)/2}}{\Gamma(\frac{n-1}{2})}v^{n-2}e^{-v^2/2}dv \\ &= \frac{1}{(n-1)\sqrt{n+1}}\int_0^\infty v^3\frac{1}{\sqrt{2\pi}}\exp\left\{-\frac{1}{2}\frac{v^2}{n+1}z^2\right\}\frac{2^{1-(n-1)/2}}{\Gamma(\frac{n-1}{2})}v^{n-2}e^{-v^2/2}dv \\ &= \frac{1}{(n-1)\sqrt{(n+1)\pi}}\int_0^\infty \frac{2^{1-n/2}}{\Gamma(\frac{n-1}{2})}v^{n+1}\exp\left\{-\frac{1}{2}v^2\left(\frac{v^2}{n+1}+1\right)\right\}dv \\ &= \frac{1}{(n-1)\sqrt{(n+1)\pi}}\int_0^\infty \frac{2^{1-n/2}}{\Gamma(\frac{n-1}{2})}\frac{\Gamma(\frac{n+2}{2})}{2^{1-(n+2)/2}}\frac{2^{1-(n+2)/2}}{\Gamma(\frac{n+2}{2})} \end{aligned}$$

$$\times \left[t\left(\frac{z^2}{n+1}+1\right)^{-1/2}\right]^{n+1} e^{-t^2/2}\left(\frac{z^2}{n+1}+1\right)^{-1/2} dt$$

$$= \frac{2}{(n-1)\sqrt{(n+1)\pi}}\frac{\Gamma(\frac{n+2}{2})}{\Gamma(\frac{n-1}{2})}\left(\frac{z^2}{n+1}+1\right)^{-(n+2)/2}\int_0^\infty \frac{2^{1-(n+2)/2}}{\Gamma(\frac{n+2}{2})}t^{n+1}e^{-t^2/2}dt$$

$$= \frac{1}{\sqrt{(n+1)\pi}}\frac{\Gamma(\frac{n+2}{2})}{\Gamma(\frac{n+1}{2})}\left(\frac{z^2}{n+1}+1\right)^{-(n+2)/2}.$$

which follows by noting that the last integrand is the pdf of a χ_{n+2} random variable. Thus, $h(z)$is the pdf of the t-distribution with $n+1$ degrees of freedom. Then, the derivative of $L(c)$ yields

$$L'(c) = -\frac{2(n-1)\sigma^2}{n}\sum_{j=1}^{k}\int_{c(z_{j-1}+z_j)/2}^{c(z_j+z_{j+1})/2} z_j(z-cz_j)h(z)dz.$$

From $L'(c)=0$ we can find the solution of c

$$c = \frac{\sum_{j=1}^k \int_{c(z_{j-1}+z_j)/2}^{c(z_j+z_{j+1})/2} z_j t h_(t)dt}{\sum_{j=1}^k \int_{c(z_{j-1}+z_j)/2}^{c(z_j+z_{j+1})/2} z_j^2 h(t)dt},$$

i.e., (5.15).

5.8.4 Proof of Theorem 5.5

Proof

Let $L(a_1,\ldots,a_k) = E[MSD(\hat{\mu}+a_1\hat{\sigma},\ldots,\hat{\mu}+a_k\hat{\sigma}]$. It follows the proof for Theorem 5.4 we have

$$\begin{aligned} L(a_1,\ldots,a_k) &= \frac{(n-1)\sigma^2}{n}\sum_{j=1}^{k}\int_{(a_{j-1}+a_j)/2}^{(a_j+a_{j+1})/2} h(z)(z-a_j)^2dz \\ &= \frac{(n-1)\sigma^2}{n}L(a_1,\ldots,a_k), \end{aligned}$$

where $h(z)$ is the pdf of the t-distribution with n+1 degrees of freedom and Z is distributed as the t-distribution with n+1 degrees of freedom. Hence, minimizing $L(a_1,\ldots,a_k)$ is equivalent to minimizing

$$E[\min_{i=1,\ldots,k}(Z-a_i)^2],$$

from which it follows that $a_1.\ldots,a_k$ minimizing the EMSD (5.14) are given by k-MSE-RPs of the t-distribution with $n+1$ degrees of freedom.

6

RPs of Multivariate Distributions

This chapter focuses on three kinds of representative points, i.e., MSE-RPs, QMC-RPs, and MSE-RPs of continuous multivariate distributions $\boldsymbol{X} \sim F(\boldsymbol{x})$ and construction of approximation distributions to $F(\boldsymbol{x})$ by these RPs. Various algorithms are introduced to generate these three RPs based on the stochastic representation of $\boldsymbol{X}$. The elliptical distribution, which is the attention of the multivariate normal distribution, has a simple stochastic representation, and its MSE-RPs, QMC-RPs, and MSE-RPs can be generated. Related algorithms are presented. Other multivariate distributions, such as the skew-normal distribution and strongly symmetric distributions, are discussed. Comparisons of estimators of the mean vector and covariance matrix among approximations by MSE-RPs, QMC-RPs, and MSE-RPs of continuous multivariate distributions are discussed. The performance of its resampling in estimating the mean vector and covariance matrix is studied. The fitting distribution of MSE of MC-RPs is also considered.

6.1 Generation of a Multivariate Distribution by Monte Carlo Methods

Let the random vector $\boldsymbol{X} \in \mathbb{R}^p$ follow a continuous multivariate distribution $F(\boldsymbol{x}) = F(x_1, \ldots, x_p)$ with pdf $p(\boldsymbol{x}) = p(x_1, \ldots, x_p)$. For statistical simulation, we need the generation of random samples from $\boldsymbol{X}$ by computer software. There are many useful multivariate distributions, among which some multivariate distributions have an analytic stochastic representation (SR) based on several independent univariate random variables, such as the uniform distribution on a rectangle of $\mathbb{R}^p$, spherical distributions, elliptical distributions, skew-normal distributions, and skew-elliptical distributions.

There is a rich literature on the generation of univariate distributions, including the inverse transformation method and stochastic representation. The inverse transformation method has been introduced in Theorem 2.1. However, this method cannot be extended to multivariate distributions, but many generating methods are proposed. Johnson (1987) comprehensively introduces the generation of multivariate statistical simulations. Palmitesta and Provasi (2015) reviewed various multivariate distributions and their computer generation. Their list includes the Dirichlet distribution, the multivariate Johnson distribution, multivariate beta and gamma distributions, the multivariate extreme value distribution, the inverse Gaussian multivariate distribution, the Cook-Johnson family of multivariate uniform distributions, the multivariate Johnson distribution, and elliptically contoured distributions. They also review some techniques for generating random vectors with given marginal distributions and correlation matrices. They said, "The distributions taken into consideration can be divided into two groups. The first group refers to specific distributions for which it is possible to analytically write the density function or the moment-generating function and, as a consequence, to derive the properties of the marginals. Such marginals,

DOI: 10.1201/9781003589389-6

in general, are an extension to the multivariate case of known random variables, which are used very often in statistical applications with unidimensional samples. On the contrary, a second group refers to distributions with a given dependence structure of the components and is defined in indirect form." Sarabia and Gómez-Déniz (2008) gave a review of some recent results on the construction of multivariate distributions with discussion.

6.1.1 Stochastic representation method

Suppose that random vector $\boldsymbol{X} \sim F(\boldsymbol{x})$ has a stochastic representation

$$\boldsymbol{X} \stackrel{d}{=} h(Z_1, \ldots, Z_q) \equiv h(\boldsymbol{Z}), \tag{6.1}$$

where $Z_1, \ldots, Z_q$ are independent. One can independently generate a random sample from each distribution of $Z_i, i = 1, \ldots, q$ and denote them by $z_1, \ldots, z_q$. Then $x = h(z_1, \ldots, z_q)$ is a random sample from $F(\boldsymbol{x})$. This method is called *stochastic representation method* (SRM). Section 1.12 gives the SR algorithm in detail and several examples for generating NT-nets on some finite domains in $\mathbb{R}^p$, such as rectangle, $A^p(r), B^p(r), V^p(r)$ and $T^p(r)$.

Example 6.1 *Let $\boldsymbol{X} \sim N_p(\boldsymbol{\mu}, \boldsymbol{\Sigma})$ follow the multivariate normal distribution. There is an SR for $\boldsymbol{X}$*

$$\boldsymbol{X} \stackrel{d}{=} \boldsymbol{\mu} + \boldsymbol{A}\boldsymbol{Z}, \tag{6.2}$$

where $\boldsymbol{Z} = (Z_1, \ldots, Z_p)^T$ with iid elements $Z_i \sim N(0, 1)$ and $\boldsymbol{A}\boldsymbol{A}^T = \boldsymbol{\Sigma}$. Often, $\boldsymbol{A}$ is chosen as the positive definite square root of $\boldsymbol{\Sigma}$, i.e., $\boldsymbol{\Sigma}^{1/2}$ is positive definite. By SR method we generate p iid samples $z_1, \ldots, z_p$ from $N(0, 1)$ and let $\boldsymbol{z} = (z_1, \ldots, z_p)^T$ and let

$$\boldsymbol{x} = \boldsymbol{\mu} + \boldsymbol{A}\boldsymbol{z}.$$

Then, $\boldsymbol{x}$ is a random sample from $N_p(\boldsymbol{\mu}, \boldsymbol{\Sigma})$.

The Box-Muller method (Hogg et al. (2005)) for the generation of $N_2(\boldsymbol{0}, \boldsymbol{I}_2)$ is based on the following SR

$$\begin{bmatrix} X_1 \\ X_2 \end{bmatrix} \stackrel{d}{=} (-2\log(U_1))^{1/2} \begin{bmatrix} \cos(2\pi U_2) \\ \sin(2\pi U_2) \end{bmatrix}, \tag{6.3}$$

where U_1 and U_2 are iid $U(0, 1)$.

Example 6.2 *Let $\boldsymbol{U}^{(p)}$ follow the uniform distribution on the unit sphere $S^p = \{\boldsymbol{y} : \boldsymbol{y}^T\boldsymbol{y} = 1\}$, denoted by $U(S^P)$. It is known (Fang and Wang (1994)) that*

$$\boldsymbol{U}^{(p)} \stackrel{d}{=} \frac{\boldsymbol{Z}}{||\boldsymbol{Z}||}, \tag{6.4}$$

where $\boldsymbol{Z} \sim N_p(\boldsymbol{0}, \boldsymbol{I}_p)$, $||\boldsymbol{Z}||$ is the L_2-norm of $\boldsymbol{Z}$. Let $z_1, \ldots, z_p$ be iid random samples from $N(0, 1)$ by the MC method and $u_i = z_i/||\boldsymbol{z}||, i = 1, \ldots, z_p$, where $\boldsymbol{z} = (z_1, \ldots, z_p)^T$. Then $\boldsymbol{u} = (u_1, \ldots, u_p)^T$ is a random sample from $U(S^P)$.

6.1.2 Conditional distribution method

The conditional distribution method changes the generation for a multivariate distribution into generation for several marginal/conditional univariate distributions. Suppose random vector $\boldsymbol{X} = (X_1, \cdots, X_p) \sim F(x_1, \ldots, x_p)$. Let $F_1(x_1)$ be the cumulative distribution function of X_1 and let $F_i(x_i|x_1, \cdots, x_{i-1})$ be the conditional distribution of X_i given $\{X_1 = x_1, \cdots, X_{i-1} = x_{i-1}\}$. The theory of probability shows

$$F(x_1, \ldots, x_p) = F_1(x_1)F_2(x_2|x_1) \cdots F_p(x_p|x_1, \cdots, x_{p-1}). \tag{6.5}$$

Note that each of $F_1(x), F_2(x_2|x_1) \cdots$ and $F_p(x_p|x_1, \cdots, x_{p-1})$ is a univariate (conditional) distributions. We can apply some methods, including the inverse transformation method to generate a random sample from these distributions. Let $U_1, \cdots, U_s$ are *iid* following $U_i \sim U(0,1)$. The following relationships

$$\begin{cases} U_1 = F_1(X_1), \\ U_i = F_i(X_i|X_1, \cdots, X_{i-1}), 2 \leq i \leq p \end{cases} \tag{6.6}$$

can obtain a random sample from respective distributions: $F_1(x_1), F_2(x_2|x_1), \cdots, F_p(x_p|x_1, \cdots, x_{p-1})$. From the conditional formula (6.5) we can generate a random sample from $F(x_1, \ldots, x_p)$.

Algorithm 6.1 (Algorithm for Conditional distribution method)

Step 1 Generate p independently random numbers $u_1, \ldots, u_p$.

Step 2 Compute $x_1 = F_1^{-1}(u_1), x_{2.1} = F_2^{-1}(u_2), \ldots, x_{p.1\cdots p-1} = F_p^{-1}(u_p)$.

Step 3 Deliver $\boldsymbol{x} = (x_1, x_{2.1}, \cdots, x_{p.1\cdots p-1})$ *as a random sample from* $F(x_1, \ldots, x_p)$.

When $X_1, \ldots, X_p$ are independent, $F_1(x_1), F_2(x_2|x_1), \cdots, F_p(x_p|x_1, \cdots, x_{p-1})$ are p marginal distributions and the conditional distribution method reduces the SR method discussed in the previous subsection. An interesting application of the conditional distribution method can be applied in the so-called *vertical density representation.*

6.1.3 Vertical density representation

Proposed the vertical density representation (VDR) was by Troutt (1991) and Troutt (1993). Kotz et al. (1997) and Fang et al. (2001) gave a systematic study on VDR of multivariate distributions. This subsection is based on their studies.

In this subsection, we use a slightly different notation from what is in the other sections. Let random vector $\boldsymbol{X} = (X_1, \cdots, X_p)$ have a probability density function (pdf) $f(x_1, \cdots, x_p)$. Then $V = f(X_1, \cdots, X_p)$ is a random variable with a density, $g(v)$ say. It can be proved by

$$g(v) = v\frac{d}{dv}A_f(v).$$

where $A_f(v) = L(S_f(v)), S_f(v) = \{\boldsymbol{x} : f(\boldsymbol{x}) \geq v\}$, and L is the Lebesgue measure in $\mathbb{R}^p$. The conditional distribution of $\boldsymbol{X}$ for given $V = v$ is $U(S_f(v))$, the uniform distribution on $S_f(v)$. The decomposition $\{V, (\boldsymbol{X}|V)\}$ is called the *vertical density representation of* $\boldsymbol{X}$ proposed by Troutt (1991). Kotz et al. (1997) gave a systematic study on various multivariate distributions. If we can generate V and the conditional distribution $(\boldsymbol{X}|V = v)$, we can obtain samples from the population $\boldsymbol{X}$. Fang et al. (2001) proposed another vertical density representation as follows. Let

$$D_{p+1}(f) = \{\boldsymbol{x} = (x_1, \cdots, x_{p+1}) : 0 \leq x_{p+1} \leq f(\boldsymbol{x}_p)\},$$

where $\boldsymbol{x}_p = (x_1, \cdots, x_p)$. Assume $\boldsymbol{X}_{p+1} = (\boldsymbol{X}_p, X_{p+1}) \sim U(D_{p+1}(f))$. We have

i) $\boldsymbol{X}_p$ has a pdf $f(\boldsymbol{x})$;

ii) The pdf of X_{p+1} is $L_p(S_f(x))$;

iii) $(\boldsymbol{X}_p|X_{p+1} = v) \sim U(S_f(v))$.

With the vertical density representation, we have the following necessary and sufficient conditions: Assume that $\boldsymbol{x} \sim U(G)$ has a finite Lebesgue measure, i.e., $L_p(G) < \infty$. Denote

$$G_{p-1}(v) = \{\boldsymbol{x} = (x_1, \cdots, x_{p-1}) : (\boldsymbol{x}, v) \in G\}.$$

Assume $L_{p-1}(D_{p-1}(v))$ is continuous in v. Then, $\boldsymbol{x} \sim U(G)$ iff

i) The pdf of $\boldsymbol{X}_p$ is $\frac{L_{p-1}(G_{p-1}(v))}{L_p(G)}$;

ii) $(\boldsymbol{X}_{p-1}|X_p = v_1) \sim U(G_{p-1}(v_1))$.

The above theory can be similarly applied to $(\boldsymbol{X}_{p-1}|X_p = v_1)$, $(\boldsymbol{x}_{p-2}|X_p = v_1, X_{p-1} = v_2)$, and so on.

Fang et al. (2001) applied the above theory to many subclasses of elliptical distributions. In general, generating MC-RPs and QMC-RPs for most multivariate distributions is difficult. Still, it is available for some kinds of distributions with a simple SR, like elliptically symmetric and skew-normal distributions.

6.1.4 Global likelihood sampling

To draw samples from multimodal and high-dimensional target distributions more conveniently and efficiently, Yi et al. (2023) and Yi and Zhou (2023) developed the global likelihood sampler (GLS), which is based on good lattice points. GLS can be regarded as a randomized QMC sampling method. GLS can leverage the excellent property of the low-discrepancy points instead of the random points to obtain representative samples. Compared with the sampling methods in the literature, GLS does not need to choose a suitable proposal distribution for a specific target distribution, and it is a problem-independent sampling method. Moreover, GLS can avoid falling into a local mode for multimodal target distributions. There are fewer parameters to set in GLS, and the computational complexity of GLS can be lower without an iterative process.

Suppose the target d-dimensional PDF $\pi(\theta)$ has the support $\mathcal{S}$, its CDF is $\Pi(\theta)$. The procedure of GLS is as in Algorithm 6.2.

Algorithm 6.2 (The procedure of GLS)

Step 0: **Input.** *The target d-dimensional PDF $\pi(\theta)$ with support $\mathcal{S}$, an M-point GLP G_0 on $[0,1]^d$, a predefined threshold ϵ, the sample size N.*

Step 1: **Truncation.** *Choose $a_t < b_t$, $t = 1, \ldots, d$ and let $\mathcal{D} = [\boldsymbol{a}, \boldsymbol{b}]$ with $\boldsymbol{a} = (a_1, \ldots, a_d)^\top$, $\boldsymbol{b} = (b_1, \ldots, b_d)^\top$, such that $\mathcal{D} \subseteq \mathcal{S}$ is a compact subregion containing most of the probabilities of interest, i.e., $\int_{\mathcal{S}\setminus\mathcal{D}} \pi(\theta)d\theta < \epsilon$. Linearly transform G_0 onto $\mathcal{D}$ and denote as $G = \{\boldsymbol{g}_k = (g_{k1}, \ldots, g_{kd}), k = 1, \ldots, M\}$. Set n, m such that $N = nm$ and let $i = 1$.*

Step 2: **Random shift.** *Generate a random shift vector $\boldsymbol{u}_i = (u_{i1}, \ldots, u_{id})^\top \in [\boldsymbol{0}, \boldsymbol{b} - \boldsymbol{a}]$, where $u_{it} \sim U(0, b_t - a_t), t = 1, \ldots, d$. Shift G by $\boldsymbol{g}_k \oplus \boldsymbol{u}_i$, where the tth entry of $\boldsymbol{g}_k \oplus \boldsymbol{u}_i$ is*

$$(\boldsymbol{g}_k \oplus \boldsymbol{u}_i)_t = \begin{cases} g_{kt} + u_{it} & \text{if } a_t \le g_{kt} + u_{it} \le b_t \\ g_{kt} + u_{it} + a_t - b_t, & \text{otherwise.} \end{cases}$$

$G \oplus \boldsymbol{u}_i = \{\boldsymbol{g}_k \oplus \boldsymbol{u}_i : k = 1, \ldots, M\}$ is the randomly-shifted GLP in the i-th batch.

Step 3: **Likelihood.** *Compute $q_{ik} = p(\boldsymbol{g}_k \oplus \boldsymbol{u}_i)/\sum_{j=1}^M p(\boldsymbol{y}_j \oplus \boldsymbol{u}_i)$ and let $\hat{F}_i$ be the multinomial distribution over $\{\boldsymbol{g}_k \oplus \boldsymbol{u}_i : k = 1, \ldots, M\}$ with weight q_{ik} for $k = 1, \ldots, M$.*

Step 4: **Batch sampling.** *Sample from $\hat{F}_i$ m times to obtain the i-th batch of samples $B_i = \{\boldsymbol{x}_{i,1}, \ldots, \boldsymbol{x}_{i,m}\}$. If $i < n$, set $i = i + 1$ and go to Step 2.*

Step 5: **Output.** *Drawn sample set $\mathbb{X} = \{\boldsymbol{x}_{i,j} \in \mathcal{D} : i = 1, \ldots, n, j = 1, \ldots, m\}$.*

In Step 2 of Algorithm 6.2, the random shifts of a GLP set produce the representative weighted points for the target distribution because of the good space-filling property of the GLP set. Based on the random shifts, multiple multinomial samplings generate multiple batches of samples, which increases the diversity and randomness of the sampling. Algorithm 6.2 generates n batches of m-size random samples, and totally $N = nm$ samples. It can be easily known that each $\boldsymbol{x}_{i,j}$ follows the same distribution, denoted by $F_M(\theta)$. Moreover, the drawn samples in different batches are independent, while those in the same batch are dependent. Regarding computational complexity for complicated distribution, To reduce the running time for a fixed sample size, we can increase the batch size by m and decrease the number of batches by n. When the size of GLP is less than 500, and the sample size is less than 400, $m = 1$ is recommended with good performance and acceptable computational cost; otherwise, $m > 1$, such as $m = \sqrt{N}$, may be a good choice.

Denote $\epsilon_{\mathrm{I}} = \sup_{\theta \in \mathcal{D}} V_{\mathrm{HK}}(f \cdot I_{[\boldsymbol{a},\theta]}) D^*(Y_M \oplus \boldsymbol{u}) v(\mathcal{D})$, where $V_{\mathrm{HK}}(f \cdot I_{[\boldsymbol{a},\theta]})$ is the total variation of $f \cdot I_{[\boldsymbol{a},\theta]}$ on $\mathcal{D}$ in the sense of Hardy and Krause and $D^*(Y_M \oplus \boldsymbol{u})$ is the star discrepancy of $Y_M \oplus \boldsymbol{u}$ and $v(\mathcal{D})$ is the volume of $\mathcal{D}$. Moreover, denote $\kappa(M) = 1/M$ when $d = 1$ and $(\log(M))^d/M$, otherwise. Moreover, we consider the following assumptions.

Assumption 1. $f(\boldsymbol{x}) I_{[\boldsymbol{a},\theta]}(\boldsymbol{x})$ is of uniformly bounded variation in the sense of Hardy and Krause for any $\theta \in \mathcal{D}$.
Assumption 2. M is sufficiently large such that $\epsilon_{\mathrm{I}} \in (0, 1/2)$.
Assumption 3. m is fixed.
Assumption 4. There exists a constant $\alpha > 0$, such that $\kappa(M)\sqrt{m}/(\log(m))^\alpha = O(1)$.

Assumption 1 is required to guarantee that the error bound in Koksma-Hlawka inequality is bounded and serves to achieve uniform asymptotic results. Commonly, monotone functions and absolutely continuous functions are of bounded variation. Assumptions 2–4 impose some assumptions on the parameters M and m in GLS to achieve the convergence in probability or with probability 1. When Assumption 1 holds, $V_{\mathrm{HK}}(f \cdot I_{[\boldsymbol{a},\theta]})$ can be bounded. The region $\mathcal{D}$ is bounded and closed; thus, $v(\mathcal{D})$ is bounded. $D^*(Y_M \oplus \boldsymbol{u}) = O(\kappa(M))$ and $\kappa(M)$ is decreasing with respect to M. Hence, Assumption 2 can be satisfied when M is enlarged to some degree. Moreover, when the sample size N is given, the batch size $m(\leq N)$ can be set based on the user's requirement for the performance and the computational complexity of GLS. Assumption 4 can also be easily satisfied when M is set larger than m to a certain extent.

Yi et al. (2023) established the asymptotic results of the GLS samples as follows.

Theorem 6.1 *For any $\epsilon_0 > 0$, we can cut out a closed and bounded hypercube $\mathcal{D}$ and perform the GLS algorithm on $\mathcal{D}$. If Assumptions 1,2,3 or Assumptions 1,2,4 with $\alpha < \beta$ holds for any $\beta > 0$, then the empirical CDF of the GLS samples, denoted by $F_{M,n,m}(\theta)$, converges to the target CDF $\Pi(\theta)$, that is,*

$$\sup_{\theta \in \mathcal{S}} |F_{M,n,m}(\theta) - \Pi(\theta)| < \epsilon_0,$$

when M and nm are sufficiently large.

The proof of Theorem 6.1 can be seen in Yi et al. (2023). From Theorem 6.1, it is shown that the rate of convergence of the drawn samples $\mathbb{X}$ is relative to the size of the used GLP M, the batch size m, and the number of batches n. Many simulation results in Yi et al. (2023) have shown that GLS performs well and is stable for multimodal distributions, even in high-dimensional spaces. Hence, GLS is well-suited to exploring thin modes and high-dimensional spaces using GLP and random shift techniques.

A suitable property of GLS is that it does not need to set the proposal distribution in advance. When the proposal distribution is well set for the methods that need proposals, GLS gives comparable performance; when the proposal distribution cannot be given suitably, GLS outperforms those methods. Then, GLS is feasible and applicable when the target distribution is complex, which is common in practice. GLS also performs well for unimodal distributions. Because of its effectiveness, robustness, and speediness, it can be a promising and competitive alternative.

6.2 Elliptically Symmetric Distributions

The class of elliptically symmetric distributions extends the multivariate normal distribution. It includes multivariate t, multiuniform on a ball, multiuniform on the unit sphere, multivariate Kotz type, multivariate Cauchy, multivariate logistics, multivariate Person types II and VII, and a mix of normal distributions on scale. The main reference for the elliptical distribution is Fang et al. (1990). The following basic knowledge is from their monograph.

The spherically symmetric distribution plays a role like the standard normal distribution $N(0, 1)$, and the elliptically symmetric distribution has its extension like $N(\mu, \sigma^2)$.

Definition 6.1 *Spherically contoured distribution*
An p dimensional random vector $\boldsymbol{X}$ *is said to have a spherically symmetric distribution (SSD), or a spherical distribution (SD) for short, if* $\boldsymbol{X} \stackrel{d}{=} \boldsymbol{P}\boldsymbol{X}$ *for every orthogonal matrix* $p \times p$ $\boldsymbol{P}$. *We write* $\boldsymbol{X} \sim S_p(\phi)$, *where* ϕ *belongs to the class of characteristic function (cf)*

$$\Phi_p = \{\phi(\cdot) : \phi(t_1^2 + \cdots t_p^2) \text{ is a p-dimensional cf }\},$$

and ϕ *is called generating cf for* $\boldsymbol{X}$.

It is unnecessary that $\boldsymbol{X} \sim S_p(\phi)$ possesses a density. For example, the uniform distribution on the unit sphere S^p does not have a density function in $\mathbb{R}^p$.

Theorem 6.2 *If* $\boldsymbol{X} \sim S_p(\phi)$ *and* $P(\boldsymbol{X} = \boldsymbol{0}) = 0$, *then*

1. $\boldsymbol{X}$ *has a stochastic representation*

$$\boldsymbol{X} \stackrel{d}{=} R\boldsymbol{U}^{(p)}, \tag{6.7}$$

where $\|\boldsymbol{X}\| \stackrel{d}{=} R$ *and* $\boldsymbol{X}/\|\boldsymbol{X}\| \stackrel{d}{=} \boldsymbol{U}^{(p)}$ *are independent, and* $\boldsymbol{U}^{(p)}$ *is uniformly distributed on the unit sphere* $S^p = \{\boldsymbol{y} : \boldsymbol{y}^T\boldsymbol{y} = 1\}$.

2. *If* $\boldsymbol{X}$ *has a density, its density must be of the form* $g(\boldsymbol{x}^T\boldsymbol{x})$. *We can write* $\boldsymbol{X} \sim S_p(g)$. *The function* g *is called the density generator.*

3. $\boldsymbol{X}$ *has a density iff* R *in (6.7) has a density denoted by* $f(r), r \geq 0$. *The relationship between* $g(\cdot)$ *and* $f(\cdot)$ *is given by*

$$f(r) = \frac{2\pi^{p/2}}{\Gamma(p/2)} r^{p-1} g(r^2). \tag{6.8}$$

Definition 6.2 *Elliptically symmetric distribution*
A random vector $\boldsymbol{X} \in \mathbb{R}^p$ *is said to have an elliptically symmetric distribution (ESD), or elliptical distribution (ED) for short, if*

$$\boldsymbol{X} \stackrel{d}{=} \boldsymbol{\mu} + R\boldsymbol{\Psi}^{1/2}\boldsymbol{U}^{(p)}, \tag{6.9}$$

where $\boldsymbol{\mu} \in \mathbb{R}^p$, $R \geq 0$ *is independent of* $\boldsymbol{U}^{(p)}$, $\boldsymbol{\Psi}$ *is a positive definite matrix of order* p, *and* $\boldsymbol{\Psi}^{1/2}$ *is the positive definite square root of* $\boldsymbol{\Psi}$.

An ESD has a cf of the form

$$\exp(i\boldsymbol{t}^T\boldsymbol{\mu})\phi(\boldsymbol{t}^T\boldsymbol{\Psi}\boldsymbol{t}),$$

and we write $\boldsymbol{X} \sim ESD_p(\boldsymbol{\mu}, \boldsymbol{\Psi}, \phi)$. It is not necessary that $\boldsymbol{X} \sim ECD_p(\boldsymbol{\mu}, \boldsymbol{\Psi}, \phi)$ possess a density. If $\boldsymbol{X}$ has a density, it has the form

$$|\boldsymbol{\Sigma}|^{1/2} g((\boldsymbol{x}-\boldsymbol{\mu})^T\boldsymbol{\Psi}^{-1}(\boldsymbol{x}-\boldsymbol{\mu})),$$

and we denote $\boldsymbol{X} \sim ESD_p(\boldsymbol{\mu}, \boldsymbol{\Psi}, g)$ or $\boldsymbol{X} \sim ECD_p(\boldsymbol{\mu}, \boldsymbol{\Psi}, g)$, as the contours of constant density are ellipsoids. The function $g(\cdot)$ is the density generator of $\boldsymbol{X}$. Note that $\boldsymbol{\Psi}$ is proportional to $\boldsymbol{\Sigma} = \text{Cov}(\boldsymbol{X})$. Different distributions of R generate different subclasses of elliptical distributions.

Example 6.3 *Let random vector $\boldsymbol{X}$ be uniformly distributed on the unit sphere in $\mathbb{R}^p$. The pdf, in a general sense, is given by*

$$p(\boldsymbol{x}) = p(x_1, \ldots, x_p) = \frac{\Gamma(p/2)}{\pi^{p/2}}(1 - \sum_{i=1}^p x_i^2), \ \sum_{i=1}^p x_i^2 < 1, \boldsymbol{x} \in \mathbb{R}_+^p.$$

So its density generator and density of R are

$$g(u) = \frac{\Gamma(p/2)}{\pi^{p/2}}(1-u), \ 0 < u < 1,$$
$$f(r) = 2r^{p-1}(1-r^2), \ 0 < r < 1,$$

respectively.

The following are some subclasses of ESD and their basic knowledge:

A. Multivariate normal distribution
$\boldsymbol{X} \sim N_p(\boldsymbol{\mu}, \boldsymbol{\Sigma})$ with the density generator g

$$g(u) = (2\pi)^{-p/2} \exp\left\{-\frac{1}{2}u\right\} \tag{6.10}$$

and the random variable R has a chi-distribution with the density

$$f(r) = \frac{1}{\Gamma(p/2)2^{p/2-1}} r^{p-1} e^{-r^2/2}. \tag{6.11}$$

The mean vector and covariance matrix of $\boldsymbol{X}$ are

$$E(\boldsymbol{X}) = \boldsymbol{\mu}, \quad \text{Cov}(\boldsymbol{X}) = \boldsymbol{\Sigma}, \tag{6.12}$$

respectively.

B. Multivariate Kotz type distribution
$\boldsymbol{X} \sim Kotz_p(\boldsymbol{\mu}, \boldsymbol{\Psi} g)$ with the density generator g

$$g(u) = \frac{s\Gamma(p/2)b^{(2N+p-2)/(2s)}}{\pi^{p/2}\Gamma((2N+p-2)/(2s))} u^{N-1} \exp\{-bu^s\} \tag{6.13}$$

with parameters b, s, N satisfying $b, s > 0$ and $2N + p > 2$, and the density of R is given by

$$f(r) = \frac{2sb^{(2N+p-2)/(2s)}}{\Gamma((2N+p-2)/(2s))} r^{2N+p-3} \exp\left\{-br^{2s}\right\}. \tag{6.14}$$

The mean vector and variance of $\boldsymbol{X}$ is given by

$$E(\boldsymbol{X}) = \boldsymbol{\mu},\ \boldsymbol{\Sigma} = \mathrm{Cov}(\boldsymbol{X}) = b^{-1/s}\frac{\Gamma((2N+p)/(2s))}{r\Gamma((2N+p-2)/(2s))}\boldsymbol{\Psi}, \tag{6.15}$$

where r is the rank of $\boldsymbol{\Psi}$.

C. Multivariate Pearson type II distribution
$\boldsymbol{X} \sim MPII_p(\boldsymbol{\mu}, \boldsymbol{\Psi}, g)$ with the density generator is

$$g(u) = \frac{\Gamma(p/2+m+1)}{\Gamma(m+1)\pi^{p/2}}(1-u)^m, \quad 0 \le u \le 1,\ m > -1, \tag{6.16}$$

and the density of R is

$$f(r) = \frac{2\Gamma(p/2+m+1)}{\Gamma(p/2)\Gamma(m+1)}r^{p-1}(1-r^2)^m,\ 0 \le r \le 1. \tag{6.17}$$

The mean vector and variance of $\boldsymbol{x}$ are

$$E(\boldsymbol{X}) = \boldsymbol{\mu},\ \boldsymbol{\Sigma} = \mathrm{Cov}(\boldsymbol{X}) = (p+2m+2)^{-1}\boldsymbol{\Psi}. \tag{6.18}$$

As $R^2 \sim Beta(p/2, m+1)$, a beta distribution, so the sample of R can be generated by the beta distribution, which takes the square root or applies the inverse transformation method directly to R.

D. Multivariate Pearson type VII distribution
$\boldsymbol{X} \sim MPVII_p(\boldsymbol{\mu}, \boldsymbol{\Psi}, g)$ with the density generator g

$$g(u) = (\pi m)^{-p/2}\frac{\Gamma(N)}{\Gamma(N-p/2)}(1+u/m)^{-N}, N > p/2, m > 0, \tag{6.19}$$

and the density of R is

$$f(r) = \frac{2}{B(p/2, N-p/2)}m^{-p/2}r^{p-1}\left(1+\frac{r^2}{m}\right)^{-N}, \tag{6.20}$$

where B is the beta function. The mean and variance of $\boldsymbol{x}$ are

$$E(\boldsymbol{X}) = \boldsymbol{\mu},\ \boldsymbol{\Sigma} = Cov(\boldsymbol{X}) = \frac{m}{2N-p-2}\boldsymbol{\Psi},\ \text{respectively.} \tag{6.21}$$

As the random variable $R^2/m \sim BeII(p/2, N-p/2)$, which is a beta type II distribution. Fang et al. (1990) gave the details of beta type II distribution (or beta prime in Table 1.4), and they also pointed out the distribution of $R^2/(m+R^2)$ which is a beta distribution with parameters $p/2$ and $(N-p/2)/2$. When $N = \frac{1}{2}(p+m)$ and m is an integer in (6.19), the corresponding distribution is called a multivariate t-distribution and is denoted by $\boldsymbol{X} \sim MT_p(m, \boldsymbol{\mu}, \boldsymbol{\Sigma})$.

Table 6.1 lists the valuable subclasses of ESDs with their density generators. From the stochastic representation of the spherical distributions (6.7), the different subclasses have different distributions of R. Figure 6.2 gives a plot of densities of R for the four subclasses of ESD, that is, $R_{normal}, R_{Kotz}, R_{P-II}$ and R_{P-VII}. The density of R_{normal} looks to have a long tail; the density of R_{Kotz} spreads more in the interval $(0.25, 1.0)$; densities of R_{P-II} and R_{P-VII} are close to each other, and the density of R_{P-II} spreads in the interval $(0, 1)$.

TABLE 6.1
Density function of some useful ECDs

Type	Density Function $g(\boldsymbol{x})$ in $\mathbb{R}^p$
Kotz type	$g(\boldsymbol{x}) = c(\boldsymbol{x}^T\boldsymbol{x})^{N-1}\exp[-b(\boldsymbol{x}^T\boldsymbol{x})^s], b, s > 0, 2N + p > 2$
Multinormal	$g(\boldsymbol{x}) = c\ \exp(-\frac{1}{2}\boldsymbol{x}^T\boldsymbol{x})$
Person Type VII	$g(\boldsymbol{x}) = c(1 + \boldsymbol{x}^T\boldsymbol{x}/m)^{-N}, N > p/2, m > 0$
Multivariate t	$g(\boldsymbol{x}) = c(1 + \boldsymbol{x}^T\boldsymbol{x}/s)^{-(p+m)/2}, m > 0$, an integer, $s > 0$
Multivariate Cauchy	$g(\boldsymbol{x}) = c(1 + \boldsymbol{x}^T\boldsymbol{x})^{-(p+1)/2}$
Person Type II	$g(\boldsymbol{x}) = c(1 - \boldsymbol{x}^T\boldsymbol{x})^m, m > -1$
Logistics	$g(\boldsymbol{x}) = c\ \exp(-\boldsymbol{x}^T\boldsymbol{x})/[1 + \exp(-\boldsymbol{x}^T\boldsymbol{x})]^2$

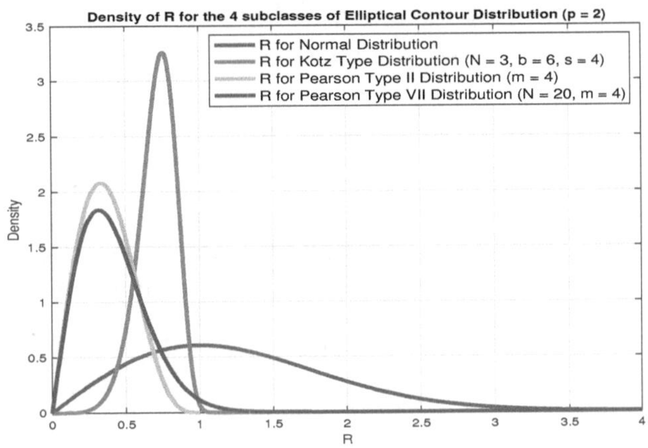

FIGURE 6.1
Densities of R for the four subclasses of ECD.

6.3 Generation of QMC-RPs for Elliptical Distributions

From Definition 6.1, it is easy to note that a critical issue in generating a random sample from an elliptical distribution is to find a feasible SR of the uniform distribution on the unit sphere $S^p = \{\boldsymbol{y} : \boldsymbol{y}^T\boldsymbol{y} = 1\}$. Then, by the SR of the elliptical distribution (6.9), we can find the QMC-RP of the elliptical distribution. There are many ways to generate a random sample from $U(S^p)$.

One method is to use the SR $\boldsymbol{U}^{(p)} \stackrel{d}{=} \boldsymbol{X}/\|\boldsymbol{X}\|$, where $\boldsymbol{X}$ can be chosen as any spherical distribution with $P(\boldsymbol{X} = \boldsymbol{0}) = 0$. In particular, we can choose the standard normal distribution $N_p(\boldsymbol{0}, \boldsymbol{I}_p)$.

Algorithm 6.3 (Generating a sample from $U(S^p)$ based on $N_p(\boldsymbol{0}, \boldsymbol{I}_p)$)

Step 1: Generate p random samples $z_1, \ldots, z_p$ from the standard normal distribution $N(0, 1)$.

Step 2: Calculate $z = (z_1^2 + \cdots + z_p^2)^{1/2}$ and $u_i = z_i/z, i = 1, \ldots, p$. The $\boldsymbol{u} = (u_1, \ldots, u_p)$ is a random sample from $U(S^p)$.

Step 3: Repeat n times of Steps 1 and 2, we get a set of n random samples from $U(S^p)$.

Algorithm 6.4 (Generation of NT-Net on S^p based on $N_p(\mathbf{0}, \boldsymbol{I}_p)$)

Step 1: Produce a NT-net on the unit cube C^p with n points by some QMC methods. Denote this net as an $n \times p$ matrix $\boldsymbol{C} = (c_{ij})$.

Step 2: calculate $z_{ij} = \Phi^{-1}(c_{ij}), i = 1, \ldots, n; j = i, \ldots, p$, where $\Phi(x)$ is the cdf of $N(0, 1)$. Now each $z_{ij} \sim N(0, 1)$.

Step 3: Calculate $z_i = (z_{i1}^2 + \cdots + z_{ip}^2)^{1/2}, i = 1, \ldots, n$ and produce the matrix $\boldsymbol{E} = (e_{ij} = \frac{z_{ij}}{z_i}), i = 1, \ldots, n; j = 1, \ldots, p$. Deliver the n rows of $\boldsymbol{E}$ as a set of QMC-RPs of $U(S^p)$.

6.3.1 An efficient algorithm for generation of NT-Net on the unit sphere S^p

Another method is the so-called TFWW algorithm that was independently proposed by Tashiro (1977) and Fang et al. (1992). Its QMC version is called the NTSR algorithm (number-theoretic stochastic representation algorithm) for generating an NT-net on S^p (Fang and Wang (1994)). The NTSR algorithm follows up the following stochastic representation.

Theorem 6.3 *Suppose $\boldsymbol{U} \sim U(S^p)$. Then $\boldsymbol{U}$ has a SR*

$$\boldsymbol{U} = \begin{bmatrix} \boldsymbol{U}_1 \\ \vdots \\ \boldsymbol{U}_m \end{bmatrix} \stackrel{d}{=} \begin{bmatrix} D_1 \boldsymbol{U}^{(1)} \\ \vdots \\ D_m \boldsymbol{U}^{(m)} \end{bmatrix}, \tag{6.22}$$

where $\boldsymbol{U}_j : t_j \times 1,\ t_1 + \cdots + t_m = p$, and

1. *$\boldsymbol{U}^{(j)} \sim U(S^{t_j})$, the uniform distribution on $S^{t_j}, j = 1, \cdots, m$.*
2. *$\boldsymbol{U}^{(1)}, \cdots, \boldsymbol{U}^{(m)}$ and $(D_1, \cdots, D_m)$ are mutually independent.*
3. *$D_j > 0, j = 1, \cdots, m$, and $(D_1^2, \cdots, D_m^2) \sim Diri_m(t_1/2, \cdots, t_m/2)$ which is a Dirichlet distribution (See Definition 6.3) with parameters $\{t_1/2, \cdots, t_m/2\}$.*

The proof of Theorem 6.3 can refer to Theorem 2.6, Fang et al. (1990).

Definition 6.3 *Let $X_1, ..., X_m$ and X_{m+1} be independent random variables, and $X_i \sim Gamma(\alpha_i, 1))$, $i = 1, ..., m + 1$. Set*

$$Y = X_1 + \cdots X_{m+1}, \quad Y_j = \frac{X_j}{Y}, j = 1, ..., m.$$

The distribution of $(Y_1, ..., Y_m)$ is called the Dirichlet distribution with parameters $\alpha_1, ..., \alpha_{m+1}$ and write

$$(Y_1, ..., Y_m) \sim Diri_m(\alpha_1, ..., \alpha_m; \alpha_{m+1}),$$
or
$$(Y_1, ..., Y_{m+1}) \sim Diri_{m+1}(\alpha_1, ..., \alpha_{m+1}).$$

The Dirichlet distribution is an extension of the beta distribution.

Notably, the uniform distribution $U(T^p)$ represents a specific case of the Dirichlet distribution with $\alpha_i = 1, i = 1, \ldots, p$. The domain T^p is defined as $T^p(r) = \{\boldsymbol{x} : \boldsymbol{x} \in R_+^p, x_1 + \ldots + x_p = r, r > 0\}$ with $r = 1$, see (1.60). The distribution $U(T^p)$ is widely recognized as a popular and essential compositional distribution in T^p. When $\alpha = 1$, the gamma distribution $Ga(1, \lambda)$ is just the exponential distribution $EP(\lambda)$. Lemma 6.1 lists three beautiful properties of Dirichlet distribution. Readers may refer to Fang et al. (1990) for further details.

Lemma 6.1 *Suppose* $(Y_1, \cdots, Y_p) \sim Diri_p(\alpha_1, \cdots, \alpha_p)$, *where* $p \geq 2$. *Then* $(Y_1, \cdots, Y_p) \in T^p$, *and the following properties hold.*

(a). There is a stochastic representation of Y_i*'s, that is,*

$$Y_i = \frac{X_i}{\sum_{j=1}^p X_j}, \quad i = 1, \cdots, p,$$

where $X_1, \cdots, X_p$ *are independent random variables following gamma distributions,* $X_i \sim Ga(\alpha_i, \lambda), \alpha_i > 0, \lambda > 0, i = 1, \cdots, p$. *The distribution of* $(Y_1, \cdots, Y_p)$ *is independent of* λ.

(b). The lower moments of $(Y_1, \cdots, Y_p)$ *are*

$$E(Y_j) = \frac{\alpha_j}{\alpha}, \quad \mathrm{Var}(Y_j) = \frac{\alpha_j(\alpha - \alpha_j)}{\alpha^2(\alpha+1)}, \quad \mathrm{Cov}(Y_i, Y_j) = \frac{-\alpha_i \alpha_j}{\alpha^2(\alpha+1)}. \tag{6.23}$$

(c). The $Diri_p(1, \cdots, 1)$ *reduces to the uniform distribution* $U(T^p)$.

Lemma 6.2 *Suppose* $\boldsymbol{X} = (X_1, \ldots, X_p)$ *forms by p iid samples from the standard exponential distribution* $EP(1)$. *Let* $R = X_1 + \cdots + X_p$ *and* $\boldsymbol{U} = (U_1, \ldots, U_p) = \boldsymbol{X}/R$. *Then,*

(a) $\boldsymbol{U} \sim U(T^p)$;

(b) R *and* $\boldsymbol{U}$ *are independent;*

(c) $\boldsymbol{X}$ *has a stochastic representation:* $\boldsymbol{X} \stackrel{d}{=} R\boldsymbol{U}$.

When we choose $t_j = 2, j = 1, \ldots, m$ for $p = 2m$ and $t_1 = 3, t_j = 2, 2 \leq jm$ for $p = 2m + 1$, the corresponding algorithm is simpler. There is a minor difference for even p or odd p.

6.3.2 The TFWW algorithm

The TFWW algorithm for even p

When $p = 2m$ is an even number and $t_1 = \cdots = t_m = 2$, we have $\boldsymbol{U}^{(j)} \sim U(S^2)$ and $(D_1^2, \cdots, D_m^2) \sim Diri_m(1, \cdots, 1) = U(T_m)$ the uniform distribution on T_m where $T_m = \{(x_1, \ldots, x_m) : x_i \geq 0, x_1 = \ldots + x_m = 1\}$ is a simplex, so the stochastic representation (6.22) can be written as the form

$$\boldsymbol{U} \stackrel{d}{=} \begin{pmatrix} D_1 \cos(2\pi\phi_1) \\ D_1 \sin(2\pi\phi_1) \\ D_2 \cos(2\pi\phi_2) \\ D_2 \sin(2\pi\phi_2) \\ \vdots \\ D_m \cos(2\pi\phi_m) \\ D_m \sin(2\pi\phi_m) \end{pmatrix}. \tag{6.24}$$

where

1. $(D_1^2, \cdots, D_m^2) \sim U(T_m)$ and $D_j > 0$, $j = 1, \cdots, m$.
2. $(\phi_1, \cdots, \phi_m)$ and $(D_1, \cdots, D_m)$ are mutually independent.
3. $(\phi_1, \cdots, \phi_m) \sim U(C^m)$, where C^m is the m-dimensional cube.

The TFWW algorithm for odd p

When $p = 2m+1$ is odd, we can choose $t_1 = 3, t_2 = \cdots = t_m = 2$, then $(D_1^2, \cdots, D_m^2) \sim Diri_m(3/2, 1, \cdots, 1)$, and the stochastic representation (6.22) can be written as

$$\boldsymbol{U} \stackrel{d}{=} \begin{pmatrix} D_1(1-2\phi_1) \\ 2D_1\sqrt{\phi_1(1-\phi_1)}\cos(2\pi\phi_2) \\ 2D_1\sqrt{\phi_1(1-\phi_1)}\sin(2\pi\phi_2) \\ D_2\cos(2\pi\phi_3) \\ D_2\sin(2\pi\phi_3) \\ \vdots \\ D_m\cos(2\pi\phi_{m+1}) \\ D_m\sin(2\pi\phi_{m+1}) \end{pmatrix} \tag{6.25}$$

where

1. $(D_1^2, \cdots, D_m^2) \sim D_m(\frac{3}{2}, 1, \cdots, 1)$.
2. $(\phi_1, \cdots, \phi_{m+1})$ and $(D_1, \cdots, D_m)$ are mutually independent.
3. $(\phi_1, \cdots, \phi_{m+1}) \sim U(C^{m+1})$.

Note that from the stochastic representations (6.24) and (6.25), one can quickly generate random samples from the uniform distribution in S^p. More details of the TFWW algorithm can be found in Section 1.5.5 of Fang and Wang (1994). The weakness of the TFWW algorithm is more complicated. Using Algorithm 6.4, we can have a simple algorithm to generate $U(S^p)$.

6.3.3 Algorithm for generation of QMC-RPs of an elliptical distribution

Let $\boldsymbol{X} \sim ESD_p(\boldsymbol{\mu}, \boldsymbol{\Psi}, g)$ follow an elliptical distribution with the density generator g. Its SR is given by $\boldsymbol{X} \stackrel{d}{=} \boldsymbol{\mu} + R\boldsymbol{\Psi}^{1/2}\boldsymbol{U}^{(p)}$.

Algorithm 6.5 (The NTSR algorithm for an elliptical distribution)

Step 1 Generate a set of quasirandom numbers $\{\boldsymbol{c}_i = (c_{i1}, \cdots, c_{ip}), i = 1, \cdots, k\}$ *on* C^p.

Step 2 Denote the cdf of R *by* $F_R(r)$ *and let* F_R^{-1} *be its inverse function. Compute* $r_i = F_R^{-1}(c_{ip})$, $i = 1, \cdots, k$.

Step 3 By the TFWW algorithm or Algorithm 6.4 to generate a set of quasirandom F-numbers $\{\boldsymbol{u}_i, i = 1, \cdots, k\}$ *of the uniform distribution on* S^p *with the first* $(p-1)$*-components of* $\boldsymbol{c}_i$*'s.*

Step 4 Then $\{\boldsymbol{y}_i = \boldsymbol{\mu} + r_i\boldsymbol{\Psi}^{1/2}\boldsymbol{u}_i, i = 1, \cdots, k\}$ *is a set of quasirandom F-numbers of the given elliptical distribution* $F(\boldsymbol{x})$.

6.4 Generation of MSE-RPs for Multivariate Distributions

The MSE-RPs of a continuous multivariate distribution $\boldsymbol{X} \sim F(\boldsymbol{x})$ in the sense of minimum mean square error and approximation distribution based on MSE-RPs have been introduced in Definitions 2.2 and 2.3 in Section 2.3. The MSE-RPs are also called *optimal vector quantizers* or *principal points* in the literature. The MSE-RPs $\boldsymbol{\xi} = \{\boldsymbol{\xi}_1, \ldots, \boldsymbol{\xi}_k\}$ minimizes

$$\text{MSE}(\mathcal{Y}) = E[\min_{1\le j\le k} (\boldsymbol{X} - \boldsymbol{Y}_j)^T(\boldsymbol{X} - \boldsymbol{Y}_j)],\ \mathcal{Y} = \{\boldsymbol{Y}_1, \ldots, \boldsymbol{Y}_k\}. \tag{6.26}$$

Suppose that we can generate a continuous multivariate distribution $F(\boldsymbol{x}) = F(x_1, \ldots, x_p)$ by the Monte Carlo method. Linde et al. (1980) proposed an iterative algorithm (LBG algorithm) for generating a set of RPs for $F(\boldsymbol{x})$ based on the well-known k-means method.

6.4.1 The LBG and NTLBG algorithms

Algorithm 6.6 (LBG Algorithm)

Step 1: Generating a training sequence $\mathcal{X} = \{\boldsymbol{x}_1, \ldots, \boldsymbol{x}_N\}$ from the given distribution $F(\boldsymbol{x})$ by a Monte Carlo method, where N is much larger than k, and k is the number of MSE-RPs for $F(\boldsymbol{x})$. Set $t = 0$.

Step 2: Give an initial set of output vectors $\mathcal{Y}_t = \{\boldsymbol{y}_{t1}, \ldots, \boldsymbol{y}_{tk}\}$ and find the associated Voronoi partition $\mathcal{S}_t = \{\mathcal{S}_{t1}, \ldots, \mathcal{S}_{tk}\}$ by assigning each $\boldsymbol{x}_i$ to the nearest region of the partition $\mathcal{S}_t$ in the sense of L_2-distance, i.e.,

$$\mathcal{S}_{ti} = \{\boldsymbol{x} : ||\boldsymbol{x} - \boldsymbol{y}_{ti}|| < ||\boldsymbol{x} - \boldsymbol{y}_{tj}||, j \neq i\}, i = 1, \ldots, k.$$

Step 3: Calculate the sample mean, denoted by $\boldsymbol{y}_{t+1,j}$, of $\boldsymbol{x}_i$ which falling in the region $\mathcal{S}_{tj}, j = 1, \ldots, k$ and denote $\mathcal{Y}_{t+1} = \{\boldsymbol{y}_{t+1,1}, \ldots, \boldsymbol{y}_{t+1,k}\}$. If $\mathcal{Y}_{t+1} = \mathcal{Y}_t$, deliver $\mathcal{Y}_t$ as MSE-RPs, and account the number of points, denoted by $n_1, \ldots, n_k$, of $\mathcal{X}$ falling in $\mathcal{S}_{tj}, j = 1, \ldots, k$, and deliver $(n_i/N, i = 1, \ldots, k)$ as estimate of probabilities of $(p_i = P(\boldsymbol{X} \in \mathcal{S}_i), i = 1, \ldots, k)$; otherwise to to next step.

Step 4: Let $t := t + 1$ and go to step 2).

Although the LBG algorithm can reach the local optimal output of $\mathcal{Y}$ with a non-increasing MSE, Fang et al. (1994b) pointed out two problems of this algorithm:

(a) The algorithm provides a local optimum that is strongly dependent on the initial points;

(b) The generation of samples of $F(\boldsymbol{x})$ and the calculation of MSE are based on Monte Carlo method, which is slow with the convergence rate $O_p(n^{-1/2})$.

Fang et al. (1994b) proposed a revised algorithm for finding MSE-RPs of $F(\boldsymbol{x})$ based on the LBG algorithm via the numeric-theoretical methods. The algorithm is called as the NTLBG algorithm.

Algorithm 6.7 (NTLBG Algorithm)

Step 1: For given $\boldsymbol{X} \sim F(\boldsymbol{x})$ to generate a set of quasirandom F-number $\boldsymbol{x}_1, \cdots, \boldsymbol{x}_N$ as a training sequence by the NTSR algorithm 6.6 with a large N. Set t = 0.

Step 2: For given k, generate a set of quasirandom F-numbers $\mathcal{Y}_t = \{\boldsymbol{y}_{t1}, \cdots, \boldsymbol{y}_{tk}\}$ *of* $F(\boldsymbol{x})$ *as an initial set of output vectors.*

Step 3: Form a partition $\{S_i\}$ *of* $\{\boldsymbol{x}_j, j = 1, \cdots, N\}$ *such that each* $\boldsymbol{x}_i$ *is assigned to the nearest region of the partition, i.e.,* $\boldsymbol{x}_i \in S_m$ *if* $||\boldsymbol{x}_i - \boldsymbol{y}_{tm}|| \le ||\boldsymbol{x}_i - \boldsymbol{y}_{tj}||$, $j \neq m$.

Step 4: Estimate the sample conditional means $\boldsymbol{y}_m = E[\boldsymbol{X}|\boldsymbol{x} \in S_m]$ *by*

$$\boldsymbol{y}_{t+1,m} = \frac{1}{n_{t+1,m}} \sum_{\boldsymbol{x}_j \in S_m} \boldsymbol{x}_j \tag{6.27}$$

where $n_{t+1,m}$ *is the number of* $\boldsymbol{x}_j$ *falling in* S_m *and forms a new set of output vector vectors* $\mathcal{Y}_{t+1} = \{\boldsymbol{y}_{t+1,1}, \cdots, \boldsymbol{y}_{t+1,k}\}$. *If* $\mathcal{Y}_{t+1} = \mathcal{Y}_t$, *deliver* $\mathcal{Y}_t$ *as MSE-RPs and go to Step 6 and* $n_{t+1,m}/N$ *as estimated probability of* $p_m = P(\boldsymbol{x} \in S_m)$, *otherwise to next step.*

Step 5: Let $t := t + 1$ *and go to step 3.*

Step 6: Calculate and deliver estimated MSE

$$MSE(\mathcal{Y}_t) = \sum_{i=1}^{N} \frac{n_{t+1,m}}{N} \sum_{m=1}^{k} (\boldsymbol{x}_i - \boldsymbol{y}_{t,m})^T (\boldsymbol{x}_i - \boldsymbol{y}_{t,m}) I(\boldsymbol{x}_i \in S_m).$$

When applying the k-means++ algorithm (refer to Subsection 3.6.4) to generate MSE-RPs of a given $F(\mathbf{x})$, we only need to replace Step 2 in the NTLBG algorithm by the k-means++ seeding method and other steps remain the same, that is, the training set is still generated by the QMC method instead of the MC method because the samples generated by the QMC method have better representativeness. For numerical examples, one can apply both the NTSR algorithm and k-means++ seeding method to generate the set of initial points and choose.

6.4.2 Convergence of RPs

Abaya and Wise (1984) studied the convergence of vector quantizers to optimal quantization. Assume $\boldsymbol{X} \sim F(\boldsymbol{x}), \boldsymbol{x} \in \mathbb{R}^p$ with optimal vector quantizers $Q(\boldsymbol{x})$ (Definition 2.2). Suppose a sequence of probability distribution functions G_j converges weakly to a distribution function F. Does the sequence of optimal quantizers for the G_n converge to an optimal quantizer for F? If so, do the respective distortions converge to the optimal distortion for F? Sufficient conditions are given to guarantee the convergence of scalar and vector quantizers with a general class of distortion measures. These results prove the existence of minimum rth power distortion vector quantizers and the convergence of a proposed algorithm for constructing optimal quantizers.

We pick up the notation from Abaya and Wise (1984). Let Q_n be an optimal k-level quantizer for G_n. Two interesting questions may be posed:

(Q1) Do the mean distortions $D(Q_n, G_n)$ converge to $D(Q, F)$, where Q is an optimal k-level quantizer for F?

(Q2) Does Q_n converge weakly to some quantizer Q; and if so, is Q optimal for F?

Two Lemmas are needed for the main result.

Lemma 6.3 *Assume that* $f_0(t)$ *is nonnegative, nondecreasing on* $[0, \infty)$ *and continuous. Let* $C(\boldsymbol{x}) = f_0(||\boldsymbol{x}||)$. *Suppose that* $F_n \xrightarrow{W} F$ *and that* $C(x - y)$ *is uniformly integrable* *concerning F_n for each y. If a sequence of k-level quantizers Q_n converges weakly to a quantizer Q, then $D(Q_n, F_n) \to D(Q, F)$.*

Lemma 6.4 *Under the conditions of Lemma 6.3, if Q_n is an optimal k-level quantizer for F_n, then Q is optimal for F among quantizers with k level or less.*

We now address the first part of (Q2): Does a sequence of optimal quantizers converge? In general, the answer is no. However, we have the following somewhat weaker property.

Lemma 6.5 *Under the conditions of Lemma 6.4, every subsequence of Q_n contains a further subsequence that converges weakly to some quantizer unless $C(x)$ is constant with F-probability one.*

The following theorem gives the main result in Abaya and Wise (1984).

Theorem 6.4 *Assume that $f_0(t)$ is nonnegative, nondecreasing on $[0, c_0)$ and continuous. Suppose that $F_n \overset{W}{\to} F$, and that Q_n is an optimum k-level quantizer for F_n. If $f(x-y)$ is uniformly integrable concerning o F_n for each y, then every convergent subsequence of Q_n converges weakly to an optimal quantizer for F. Such a subsequence exists unless $f(x)$ is constant with F-probability 1. Moreover, $D(Q_n F_n)$ converges to the optimal mean distortion for quantizing F with k levels.*

For some classes of multivariate distributions, like elliptical distributions, one can find more beautiful results on their approximation distributions.

6.5 More Symmetric Multivariate Distributions

For univariate random variable X, we say X is symmetric about 0 if $X \overset{d}{=} -X$ and X is symmetric about μ if $X - \mu \overset{d}{=} -(X - \mu)$. Theorem 3.2 gives a symmetric structure for MSE-RPs of univariate distributions. There are many different extensions of "symmetry" in multivariate cases.

Definition 6.4 *A random vector $\boldsymbol{X} \in \mathbb{R}^p$ is symmetric to $\mathbf{0}$ if $\boldsymbol{X} \overset{d}{=} -\boldsymbol{X}$. $\boldsymbol{X}$ is symmetric to its mean vector $\boldsymbol{\mu}$ if $\boldsymbol{X} - \boldsymbol{\mu} \overset{d}{=} -(\boldsymbol{X} - \boldsymbol{\mu})$.*

There are many kinds of symmetry; Section 6.2 introduces spherical and elliptical symmetry concepts. More studies on symmetric multivariate distributions can refer to Fang et al. (1990). This section introduces other symmetric multivariate distributions and their RPs discussed by Tarpey (1995).

6.5.1 Strongly symmetric distribution

Definition 6.5 *A random vector $\boldsymbol{X}$ is strongly symmetric with its mean $\boldsymbol{\mu} = E(\boldsymbol{X})$ if there exists an orthogonal $p \times p$ matrix $\boldsymbol{H}$ such that*

$$\boldsymbol{X} \overset{d}{=} \boldsymbol{\mu} + \boldsymbol{H}\boldsymbol{Y},$$

where $\boldsymbol{Y} = (Y_1, \ldots, Y_p)^T$ satisfies the property:

$$(Y_1, \ldots, Y_p) \overset{d}{=} (\pm Y_1, \ldots, \pm Y_p). \tag{6.28}$$

Lemma 6.6 *If $\boldsymbol{X} \in \mathbb{R}^p$ is a random vector whose MSE-RPs are mutually independent and have symmetric distributions. Then, $\boldsymbol{x}$ is strongly symmetric.*

Proof *$\boldsymbol{X}$ can be expressed as its SR $\boldsymbol{X} = \boldsymbol{\mu} + \boldsymbol{H}^T\boldsymbol{\xi}$, where $\boldsymbol{H}$ is orthogonal and $\boldsymbol{\xi}^T = (Z_1, \ldots, Z_p)$ is a vector of MSE-RPs, independently distributed and symmetric. Therefore, $(Z_1, \ldots, Z_p) \stackrel{d}{=} (\pm Z_1, \ldots, \pm Z_p)$ that implies that $\boldsymbol{X}$ is strongly symmetric.*

Lemma 6.7 *A random vector $\boldsymbol{X}$ is symmetric to its mean vector $\boldsymbol{\mu}$ if and only if $\boldsymbol{a}^T(\boldsymbol{X}-\boldsymbol{\mu})$ is symmetric for any $\boldsymbol{a} \in \mathbb{R}^p$.*

Suppose $\boldsymbol{X}$ is a p-variate strongly symmetric random vector. The next theorem given by Tarpey (1995) shows that a set of k self-consistent points for the marginal distribution in a principal component subspace determines a set of k self-consistent points of $\boldsymbol{x}$.

Theorem 6.5 *Let $\boldsymbol{X} = \boldsymbol{\mu} + \boldsymbol{H}\boldsymbol{Y}$ be a strongly symmetric p-variate random vector where $\boldsymbol{H} = [\boldsymbol{\beta}_1, \ldots, \boldsymbol{\beta}_p]$ is orthogonal and $\boldsymbol{Y}^T = (Y_1, \ldots, Y_p) \stackrel{d}{=} (\pm Y_1, \ldots, \pm Y_p)$. If k points $\{\boldsymbol{y}_1 - \boldsymbol{\mu}, \ldots, \boldsymbol{y}_k - \boldsymbol{\mu}\}$ span the same subspace as q eigenvectors, say $\{\boldsymbol{\beta}_1, \ldots, \boldsymbol{\beta}_q\}, q < p$, and if $\{\boldsymbol{H}_1^T\boldsymbol{y}_j, j = 1, \ldots, k\}$ are k self-consistent points for $\boldsymbol{H}_1^T\boldsymbol{X}$, where $\boldsymbol{H}_1 = [\boldsymbol{\beta}_1, \ldots, \boldsymbol{\beta}_q]$, then $\boldsymbol{y}_j, j = 1, \ldots, k$ are k self-consistent points for $\boldsymbol{X}$.*

The proof can refer to Subsection 6.9.1.

6.5.2 Skew-normal distribution

The density function defines the standard skew-normal distribution as

$$p(x) = 2\phi(x)\Phi(\alpha x),$$

where α is the shape parameter and $\phi(\cdot)$ and $\Phi(\cdot)$ are pdf and cdf of $N(0,1)$. Insert location parameter μ and scale parameter σ, the density function of the skew-normal distribution $X \sim SN(\mu, \sigma, \alpha)$ is

$$p(x; \mu, \sigma, \alpha) = \frac{1}{\sigma\pi} e^{-\frac{(x-\mu)^2}{2\sigma^2}} \int_{-\infty}^{\alpha(\frac{x-\mu}{\sigma})} e^{-t^2/2} dt. \tag{6.29}$$

The first four moments are

$$\begin{aligned}
&\text{mean: } \mu + \sigma\delta\sqrt{\frac{2}{\pi}}, \text{ with } \delta = \frac{\alpha}{\sqrt{1+\alpha^2}};\\
&\text{variance: } \sigma^2(1 - \frac{2\delta^2}{\pi});\\
&\text{skewness: } \frac{4-\pi}{2\frac{(\delta\sqrt{2/\pi})^3}{(1-2\delta^2/\boldsymbol{i})^{3/2}}}; \text{ and}\\
&\text{kurtoss: } 2(\pi-3)\frac{(\delta\sqrt{2/\pi})^4}{(1-2\delta^2/\boldsymbol{i})^2}.
\end{aligned}$$

The multivariate skew-normal distribution is an extension of the univariate one and can be applied to approximate the distribution of real data in many situations. Suppose $\boldsymbol{X}$ be a p-variate random vector of skew-normal distribution with location vector $\boldsymbol{a} \in \mathbb{R}^p$, scale matrix $\boldsymbol{G}$ to be a positive definite $p \times p$ matrix and skewness vector $\boldsymbol{\eta} \in \mathbb{R}^p$, we write as $\boldsymbol{X} \sim SN_p(\boldsymbol{a}, \boldsymbol{G}, \boldsymbol{\eta})$. The density of $\boldsymbol{X}$ is given by

$$p(\boldsymbol{x}) = 2\phi_p(\boldsymbol{x}|\boldsymbol{a}, \boldsymbol{G})\Phi(\boldsymbol{\eta}^T\boldsymbol{G}^{-1/2}(\boldsymbol{x} - \boldsymbol{a})), \tag{6.30}$$

where $\phi_p(\boldsymbol{x}|\boldsymbol{a}, \boldsymbol{G})$ and $\Phi(\boldsymbol{x}|\boldsymbol{a}, \boldsymbol{G})$ are the density function and cdf of $N_p(\boldsymbol{a}, \boldsymbol{G})$, respectively. See Azzalini and Valle (1996) and Arellano-Valle and Genton (2005) for a comprehensive review. If $\boldsymbol{a} = \boldsymbol{0}$ and $\boldsymbol{G} = \boldsymbol{I}_p$, then $SN_p(\boldsymbol{a}, \boldsymbol{G}, \boldsymbol{\eta})$ is denoted as $SN_p(\boldsymbol{\eta})$. Let $\boldsymbol{W} \sim SN_p(\boldsymbol{\eta})$, the stochastic representation of $\boldsymbol{W}$ is given by

$$\boldsymbol{W} \stackrel{d}{=} \delta|Z_0| + (\boldsymbol{I}_p - \boldsymbol{\delta}\boldsymbol{\delta}^T)^{1/2}\boldsymbol{Z}, \tag{6.31}$$

where $Z_0 \sim N(0, 1)$ is independent of $\boldsymbol{Z} \sim N_p(\boldsymbol{0}, \boldsymbol{I}_d)$ and $\boldsymbol{\delta} = \boldsymbol{\eta}/\sqrt{1 + \boldsymbol{\eta}^T\boldsymbol{\eta}}$. For $\boldsymbol{X} \sim SN_d(\mathbf{a}, \boldsymbol{G}, \boldsymbol{\eta})$, the relation between $\boldsymbol{W} \sim SN_p(\boldsymbol{\eta})$ and $\boldsymbol{X}$ is

$$\boldsymbol{X} \stackrel{d}{=} \boldsymbol{a} + \boldsymbol{G}^{1/2}\boldsymbol{W}, \tag{6.32}$$

and the mean vector $\boldsymbol{\mu}$ and covariance matrix $\boldsymbol{\Sigma}$ of $\boldsymbol{X}$ are

$$\boldsymbol{\mu} = \mathbf{a} + \sqrt{\frac{2}{\pi}}\boldsymbol{G}^{1/2}\boldsymbol{\eta} \quad \textit{and} \quad \boldsymbol{\Sigma} = \boldsymbol{G} - \frac{2}{\pi}\boldsymbol{G}^{1/2}\boldsymbol{\eta}\boldsymbol{\eta}^T\boldsymbol{G}^{1/2}, \tag{6.33}$$

respectively.

Algorithm 6.8 (The NTSR algorithm for QMC-RPs of the SN distribution)
Step 1 Generate a set of quasi-random numbers $\{\boldsymbol{c}_i = (c_{i1}, \cdots, c_{i,p+1}), i = 1, \cdots, k\}$.

Step 2 Denote the cdf of R *by* $F_R(r)$ *of the* χ*-distribution (6.11) and let* F_R^{-1} *be its inverse function. Compute* $r_i = F_R^{-1}(c_{id})$, $i = 1, \cdots, k$.

Step 3 Generate a set of quasi-random F-numbers $\{\boldsymbol{u}_i, i = 1, \cdots, k\}$ *of the uniform distribution on* S^p *with the first* $(p-1)$*-components of* $\boldsymbol{c}_i$*'s. Then* $\{(\boldsymbol{z}_1)_i = r_i\boldsymbol{u}_i, i = 1, \cdots, k\}$ *is a set of quasi-random F-numbers of* $\boldsymbol{Z}$.

Step 4 Let Φ^{-1} *be the inverse function of* Φ. *Compute* $(z_0)_i = \Phi_1^{-1}(c_{i,p+1})$, $i = 1, \cdots, k$.

Step 5 Then, $\{\boldsymbol{w}_i = \boldsymbol{\delta}|(z_0)_i| + (\boldsymbol{I}_d - \boldsymbol{\delta}\boldsymbol{\delta}^T)^{1/2}(\boldsymbol{z}_1)_i\}, i = 1, \ldots, k$ *is a set of quasi-random F-numbers of* $\boldsymbol{w}$, *and* $\boldsymbol{\xi}_i = \boldsymbol{a} + \boldsymbol{G}^{1/2}\boldsymbol{w}_i, i = 1, \ldots, k$ *is a set of QMC-RPs of* $SN_p(\boldsymbol{a}, \boldsymbol{G}, \boldsymbol{\eta})$.

Similarly to Algorithm 6.7, we can find an algorithm for generating MSE-RPs of the skew-normal distribution. Tarpey and Loperfido (2015) provided a generalized principal subspace theorem for multivariate location mixtures of spherically symmetric distribution based on the notion of self-consistency, including the following result for skew-normal distribution, where the skew-normal distribution can be regarded as an infinite location mixture of multivariate standard normal distributions.

Theorem 6.6 *Let* $\mathbf{X} \in \mathbb{R}^p$ *be a random vector of skew-normal distribution with location parameter* $\boldsymbol{a} = \boldsymbol{0}$, *scale matrix* $\boldsymbol{I}_p + \boldsymbol{\eta}\boldsymbol{\eta}^T$ *and shape parameter* $\boldsymbol{G}^{-1/2}\boldsymbol{\eta}/\sqrt{1 + \boldsymbol{\eta}^T\boldsymbol{G}^{-1}\boldsymbol{\eta}}$, *where the dispersion matrix* $\boldsymbol{G} = \boldsymbol{I}_p$ *and skewness vector* $\boldsymbol{\eta} \in \mathbb{R}^p$. *If a set of k MSE-RPs of* $\boldsymbol{X}$ *lie on a line, then the k MSE-RPs lie on the line spanned by* $\boldsymbol{\eta}$.

Table 6.2 shows the estimation bias of the mean vector and the covariance by the approximation distribution based on three types of RP of $\mathrm{SN}_3(\boldsymbol{0}, \boldsymbol{G}_p, \eta)$ with $\eta = (2, 3, 1)^T$. The estimation bias of the mean vector is used by $||\boldsymbol{\mu} - \hat{\boldsymbol{\mu}}||$ and the estimation bias of the covariance matrix is used by $||\boldsymbol{\Sigma} - \hat{\boldsymbol{\Sigma}}||_F$, where the Frobenius norm of a matrix $\boldsymbol{A}$ is defined by

$$||\boldsymbol{A}||_F = \sqrt{\mathrm{tr}(\boldsymbol{A}^T\boldsymbol{A})}. \tag{6.34}$$

We can see that MSE-RPs perform best among the three types of RPs.

TABLE 6.2
Estimation bias for $\mathrm{SN}_3(\mathbf{0}, \boldsymbol{G}_p, (2,3,1)^T)$

$p = 3$	Mean Vector			Covariance Matrix		
k	MC-RPs	QMC-RPs	MSE-RPs	MC-RPs	QMC-RPs	MSE-RPs
15	0.524670	0.156166	**0.000047**	0.798052	0.468315	**0.215005**
25	0.341151	0.039600	**0.000047**	0.355807	0.335800	**0.157446**
31	0.187494	0.059448	**0.000047**	0.869429	0.324445	**0.138094**
307	0.053655	0.011552	**0.000047**	0.100928	0.072612	**0.031686**
562	0.066912	0.010646	**0.000047**	0.133469	0.053313	**0.020852**
701	0.045041	0.004708	**0.000047**	0.212273	0.068716	**0.017842**
1019	0.055986	0.035069	**0.000047**	0.081607	0.017985	**0.013711**
2129	0.043059	0.001361	**0.000047**	0.050595	0.014921	**0.007593**
3001	0.017078	0.002863	**0.000047**	0.037062	0.010527	**0.005688**
4001	0.014384	0.000833	**0.000047**	0.036917	0.010418	**0.004408**

6.5.3 The skew-elliptical distribution

As a natural extension of the skew-normal distribution, the skew-elliptical (SE) family of distributions (including the skew-t distribution) is an extension of the skew-normal distribution and is proposed by Branco and Dey (2001) and studied by many authors including Fang (2003). Recent development can refer to **?**.

The skew-elliptical density function takes the form

$$p(\boldsymbol{x}) = 2g(\boldsymbol{x})G(\lambda \boldsymbol{x}), \tag{6.35}$$

where $g(\cdot)$ denotes a density function of a random variable X that is elliptical distributed about zero, the skewing term $G(\cdot)$ denotes the distribution function of a random variable related to X that is also elliptical distributed about zero and λ is the shape or skewness parameter (Adcock (2023)). A special case is the skew-t distribution.

Definition 6.6 *A random variable X with a pdf*

$$p(x) = 2t_\nu(x)T_{\nu+1}(\lambda x), \tag{6.36}$$

where $t_\nu(x)$ is the pdf of the t-distribution with the degrees of freedom ν, and $T_{\nu+1}(\lambda x)$ is the cdf of the t-distribution with the degrees of freedom $\nu+1$. Adcock (2023) defined the skew-t distribution by

$$p(x) = 2t_\nu(x)T_{\nu+1}\left(\sqrt{\frac{1+\nu}{x^2+\nu}}\lambda x\right), \tag{6.37}$$

and denote $X \sim ST_\nu(\lambda, \nu)$.

From Adelchi (2003) the first two moments of $ST_\nu(\lambda, \nu)$ are given as follows with $\delta = \frac{\lambda}{\sqrt{1+\lambda^2}}$.

$$\mu = E(X) = \delta\left(\frac{\nu}{\pi}\right)^{1/2} \frac{\Gamma(\frac{1}{2}(\nu-1))}{\Gamma(\frac{1}{2}\nu)},$$

$$\mathrm{Var}(X) = \frac{\nu}{\nu-2} - \mu^2,$$

Adcock (2023) obtained the following interesting result.

Theorem 6.7 *Let two random variables* $Y \sim N(\lambda x\sqrt{1+\frac{x^2}{\nu}}, \frac{1}{v})$ *and* $V/\nu \sim \chi^2_\nu$. *Assume the conditional distribution* $(X|V=v) \sim N(0, \frac{1}{v})$. *Then* $(X|Y>0) \sim ST_\nu(\lambda, \nu)$.

The proof can be found in P.407 of Adcock (2023). Jones and Faddy (2003) gave different consideration to the skew-t distribution and applications.

Zhou et al. (2024a) studied the generalized alpha skew-t (GAST) distribution where it has a pdf

$$f(x;\alpha,s,\nu) = \frac{(1-\alpha x)^2+1}{c(\alpha,s,\nu)} t(x;\nu)T\left(\sqrt{\frac{1+\nu}{x^2+\nu}}sx;\nu+1\right), \nu>2, x\in\mathbb{R},$$

where

$$c(\alpha,s,\nu) = 1-\alpha\left[\frac{s}{\sqrt{1+s^2}}\left(\frac{\nu}{\pi}\right)^{1/2}\Gamma\left(\frac{\nu-1}{2}\right)/\Gamma\left(\frac{\nu}{2}\right)\right]+\frac{\alpha^2}{2}\left(\frac{\nu}{\nu-2}\right).$$

They studied some properties of the GAST distribution, including calculating the moments and the relationship between the number of peaks and the GAST parameters. The three types of RPs are utilized in estimating moments and densities of the GAST distribution with known and unknown parameters. The MSE-RPs perform the best among all case studies. For unknown parameter cases, a revised maximum likelihood estimation (MLE) parameter estimation method is compared with the plain MLE method. This indicates that the revised MLE method suits GAST distribution with unimodal or unobvious bimodal patterns. This paper includes two real-data applications in which the GAST model appears adaptable to various data types.

6.6 Three Types of Approximation Distributions

For a given continuous distribution $\boldsymbol{X} \sim F(\boldsymbol{x}) \in \mathbb{R}^p$ with a finite mean vector $\boldsymbol{\mu}$ and a covariance matrix $\boldsymbol{\Sigma}$, a discrete approximation $\hat{F}_k(\mathbf{x})$ with
support points $\{\boldsymbol{y}_1,\ldots,\boldsymbol{y}_k\}$,
related Voronoi regions $S_1,\ldots,S_p$, and
probabilities $p_j = \Pr(\boldsymbol{X}=S_j)$, $j=1,\ldots,k$,
can be constructed by MC-RPs, QMC-RPs, and MSE-RPs, respectively. For the methods of MC-RP and QMC-RP, all the $p_j = \frac{1}{k}$, but the probabilities of the MSE-RP method are different. We denote these three approximations by $Y^k_{MC} \sim F^k_{MC}$, $Y^k_{QMC} \sim F^k_{QMC}$ and $Y^k_{MSE} \sim F^k_{MSE}$, respectively. We can often omit k in the notation without confusion. The corresponding mean vector and covariance of $\hat{F}_k(\mathbf{x})$ are

$$\hat{\boldsymbol{\mu}} = \sum_{j=1}^k \mathbf{y}_j p_j \quad \text{and} \quad \hat{\boldsymbol{\Sigma}} = \sum_{j=1}^k (\mathbf{y}_j-\hat{\mu})(\mathbf{y}_j-\hat{\mu})^\top p_j, \tag{6.38}$$

which are the estimations of $\boldsymbol{\mu}$ and $\boldsymbol{\Sigma}$, respectively. The L_2 norm is applied to evaluate the estimation bias of the mean vector $\boldsymbol{\mu}$ by $||\boldsymbol{\mu}-\hat{\boldsymbol{\mu}}||$, and a Frobenius norm $||\boldsymbol{\Sigma}-\hat{\boldsymbol{\Sigma}}||_F$ for the covariance matrix $\boldsymbol{\Sigma}$, where the Frobenius norm of a matrix $\boldsymbol{A}$ is defined by (6.34).

6.6.1 Numerical simulation

Following Yang et al. (2021), we try to estimate the mean vector and covariance matrix of some elliptically contoured distributions and the skew-normal distribution of dimension

TABLE 6.3
Estimation bias for $\text{Kotz}_3(3,3,1)$, with $\mathbf{\Sigma}_i, i = 1,2,3$

$\mathbf{\Sigma} = \mathbf{\Sigma}_1$	Estimation bias of mean vector μ			Estimation bias of covariance matrix $\mathbf{\Sigma}$		
k	MC-RPs	QMC-RPs	MSE-RPs	MC-RPs	QMC-RPs	MSE-RPs
15	0.332666	0.014961	**0.000040**	2.132877	0.451716	**0.107800**
35	0.242490	0.021176	**0.000040**	0.516929	0.564609	**0.039294**
185	0.187304	0.020161	**0.000040**	0.137735	0.130086	**0.007840**
418	0.079177	0.002024	**0.000040**	0.418887	0.008194	**0.003777**
1459	0.004132	0.005112	**0.000040**	0.174390	0.014945	**0.001427**
2440	0.009230	0.001968	**0.000040**	0.225844	0.026886	**0.001063**
$\mathbf{\Sigma} = \mathbf{\Sigma}_2$	Estimation bias of mean vector μ			Estimation bias of covariance matrix $\mathbf{\Sigma}$		
k	MC-RPs	QMC-RPs	MSE-RPs	MC-RPs	QMC-RPs	MSE-RPs
15	0.717651	0.007931	**0.000033**	1.863166	0.276473	**0.039740**
35	0.198866	0.014331	**0.000033**	0.272105	**0.165903**	0.017020
185	0.120270	0.011898	**0.000033**	0.079402	0.029729	**0.004190**
418	0.173440	0.001293	**0.000033**	0.049242	**0.001864**	0.002387
1459	0.031152	0.003180	**0.000033**	0.141750	0.004449	**0.001050**
2440	0.006956	0.000998	**0.000033**	0.057752	0.006836	**0.000764**
$\mathbf{\Sigma} = \mathbf{\Sigma}_3$	Estimation bias of mean vector μ			Estimation bias of covariance matrix $\mathbf{\Sigma}$		
k	MC-RPs	QMC-RPs	MSE-RPs	MC-RPs	QMC-RPs	MSE-RPs
15	0.599516	0.026194	**0.000088**	0.527672	0.350963	**0.252247**
35	0.359905	0.019325	**0.000088**	1.105650	0.262044	**0.143330**
185	0.268338	0.014523	**0.000088**	0.137162	**0.042243**	0.051571
418	0.084572	0.002398	**0.000088**	0.247787	**0.007914**	0.030914
1459	0.038571	0.004452	**0.000088**	0.217465	**0.005527**	0.013885
2440	0.092121	0.001066	**0.000088**	0.068231	**0.005999**	0.009895

$d = 3$ by the three types of RP, and this is a preliminary comparison among the MC-RPs, QMC-RPs, and MSE-RPs. Assume that the mean vector $\boldsymbol{\mu} = \mathbf{0}$, and three covariance matrices $\mathbf{\Sigma}_i, i = 1, \ldots, 3$ are randomly generated as following

$$\mathbf{\Sigma}_1 = \begin{bmatrix} 1.5003 & 1.3294 & 0.8439 \\ 1.3294 & 1.2437 & 0.6936 \\ 0.8439 & 0.6936 & 1.2935 \end{bmatrix}, \quad \mathbf{\Sigma}_2 = \begin{bmatrix} 1.1707 & 0.8369 & 1.4413 \\ 0.8369 & 0.6368 & 1.0610 \\ 1.4413 & 1.0610 & 1.8034 \end{bmatrix},$$

$$\mathbf{\Sigma}_3 = \begin{bmatrix} 2.2729 & 1.4294 & 1.0281 \\ 1.4294 & 1.2994 & 0.9218 \\ 1.0281 & 0.9218 & 1.0986 \end{bmatrix}.$$

We have carried out the numerical experiments for normal and Kotz-type distributions. Assume that the normal distribution $\text{N}_3(\mathbf{0}, \mathbf{\Sigma})$ and the Kotz type distribution $\text{Kotz}_3(3,3,1)$ share the same mean vector $\boldsymbol{\mu} = \mathbf{0}$ and covariance matrix $\mathbf{\Sigma} = \mathbf{\Sigma}_i$ in the experiments. However, the numerical results of the normal distribution are quite similar to that of the Kotz type distribution, so we omit the tables of results for the normal distribution. The results of estimation bias for $\text{Kotz}_3(3,3,1)$ are shown in Table 6.3 with the counts of the smallest estimation bias for each set of RPs in Table 6.4. For the skew-normal distribution $\text{SN}_3(\mathbf{a}, \mathbf{G}, \eta)$, suppose that the location vector $\mathbf{a} = \mathbf{0}$, the scale matrix $\mathbf{G} = \mathbf{\Sigma}_i$ and the skewness vector $\eta = (2,3,1)^\top$, then the mean vector and covariance matrix of skew-normal distribution with these parameters can be calculated by (6.33). Tables 6.5 and 6.6 show the results for $\text{SN}_3(\mathbf{0}, \boldsymbol{G}, (2,3,1)^\top)$.

The numbers of QMC-RPs (k) are chosen from the tables of *glp* set in Appendix A of (Fang and Wang, 1994). In general, the MSE-RPs have a better performance, but the

TABLE 6.4
Number of smallest estimation bias for $\text{Kotz}_3(3,3,1)$ with $\boldsymbol{\Sigma}_i, i=1,2,3$

	Estimation of μ			Estimation of $\boldsymbol{\Sigma}$		
$\boldsymbol{\Sigma}=\boldsymbol{\Sigma}_i$	MC-RPs	QMC-RPs	MSE-RPs	MC-RPs	QMC-RPs	MSE-RPs
$\boldsymbol{\Sigma}_1$	0	0	6	0	0	6
$\boldsymbol{\Sigma}_2$	0	0	6	0	2	4
$\boldsymbol{\Sigma}_3$	0	0	6	0	4	2
Total	0	0	18	0	6	12

TABLE 6.5
Estimation bias for $\text{SN}_3(\mathbf{0},\mathbf{G},(2,3,1)^\top)$, where $\mathbf{G}=\boldsymbol{\Sigma}_i$, $i=1,2,3$

$\mathbf{G}=\boldsymbol{\Sigma}_1$	Estimation bias of mean vector μ			Estimation bias of covariance matrix $\boldsymbol{\Sigma}$		
k	MC-RPs	QMC-RPs	MSE-RPs	MC-RPs	QMC-RPs	MSE-RPs
15	0.167373	0.066652	**0.000073**	0.466210	0.450941	**0.159836**
31	0.387120	0.057031	**0.000073**	0.367563	0.285593	**0.086323**
307	0.031687	0.019381	**0.000073**	0.077712	0.031647	**0.018398**
562	0.070129	0.009711	**0.000073**	0.057653	0.027562	**0.012376**
1019	0.021079	0.005285	**0.000073**	0.090031	0.016345	**0.008202**
3001	0.014740	0.001113	**0.000073**	0.024302	0.006504	**0.003716**
$\mathbf{G}=\boldsymbol{\Sigma}_2$	Estimation bias of mean vector μ			Estimation bias of covariance matrix $\boldsymbol{\Sigma}$		
k	MC-RPs	QMC-RPs	MSE-RPs	MC-RPs	QMC-RPs	MSE-RPs
15	0.442373	0.086722	**0.000036**	1.128469	0.581315	**0.033676**
31	0.203203	0.024511	**0.000036**	0.632891	0.419159	**0.017286**
307	0.076553	0.016586	**0.000036**	0.189267	0.036246	**0.002495**
562	0.017958	0.010257	**0.000036**	0.029340	0.009044	**0.001648**
1019	0.058045	0.006286	**0.000036**	0.026544	0.021020	**0.001106**
3001	0.002164	0.001219	**0.000036**	0.052943	0.006965	**0.000513**
$\mathbf{G}=\boldsymbol{\Sigma}_3$	Estimation bias of mean vector μ			Estimation bias of covariance matrix $\boldsymbol{\Sigma}$		
k	MC-RPs	QMC-RPs	MSE-RPs	MC-RPs	QMC-RPs	MSE-RPs
15	0.471961	0.120231	**0.000101**	0.601581	0.351456	**0.217226**
31	0.139151	0.072319	**0.000101**	0.960364	0.283401	**0.139317**
307	0.159581	0.024045	**0.000101**	0.105484	0.040397	**0.033554**
562	0.078494	0.012173	**0.000101**	0.109267	0.027626	**0.022504**
1019	0.057387	0.007629	**0.000101**	0.065771	0.024161	**0.015044**
3001	0.031081	0.002069	**0.000101**	0.011811	**0.005175**	0.006812

TABLE 6.6
Number of smallest estimation bias for $\text{SN}_3(\mathbf{0},\mathbf{G},(2,3,1)^\top)$, where $\mathbf{G}=\boldsymbol{\Sigma}_i$, $i=1,2,3$

	Estimation of μ			Estimation of $\boldsymbol{\Sigma}$		
$\mathbf{G}=\boldsymbol{\Sigma}_i$	MC-RPs	QMC-RPs	MSE-RPs	MC-RPs	QMC-RPs	MSE-RPs
$\boldsymbol{\Sigma}_1$	0	0	6	0	0	6
$\boldsymbol{\Sigma}_2$	0	0	6	0	0	6
$\boldsymbol{\Sigma}_3$	0	0	6	0	1	5
Total	0	0	18	0	1	17

results obtained by QMC-RPs are more accurate in some cases. Hence, the performance of QMC-RPs and MSE-RPs may depend on the underlying distribution and their parameters.

TABLE 6.7
Estimation bias for $\text{Kotz}_3(3,3,1)$ by resampling, $i = 1,2,3$

$\mathbf{\Sigma} = \mathbf{\Sigma}_1$	Estimation bias of mean vector μ			Estimation bias of covariance matrix $\mathbf{\Sigma}$		
k	MC-RPs	QMC-RPs	MSE-RPs	MC-RPs	QMC-RPs	MSE-RPs
15	0.339382	**0.011185**	0.017939	2.705093	**0.391274**	0.919508
35	0.251636	0.012450	**0.010119**	0.296953	**0.170563**	0.351828
185	0.174243	0.024130	**0.001357**	0.087749	0.238956	**0.034090**
418	0.081861	0.004841	**0.002727**	0.448038	**0.047386**	0.048232
1459	0.003603	0.005111	**0.001660**	0.182671	0.009133	**0.006990**
2440	0.007819	0.000823	**0.000213**	0.225136	0.035529	**0.005571**
$\mathbf{\Sigma} = \mathbf{\Sigma}_2$	Estimation bias of mean vector μ			Estimation bias of covariance matrix $\mathbf{\Sigma}$		
k	MC-RPs	QMC-RPs	MSE-RPs	MC-RPs	QMC-RPs	MSE-RPs
15	0.721514	0.008834	**0.001261**	1.986047	**0.090411**	0.258858
35	0.201690	0.025219	**0.006142**	0.341458	**0.043095**	0.111629
185	0.115714	0.006730	**0.000727**	0.076345	0.039866	**0.023155**
418	0.175352	**0.000405**	0.000967	0.064786	**0.003232**	0.009732
1459	0.030388	0.003551	**0.001631**	0.140709	0.002446	**0.001854**
2440	0.007688	0.001285	**0.000575**	0.056606	0.008328	**0.001932**
$\mathbf{\Sigma} = \mathbf{\Sigma}_3$	Estimation bias of mean vector μ			Estimation bias of covariance matrix $\mathbf{\Sigma}$		
k	MC-RPs	QMC-RPs	MSE-RPs	MC-RPs	QMC-RPs	MSE-RPs
15	0.590542	0.026633	**0.004299**	0.757688	**0.362763**	0.463363
35	0.355764	0.027596	**0.002901**	1.175356	**0.126148**	0.253164
185	0.269042	0.015239	**0.000977**	0.146629	0.063476	**0.062544**
418	0.082881	0.003072	**0.001442**	0.254755	**0.007814**	0.037470
1459	0.039575	0.004632	**0.000961**	0.215437	**0.004571**	0.015924
2440	0.090849	0.001651	**0.000810**	0.069431	**0.008963**	0.010297

6.6.2 Estimation of mean and covariance by resampling

The sets of MC-RPs, QMC-RPs, and MSE-RPs are also used for estimating mean vector and covariance matrix of normal, Kotz-type, and skew-normal distributions in the above experiments that were determined using Bootstrap. In the following experiments, we choose the resampling size $B = 2000$, and the numerical results are shown in the same form as that of the preliminary comparison in Tables 6.3 - 6.6. The results of $N_3(\mathbf{0}, \mathbf{\Sigma})$ by resampling are also omitted, as they are also similar to that of $\text{Kotz}_3(3,3,1)$ with the same mean vector and covariance matrix. Based on the results of $\text{Kotz}_3(3,3,1)$ (Tables 6.7 and 6.8), in most cases, the MSE-RPs give the best accuracy among the three methods, but the QMC-RPs perform slightly better than the In some cases, MSE-RPs are used to estimate the covariance matrix. Tables 6.9 and 6.10 for $\text{SN}_3(\mathbf{0}, \mathbf{G}, (2,3,1)^\top)$ show that MSE-RPs can obtain a more accurate estimation.

In the traditional bootstrap, a set of MC-RPs is used for the support of the discrete approximation. As we may see in Tables 6.7 - 6.10, there is a significant improvement in most cases if the QMC-RPs and MSE-RPs are regarded as the support points in bootstrap, so these types of RPs have a great potential for statistical inference by resampling.

6.7 Expected mean squared distance

Section 5.5.2 introduced the expected mean squared distance in univariate distributions proposed by Matsuura et al. (2015). The authors also extended the EMSD into multivariate distributions.

TABLE 6.8
Number of smallest estimation bias for Kotz3(3, 3, 1) by resampling, $i = 1, 2, 3$

	Estimation of μ			Estimation of $\mathbf{\Sigma}$		
$\mathbf{\Sigma} = \mathbf{\Sigma}_i$	MC-RPs	QMC-RPs	MSE-RPs	MC-RPs	QMC-RPs	MSE-RPs
$\mathbf{\Sigma}_1$	0	1	5	0	3	3
$\mathbf{\Sigma}_2$	0	1	5	0	3	3
$\mathbf{\Sigma}_3$	0	0	6	0	5	1
Total	0	2	16	0	11	7

TABLE 6.9
Estimation bias for $\text{SN}_3(\mathbf{0}, \mathbf{G}, (2, 3, 1)^T)$ by resampling, where $\mathbf{G} = \mathbf{\Sigma}_i$, $i = 1, 2, 3$

$\mathbf{G} = \mathbf{\Sigma}_1$	Estimation bias of mean vector μ			Estimation bias of covariance matrix $\mathbf{\Sigma}$		
k	MC-RPs	QMC-RPs	MSE-RPs	MC-RPs	QMC-RPs	MSE-RPs
15	0.169509	0.070159	**0.001665**	0.440627	0.523081	**0.250810**
31	0.382753	0.055064	**0.007285**	0.368922	0.308939	**0.124137**
307	0.030967	0.020740	**0.001416**	0.071353	0.025172	**0.021591**
701	0.013907	0.006590	**0.000480**	0.076101	0.078252	**0.013565**
2129	0.024097	0.001941	**0.001034**	0.021186	0.009948	**0.006082**
3001	0.014637	0.000993	**0.000588**	0.024313	0.006227	**0.004081**
$\mathbf{G} = \mathbf{\Sigma}_2$	Estimation bias of mean vector μ			Estimation bias of covariance matrix $\mathbf{\Sigma}$		
k	MC-RPs	QMC-RPs	MSE-RPs	MC-RPs	QMC-RPs	MSE-RPs
15	0.449962	0.085720	**0.008419**	1.201827	0.680859	**0.161385**
31	0.216299	0.024945	**0.000724**	0.537853	0.472456	**0.058401**
307	0.072198	0.017018	**0.003739**	0.198651	0.026776	**0.011563**
701	0.010601	0.006691	**0.001769**	0.130132	0.019875	**0.005393**
2129	0.047615	0.001996	**0.000658**	0.043335	0.010142	**0.001982**
3001	0.001911	0.000743	**0.000562**	0.052557	0.007514	**0.001208**
$\mathbf{G} = \mathbf{\Sigma}_3$	Estimation bias of mean vector μ			Estimation bias of covariance matrix $\mathbf{\Sigma}$		
k	MC-RPs	QMC-RPs	MSE-RPs	MC-RPs	QMC-RPs	MSE-RPs
15	0.467046	0.123618	**0.002434**	0.640243	0.428801	**0.306574**
31	0.142672	0.073466	**0.010198**	0.879896	0.320486	**0.203242**
307	0.157332	0.022044	**0.001468**	0.104098	0.038326	**0.036394**
701	0.049189	0.009260	**0.000397**	0.109363	0.088129	**0.019277**
2129	0.021702	0.001779	**0.001540**	0.038961	0.012144	**0.011136**
3001	0.031076	0.001767	**0.000468**	0.011745	**0.006485**	0.008143

TABLE 6.10
Number of smallest estimation bias for $\text{SN}_3(\mathbf{0}, \mathbf{G}, (2, 3, 1)^T)$ by resampling, where $\mathbf{G} = \mathbf{\Sigma}_i$, $i = 1, 2, 3$

	Estimation of μ			Estimation of $\mathbf{\Sigma}$		
$\mathbf{G} = \mathbf{\Sigma}_i$	MC-RPs	QMC-RPs	MSE-RPs	MC-RPs	QMC-RPs	MSE-RPs
$\mathbf{\Sigma}_1$	0	0	6	0	0	6
$\mathbf{\Sigma}_2$	0	0	6	0	0	6
$\mathbf{\Sigma}_3$	0	0	6	0	1	5
Total	0	0	18	0	1	17

First, we consider the multivariate normal case. Let $\boldsymbol{X} \sim N_p(\boldsymbol{\mu}, \mathbf{\Sigma})$, where $\mathbf{\Sigma} = \sigma^2\Psi$ with $\boldsymbol{\mu}$ and σ^2 unknown, the matrix Ψ is known. Let $\boldsymbol{x}_1, \ldots, \boldsymbol{x}_n$ are i.i.d. samples from the population. We want to estimate k-MSE-RPs of $N_p(\boldsymbol{\mu}, \mathbf{\Sigma})$. When the largest eigenvalue

λ_1 of Ψ is sufficiently larger than the second largest eigenvalue, k-MSE-RPs of $\boldsymbol{X}$ will be collinear, lying on the first principal component axis (as demonstrated in the bivariate normal case by Tarpey (1998)). Let $\boldsymbol{\beta}$ be the normalized eigenvector corresponding to λ_1. Consider the k-MSE-RPs are of the form

$$\hat{\boldsymbol{\xi}}_j = \hat{\boldsymbol{\mu}} + a_j \hat{\sigma} \sqrt{\lambda_1} \boldsymbol{\beta}, \ j = 1, \dots, k, \tag{6.39}$$

where the maximum likelihood estimators of $\boldsymbol{\mu}$ and σ^2 are

$$\hat{\boldsymbol{\mu}} = \bar{\boldsymbol{x}}, \hat{\sigma^2} = \frac{1}{np} \sum_{i=1}^{n} (\boldsymbol{x}_i - \bar{\boldsymbol{x}})^T \Psi^{-1} (\boldsymbol{x}_i - \bar{\boldsymbol{x}}),$$

respectively.

Theorem 6.8 *Under the above assumption,* $a_1, \dots, a_k$ *defining the collinear points in (6.39) that minimize the*

$$EMSD = E[MSD(\hat{\boldsymbol{\xi}}_1, \dots, \hat{\boldsymbol{\xi}}_k)] = E[\min_j ||\boldsymbol{x} - \hat{\boldsymbol{\xi}}_j||^2]$$

are k-MSE-RPs of

$$\sqrt{\frac{(n+1)p}{(n-1)p+2}}\, t_{(n-1)p+2},$$

where t_v *is the t-distribution with* v *degrees of freedom.*

The findings from this theorem are summarized as the following (i)-(iii):

(i) For any dimension p of the distribution, the minimum EMSD estimator is given based on k-MSE-RPs of a scaled t-distribution with mean 0 and variance $\frac{n+1}{n-1}$.

(ii) For any dimension p of the distribution, as the sample size n goes to infinity, the minimum EMSD estimator is given based on the k-MSE-RPs of $N(0, 1)$.

(iii) For a given sample size n, as the dimension p goes to infinity, then the minimum EMSD estimator is given based on the k-MSE-RPs of $N(0, \frac{n+1}{n-1})$.

The extension of the result to elliptically symmetric distributions is straightforward.

Theorem 6.9 *Let* $\boldsymbol{X}$ *be distributed as a p-dimensional elliptically symmetric distribution with mean* $\boldsymbol{\mu}$ *and covariance matrix* $\boldsymbol{\Sigma} = \sigma^2 \Psi$*, where* σ *is a scalar. Then its pdf is of the form*

$$p(\boldsymbol{x}) = \frac{1}{\sigma^p} |\Psi|^{-1/2} g[(\boldsymbol{x} - \boldsymbol{\mu})^T \Psi^{-1} (\boldsymbol{x} - \boldsymbol{\mu})/\sigma], \tag{6.40}$$

where $\boldsymbol{\mu}$ *and* σ *are unknown and* Ψ *and* g *are known. Let* $\boldsymbol{x}_1, \dots, \boldsymbol{x}_n$ *be a random sample from the distribution of* $\boldsymbol{X}$*. Let* $\hat{\boldsymbol{\mu}}$ *and* $\hat{\sigma}$ *are MLE of* $\boldsymbol{\mu}$ *and* σ*, respectively. Let* $\boldsymbol{\beta}_1$ *denote a normalized eigenvector of* Ψ *corresponding to its largest eigenvalue* λ_1 *and define*

$$\boldsymbol{V} = \frac{\boldsymbol{\beta}_1^T (\boldsymbol{X} - \hat{\boldsymbol{\mu}})}{\sigma \sqrt{\lambda_1}}, \ W = \frac{\hat{\sigma}}{\sigma},$$

where $\boldsymbol{X}$ *is independent with the random sample* $\boldsymbol{x}_1, \dots, \boldsymbol{x}_n$*. Denote the pdf of* $\boldsymbol{V}|W$ *by* $\psi(u|w)$ *and let*

$$\hat{\boldsymbol{\xi}}_j = \hat{\boldsymbol{\mu}} + a_j \hat{\sigma} \sqrt{\lambda_1} \boldsymbol{\beta}, \ j = 1, \dots, k.$$

Then, $a_1, \dots, a_k$ *minimizing the* $EMSD = E[MSD(\hat{\boldsymbol{\xi}}_1, \dots, \hat{\boldsymbol{\xi}}_k)]$ *are given by k-MSE-RPs of the distribution with pdf* $h(z)$*, where*

$$h(z) = \frac{E[W^3 \psi(zW|W)]}{E(W^2)}. \tag{6.41}$$

Antle and Bain (1969) pointed out that the distributions of U and W do not depend on $\boldsymbol{\mu}$ and σ. Hence, the pdf (6.41) does not depend on the unknown parameters $\boldsymbol{\mu}$ and σ, and thus k-MSE-RPs of the distribution with pdf (6.41) can be calculated.

For the spherically symmetric distribution with location parameter $\boldsymbol{\mu}$ in $\mathbb{R}^p$, its pdf has of the form (refer to (6.40)

$$p(\boldsymbol{x}) = \frac{1}{\sigma^p} g\left(\frac{(\boldsymbol{x}-\boldsymbol{\mu})^T(\boldsymbol{x}-\boldsymbol{\mu})}{\sigma}\right). \tag{6.42}$$

Let $S_p = \{\boldsymbol{\mu} + \delta\sigma \boldsymbol{U}^{(p)}\}$ (refer to (6.7)). Then the mean squared distance (MSD) between $\boldsymbol{X}$ and S_p is

$$E[\min_{\boldsymbol{Y}\in S_p} \|\boldsymbol{X}-\boldsymbol{Y}\|^2$$

that is minimized when $\delta = E[\|\boldsymbol{X}-\boldsymbol{\mu}\|^2]$. This result does not depend on $\boldsymbol{\mu}$ and g. Then we can choose $\boldsymbol{X} \sim N_p(\boldsymbol{0}, \sigma^2 \boldsymbol{I}_p)$ and find

$$\delta = E(\chi_p) = \frac{\sqrt{2}\Gamma(\frac{p+1}{2})}{\Gamma(\frac{p}{2})}. \tag{6.43}$$

Theorem 6.10 *Let $\boldsymbol{X}$ be distributed as a p-dimensional spherical distribution with pdf (6.42), where $\boldsymbol{\mu}$ and σ are unknown. Let $\hat{\boldsymbol{\mu}}$ and $\hat{\sigma}$ be MLE of $\boldsymbol{\mu}$ and σ based on a random sample, respectively.*

$$\boldsymbol{V} = \frac{(\boldsymbol{X}-\hat{\boldsymbol{\mu}})}{\sigma}, \ W = \frac{\hat{\sigma}}{\sigma},$$

where $\boldsymbol{X}$ is independent with the random sample $\boldsymbol{x}_1, \ldots, \boldsymbol{x}_n$. Let

$$\hat{S}_p = \{\hat{\boldsymbol{\mu}} + a\hat{\sigma}\boldsymbol{U}^{(p)}\}.$$

Then, the value of a minimizing

$$E[\min_{\boldsymbol{Y}\in \hat{S}_p} \|\boldsymbol{X}-\boldsymbol{Y}\|^2$$

with respect to a is given by

$$a = \frac{E(W\boldsymbol{V}^T\boldsymbol{V})}{E(W^2)}. \tag{6.44}$$

Since the distributions of $\boldsymbol{V}$ and W do not depend on the unknown parameters $\boldsymbol{\mu}$ and σ, the value of a can be calculated for a multivariate normal distribution $N_p(\boldsymbol{\mu}, \sigma^2\boldsymbol{I}_p)$ and find

$$\begin{aligned} a &= \sqrt{\frac{(n+1)p}{2}}\frac{\Gamma(\frac{(n-1)p+1}{2})}{\Gamma(\frac{(n-1)p+2}{2})}\delta \\ &= \sqrt{(n+1)p}\frac{\Gamma(\frac{p+1}{2})}{\Gamma(\frac{p}{2})}\frac{\Gamma(\frac{(n-1)p+1}{2})}{\Gamma(\frac{(n-1)p+2}{2})}, \end{aligned}$$

where δ is given by (6.43).

Matsuura et al. (2015) pointed out:

(1) For a given dimension p, $a \to \delta$ as the sample size n goes to infinity;

(2) For a given sample size n, $a \to \sqrt{\frac{n+1}{n-1}}\delta$ as the dimension p of the distribution goes to infinity.

Matsuura and Tarpey (2018) used the research line in Matsuura et al. (2015) to focus on RPs of location-scale and location-scale-rotation multivariate families. They also extended Theorem 5.6 into the multivariate case. Let $\boldsymbol{X} \in \mathbb{R}^p$ follow a location-scale multivariate family with pdf (5.1).

Theorem 6.11 *Let*

$$h(\boldsymbol{z}) = \frac{E[W^{p+2}\psi(\boldsymbol{z}W|W)]}{E(W^2)}. \tag{6.45}$$

The values of $\{\boldsymbol{a}_1, \ldots, \boldsymbol{a}_k\}$ *minimizing*

$$\text{EMSD} = E[\text{MSD}(\hat{\boldsymbol{\mu}} + \hat{\sigma}\boldsymbol{a}_1, \ldots, \hat{\boldsymbol{\mu}} + \hat{\sigma}\boldsymbol{a}_k)]$$

are given by k-MSE-RPs of the distribution with pdf (6.45).

The proof is similar to that for Theorem 5.6. It depends on the independence of $\boldsymbol{V} = \frac{\boldsymbol{X}-\hat{\boldsymbol{\mu}}}{\sigma}$ and $W = \frac{\hat{\sigma}}{\sigma}$.

Example 6.4 *Let* $\boldsymbol{X} \sim N_p(\boldsymbol{\mu}, \sigma\boldsymbol{\Psi})$ *where* $\boldsymbol{\mu}$ *and* σ *are unknown and* $\boldsymbol{\Psi}$ *is known. Let* $\boldsymbol{x}_1, \ldots, \boldsymbol{x}_n$ *be a random sample from* $\boldsymbol{X}$ *and the MLEs of* $\boldsymbol{\mu}$ *and* σ *are* $\bar{\boldsymbol{x}}$ *(the sample mean vector) and*

$$\hat{\sigma} = \sqrt{\frac{\sum_{i=1}^{n}(\boldsymbol{x}_i - \bar{\boldsymbol{x}})^T\boldsymbol{\Psi}^{-1}(\boldsymbol{x}_i - \bar{\boldsymbol{x}})}{np}},$$

respectively. Theorem 6.11 shows that the density function of $h(\boldsymbol{z})$ *is just related to*

$$\sqrt{\frac{(n+1)p}{(n-1)p+2}}t_{(n-1)p+2}(\boldsymbol{0}, \boldsymbol{\Psi}),$$

where $t_\nu(\boldsymbol{0}, \boldsymbol{\Psi})$ *is the density function of the multivariate t-distribution with* ν *degrees of freedom, location* $\boldsymbol{0}$ *and scale matrix* $\boldsymbol{\Psi}$ *having pdf*

$$p_\nu(\boldsymbol{t}) = \frac{\Gamma(\frac{\nu+p}{2})}{\Gamma_{\frac{\nu}{2}}\nu^{p\pi/2}}|\boldsymbol{\Psi}|^{-1/2}\left(1 + \frac{\boldsymbol{t}^T\boldsymbol{\Psi}^{-1}\boldsymbol{t}}{\nu}\right)^{-(\nu+p)/2}.$$

The RPs of $p_\nu(\boldsymbol{t})$ *can be generated by the NTLBG algorithm in Subsection 6.4.1 with pdf of* R *related to pdf of* $p_\nu(\boldsymbol{t})$ *in Table 6.1. Multiplying by* $\sqrt{\frac{(n+1)p}{(n-1)p+2}}$ *for each of the RPs of* $p_\nu(\boldsymbol{t})$ *can obtain* $\boldsymbol{a}_j, j = 1, \ldots, k$.

More discussion aline this direction can refer to Matsuura and Tarpey (2018). Their approach can work only for very few multivariate distributions. A more general approach is necessary.

6.8 Distribution of MSE of MC-RPs

In the previous section, we show that MSE-RPs and F_{MSE}^k have a better performance in the estimation of the population mean vector and covariance matrix than the MC and QMC methods in most cases. It shows that the criterion of the minimum square error is reasonable. We might be interested in the difference in MSE between MC-RPs and MSE-RPs.

Let $\boldsymbol{X} \sim F(\boldsymbol{x})$ in $\mathbb{R}^p$ with mean vector $\boldsymbol{\mu}$ and covariance matrix $\boldsymbol{\Sigma}$. In this study, we focus only on the elliptical distribution. Suppose $\mathcal{Y} = \{\boldsymbol{y}_1, \cdots, \boldsymbol{y}_k\}$ is a set of k random samples from $F(\boldsymbol{x})$ and the corresponding $\boldsymbol{Y}_{MC} \sim F_{MC}$ that is uniformly distributed on $\mathcal{Y}$. The MSE($\mathcal{Y}$) is given by

$$\text{MSE}(\mathcal{Y}) = E||\boldsymbol{X} - \boldsymbol{Y}_{MC}(\mathcal{Y})||^2.$$

The MSE($\mathcal{Y}$) is a random variable. Unfortunately, the above formula is difficult to follow up. Instead, we can generate a set of training samples $\{\boldsymbol{x}_1, \ldots, \boldsymbol{x}_N\}$ from $F(\boldsymbol{x})$. By the training sample and k-means technique, we can easily estimate MSE($\mathcal{Y}$). Fang et al. (2020) found that the distribution of MSE($\mathcal{Y}$) can be fitted by the generalized extreme value distribution when $\boldsymbol{X}$ has an elliptical distribution.

The extreme value distribution and generalized extreme value distribution (GEVD) are often used to model the smallest or largest value among a large set of independently identically distributed random variables representing measurements or observations. The density function of a generalized extreme value distribution (GEVD) is defined as

$$f(x|\kappa, \mu, \sigma) = \frac{1}{\sigma} \exp\left\{ -\left(1 + \kappa \cdot \frac{x-\mu}{\sigma}\right)^{1/\kappa} \right\} \left(1 + \kappa \cdot \frac{x-\mu}{\sigma}\right)^{-1-1/\kappa} \tag{6.46}$$

for $1 + \kappa \cdot \frac{x-\mu}{\sigma} > 0$, where the location parameter $\mu \in (-\infty, \infty)$, the scale parameter $\sigma > 0$ and the shape parameter $\kappa \in (-\infty, \infty)$. There are three types of GEVD,

- Type I: Gumbel distribution, $\kappa = 0$, the domain $-\infty < x < \infty$.
- Type II: Fráchet distribution, $\kappa > 0$, the domain $0 < x < \infty$.
- Type III: Weibull distribution, $\kappa < 0$, the domain $-\infty < x < 0$.

As MSE($\mathcal{Y}$) > 0, it is appropriate to fit the distribution of MSE($\mathcal{Y}$) by Freáchet distribution (Type II GEVD).

In the study, $F(\boldsymbol{x})$ is chosen $ESD_p(\boldsymbol{\mu}, \boldsymbol{\Sigma}, g)$ where $\boldsymbol{\Sigma}$ is of the type

$$\boldsymbol{\Sigma}_p(\rho) = \begin{bmatrix} 1 & \rho & \cdots & \rho \\ \rho & 1 & \cdots & \rho \\ \vdots & \vdots & & \vdots \\ \rho & \rho & \cdots & 1 \end{bmatrix}, \tag{6.47}$$

where $\rho = 0, 0.3, 0.6, 0.9$, $p = 2, 3$ and choose four subclasses of multivariate elliptical distribution: Kotz type distribution with parameters are $N = 3$, $b = 6$, $s = 4$, Pearson type II distribution with $m = 4$, and Pearson type VII distribution with $N = 20$, $m = 4$.

Randomly generate k-samples from the $F(\boldsymbol{x})$ for $M = 2000$ times, then we have MSE($\mathcal{Y}_j$), $j = 1, \cdots, M$ for the distribution fitting. Table 6.11 shows the fitting parameters κ, σ and μ of the GEVD which is applied to fit the distribution of MSE for the four subclasses of the elliptical distribution and the minimum number of MSE samples from its distribution. The MSE of MC-RPs obtained by the NTLBG algorithm are also listed in the table. The histogram of MSE and the fitting density of GEVD for the MSE for $\rho = 0.6$ and $p = 3$ are shown in Figure 6.2. It could be observed that the GEVD could fit the MSE well. Similar figures can also be obtained using GEVD to fit other cases with different values of ρ and p. Hence, the GEVD might be appropriate for the approximate distribution of MSE of MC-RPs. In addition, all of the MSE of RPs by the NTLBG algorithm are less than the minimum of the samples of MSE, which implies that the NTLBG algorithm is robust for finding the MSE-RPs of $F(\boldsymbol{x})$, where $F(\boldsymbol{x})$ is an elliptical distribution.

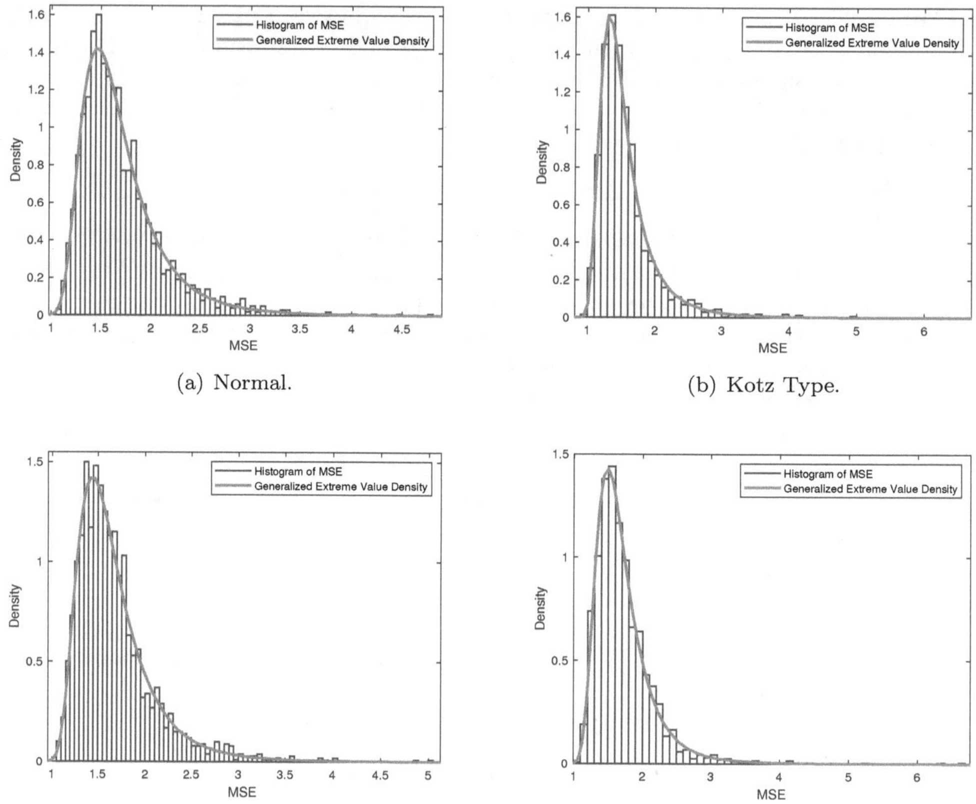

(a) Normal. (b) Kotz Type.

(c) Pearson Type II. (d) Pearson Type VII.

FIGURE 6.2
Distribution fitting for the four subclasses of elliptical distribution ($\rho = 0.6$, $p = 3$).

6.9 Proof of Theorems in Chapter 6

6.9.1 Proof of Theorem 6.5

Proof *Without loss of generality, assume that* $\boldsymbol{\mu} = \mathbf{0}$. *Sprit* $\boldsymbol{H} = [\boldsymbol{H}_1 : \boldsymbol{H}_2]$, *where* $\boldsymbol{H}_1 = [\boldsymbol{\beta}_1, \ldots, \boldsymbol{\beta}_q]$ *and* $\boldsymbol{H}_2 = [\boldsymbol{\beta}_{q+1}, \ldots, \boldsymbol{\beta}_p]$. *Set*

$$\boldsymbol{Y} = \begin{bmatrix} \boldsymbol{Y}_1 \\ \boldsymbol{Y}_2 \end{bmatrix} = \begin{bmatrix} \boldsymbol{H}_1^T \boldsymbol{X} \\ \boldsymbol{H}_2^T \boldsymbol{X} \end{bmatrix}$$

and $\boldsymbol{y}_j^* = \boldsymbol{H}_1^T \boldsymbol{y}_j, j = 1, \ldots, k$. *Then*

$$\begin{aligned} \|\boldsymbol{X} - \boldsymbol{y}_j\|^2 &= \|\boldsymbol{H}^T \boldsymbol{X} - \boldsymbol{H}^T \boldsymbol{y}_j\|^2 \\ &= \|\boldsymbol{H}_1^T \boldsymbol{X} - \boldsymbol{H}_1^T \boldsymbol{y}_j\|^2 + \|\boldsymbol{H}_2^T \boldsymbol{x} - \boldsymbol{H}_2^T \boldsymbol{y}_j\|^2 \\ &= \|\boldsymbol{H}_1^T \boldsymbol{X} - \boldsymbol{H}_1^T \boldsymbol{y}_j\|^2 + \|\boldsymbol{H}_2^T \boldsymbol{x}\|^2 \\ &= \|\boldsymbol{Y}_1 - \boldsymbol{y}_j^*\|^2 + \|\boldsymbol{Y}_2\|^2. \end{aligned}$$

TABLE 6.11
Distribution fitting for MSE for the 4 subclasses of elliptical distribution with generalized extreme value distribution ($k = 5$)

Distribution	p	ρ	Parameters by MLE κ	σ	μ	$\min_j(MSE(\mathcal{Y}_j))$	MSE by NTLBG
Normal	2	0	0.1411	0.1712	0.9525	0.6696	0.6221
Normal	2	0.3	0.1455	0.1651	0.9037	0.6403	0.5698
Normal	2	0.6	0.1977	0.1693	0.7917	0.4988	0.4711
Normal	2	0.9	0.2874	0.1658	0.5516	0.2894	0.2520
Kotz Type	2	0	0.2452	0.2301	0.6840	0.3440	0.3041
Kotz Type	2	0.3	0.2770	0.2107	0.6549	0.3450	0.2968
Kotz Type	2	0.6	0.2671	0.1807	0.5763	0.2988	0.2664
Kotz Type	2	0.9	0.5444	0.1156	0.3580	0.2066	0.1725
Pearson Type II	2	0	0.1421	0.1856	0.9212	0.6187	0.5664
Pearson Type II	2	0.3	0.1550	0.1823	0.8860	0.5884	0.5226
Pearson Type II	2	0.6	0.2087	0.1814	0.7687	0.4627	0.4344
Pearson Type II	2	0.9	0.2766	0.1642	0.5143	0.2669	0.2298
Pearson Type VII	2	0	0.1317	0.1663	0.9661	0.6682	0.6396
Pearson Type VII	2	0.3	0.1584	0.1624	0.9215	0.6345	0.5850
Pearson Type VII	2	0.6	0.1822	0.1694	0.8093	0.5322	0.4830
Pearson Type VII	2	0.9	0.2603	0.1773	0.5768	0.3283	0.2597
Normal	3	0	0.1227	0.2444	1.9515	1.5270	1.4093
Normal	3	0.3	0.1151	0.2543	1.8338	1.3922	1.2478
Normal	3	0.6	0.1598	0.2618	1.5050	1.0773	0.9517
Normal	3	0.9	0.2570	0.2502	0.9124	0.5331	0.4236
Kotz Type	3	0	0.0555	0.2997	1.8531	1.2644	1.1747
Kotz Type	3	0.3	0.0875	0.2791	1.7277	1.2152	1.0284
Kotz Type	3	0.6	0.2242	0.2343	1.3650	0.9127	0.7895
Kotz Type	3	0.9	0.4560	0.1995	0.7362	0.4419	0.3371
Pearson Type II	3	0	0.1107	0.2663	1.9470	1.4680	1.3473
Pearson Type II	3	0.3	0.1089	0.2587	1.8088	1.3373	1.1897
Pearson Type II	3	0.6	0.1639	0.2618	1.4792	1.0587	0.9137
Pearson Type II	3	0.9	0.3170	0.2416	0.8641	0.5299	0.3936
Pearson Type VII	3	0	0.1157	0.2299	1.9596	1.5706	1.4317
Pearson Type VII	3	0.3	0.1195	0.2396	1.8360	1.4186	1.2687
Pearson Type VII	3	0.6	0.1435	0.2595	1.5208	1.0936	0.9659
Pearson Type VII	3	0.9	0.2474	0.2585	0.9457	0.5331	0.4353

Let D_j and D_j^* be the domains of attraction for $\boldsymbol{y}_j$ and $\boldsymbol{y}_j^*$, respectively, $j = 1, \ldots, k$. It is easy to see that if $\|\boldsymbol{x} - \boldsymbol{y}_j\|^2 < \|\boldsymbol{x} - \boldsymbol{y}_l\|^2$ then $\|\boldsymbol{Y}_1 - \boldsymbol{y}_j^*\|^2 < \|\boldsymbol{Y}_1 - \boldsymbol{y}_l^*\|^2$, and $\{\boldsymbol{x} \in D_j\} = \{\boldsymbol{Y}_1 \in D_j^*\}$. By strong symmetry

$$\begin{bmatrix} \boldsymbol{Y}_1 \\ \boldsymbol{Y}_2 \end{bmatrix} \stackrel{d}{=} \begin{bmatrix} \boldsymbol{Y}_1 \\ -\boldsymbol{Y}_2 \end{bmatrix}$$

and

$$E[\boldsymbol{Y}_2|\boldsymbol{Y}_1 \in D_j^*] = E[-\boldsymbol{Y}_2|\boldsymbol{Y}_1 \in D_j^*], j = 1, \ldots, k,$$

which implies $E[\boldsymbol{Y}_2|\boldsymbol{Y}_1 \in D_j^*] = \mathbf{0}$. By the self-consistency assumption, $E[\boldsymbol{Y}|\boldsymbol{Y}_1 \in D_j] =$

$\boldsymbol{y}_j^*$. *Therefore, by the use of the orthogonal* $\boldsymbol{H}$, *we have*

$$\begin{aligned} E[\boldsymbol{X}|\boldsymbol{X} \in D_j] &= \boldsymbol{H}\boldsymbol{H}^T E[\boldsymbol{x}|\boldsymbol{H}_1^T\boldsymbol{x} \in D_j^*] \\ &= \boldsymbol{H}\begin{bmatrix} E[\boldsymbol{H}_1^T\boldsymbol{x}|\boldsymbol{H}_1^T\boldsymbol{x} \in D_j^*] \\ E[\boldsymbol{H}_2^T\boldsymbol{x}|\boldsymbol{H}_1^T\boldsymbol{x} \in D_j^*] \end{bmatrix} \\ &= \boldsymbol{H}\begin{bmatrix} E[\boldsymbol{Y}_1|\boldsymbol{Y}_1 \in D_j^*] \\ E[\boldsymbol{Y}_2|\boldsymbol{Y}_1 \in D_j^*] \end{bmatrix} \\ &= \boldsymbol{H}\begin{bmatrix} \boldsymbol{y}_j^* \\ \boldsymbol{0} \end{bmatrix} = \boldsymbol{H}\begin{bmatrix} \boldsymbol{H}_1^T\boldsymbol{y}_j \\ \boldsymbol{H}_2^T\boldsymbol{y}_j \end{bmatrix} \\ &= \boldsymbol{H}\boldsymbol{H}^T\boldsymbol{y}_j = \boldsymbol{y}_j, \end{aligned}$$

and the points $\boldsymbol{y}_1, \ldots, \boldsymbol{y}_k$ *are self-consistent points of* $\boldsymbol{X}$.

7

Properties of MSE-RPs of the Multivariate Distributions

This chapter focuses on MSE-RPs of the multivariate distributions and related discrete approximation distribution $F_{MSE}(\boldsymbol{x})$ to $F(\boldsymbol{x})$. However, a more general concept than MSE-RPs is the self-consistent points. The study of the properties of the self-consistent points is introduced. These properties can be extended to the MSE-RPs. A small number of MSE-RPs for the elliptical distribution is given. The relationship between MSE-RPs and the principal components of the multivariate analysis is discussed.

7.1 Self-Consistency Points of Multivariate Distributions

Let random vector $\boldsymbol{X} \in \mathbb{R}^p$ follow a multivariate distribution $F(\boldsymbol{x})$ with finite mean vector $\boldsymbol{\mu}$ and covariance matrix $\boldsymbol{\Sigma}$. A more general concept than MSE-RPs is the *self-consistent points* (SC-RPs) defined in Definition 2.4. The term "self-consistency" was inspired by Hastie and Stuetzle (1989), who defined self-consistent curves and principal curves. An earlier definition of self-consistency of estimators is due to Efron (1967). Tarpey and Flury (1996) gave a comprehensive study on self-consistency. We recall the concept and discuss it in more detail in this section. The content of this section is from Tarpey and Flury (1996).

7.1.1 Definition and basic properties

Suppose that k points $\boldsymbol{y}_1, \ldots, \boldsymbol{y}_k$ in $\mathbb{R}^p$ are given, and $\mathbb{R}^p$ is partitioned into k regions (called 'domains of attraction' or 'Voronoi regions') $S_1, \ldots, S_k$, where S_j are defined as

$$S_j = \{\boldsymbol{x} : (\boldsymbol{x} - \boldsymbol{y}_j)^T(\boldsymbol{x} - \boldsymbol{y}_j) < (\boldsymbol{x} - \boldsymbol{y}_h)^T(\boldsymbol{x} - \boldsymbol{y}_h), \forall h \neq j\}, j = 1, ..., k. \tag{7.1}$$

Graf and Luschgy (2007) gave more concepts: Voronoi regions, Voronoi diagrams, and Voronoi partitions, and a deeper study of these concepts. The self-consistent points are defined as follows.

Definition 7.1 *A set of points* $\{\boldsymbol{y}_1, \ldots, \boldsymbol{y}_k\}$ *is called self-consistent for* $\boldsymbol{X}$ *if*

$$E(\boldsymbol{X}|\boldsymbol{X} \in S_j) = \boldsymbol{y}_j, j = 1, \ldots, k, \tag{7.2}$$

where S_j *is the domain of attraction of* $\boldsymbol{y}_j$ *defined in (7.1). Define a discrete random vector and related distribution* $\mathbf{Y}^k_{SC} \sim F^k_{SC}$ *by*

$$\mathbf{Y}^k_{SC} = \boldsymbol{y}_j, \quad \text{if } \boldsymbol{X} \in S_j, j = 1, \ldots, k. \tag{7.3}$$

Therefore,

$$P(\mathbf{Y}^k_{SC} = \boldsymbol{y}_j) = P(\boldsymbol{X} \in S_j) \equiv p_j, j = 1, \ldots, k. \tag{7.4}$$

DOI: 10.1201/9781003589389-7

An important property of $\boldsymbol{Y}_{SC}^k$ is unbiased (McQueen (1967)).

Theorem 7.1 *Let $\boldsymbol{X} \in \mathbb{R}^p$ be a continuous random vector and $\boldsymbol{Y}_{SC}^k$ be a self-consistent vector for $\boldsymbol{X}$ defined in (7.3). Then*

1. *$E(\boldsymbol{Y}_{SC}^k) = E(\boldsymbol{X})$ and $E(\boldsymbol{X})$ is located in the convex hull of $\boldsymbol{y}_1, \ldots, \boldsymbol{y}_k$.*
2. *Suppose $\boldsymbol{X}$ has the mean vector $\mathbf{0}$ and covariance matrix $\boldsymbol{\Sigma}$, then*

$$\text{MSE}(\mathcal{Y}) = \text{MSE}(\boldsymbol{y}_1, \ldots, \boldsymbol{y}_k) = tr(\boldsymbol{\Sigma}) - \sum_{j=1}^{k} p_i \|\boldsymbol{y}_j\|^2. \tag{7.5}$$

The proof of Theorem 7.1 is given in Subsection 7.5.1. The following theorem is from Flury (1993).

Theorem 7.2 *MSE-RPs $\boldsymbol{\xi} = \{\boldsymbol{\xi}_j \in \mathbb{R}^p, 1 \le j \le k\}$ for $\boldsymbol{X}$ are self-consistent and the related MSE is given by*

$$\text{MSE}(\boldsymbol{\xi}) = \sum_{j=1}^{k} p_j tr(\boldsymbol{\Sigma}_j), \tag{7.6}$$

where $\boldsymbol{\Sigma}_j = \text{Cov}(\boldsymbol{X}|\boldsymbol{X} \in S_j)$.

The proofs of Theorem 7.2 can refer to Subsection 7.5.2. Combining the above two theorems, we can give another definition for MSE-RPs.

Definition 7.2 *A self-consistent set of $\{\boldsymbol{\xi}_1, \ldots, \boldsymbol{\xi}_k\}$ for $\boldsymbol{X} \in \mathbb{R}^p$ is a set of MSE-RPs of a random vector $\boldsymbol{X}$ if*

$$\text{MSE}(\boldsymbol{\xi}_1, \ldots, \boldsymbol{\xi}_k) \le \text{MSE}(\boldsymbol{y}_1, \ldots, \boldsymbol{y}_k)$$

for all self-consistent sets $\{\boldsymbol{y}_1, \ldots, \boldsymbol{y}_k\}$. Define a discrete random vector $\boldsymbol{\xi}^k = \boldsymbol{y}_{MSE}^k$ with support points $\{\boldsymbol{\xi}_1, \ldots, \boldsymbol{\xi}_k\}$ and $P(\boldsymbol{y}_{MSE} = \boldsymbol{\xi}_j) = P(\boldsymbol{X} \in S_j) = p_j, j = 1, \ldots, k$. When the k is fixed in the case, we may omit k from the notation $\boldsymbol{Y}_{SC}^k \sim F_{MC}^k$ and $\boldsymbol{\xi}^k = \boldsymbol{Y}_{MSE}^k \sim F_{MSE}^k$.

The MSE of MSE-RPs can be expressed as

$$\text{MSE}(\boldsymbol{\xi}) = E(\boldsymbol{X} - \boldsymbol{Y}_{MSE})^T(\boldsymbol{X} - \boldsymbol{Y}_{MSE}) = E\|(\boldsymbol{X} - \boldsymbol{Y}_{MSE})\|^2. \tag{7.7}$$

As a set of MSE-RPs belongs to the class of self-consistent sets, many properties for the set of SC-RPs that can be applied to the set of MSE-RPs properties are good examples.

Theorem 7.3 *Let $\boldsymbol{X}_1$ and $\boldsymbol{X}_2$ be two random vectors in $\mathbb{R}^p$ with relation $\boldsymbol{X}_2 = \boldsymbol{a} + b\boldsymbol{H}\boldsymbol{X}_1$, where $\boldsymbol{a} \in \mathbb{R}^p$, $b \in \mathbb{R}$ and $\boldsymbol{H}$ is an orthogonal matrix of order p. We have*

(a) If $\{\boldsymbol{y}_1, \ldots, \boldsymbol{y}_k\}$ is a set of SC-RPs of $\boldsymbol{X}_1$, then $\{\boldsymbol{a} + b\boldsymbol{H}\boldsymbol{y}_1, \ldots, \boldsymbol{a} + b\boldsymbol{H}\boldsymbol{y}_k\}$ is a set of SC-RPs of $\boldsymbol{X}_2$;

(b) If $\{\boldsymbol{\xi}_1, \ldots, \boldsymbol{\xi}_k\}$ is a set of MSE-RPs of $\boldsymbol{X}_1$, then $\{\boldsymbol{a} + b\boldsymbol{H}\boldsymbol{\xi}_1, \ldots, \boldsymbol{a} + b\boldsymbol{H}\boldsymbol{\xi}_k\}$ is a set of MSE-RPs of $\boldsymbol{X}_2$.

Matsuura and Kurata (2013) extended the fact (b) into the case: $\boldsymbol{H}$ is $p \times m$ such that $\boldsymbol{H}^T\boldsymbol{H} = \boldsymbol{I}_m$.

Theorem 7.4 *Let $\boldsymbol{X}$ be a random vector in $\mathbb{R}^p$ with mean vector $\mathbf{0}$. Suppose $\boldsymbol{y}_1, \ldots, \boldsymbol{y}_k$ are k SC-RPs for $\boldsymbol{X}$, and $\mathcal{P}$ is the liner subspace spanned by $\boldsymbol{y}_1, \ldots, \boldsymbol{y}_k$ with dimension $q < p$. Let $\boldsymbol{a}_1, \ldots, \boldsymbol{a}_q \in \mathbb{R}^p$ be an orthonormal basis of $\mathcal{P}$ and denote $\boldsymbol{A} = [\boldsymbol{a}_1 : \ldots : \boldsymbol{a}_q]$. Then the random vector $\boldsymbol{A}^T\boldsymbol{X}$ has a set of k SC-RPs $\boldsymbol{A}^T\boldsymbol{y}_1, \ldots, \boldsymbol{A}^T\boldsymbol{y}_k$.*

Proof *Let $S_j \in \mathbb{R}^p$ be the domain of attraction of $\boldsymbol{y}_j$, and let $S_j^* \in \mathbb{R}^p$ denote the domain of attraction of $A^T\boldsymbol{y}_j, j = 1, \ldots, k$. Then $\boldsymbol{X} \in S_j$ is equivalent to $\boldsymbol{A}^T\boldsymbol{X} \in S_j^*$. We then have $E[\boldsymbol{A}^T\boldsymbol{X}|\boldsymbol{A}^T\boldsymbol{X} \in D_j^*] = \boldsymbol{A}^T\boldsymbol{y}_j$.*

When $k = 1$, we have a similar result with Theorem 3.5.

Theorem 7.5 *Let $\boldsymbol{X} \sim F(\boldsymbol{x})$ in $\mathbb{R}^p$ with mean vector $E(\boldsymbol{X}) = \boldsymbol{\mu}$ and covariance matrix $Cov(\boldsymbol{X}) = \boldsymbol{\Sigma}$. Then the MSE-RP with $k = 1$ is $\boldsymbol{\xi} = \boldsymbol{\mu}$ and the related* $\mathrm{MSE}(\boldsymbol{\xi}) = tr(\boldsymbol{\Sigma})$.

The proof of the theorem gives in Subsection 7.5.3. This theorem is a special case of (7.6). But it seems to contradict (7.5). For the case of $k = 1$ and $\boldsymbol{\mu} = \mathbf{0}$, $\boldsymbol{y}_1$ is only one self-consistent point of $\boldsymbol{X}$, from Theorem 7.1 $E(\boldsymbol{y}_1) = E(\boldsymbol{Y}_{SC}) = E(\boldsymbol{X}) = \mathbf{0}$, and $\mathrm{MSE}(\boldsymbol{y}_1) = tr(\boldsymbol{\Sigma}) - ||\boldsymbol{y}_1||^2$. In this case, the MSE-RP $\boldsymbol{\xi} = \mathbf{0}$ and

$$\mathrm{MSE}(\boldsymbol{\xi}) = tr(\boldsymbol{\Sigma}) \leq \mathrm{MSE}(\boldsymbol{y}_1) = tr(\boldsymbol{\Sigma}) - ||\boldsymbol{y}_1||^2,$$

the equal holds if and only if $||\boldsymbol{y}_1|| = 0$, and if and only if $\boldsymbol{y}_1 = \mathbf{0}$. Note that $E(\boldsymbol{y}_1) = \mathbf{0}$ does not imply $\boldsymbol{y}_1 = \mathbf{0}$.

7.1.2 Advanced study on the self-consistency

The contents of this subsection cite Tarpey and Flury (1996). Suppose we want to approximate the distribution of a random vector $\boldsymbol{X}$ by a random vector $\boldsymbol{Y}$ with minimum mean squared error

$$MSE(\boldsymbol{Y}; \boldsymbol{X}) = E||\boldsymbol{Y} - \boldsymbol{X}||^2 \tag{7.8}$$

if both $\boldsymbol{X}$ and $\boldsymbol{Y}$ have finite second moments. In terms of mean squared error, the approximation of $\boldsymbol{X}$ by $\boldsymbol{Y}$ can always be improved using $E[\boldsymbol{X}|\boldsymbol{Y}]$ since for any measurable function g,

$$E||\boldsymbol{X} - E(\boldsymbol{X}|\boldsymbol{Y})||^2 \leq E||\boldsymbol{X} - g(\boldsymbol{Y})||^2.$$

In particular, Bickel and Doksum (1977) pointed out

$$E||\boldsymbol{X} - E(\boldsymbol{X}|\boldsymbol{Y})||^2 \leq E||\boldsymbol{X} - \boldsymbol{Y}||^2.$$

Thus the random vector $\boldsymbol{Y}$ is locally optimal for approximating $\boldsymbol{X}$ if $\boldsymbol{Y} = E[\boldsymbol{X}|\boldsymbol{Y}]$, in which case Tarpey and Flury (1996) called $\boldsymbol{Y}$ self-consistent for $\boldsymbol{X}$. The following definition of self-consistent is more general.

Definition 7.3 *For two jointly distributed random vectors $\boldsymbol{X}$ and $\boldsymbol{Y}$ with finite second moments, we say that $\boldsymbol{Y}$ is self-consistent for $\boldsymbol{X}$ if $E(\boldsymbol{X}|\boldsymbol{Y}) = \boldsymbol{Y}$ almost surely.*

Note that this definition does not restrict $\boldsymbol{Y}$ to be discrete. Two extreme cases: (a) $\boldsymbol{Y} = \boldsymbol{X}$, i.e., $\boldsymbol{X}$ is self-consistent for $\boldsymbol{X}$; (b) $\boldsymbol{Y} = E(\boldsymbol{X})$ is also self-consistent for $\boldsymbol{X}$.

Lemma 7.1 *If $\boldsymbol{Y}$ is self-consistent for $\boldsymbol{X}$ then $E(\boldsymbol{Y}) = E(\boldsymbol{X})$.*

Proof *The lemma follows from* $E(E(\boldsymbol{X}|\boldsymbol{Y})) = E(\boldsymbol{X})$.

This Lemma is more general than Theorem 7.1. Some examples given by Tarpey and Flury (1996) are as follows:

Example 7.1 *Let* $X \sim N(0,1)$ *and*

$$Y = \begin{cases} -\sqrt{2/\pi}, & X < 0, \\ \sqrt{2/\pi}, & X \geq 0. \end{cases}$$

Then Y *is self-consistent for* X*. In fact,* Y *is constructed by two MSE-RPs,* $\pm\sqrt{2/\pi}$*, of* $N(0,1)$.

Example 7.2 *Let* $\boldsymbol{X} = (X_1, X_2)^T \sim N_2(\boldsymbol{0}, \boldsymbol{I}_2)$*. It is known that* $E(X_1|X_1 \geq 0) = \sqrt{2/\pi}$.

$$\boldsymbol{Y} = \begin{cases} (X_1, -\sqrt{2/\pi})^T, & \text{if } X_2 < 0, \\ (X_1, \ \sqrt{2/\pi})^T, & \text{if } X_2 \geq 0. \end{cases}$$

Then $\boldsymbol{Y}$ *is self-consistent for* $\boldsymbol{X}$*. This example is an extension of Example 7.1.*

Example 7.3 *For a random vector* $\boldsymbol{X} \in \mathbb{R}^p$*, suppose* $\mathcal{L}$ *is a measurable set in* $\mathbb{R}^p$ *such that for any* $\boldsymbol{y} \in \mathcal{L}$

$$\boldsymbol{y} = E(\boldsymbol{X}|\boldsymbol{X} \in \mathcal{D}_{\boldsymbol{y}}),$$

where $\mathcal{D}_{\boldsymbol{y}}$ *is domain of attraction of* $\boldsymbol{y}$*, that is*

$$\mathcal{D}_{\boldsymbol{y}} = \{\boldsymbol{x} \in \mathbb{R}^p : ||\boldsymbol{x} - \boldsymbol{y}|| < ||x - \boldsymbol{y}^*||, \forall \boldsymbol{y}^* \in \mathcal{L}\}.$$

Define $\boldsymbol{Y} = \boldsymbol{y}$ *if* $\boldsymbol{X} \in \mathcal{D}_{\boldsymbol{y}}$*, Then* $\boldsymbol{Y}$ *is self-consistent for* $\boldsymbol{X}$ *as*

$$E(\boldsymbol{X}|\boldsymbol{Y}) = E(\boldsymbol{X}|\boldsymbol{X} \in \mathcal{D}_{\boldsymbol{y}}) = \boldsymbol{y}.$$

Denote the covariance matrices of $\boldsymbol{X}$ and $\boldsymbol{Y}$ by $\boldsymbol{\Sigma}_{\boldsymbol{x}}$ and $\boldsymbol{\Sigma}_{\boldsymbol{y}}$, respectively. Let $\boldsymbol{X} \in \mathbb{R}^p$ be a continuous random vector, and let $\boldsymbol{Y}$ be $\boldsymbol{Y}_{SC}$. Then

$$\boldsymbol{\Sigma}_{\boldsymbol{y}} = E(\boldsymbol{Y} - E(\boldsymbol{Y}))(\boldsymbol{Y} - E(\boldsymbol{Y}))^T = \sum_{j=1}^{k} p_j(\boldsymbol{y}_j - \boldsymbol{\mu})(\boldsymbol{y}_j - \boldsymbol{\mu})^T. \tag{7.9}$$

When $\boldsymbol{\mu} = \boldsymbol{0}$, then $\boldsymbol{\Sigma}_y = \sum_{j=1}^k p_j \boldsymbol{y}_j \boldsymbol{y}_j^T$.

Theorem 7.6 *If* $\boldsymbol{Y}$ *is self-consistent for* $\boldsymbol{X}$*, then*
(i) $\boldsymbol{\Sigma}_{\boldsymbol{X}} \geq \boldsymbol{\Sigma}_{\boldsymbol{Y}}$;
(ii) $MSE(\boldsymbol{Y}; \boldsymbol{X}) = tr(\boldsymbol{\Sigma}_{\boldsymbol{X}}) - tr(\boldsymbol{\Sigma}_{\boldsymbol{Y}})$.

The proof can be found in Subsection 7.5.4. It follows from Theorem 7.1 that Cov$[\boldsymbol{Y}] =$ Cov$[\boldsymbol{X}]$ if Cov$[\boldsymbol{X}|\boldsymbol{Y}] = \boldsymbol{0}$ *a.s.*, i.e., $\boldsymbol{Y} = \boldsymbol{X}$ *a.s.* Applying this theorem to the univariate case where X and Y are random variables.

Corollary 7.1 *If* Y *is self-consistent for* X*, then* $\text{Var}(Y) \leq \text{Var}(X)$.

From Theorem 7.1 for $k > 1$

$$MSE(\boldsymbol{Y};\boldsymbol{X}) = tr(\boldsymbol{\Sigma}_{\boldsymbol{X}}) - \sum_{j=1}^{k} p_j(\boldsymbol{y}_j - \boldsymbol{\mu})^T(\boldsymbol{y}_j - \boldsymbol{\mu}).$$

When $E(\boldsymbol{X}) = \boldsymbol{0}$ and $k > 1$ we have

$$MSE(\boldsymbol{Y};\boldsymbol{X}) = tr(\boldsymbol{\Sigma}_x) - \sum_{j=1}^{k} p_j||\boldsymbol{y}_j||^2.$$

When $k = 1$, $tr(\boldsymbol{\Sigma}_{\boldsymbol{Y}}) = 0$ and $MSE(\boldsymbol{Y};\boldsymbol{X}) = tr(\boldsymbol{\Sigma}_{\boldsymbol{X}})$.

Theorem 7.7 *Suppose* $\boldsymbol{Y}$ *is self-consistent for* $\boldsymbol{X} \in \mathbb{R}^p$ *with* $E(\boldsymbol{X}) = \boldsymbol{0}$, *and* $\mathcal{L}(\boldsymbol{y})$ *is the linear subspace spanned by* q *orthonormal column vectors in the* $q \times p$ *matrix* $\boldsymbol{A}$. *Let* $\boldsymbol{P} = \boldsymbol{A}\boldsymbol{A}^T$ *denote the associated projection matrix. Then* $\boldsymbol{Y}$ *and* $\boldsymbol{A}^T\boldsymbol{Y}$ *are self-consistent for* $\boldsymbol{P}\boldsymbol{X}$ *and* $\boldsymbol{A}^T\boldsymbol{X}$, *respectively.*

Theorem 7.8 *Suppose random vector* $\boldsymbol{X} \in \mathbb{R}^p$ *is partitioned into* q *and* $p - q$ *components as*

$$\boldsymbol{X} = \begin{bmatrix} \boldsymbol{X}_1 \\ \boldsymbol{X}_2 \end{bmatrix}.$$

Then, the random vector

$$\boldsymbol{Y} = \begin{bmatrix} \boldsymbol{Y}_1 \\ \boldsymbol{Y}_2 \end{bmatrix} = \begin{bmatrix} \boldsymbol{X}_1 \\ E(\boldsymbol{X}_2|\boldsymbol{X}_1) \end{bmatrix}$$

is self-consistent for $\boldsymbol{X}$.

Proof *As*

$$\begin{aligned} E[\boldsymbol{X}_1|\boldsymbol{Y}] &= E[\boldsymbol{X}_1|\boldsymbol{X}_1, E(\boldsymbol{X}_2|\boldsymbol{X}_1)] = E[\boldsymbol{X}_1|\boldsymbol{X}_1] = \boldsymbol{X}_1 = \boldsymbol{Y}_1, \\ E[\boldsymbol{X}_2|\boldsymbol{Y}] &= E[\boldsymbol{X}_2|\boldsymbol{X}_1] = \boldsymbol{Y}_2. \end{aligned}$$

Hence

$$E[\boldsymbol{X}|\boldsymbol{Y}] = \begin{bmatrix} E[\boldsymbol{X}_1|\boldsymbol{Y}] \\ E[\boldsymbol{X}_2|\boldsymbol{Y}] \end{bmatrix} = \begin{bmatrix} \boldsymbol{Y}_1 \\ \boldsymbol{Y}_2 \end{bmatrix}.$$

Example 7.4 *Suppose* $\boldsymbol{X}$ *has a bivariate normal distribution with mean* $\boldsymbol{0}$ *and covariance matrix* $\boldsymbol{\Sigma} = \begin{pmatrix} 1 & \rho \\ \rho & 1 \end{pmatrix}$. *Then*

$$\boldsymbol{Y} = \begin{bmatrix} X_1 \\ E(X_2|X_1) \end{bmatrix} = \begin{bmatrix} X_1 \\ \rho X_1 \end{bmatrix}$$

is self-consistent for $\boldsymbol{X}$ *with* $MSE[\boldsymbol{Y};\boldsymbol{X}] = 1 - \rho^2$.

For high-dimensional random variables, it is often desirable to find a low-dimensional approximation. The following Theorem 7.9 is called the *principal subspace theorem* proposed by Tarpey et al. (1995) for the special case of self-consistent approximations whose support consists of k distinct points.

Theorem 7.9 *Suppose $\boldsymbol{X}$ is a vector in $\mathbb{R}^p$ with mean $\mathbf{0}$ and covariance matrix $\boldsymbol{\Sigma} > 0$.*

Suppose $\boldsymbol{Y}$ is self-consistent for $\boldsymbol{X}$ and the support of $\boldsymbol{Y}$ spans a linear subspace $\mathcal{A}$ of dimension $q < p$. Suppose, furthermore, that $\boldsymbol{Y}$ is measurable concerning the orthogonal projection of $\boldsymbol{X}$ on $\mathcal{A}$. Then $\mathcal{A}$ is spanned by q eigenvectors of $\boldsymbol{\Sigma}$.

The proof can refer to Subsection 7.5.5. Theorem 7.9 shows that for certain types of self-consistent. Approximation attention may be restricted to subspaces spanned by eigenvectors of the covariance matrix, including principal components.

Next, we show how principal components can be based on self-consistency. Suppose $\boldsymbol{A}_1$ is a $p \times q$ matrix ($q < p$) such that all columns of $\boldsymbol{A}_1$ have unit length and are mutually orthogonal. Let $\boldsymbol{P} = \boldsymbol{A}\boldsymbol{A}^T$ denote the projection matrix associated with the orthogonal projection from $\mathbb{R}^p$ into the subspace $\mathcal{M}$ spanned by the columns of $\boldsymbol{A}_1$. Let $\boldsymbol{b} \in \mathbb{R}^p$ and consider the transformation $\boldsymbol{Y} = \boldsymbol{b} + \boldsymbol{P}\boldsymbol{X}$. If $\boldsymbol{Y}$ is self-consistent for $\boldsymbol{X}$, then $E(\boldsymbol{Y}) = E(\boldsymbol{X})$ by Lemma 7.1. That is $\boldsymbol{b} = (\boldsymbol{I}_p - \boldsymbol{P})\boldsymbol{\mu}$. Thus, if $\boldsymbol{Y}$ is self-consistent for $\boldsymbol{X}$ into a linear space, then $\boldsymbol{Y} = (\boldsymbol{I}_p - \boldsymbol{P})\boldsymbol{\mu} + \boldsymbol{P}\boldsymbol{X}$. Assume, without loss of generality, that $\boldsymbol{\mu} = \mathbf{0}$. Let $\boldsymbol{A}_2$ denote a $(p-q) \times p$ matrix such that $[\boldsymbol{A}_1 : \boldsymbol{A}_2]$ is orthogonal and let $\boldsymbol{Q} = \boldsymbol{I}_p - \boldsymbol{P}$. Then $\boldsymbol{Y} = \boldsymbol{P}\boldsymbol{X}$ is self-consistency for $\boldsymbol{X}$ implies $E[\boldsymbol{Q}\boldsymbol{x}(\boldsymbol{P}\boldsymbol{x}] = \mathbf{0}$ a.s. However, for any random variables U and V, $\text{Cov}[U, V] = 0$ implies $\text{Cov}(U, V) = 0$. Thus $\boldsymbol{A}_2^T \boldsymbol{\Sigma} \boldsymbol{A}_1 = \mathbf{0}$. Tarpey et al. (1995) gave the following theorem.

Theorem 7.10 *If $\boldsymbol{Y} = (\boldsymbol{I}_p - \boldsymbol{P})\boldsymbol{\mu} + \boldsymbol{P}\boldsymbol{X}$ is self-consistent for $\boldsymbol{X} \in \mathbb{R}^p$ where $\boldsymbol{\mu} = E[\boldsymbol{X}]$ and $\boldsymbol{P}$ is the projection matrix associated with the orthogonal projection from $\mathbb{R}^P$ into a linear subspace $\mathcal{M}$ of dimension $q < p$, then $\mathcal{M}$ is spanned by some q eigenvectors of $\boldsymbol{\Sigma} = Cov(\boldsymbol{X})$.*

Example 7.5 *Suppose $\boldsymbol{X}$ is bivariate normal with mean $\mathbf{0}$ and covariance matrix $\boldsymbol{\Sigma} = \begin{pmatrix} 1 & \rho \\ \rho & 1 \end{pmatrix}$ discussed in Example 7.4. The two eigenvalues of $\boldsymbol{\Sigma}$ are $1+\rho$ and $1-\rho$ with associated eigenvectors $\boldsymbol{\beta}_1 = \frac{1}{\sqrt{2}}(1,1)^T$ and $\boldsymbol{\beta}_2 = \frac{1}{\sqrt{2}}(-1,1)^T$, respectively. The corresponding*

$$\boldsymbol{P} = \boldsymbol{\beta}_1\boldsymbol{\beta}_1^T = \frac{1}{2}\begin{bmatrix} 1 & 1 \\ 1 & 1 \end{bmatrix}, \text{ and } \boldsymbol{Y} = \frac{1}{2}\begin{bmatrix} X_1 + X_2 \\ X_1 + X_2 \end{bmatrix}.$$

$\boldsymbol{Y}$ is self-consistent for $\boldsymbol{X}$ with MSE$=1-\rho$. Compare $\boldsymbol{Y}$ in Example 7.5 with MSE$=1-\rho^2$ whose support is along the line of X_2 on X_1 which is not spanned by an eigenvector of the covariance matrix,

7.2 Properties of MSE-RPs of Multivariate Distributions

This section assumes that the random vector $\boldsymbol{X}$ has a finite mean vector $\boldsymbol{\mu}$ and covariance matrix $\boldsymbol{\Sigma}$. A set of MSE-RPs of $\boldsymbol{X}$ is denoted by $\{\boldsymbol{\xi}_1, \ldots, \boldsymbol{\xi}_k\}$ and the corresponding vector is $\boldsymbol{\xi} = \boldsymbol{Y}_{MSE}$ with Voronoi regions $S_j, j = 1, \ldots, k$ and probabilities $p_j = P(\boldsymbol{X} \in S_j), j = 1, \ldots, k$.

Definition 7.2 indicates that a set of MSE-RPs must be a set of self-consistent points. Therefore, some properties mentioned in Section 7.1 can be applied to MSE-RPs. Flury (1990) obtained the following theorem.

Theorem 7.11 *Under the assumption on $\boldsymbol{X}$ we have*

1. *When $k = 1$ the MSE-RP is given by $\boldsymbol{\xi} = \boldsymbol{\mu}$;*

2. $E(\boldsymbol{Y}_{MSE}) = E(\boldsymbol{X}) = \sum_{j=1}^k p_j \boldsymbol{\xi}_j$, *i.e.,* $E(\boldsymbol{X})$ *is in the convex hull of* $\boldsymbol{\xi}_1, \ldots, \boldsymbol{\xi}_k$*;*

3. $\text{MSE}(\boldsymbol{Y}_{MSE}) = tr(\boldsymbol{\Sigma}) - tr(\boldsymbol{\Sigma}_{\boldsymbol{\xi}}) = tr(\boldsymbol{\Sigma}) - \sum_{j=1}^k p_j(\boldsymbol{\xi}_j - \boldsymbol{\mu})(\boldsymbol{\xi}_j - \boldsymbol{\mu})^T$.

4. For $k = 2$*, the minimum value of* $\text{MSE}(\boldsymbol{\xi}_1, \boldsymbol{\xi}_2)$ *is attained if the two points are projected onto a line through the mean vector* $\boldsymbol{\mu}$ *with the same direction as* $\boldsymbol{\xi}_2 - \boldsymbol{\xi}_1$.

5. The rank of space spading by $\{\boldsymbol{\xi}_1 - \boldsymbol{\mu}, \ldots, \boldsymbol{\xi}_k - \boldsymbol{\mu}\}$ *is less than or equal to* k.

7.3 Property of MSE-RPs of The Elliptical Distributions

It is too difficult to explore more properties of MSE-RPs for all continuous multivariate distributions. Therefore, many authors focus on the elliptical distributions due to their stochastic representation. Section 6.2 briefly introduces the elliptical distributions. Flury (1990), Fang et al. (1994b), Fang and Wang (1994), Tarpey et al. (1995), Tarpey and Flury (1996), Fang et al. (2020), and Yang et al. (2021) gave a comprehensive study on MSE-RPs of the elliptical distributions. The contents in this section are from the above publications.

7.3.1 The case of $k = 2$

In this subsection, we consider the case of $k = 2$ and $\boldsymbol{X} \sim ESD_p(\boldsymbol{\mu}, \Psi, g)$ having finite mean vector $\boldsymbol{\mu}$, and covariance matrix $\boldsymbol{\Sigma}$ that is proportional to Ψ and density generator g. Flury (1990) connected 2 MSE-RPs and the eigenvector concerning the largest eigenvalue of $\text{Cov}(\boldsymbol{X})$.

Theorem 7.12 *Under the assumption in this subsection, the two MSE-RPs of* $\boldsymbol{X}$ *have the form*

$$\boldsymbol{\xi}_1 = \boldsymbol{\mu} + \gamma_1 \boldsymbol{\beta}, \boldsymbol{\xi}_2 = \boldsymbol{\mu} - \gamma_2 \boldsymbol{\beta}, \tag{7.10}$$

where $\boldsymbol{\beta} \in \mathbb{R}^p$ *is the normalized eigenvector associated with the largest eigenvalue of* $\boldsymbol{\Sigma}$*, and* $\{\gamma_1, \gamma_2\}$ *are the two MSE-RPs of the univariate random variable* $\boldsymbol{\beta}^T(\boldsymbol{X} - \boldsymbol{\mu})$*. If the MSE-RPs are not unique, they can be chosen of the given form.*

The proof of the theorem can refer to Subsection 7.5.8. Yang et al. (2021) gave a formula for calculating the MSE of two MSE-RPs of $\boldsymbol{X}$.

Corollary 7.2 *Under the assumption in this subsection the MSE of two MSE-RPs of* $\boldsymbol{X}$ *is given by*

$$MSE(\xi_1, \xi_2) = tr(\boldsymbol{\Sigma}) - (1 - h)\boldsymbol{\beta}^T \boldsymbol{\Sigma} \boldsymbol{\beta}, \tag{7.11}$$

where

$$h = 1 - (E(|\boldsymbol{\beta}^T \boldsymbol{\Sigma}^{-1/2} \boldsymbol{X}|))^2.$$

The proof can refer to Subsection 7.5.7. Note that $\boldsymbol{\Sigma}^{-1/2}(\boldsymbol{x} - \boldsymbol{\mu})$ is a spherical distribution, and the value of

$$\left(\text{E}\left(\left|\boldsymbol{\beta}^T \boldsymbol{\Sigma}^{-1/2}(\boldsymbol{x} - \boldsymbol{\mu})\right|\right)\right)^2$$

will not change when $\boldsymbol{\Sigma}^{-1/2}(\boldsymbol{x}-\boldsymbol{\mu})$ is projected onto other directions, so $\boldsymbol{\beta}$ could be arbitrary normalized vectors. However, as the points $\gamma_1 = -\lambda^{1/2}\gamma$ and $\gamma_2 = \lambda^{1/2}\gamma$ should be projected onto the direction of $\boldsymbol{\beta}$, so it is convenient to use the eigenvector associated with the largest eigenvalue λ. The following two examples are also from Flury (1990).

Example 7.6 *Suppose $\boldsymbol{X} \sim N_p(\boldsymbol{\mu}, \boldsymbol{\Sigma})$. Let λ be the largest eigenvalue of $\boldsymbol{\Sigma}$ and $\boldsymbol{\beta}$ be its associated normalized eigenvector. Consider two cases:*

(a) if λ is simple, two MSE-RPs of $\boldsymbol{X}$ are given by $\boldsymbol{\mu}-(2\lambda/\pi)^{1/2}\boldsymbol{\beta}$ and $\boldsymbol{\mu}+(2\lambda/\pi)^{1/2}\boldsymbol{\beta}$;

(b) if λ has multiplicity r, let $\mathcal{T}$ denote the sphere of radius $(2\lambda/\pi)^{1/2}$, centered at $\boldsymbol{\mu}$, in the latent space of λ. Then any two points on $\mathcal{T}$ symmetric to $\boldsymbol{\mu}$ are two MSE-RPs of $\boldsymbol{X}$.

Example 7.7 *Consider $\boldsymbol{X}$ to follow the uniform distribution on*

$$C = \{\boldsymbol{x} \in \mathbb{R}^P : (\boldsymbol{x}-\boldsymbol{\mu})^T \boldsymbol{\Psi}^{-1}(\boldsymbol{x}-\boldsymbol{\mu}) \le 1\}.$$

Let λ and $\boldsymbol{\beta}$ denote the largest eigenvalue and associated eigenvector, respectively, of $\boldsymbol{\Psi}$, with $\|\boldsymbol{\beta}\| = 1$. Suppose that λ is simple. Then, the two MSE-RPs of $\boldsymbol{X}$ are

$$\boldsymbol{\mu} \pm \frac{2\lambda^{1/2}\Gamma(\frac{1}{2}p)}{(p+1)\Gamma(\frac{1}{2}(p+1)\Gamma(\frac{1}{2})}\boldsymbol{\beta}$$

provided that the local minimum taken at these points is the global minimum.

7.3.2 Relationship between the principal components and MSE-RPs of the elliptical distribution

Theorem 7.4 discusses the linear subspace spanning by a set of self-consistent points. When this set of self-consistent points is a set of MSE-RPs of a random vector $\boldsymbol{X}$, the corresponding linear subspace should have some particular property. Flury (1990) conjectured that this subspace is related to the principal components of $\boldsymbol{\Sigma} = \mathrm{Cov}(\boldsymbol{X})$. Tarpey (1995) pointed out that k MSE-RPs ($k < p$) of an elliptical distribution lie in the linear subspace spanned by the first several principal components.

Theorem 7.13 *(Principal subspace theorem)*
Let $\boldsymbol{X} \sim ESD_p(\boldsymbol{0}, \boldsymbol{\Psi}, g)$. If a set of k MSE-RPs of $\boldsymbol{X}$ spans a subspace $\mathcal{V}$ of dimension $q, q < p$, then $\boldsymbol{\Psi}$ has a set of eigenvectors $\boldsymbol{\beta}_1, \ldots, \boldsymbol{\beta}_p$ with associated ordered eigenvalues $\lambda_1 \ge \lambda_2 \cdots \ge \lambda_p$ such that $\mathcal{V}$ is spanned by $\boldsymbol{\beta}_1, \ldots, \boldsymbol{\beta}_q$.

As MSE-RPs of $\boldsymbol{X}$ are self-consistent for $\boldsymbol{X}$, applying Theorem 7.9, we obtain this theorem.

Theorem 7.14 *Let $\boldsymbol{X} \sim ESD_p(\boldsymbol{0}, \boldsymbol{\Psi}, g)$, where the covariance has a spectral decomposition $\boldsymbol{\Psi} = \boldsymbol{B}\boldsymbol{\Lambda}\boldsymbol{B}^T$, $\boldsymbol{B} = [\boldsymbol{\beta}_1, \ldots, \boldsymbol{\beta}_p]$ is orthogonal, and $\boldsymbol{\Lambda}$ is diagonal. Suppose that the k vectors $\boldsymbol{y}_1, \ldots, \boldsymbol{y}_k$ span the same subspace as the q eigenvectors $\boldsymbol{\beta}_1, \ldots, \boldsymbol{\beta}_q$. Let $\boldsymbol{B}_1 = [\boldsymbol{\beta}_1, \ldots, \boldsymbol{\beta}_q]$ and $\boldsymbol{X}^* = \boldsymbol{B}_1\boldsymbol{X}$. If $\boldsymbol{z}_j = \boldsymbol{B}^T\boldsymbol{y}_j, j = 1, \ldots, k$ are self-consistent points of $\boldsymbol{X}^*$, then $\boldsymbol{y}_1, \ldots, \boldsymbol{y}_k$ are self-consistent points of $\boldsymbol{X}$.*

7.3.3 Connection between MSE and correlation ρ in covariance matrix

Consider the covariance matrix of $\boldsymbol{X}$ to be $\boldsymbol{\Sigma}_p(\rho)$ shown in (6.47). Let $\boldsymbol{\xi} = \{\boldsymbol{\xi}_1, \cdots, \boldsymbol{\xi}_k\}$ be the set of MSE-RPs of $\boldsymbol{X}$ and $MSE(\boldsymbol{\xi})$ is the related MSE of $\boldsymbol{\xi}$. When the number of MSE-RPs is $k = 2$, by Corollary 7.2, the set of RPs are $\boldsymbol{\xi} = \{\boldsymbol{\xi}_1, \boldsymbol{\xi}_2\}$ has the MSE

$$\mathrm{MSE}(\boldsymbol{\xi}) = tr(\boldsymbol{\Sigma}) - \lambda\left(E\left(\left|\boldsymbol{\beta}^T\left[\boldsymbol{\Sigma}_p(\rho)\right]^{-\frac{1}{2}}(\boldsymbol{X}-\boldsymbol{\mu})\right|\right)\right)^2,$$

where λ is the largest eigenvalue of $\mathbf{\Sigma}$, $\boldsymbol{\beta}$ is the associated normalized eigenvector of λ and $\boldsymbol{\beta}^T\boldsymbol{\beta} = 1$. From Theorem 7.9, these two MSE-RPs in the subspace generated by $\boldsymbol{\beta}$. It is easy to find that the eigenvalues of $\mathbf{\Sigma}_p(\rho)$ are $\lambda_1 = 1 + (p-1)\rho, \lambda_2 = \cdots = \lambda_p = 1 - \rho$. When $\rho \geq 0$, the largest eigenvalue is $\lambda = \lambda_1$ and the associated standard eigenvector is

$$\boldsymbol{\beta} = \left[\begin{array}{cccc} \frac{1}{\sqrt{p}} & \frac{1}{\sqrt{p}} & \cdots & \frac{1}{\sqrt{p}} \end{array} \right]^T,$$

then the MSE of $k = 2$ MSE-RPs of $\boldsymbol{X}$ is given by

$$\text{MSE}(\boldsymbol{\xi}) = p - (1 + (p-1)\rho) \left(E \left(\left| \boldsymbol{\beta}^T \left[\mathbf{\Sigma}_p(\rho) \right]^{-\frac{1}{2}} (\boldsymbol{X} - \boldsymbol{\mu}) \right| \right) \right)^2. \tag{7.12}$$

It is obvious that the value of MSE($\boldsymbol{\xi}$) decreases with an increasing value of ρ. As $\rho \in [0, 1]$, we have the maximum of MSE

$$\max_{0 \leq \rho \leq 1} \text{MSE}(\boldsymbol{\xi}) = p - \left(E \left(\left| \boldsymbol{\beta}^T \left[\mathbf{\Sigma}_p(\rho) \right]^{-\frac{1}{2}} (\boldsymbol{X} - \boldsymbol{\mu}) \right| \right) \right)^2 \tag{7.13}$$

and the minimum MSE

$$\begin{aligned} \min_{0 \leq \rho \leq 1} MSE(\boldsymbol{\xi}) &= p - \left(E \left(\left| \boldsymbol{\beta}^T \left[\mathbf{\Sigma}_p(\rho) \right]^{-\frac{1}{2}} (\boldsymbol{X} - \boldsymbol{\mu}) \right| \right) \right)^2 (1 + (p-1)) \\ &= p - p \left(E \left(\left| \boldsymbol{\beta}^T \left[\mathbf{\Sigma}_p(\rho) \right]^{-\frac{1}{2}} (\boldsymbol{X} - \boldsymbol{\mu}) \right| \right) \right)^2. \end{aligned} \tag{7.14}$$

When $\rho < 0$, the largest eigenvalue is $\lambda = \lambda_2 = \cdots = \lambda_p = 1 - \rho$ instead of λ_1. Hence, let $\lambda = \lambda_j = 1 - \rho$ and the associated eigenvector $\boldsymbol{\beta} = \boldsymbol{e}_j$ for any $2 \leq j \leq p$, the MSE of $\boldsymbol{\xi} = \{\boldsymbol{\xi}_1, \boldsymbol{\xi}_2\}$ is given by

$$\text{MSE}(\boldsymbol{\xi}) = p - (1 - \rho) \left(E \left(\left| \boldsymbol{\beta}^T \left[\mathbf{\Sigma}_p(\rho) \right]^{-\frac{1}{2}} (\boldsymbol{X} - \boldsymbol{\mu}) \right| \right) \right)^2. \tag{7.15}$$

Here, we can conclude that an increasing value of ρ leads to a larger MSE of RPs. When $\rho \leq 0$, the maximum of MSE is

$$\max_{-1 \leq \rho \leq 0} MSE(\boldsymbol{\xi}) = p - \left(E \left(\left| \boldsymbol{\beta}^T \left[\mathbf{\Sigma}_p(\rho) \right]^{-\frac{1}{2}} (\boldsymbol{X} - \boldsymbol{\mu}) \right| \right) \right)^2 \tag{7.16}$$

with $\rho = 0$ which is the same with (7.13), and the minimum of MSE is

$$\min_{-1 \leq \rho \leq 0} MSE(\boldsymbol{\xi}) = p - 2 \left(E \left(\left| \boldsymbol{\beta}^T \left[\mathbf{\Sigma}_p(\rho) \right]^{-\frac{1}{2}} (\boldsymbol{X} - \boldsymbol{\mu}) \right| \right) \right)^2. \tag{7.17}$$

A special case based on (7.12) and (7.15) is that, when $p = 2$, the MSEs of the two RPs of $\boldsymbol{X}$ could be unified with the same absolute value of ρ, that is,

$$MSE(\boldsymbol{\xi}) = 2 - (1 + |\rho|) \left(E \left(\left| \boldsymbol{\beta}^T \left[\mathbf{\Sigma}_p(\rho) \right]^{-\frac{1}{2}} (\boldsymbol{X} - \boldsymbol{\mu}) \right| \right) \right)^2; \tag{7.18}$$

and when $p > 2$, we do not have this property. Therefore, when we combine the cases $\rho \geq 0$ and $\rho < 0$, we conclude that an increasing value of $|\rho|$ results in a smaller value of MSE.

Tables 7.1 and 7.2 list some numerical results to show the relationships between the MSE of MSE-RPs and the value of ρ in the covariance matrix $\mathbf{\Sigma}_p(\rho)$. The various are considered for the multivariate normal, Kotz type, Pearson type II and VII distributions, including $k = 2, 4, 8, 11$, $\rho = -0.9, -0.8, \cdots, 0.8, 0.9$, and $p = 2, 5$. In Tables 7.1 and 7.2, the values of MSE for $k = 2$ are calculated by (7.12) and (7.15), and the value of MSE for $k > 2$

TABLE 7.1
MSE of RPs with different ρ

	Multivariate Normal Distribution							
	$p=2$				$p=5$			
ρ	$k=2$	$k=4$	$k=8$	$k=11$	$k=2$	$k=4$	$k=8$	$k=11$
-0.9	0.7904	0.3231	0.1666	0.1250				
-0.7	0.9177	0.4997	0.2747	0.2101				
-0.5	1.0451	0.6062	0.3418	0.2584				
-0.3	1.1724	0.6720	0.3799	0.2872				
-0.1	1.2997	0.7167	0.3966	0.3006	4.2997	3.4455	2.6392	2.3935
0	1.3634	0.7268	0.4024	0.3031	4.3634	3.5866	2.7919	2.4624
0.1	1.2997	0.7167	0.3966	0.3006	4.1087	3.4603	2.7145	2.4145
0.3	1.1724	0.6720	0.3799	0.2872	3.5994	2.9692	2.3980	2.1405
0.5	1.0451	0.6062	0.3419	0.2585	3.0901	2.3529	1.9289	1.7239
0.7	0.9177	0.5248	0.2748	0.2102	2.5808	1.6469	1.3081	1.1839
0.9	0.7904	0.3233	0.1667	0.1251	2.0715	0.9410	0.5590	0.4889
	Multivariate Kotz Type Distribution							
	$p=2$				$p=5$			
ρ	$k=2$	$k=4$	$k=8$	$k=11$	$k=2$	$k=4$	$k=8$	$k=11$
-0.9	0.5077	0.2401	0.0978	0.0629				
-0.7	0.6648	0.3133	0.1263	0.0859				
-0.5	0.8219	0.3695	0.1445	0.1007				
-0.3	0.9790	0.4076	0.1552	0.1097				
-0.1	1.1361	0.4268	0.1592	0.1138	4.2410	3.3150	2.4419	2.1752
0	1.2146	0.4292	0.1597	0.1143	4.3100	3.4682	2.6070	2.2495
0.1	1.1361	0.4268	0.1592	0.1138	4.0340	3.3311	2.5291	2.2041
0.3	0.9790	0.4076	0.1552	0.1097	3.4820	2.8468	2.2262	1.9595
0.5	0.8219	0.3695	0.1445	0.1007	2.9301	2.2672	1.7873	1.5849
0.7	0.6648	0.3133	0.1263	0.0859	2.3781	1.5424	1.2399	1.1079
0.9	0.5077	0.2402	0.0978	0.0629	1.8261	0.8145	0.5144	0.4638

are obtained by NTLBG algorithm. The parameters of the four subclasses of the elliptical distribution are the same as those in Table 6.11. To generate the set of quasi-random F-numbers on the elliptical distribution, the generating vectors of the glp method (Definition 1.9) are

$$(n; h_1, h_2) = (750251; 1, 46368) \text{ for } p = 2 \text{ and}$$
$$(n; h_1, h_2, h_3, h_4, h_5) = (100063; 1, 33755, 65170, 12740, 6878) \text{ for } p = 5.$$

We cannot find the MSE-RPs and the related MSE with some value of ρ when $p = 5$, as $\boldsymbol{\Sigma}_p(\rho), \rho = -0.9, -0.8, \cdots, -0.3$ are not positive definite, so there exist some blanks for these values in Tables 7.1 and 7.2. From the results, we can conclude that

(i) the MSE becomes larger when ρ is getting close to 0, and the largest MSE appears if $\rho = 0$,

(ii) when $p = 2$, the MSE has the same value for positive or negative ρ with the same absolute value, but this is not applicable for the general cases when $p > 2$.

7.3.4 Relationship between the principal components and MSE-RPs under the normal distribution

Let $\boldsymbol{X} \sim N_p(\mathbf{0}, \boldsymbol{\Sigma}_\rho)$, where $\boldsymbol{\Sigma}_\rho$ is given by (6.47). Denote the eigenvalues of $\boldsymbol{\Sigma}_\rho$ by $\lambda_1 \geq \lambda_2 \geq \cdots \geq \lambda_p$ and $\boldsymbol{\beta}_1, \boldsymbol{\beta}_2, \cdots, \boldsymbol{\beta}_p$ the corresponding eigenvectors. Table 7.3 lists MSE values for $p = 2, 5$, $k = 2, 3, 4, 8, 11$ and $\rho = -0.2, -0.1, 0, 0.1, 0.3, 0.7, 0.9$. Multivariate statistical analysis defines the contribution rate of the first principal component as: $C_1 = \lambda_1 / tr(\boldsymbol{\Sigma}_\rho)$. Table 7.3 shows the following facts:

TABLE 7.2
MSE of RPs with different ρ

	Multivariate Pearson Type II Distribution							
	$p=2$				$p=5$			
ρ	$k=2$	$k=4$	$k=8$	$k=11$	$k=2$	$k=4$	$k=8$	$k=11$
-0.9	0.7390	0.2939	0.1520	0.1124				
-0.7	0.8718	0.4825	0.2469	0.1876				
-0.5	1.0045	0.5622	0.3085	0.2294				
-0.3	1.1372	0.6249	0.3429	0.2573				
-0.1	1.2700	0.6650	0.3575	0.2680	4.2760	3.3928	2.5599	2.3056
0	1.3363	0.6727	0.3591	0.2699	4.3418	3.5388	2.7172	2.3769
0.1	1.2700	0.6650	0.3575	0.2681	4.0785	3.4083	2.6400	2.3298
0.3	1.1372	0.6249	0.3429	0.2566	3.5520	2.9185	2.3300	2.0680
0.5	1.0045	0.5623	0.3085	0.2294	3.0254	2.3157	1.8703	1.6673
0.7	0.8718	0.4826	0.2469	0.1876	2.4988	1.5999	1.2748	1.1520
0.9	0.7390	0.2940	0.1520	0.1125	1.9723	0.8841	0.5375	0.4756
	Multivariate Pearson Type VII Distribution							
	$p=2$				$p=5$			
ρ	$k=2$	$k=4$	$k=8$	$k=11$	$k=2$	$k=4$	$k=8$	$k=11$
-0.9	0.8071	0.3332	0.1717	0.1290				
-0.7	0.9327	0.5087	0.2838	0.2176				
-0.5	1.0582	0.6205	0.3527	0.2679				
-0.3	1.1838	0.6872	0.3923	0.2975				
-0.1	1.3094	0.7333	0.4097	0.3116	4.3104	3.4685	2.6752	2.4325
0	1.3722	0.7443	0.4158	0.3143	4.3731	3.6078	2.8251	2.5005
0.1	1.3094	0.7333	0.4097	0.3117	4.1223	3.4834	2.7474	2.4522
0.3	1.1838	0.6873	0.3923	0.2976	3.6207	2.9918	2.4258	2.1727
0.5	1.0582	0.6206	0.3527	0.2687	3.1192	2.3703	1.9552	1.7485
0.7	0.9327	0.5088	0.2839	0.2177	2.6176	1.6690	1.3235	1.1988
0.9	0.8071	0.3334	0.1697	0.1292	2.1161	0.9678	0.5698	0.4951

(i) when k increases MSE decreases;
(ii) when p increases MSE increases;
(iii) when $|\rho|$ increases, MSE decreases.

TABLE 7.3
MSE of the multivariate normal distributions for $p=2$ and $p=5$

	Multivariate Normal Distribution									
	$p=2$					$p=5$				
ρ	$k=2$	$k=3$	$k=4$	$k=8$	$k=11$	$k=2$	$k=3$	$k=4$	$k=8$	$k=11$
-0.2	1.2361	0.9124	0.6975	0.3890	0.2954	4.2361	3.7104	3.3037	2.4247	2.1565
-0.1	1.2997	0.9227	0.7167	0.3966	0.3006	4.2997	3.8181	3.4455	2.6392	2.3935
0	1.3634	0.9257	0.7268	0.4024	0.3031	4.3634	3.9252	3.5866	2.7919	2.4624
0.1	1.2997	0.9226	0.7167	0.3966	0.3006	4.1087	3.7460	3.4603	2.7145	2.4145
0.3	1.1724	0.8938	0.6720	0.3799	0.2872	3.5994	3.2182	2.9692	2.3980	2.1405
0.5	1.0451	0.7853	0.6062	0.3419	0.2585	3.0901	2.5702	2.3529	1.9289	1.7239
0.7	0.9177	0.6233	0.5248	0.2748	0.2102	2.5808	1.9222	1.6469	1.3081	1.1839
0.9	0.7904	0.4614	0.3233	0.1667	0.1251	2.0715	1.2743	0.9410	0.5590	0.4889

To study the relationship between MSE-RPs of the multivariate normal distribution and the corresponding principal components, first consider $p=2, \mathbf{\Sigma} = \text{diag}(\sigma^2, 1) \equiv \mathbf{\Sigma}_\sigma$.

1. For $k=2$, two MSE-RPS of $\boldsymbol{X}$ locate at the line of the first component of $\mathbf{\Sigma}$.
2. For $k=3$, Fang et al. (2020) found that the three of MSE-RPs of $\boldsymbol{X}$ construct an equilateral triangle when $\sigma < \sigma_0(3) = 1.5416$. But when $\sigma \geq 1.5416$, all three MSE-RPs fall on the first principal component. This phenomenon tells us that when $\sigma \geq 1.5416$, one principal component can obtain the primary information of

$\boldsymbol{X}$ and its contribution rate is $C_1 = \frac{\sigma^2}{1+\sigma^2}$. When $\sigma = 1.5416$, the corresponding contribution rate is $C_1 = 0.7038$.

3. For $k = 4$, if $\sigma < \sigma_0(4) = 2.1335$, four MSE-RPs form a rectangle. If $\sigma \geq \sigma_0(4) = 2.1335$, all four MSE-RPs fall on the first principal component, and the contribution rate of the first principal component is $C_1 = 0.8199$.
4. For $k = 5$, the critical value of σ is $\sigma_0(5) = 2.7199$ for all five MSE-RPs fall on the first principal component. Table 7.4 lists the corresponding MSE for $3 \leq k \leq 5$ including $\sigma = \sigma_0(k)$.

TABLE 7.4
Critical σ for binormal distribution and related C_1

k	$\sigma_0(k)$	C_1	MSE
3	1.5416	0.7038	1.4520
4	2.1335	0.8199	1.5349
5	2.7199	0.8809	1.5916

TABLE 7.5
Critical $\rho_0(k)$ and related C_1 for bionormal

k	$\rho_0(k)$	C_1	MSE
3	0.4078	0.7039	0.8600
4	0.4230	0.7115	0.8520
5	0.7476	0.8724	0.3921

For the cases for $p = 2$ and $k = 3, 4, 5$, we want to know how the location of k RPs changes according to ρ increases. The immediate question arising from this problem is, under what circumstances would it be optimal to choose k points that lie precisely on the first principal axis?

Denote $\rho_0(k)$ as the critical value of ρ for all of k RPs fall on the first component direction as soon as $\rho \geq \rho_0(k)$. For example, when $k = 3$ and $\rho < \rho_0(3) = 0.4078$, the three of RPs form a triangle; but when $\rho \geq 0.4078$, the three RPs with MSE=0.8600 fall on the first component line. The contribution rate of the first component is $C_1 = (1 + \rho_0)/2 = (1 + 0.4078)/2 = 0.7039$. See Table 7.5 for the details.

We are studying the geometric structure of the representative points of the ellipsoid contour distribution, such as $p > 4$. When $k = 3$, the first and second principal components determine whether the four representative points are divided into two pairs. The relationship between the geometric structure of the principal components of different elliptically distributed subclasses.

For $k = 3$, the RPs of the distribution have two possible structures:
(1) Two RPs on the first eigenvector direction while another RP is on the second.
(2) Three RPs, all on the first eigenvector direction.

Case $p = 2, k = 3$ and $\boldsymbol{\Sigma} = \begin{bmatrix} 1 & \rho \\ \rho & 1 \end{bmatrix}$.

All the $k = 3$ RPs are in the direction of eigenvector $\boldsymbol{\beta}_1$ when $\rho \geq \rho_0(3) = 0.4078$. In this case, the RPs are obtained by the NTLBG algorithm, which is $(0.6595, 10622)^T$,

$(-1.0625, -0.6594)^T$ and $(0.3942, -0.3946)^T$. Hence, one RP is almost in the direction of $\boldsymbol{\beta}_2$, and the other two RPs are approximately symmetric about $\boldsymbol{\beta}_2$.

When $\rho_0(3) = 0.4078$ and we take the origin $(0,0)^T$ as one of the RPs, and we assume that the remaining two RPs are symmetric about the eigenvector $\boldsymbol{\beta}_2$ and search for them in the direction of $\boldsymbol{\beta}_1$, then we have the RPs $(-1.0268, -1.0268)^T$, $(0,0)^T$ and $(1.0268, -1.0268)^T$.

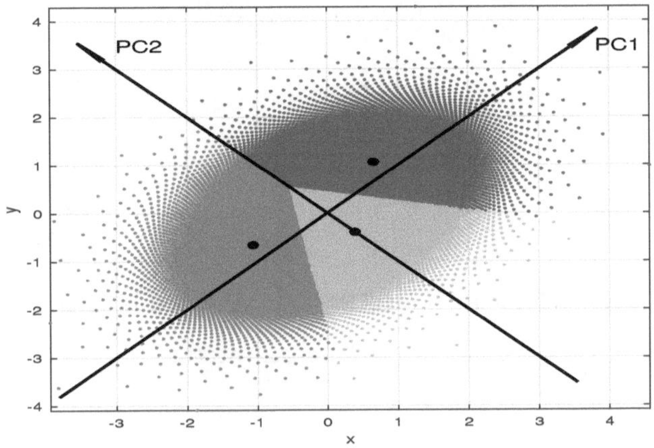

FIGURE 7.1
Patterns of $k = 3$ RPs with $\rho = 0.4077$ and $MSE = 0.86000923$.

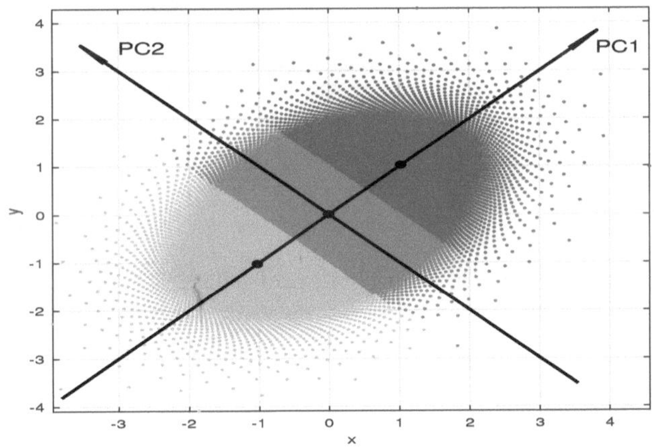

FIGURE 7.2
Patterns of $k = 3$ RPs with $\rho = 0.4078$ and $MSE = 0.85995887$.

Case $p = 2, k = 4$

All the RPs are in the direction of the eigenvector $\boldsymbol{\beta}_1$ when $\rho > \rho_0(4) = 0.7042$, and the RPs are symmetric about $\boldsymbol{\beta}_2$.

When $\rho = 0.7041 < \rho_0(4)$, two of the RPs are on the direction of $\boldsymbol{\beta}_1$ and the other two are on the direction of $\boldsymbol{\beta}_2$, and each pair of RPs in the same direction is symmetric about the origin $(0,0)^T$.

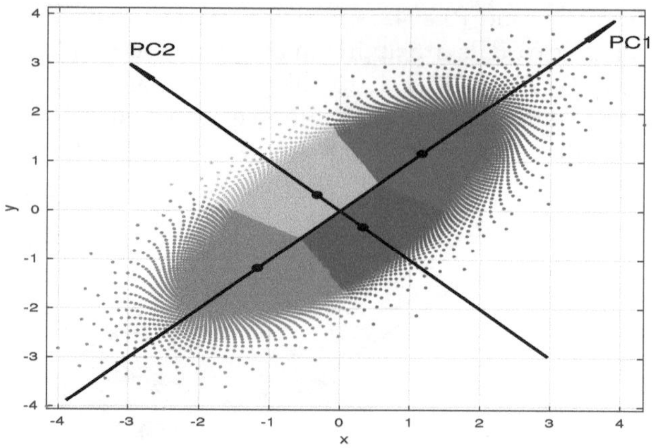

FIGURE 7.3
Patterns of $k = 4$ RPs with $\rho = 0.7041$ and $MSE = 0.5229128$.

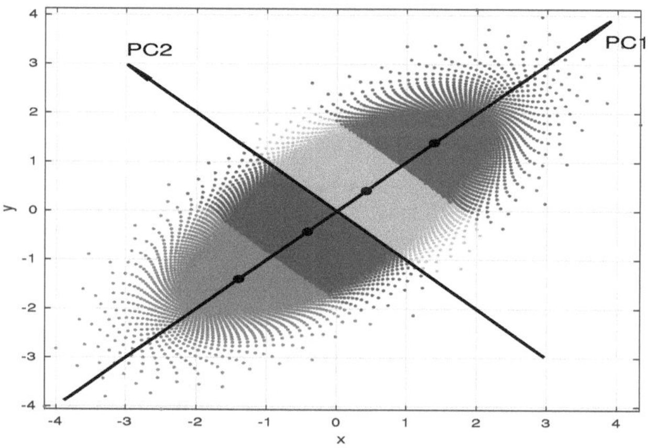

FIGURE 7.4
Patterns of $k = 4$ RPs with $\rho = 0.7042$ and $MSE = 0.49603681$.

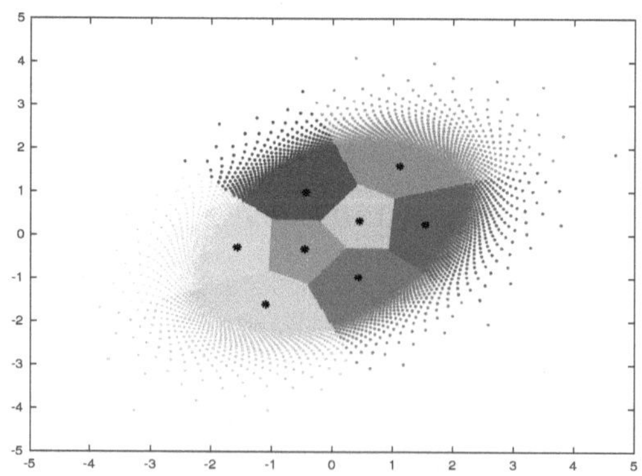

FIGURE 7.5
8 MSE-RPs of multivariate normal distribution.

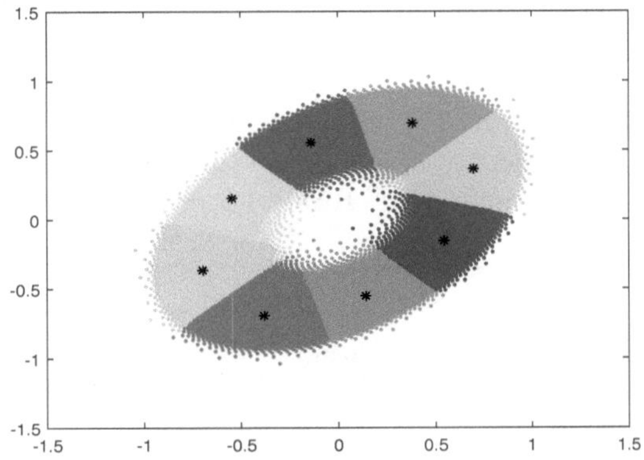

FIGURE 7.6
8 MSE-RPs of Kotz type distribution.

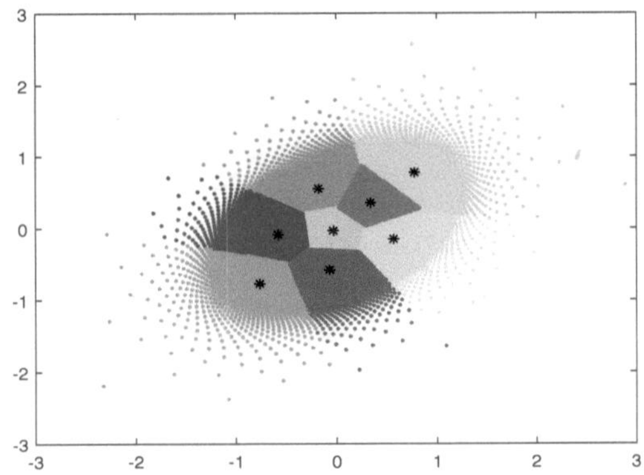

FIGURE 7.7
8 MSE-RPs of Pearson type II distribution.

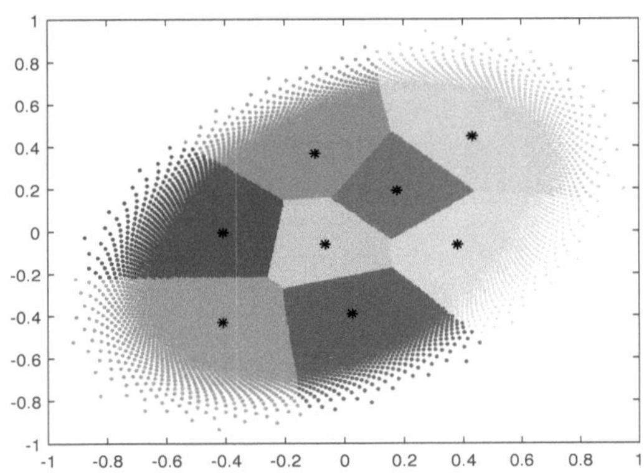

FIGURE 7.8
8 MSE-RPs of Pearson type VII distribution.

7.4 Parallel Principal Axes

Tarpey (2000) proposed an idea of the construction of MSE-RPs for the elliptical distributions. Parallel principal axes correspond to an approximation to a distribution by combining the ideas of principal points (MSE-RPs) and principal component axes.

Definition 7.4 *Let $\boldsymbol{X}$ denote a p-variate random vector. Suppose a set of k parallel lines or (hyper)planes are orthogonal to a principal component axis, intersecting this axis at the k MSE-RPs of the marginal distribution of $\boldsymbol{X}$ on this principal component axis. Then, these k parallel axes (planes) are called k parallel principal axes (or planes) of $\boldsymbol{X}$.*

Lemma 7.2 *Let $\boldsymbol{X}$ denote a p-variate random vector with finite second moments. For a given collection of k parallel axes (or hyperplanes) of dimension $p-1$, let z_1 be the axis to which the planes are perpendicular and let Z_1 be the marginal distribution of $\boldsymbol{X}$ along the z_1 axis. Then, the MSE-RP of the projection of $\boldsymbol{x}$ onto the nearest of the k parallel planes is minimized when the planes intersect the z_1 axis at the k MSE-RPs of Z_1. Furthermore, if $\boldsymbol{X}$ is elliptically distributed, then the MSE is minimized if the z_1 axis corresponds to the last principal component axis for $\boldsymbol{X}$.*

7.5 Proofs of Theorems in Chapter 7

7.5.1 Proof of Theorem 7.1

Proof *For assertion 1, we have*

$$E(\boldsymbol{X}) = \sum_{j=1}^{k} E(\boldsymbol{X}|\boldsymbol{X} \in S_j)P(\boldsymbol{X} \in S_j) = \sum_{j=1}^{k} \boldsymbol{y}_j p_j = E(\boldsymbol{y}_{SC}).$$

For assertion 2, we have

$$\begin{aligned}
\text{MSE}(\boldsymbol{y}_1,\dots,\boldsymbol{y}_k) &= \sum_{j=1}^{k} p_i E(\boldsymbol{X}-\boldsymbol{y}_j)^T(\boldsymbol{X}-\boldsymbol{y}_j) \\
&= E(\boldsymbol{X}^T\boldsymbol{X}) - \sum_{j=1}^{k} p_j \boldsymbol{y}_j^T[2E(\boldsymbol{X}|S_j) - \boldsymbol{y}_j] \\
&= tr(\boldsymbol{\Sigma}) - \sum_{j=1}^{k} p_j \boldsymbol{y}_j^T[2\boldsymbol{y}_j - \boldsymbol{y}_j] \\
&= tr(\boldsymbol{\Sigma}) - \sum_{j=1}^{k} p_j \|\boldsymbol{y}_j\|^2.
\end{aligned}$$

7.5.2 Proof of Theorem 7.2

Proof

$$\begin{aligned}
\text{MSE}(\boldsymbol{\xi}) &= E\|\boldsymbol{X}-\boldsymbol{\xi}\|^2 \\
&= \sum_{j=1}^{k} p_j E((\boldsymbol{X}-\boldsymbol{\xi})^T(\boldsymbol{X}-\boldsymbol{\xi})|\boldsymbol{X}\in S_j) \\
&= \min_{\boldsymbol{y}_j, j=1,\dots,k}[\sum_{j=1}^{k} p_j E((\boldsymbol{X}-\boldsymbol{y}_j)^T(\boldsymbol{X}-\boldsymbol{y}_j)|\boldsymbol{X}\in S_j)] \\
&= \sum_{j=1}^{k} p_j \min_{\boldsymbol{y}_j}[E((\boldsymbol{X}-\boldsymbol{y}_j)^T(\boldsymbol{X}-\boldsymbol{y}_j)|\boldsymbol{X}\in S_j)].
\end{aligned}$$

But the jth term in the above equation is minimized by choosing $\boldsymbol{y}_j = E(\boldsymbol{X}|\boldsymbol{X}\in S_j)$*, i.e., RPs are conditional means, hence self-consistent.*

7.5.3 Proof of Theorem 7.5

Proof *Let* $\boldsymbol{\xi}\in\mathbb{R}^p$ *be the RP of* $\boldsymbol{X}$ *with* $k=1$*. The MSE*

$$\begin{aligned}
\text{MSE}(\boldsymbol{\xi}) &= E\|\boldsymbol{X}-\boldsymbol{\xi}\|^2 = E(\boldsymbol{X}-\boldsymbol{\mu}+\boldsymbol{\mu}-\boldsymbol{\xi})^T(\boldsymbol{X}-\boldsymbol{\mu}+\boldsymbol{\mu}-\boldsymbol{\xi}) \\
&= E(\boldsymbol{X}-\boldsymbol{\mu})^T(\boldsymbol{X}-\boldsymbol{\mu}) + (\boldsymbol{\mu}-\boldsymbol{\xi})^T(\boldsymbol{\mu}-\boldsymbol{\xi}) \\
&\ge E(\boldsymbol{X}-\boldsymbol{\mu})^T(\boldsymbol{X}-\boldsymbol{\mu}) = E\left(tr(\boldsymbol{X}-\boldsymbol{\mu})^T(\boldsymbol{X}-\boldsymbol{\mu})\right) \\
&= tr\left(E(\boldsymbol{X}-\boldsymbol{\mu})^T(\boldsymbol{X}-\boldsymbol{\mu})\right) = tr\left(E\left((\boldsymbol{X}-\boldsymbol{\mu})(\boldsymbol{X}-\boldsymbol{\mu})^T\right)\right) \\
&= tr(\boldsymbol{\Sigma}) \qquad (7.19)
\end{aligned}$$

which shows that $\boldsymbol{\xi}=\boldsymbol{\mu}$ *and the related MSE is* $tr(\boldsymbol{\Sigma})$*.*

7.5.4 Proof of Theorem 7.6

Proof *Without loss of generality assume* $E[\boldsymbol{X}]=\boldsymbol{0}$*. For part (i), by self-consistency of* $\boldsymbol{Y}$ *for* $\boldsymbol{X}$ *and using the conditional variance formula*

$$Cov(\boldsymbol{X}) = Cov(E(\boldsymbol{X}|\boldsymbol{Y})) + E[Cov(\boldsymbol{X}|\boldsymbol{Y})], \qquad (7.20)$$

we have

$$Cov(\boldsymbol{X}) = Cov(\boldsymbol{Y}) + E[Cov(\boldsymbol{X}|\boldsymbol{Y})]. \qquad (7.21)$$

The assertion (i) follows as $Cov(\boldsymbol{X}|\boldsymbol{Y})>0$*, positive definite matrix. To prove (ii), we have*

$$\begin{aligned}
E\|\boldsymbol{X}-\boldsymbol{Y}\|^2 &= E[\boldsymbol{X}^T\boldsymbol{X}] - 2E[\boldsymbol{Y}^T\boldsymbol{X}] + E[\boldsymbol{Y}^T\boldsymbol{Y}] \\
&= tr(\boldsymbol{\Sigma}_x) - 2E[E[\boldsymbol{Y}^T\boldsymbol{X}|\boldsymbol{Y}]] + tr(\boldsymbol{\Sigma}_y) \\
&= tr(\boldsymbol{\Sigma}_x) - 2E[\boldsymbol{Y}^T E[\boldsymbol{X}|\boldsymbol{Y}]] + tr(\boldsymbol{\Sigma}_y) \\
&= tr(\boldsymbol{\Sigma}_x) - 2E[\boldsymbol{Y}^T\boldsymbol{Y}] + tr(\boldsymbol{\Sigma}_y) \\
&= tr(\boldsymbol{\Sigma}_x) - tr(\boldsymbol{\Sigma}_y).
\end{aligned}$$

7.5.5 Proof of Theorem 7.9

Proof *Let* $\boldsymbol{A} = [\boldsymbol{A}_1 : \boldsymbol{A}_2]$ *be an orthogonal* $p \times p$ *matrix, partitioned into* q *columns* $\boldsymbol{A}_1$ *and* $p - q$ *columns* $\boldsymbol{A}_2$ *such that the columns of* $\boldsymbol{A}_1$ *span* $\mathcal{A}$. *Then,* $\boldsymbol{A}_2\boldsymbol{Y} = \boldsymbol{0}$ *a.s. The covariance matrix of* $\boldsymbol{A}^T\boldsymbol{X} = (\boldsymbol{A}_1^T\boldsymbol{X}, \boldsymbol{A}_2^T\boldsymbol{X})^T$ *is*

$$\begin{bmatrix} \boldsymbol{A}_1^T\boldsymbol{\Sigma}\boldsymbol{A}_1 & \boldsymbol{A}_1^T\boldsymbol{\Sigma}\boldsymbol{A}_2 \\ \boldsymbol{A}_2^T\boldsymbol{\Sigma}\boldsymbol{A}_1 & \boldsymbol{A}_2^T\boldsymbol{\Sigma}\boldsymbol{A}_2 \end{bmatrix}.$$

By the linearity of the conditional expectation,

$$E[\boldsymbol{A}_2^T\boldsymbol{X}|\boldsymbol{A}_1^T\boldsymbol{X}] = \boldsymbol{A}_2^T\boldsymbol{\Sigma}\boldsymbol{A}_1(\boldsymbol{A}_1^T\boldsymbol{\Sigma}\boldsymbol{A}_1)^{-1}\boldsymbol{A}_1^T\boldsymbol{X}.$$

Since $\boldsymbol{Y}$ *is measurable with respect to* $\boldsymbol{A}^T\boldsymbol{X}$,

$$\begin{aligned} E[\boldsymbol{A}_2^T\boldsymbol{X}] &= E[E(E[\boldsymbol{A}_2^T\boldsymbol{X}|\boldsymbol{A}_1^T\boldsymbol{X})|\boldsymbol{Y}] \\ &= E[\boldsymbol{A}_2^T\boldsymbol{\Sigma}\boldsymbol{A}_1(\boldsymbol{A}_1^T\boldsymbol{\Sigma}\boldsymbol{A}_1)^{-1}\boldsymbol{A}_1^T\boldsymbol{X}|\boldsymbol{Y}] \\ &= \boldsymbol{A}_2^T\boldsymbol{\Sigma}\boldsymbol{A}_1(\boldsymbol{A}_1^T\boldsymbol{\Sigma}\boldsymbol{A}_1)^{-1}\boldsymbol{A}_1^T E[\boldsymbol{X}|\boldsymbol{Y}] \\ &= \boldsymbol{A}_2^T\boldsymbol{\Sigma}\boldsymbol{A}_1(\boldsymbol{A}_1^T\boldsymbol{\Sigma}\boldsymbol{A}_1)^{-1}\boldsymbol{A}_1^T\boldsymbol{Y}, \end{aligned}$$

by self-consistency of $\boldsymbol{Y}$ *for* $\boldsymbol{X}$. *Also, by self-consistency,* $E[\boldsymbol{A}_2^T\boldsymbol{X}|\boldsymbol{Y}] = \boldsymbol{A}_2^T\boldsymbol{Y} = \boldsymbol{0}$. *Therefore,* $\boldsymbol{A}_2^T\boldsymbol{\Sigma}\boldsymbol{A}_1(\boldsymbol{A}_1^T\boldsymbol{\Sigma}\boldsymbol{A}_1)^{-1}\boldsymbol{A}_1^T\boldsymbol{Y} = \boldsymbol{0}$, *which implies* $\boldsymbol{A}_2^T\boldsymbol{\Sigma}\boldsymbol{A}_1 = \boldsymbol{0}$. *That is, the columns of* $\boldsymbol{A}_2$ *are orthogonal to the columns of* $\boldsymbol{\Sigma}\boldsymbol{A}_1$, *or* $\boldsymbol{\Sigma}\boldsymbol{A}_1 = \boldsymbol{A}_1\boldsymbol{H}$ *for some orthogonal matrix* $\boldsymbol{H}$ *of dimension* $q \times q$. *This means in turn that the columns of* $\boldsymbol{A}_1$ *span the same q-dimensional subspace as some* q *eigenvectors of* $\boldsymbol{\Sigma}$.

7.5.6 Proof of Theorem 7.11

The following gives two proofs for issues 4 and 5 in Theorem 7.11.

Proof *Let* $\boldsymbol{a} = \frac{\boldsymbol{c}_2 - \boldsymbol{c}_1}{||\boldsymbol{c}_2 - \boldsymbol{c}_1||}$. *Take matrix* $\boldsymbol{B} : p \times (p-1)$ *such that* $[\boldsymbol{a} : \boldsymbol{B}]$ *is orthogonal. Let*

$$\boldsymbol{y} = \begin{bmatrix} Y_1 \\ \boldsymbol{y}_2 \end{bmatrix} = [\boldsymbol{a} : \boldsymbol{B}]^T\boldsymbol{X}, \quad \boldsymbol{v}_i = \begin{bmatrix} v_{i1} \\ \boldsymbol{v}_{i2} \end{bmatrix} = [\boldsymbol{a} : \boldsymbol{B}]^T\boldsymbol{c}_i, i = 1, 2,$$

where Y_1, v_{11} *and* v_{21} *belong to* $\mathbb{R}$, *and* $\boldsymbol{y}_2, \boldsymbol{v}_{12}$ *and* $\boldsymbol{v}_{22}$ *belong to* $\mathbb{R}^{p-1}$. *From orthogonality of* $[\boldsymbol{a} : \boldsymbol{B}]$, *we have*

$$\boldsymbol{0} = \boldsymbol{B}^T\boldsymbol{a} = \boldsymbol{B}^T\frac{\boldsymbol{c}_2 - \boldsymbol{c}_1}{||\boldsymbol{c}_2 - \boldsymbol{c}_1||}$$

that implies $\boldsymbol{B}^T\boldsymbol{c}_2 = \boldsymbol{B}^T\boldsymbol{c}_1$ *and* $\boldsymbol{v}_{12} = \boldsymbol{v}_{22} = \boldsymbol{v}_2$. *We have*

$$\begin{aligned} E\{d^2(\boldsymbol{X}|\boldsymbol{c}_l, \boldsymbol{c}_2)\} &= E\{d^2(\boldsymbol{y}|\boldsymbol{v}_1, \boldsymbol{v}_2)\} \\ &= E\{d^2(Y_1|v_{11}, v_{21})\} + E\{d^2(\boldsymbol{y}_2|\boldsymbol{v}_2, \boldsymbol{v}_2)\} \\ &= E\{d^2(Y_1|v_{11}, v_{21})\} + E\{d^2(\boldsymbol{y}_2|0, 0)\} \end{aligned}$$

with equality exactly if $\boldsymbol{v}_2 = 0$ *as* $\boldsymbol{\mu} = \boldsymbol{0}$.

Proof *If* $k > p$, *the theorem implies that the result is trivial. For* $k \leq p$, *the theorem implies that any plane that contains all* k *RPs contains also* $\boldsymbol{\mu}$ *that must be in the convex hull of* $\boldsymbol{y}_1, \ldots, \boldsymbol{y}_k$ *(see Theorem 7.1). Let* $\boldsymbol{c}_1, \ldots, \boldsymbol{c}_k \in \mathbb{R}$ *be arbitrary* k *points, and let*

$\boldsymbol{b}_i = \boldsymbol{c}_i - \boldsymbol{c}_k, i = 1, \ldots, k$. *Denote* $m = \text{rank}(\boldsymbol{b}_1, \ldots, \boldsymbol{b}_{k-1})$. *Let* $\mathcal{P}$ *be the linear subspace spanned by* $\boldsymbol{b}_1, \ldots, \boldsymbol{b}_{k-1}$, *and* $\boldsymbol{a}_1, \ldots, \boldsymbol{a}_m$ *be an orthonomal basis of* $\mathcal{P}$. *Let* $\boldsymbol{A}_1 = [\boldsymbol{a}_1, \ldots, \boldsymbol{a}_m]$ *and let* $\boldsymbol{A}_2$ *be such that* $\boldsymbol{A} = [\boldsymbol{A}_1 : \boldsymbol{A}_2]$ *is orthogonal. Denote*

$$\boldsymbol{d}_i = \boldsymbol{A}^T \boldsymbol{c}_i = \begin{bmatrix} \boldsymbol{A}_1^T \boldsymbol{c}_i \\ \boldsymbol{A}_2^T \boldsymbol{c}_i \end{bmatrix} = \begin{bmatrix} \boldsymbol{d}_i^{(1)} \\ \boldsymbol{d}_i^{(2)} \end{bmatrix}.$$

Since all $\boldsymbol{c}_i - \boldsymbol{c}_j$ *are in* $\mathcal{P}$ *spanned by the columns of* $\boldsymbol{A}_1$, *we can express* $\boldsymbol{c}_i - \boldsymbol{c}_j = \boldsymbol{A}_1 \boldsymbol{M}_{ij}$ *for some* $m \times m$ *matrices* $\boldsymbol{M}_{ij}$. *We have*

$$\boldsymbol{d}_i^{(2)} - \boldsymbol{d}_j^{(2)} = \boldsymbol{A}_2^T (\boldsymbol{c}_i - \boldsymbol{c}_j) = \boldsymbol{A}_2^T \boldsymbol{A}_1 \boldsymbol{M}_{ij} = \boldsymbol{0}, i, j = 1, \ldots, k.$$

As a result, all $\boldsymbol{d}_i^{(2)}$ *are the same and denote* $\boldsymbol{d}^{(2)} = \boldsymbol{d}_i^{(2)}$. *Without loss of generality, assume* $\boldsymbol{\mu} = \boldsymbol{0}$, *and let*

$$\boldsymbol{y} = \begin{bmatrix} \boldsymbol{y}_1 \\ \boldsymbol{y}_2 \end{bmatrix} = \boldsymbol{A}^T \boldsymbol{X} = \begin{bmatrix} \boldsymbol{A}_1^T \boldsymbol{X} \\ \boldsymbol{A}_2^T \boldsymbol{X} \end{bmatrix}.$$

As the matrix $\boldsymbol{A}$ *is orthogonal, we have*

$$\begin{aligned} E[d^2(\boldsymbol{X}|\boldsymbol{c}_1, \ldots, \boldsymbol{c}_k)] &= E[d^2(\boldsymbol{y}|\boldsymbol{d}_1, \ldots, \boldsymbol{d}_k)] \\ &= E[d^2(\boldsymbol{y}_1|\boldsymbol{d}_1^{(1)}, \ldots, \boldsymbol{d}_k^{(1)})] + E[d^2(\boldsymbol{y}_2|\boldsymbol{d}^{(2)}, \ldots, \boldsymbol{d}^{(2)})] \\ &\geq E[d^2(\boldsymbol{y}_1|\boldsymbol{d}_1^{(1)}, \ldots, \boldsymbol{d}_k^{(1)})] + E[\boldsymbol{d}^{(2)}(\boldsymbol{y}_2|\boldsymbol{0}, \ldots, \boldsymbol{0})], \end{aligned}$$

where the equal holds if $\boldsymbol{d}^{(2)} = \boldsymbol{0}$. *Define*

$$\boldsymbol{c}_i^* = [\boldsymbol{A}_1 : \boldsymbol{A}_2] \begin{bmatrix} \boldsymbol{d}_i^{(1)} \\ \boldsymbol{0} \end{bmatrix} = \boldsymbol{A}_1 \boldsymbol{d}_i^{(1)} = \boldsymbol{A}_1 \boldsymbol{A}_1^T \boldsymbol{c}_i$$

and

$$E[d^2(\boldsymbol{X}|\boldsymbol{c}_1^*, \ldots, \boldsymbol{c}_k^*)] \leq E[d^2(\boldsymbol{X}|\boldsymbol{c}_1, \ldots, \boldsymbol{c}_k)],$$

with equality if $\boldsymbol{A}_1 \boldsymbol{A}_1^T \boldsymbol{c}_i = \boldsymbol{c}_i, i = 1, \ldots, k$. *Since*

$$[\boldsymbol{c}_1^*, \ldots, \boldsymbol{c}_k^*] = \boldsymbol{A}_1 \boldsymbol{A}_1^T [\boldsymbol{c}_1, \ldots, \boldsymbol{c}_k]$$

and $\text{rank}(\boldsymbol{A}_1) = m \leq k - 1$. *The proof is complete.*

7.5.7 Proof of Corollary 7.2

Proof *Let* $\gamma = E(|\boldsymbol{\beta}^T \boldsymbol{\Sigma}^{-1/2}(\boldsymbol{X} - \boldsymbol{\mu})|)$, *then* $-\gamma$ *and* γ *are two RPs of the univariate symmetric distribution* $Z = \boldsymbol{\beta}^T \boldsymbol{\Sigma}^{-1/2}(\boldsymbol{X} - \boldsymbol{\mu})$. *From Theorem 7.12 the two RPs of* $\boldsymbol{X}$ *are*

$$\begin{aligned} \boldsymbol{\xi}_1 &= \boldsymbol{\mu} - \lambda^{1/2} \gamma \boldsymbol{\beta} = \boldsymbol{\mu} - \lambda^{1/2} E(|Z|) \boldsymbol{\beta}, \\ \boldsymbol{\xi}_2 &= \boldsymbol{\mu} + \lambda^{1/2} \gamma \boldsymbol{\beta} = \boldsymbol{\mu} + \lambda^{1/2} E(|Z|) \boldsymbol{\beta}, \end{aligned} \tag{7.22}$$

which has the same form with (7.10). The MSE of $\boldsymbol{\xi} = \{\boldsymbol{\xi}_1, \boldsymbol{\xi}_2\}$ *is*

$$MSE(\boldsymbol{\xi}_1, \boldsymbol{\xi}_2) = tr(\boldsymbol{\Sigma}) - (1 - h) \boldsymbol{\beta}^T \boldsymbol{\Sigma} \boldsymbol{\beta} \tag{7.23}$$

where $h < 1$ *and* $\lambda = \boldsymbol{\beta}^T \boldsymbol{\Sigma} \boldsymbol{\beta}$. *With some calculations, we can find that*

$$h = 1 - (E(|\boldsymbol{\beta}^T \boldsymbol{\Sigma}^{-1/2}(\boldsymbol{x} - \boldsymbol{\mu})|))^2, \tag{7.24}$$

and h *is dependent on the distribution of* $\boldsymbol{X}$. *Hence, the MSE of two RPs of* $\boldsymbol{X}$ *is*

$$MSE(\boldsymbol{\xi}_1, \boldsymbol{\xi}_2) = tr(\boldsymbol{\Sigma}) - \lambda \left(E \left(\left| \boldsymbol{\beta}^T \boldsymbol{\Sigma}^{-1/2} (\boldsymbol{x} - \boldsymbol{\mu}) \right| \right) \right)^2. \tag{7.25}$$

7.5.8 Proof of Theorem 7.12

Proof *From Theorem 7.11, two MSE-RPs should be on a line through the mean vector* $\boldsymbol{\mu}$ *with the same direction as* $\mathbf{c}_2 - \mathbf{c}_1$*, denoted by* $\boldsymbol{a}$ *with* $\boldsymbol{a}^T\boldsymbol{a} = 1$*. Let* $Z = \boldsymbol{a}^T\boldsymbol{X}$*, then* Z *follows a symmetric distribution about* $\boldsymbol{a}^T\mu$*. It is known that* $\mathrm{E}(Z) = \boldsymbol{a}^T\mu$ *and* $\mathrm{Var}(Z) = \boldsymbol{a}^T\boldsymbol{\Sigma}\boldsymbol{a}$*, where* $\boldsymbol{\Sigma} = Cov(\boldsymbol{X})$ *is proportional to* Ψ*.*

For the given multivariate elliptical distribution, let $\boldsymbol{\xi} = \{\boldsymbol{\xi}_1, \boldsymbol{\xi}_2\}$ *be the set of two MSE-RPs of* $\boldsymbol{X}$*. From the Theorem 2 in Flury (1990), the theoretical results of MSE-RPs of* $\boldsymbol{X}$ *are given, which are*

$$\boldsymbol{\xi}_1 = \boldsymbol{\mu} - \gamma_1\boldsymbol{\beta} \quad \textit{and} \quad \boldsymbol{\xi}_2 = \boldsymbol{\mu} + \gamma_2\boldsymbol{\beta}, \tag{7.26}$$

where $\boldsymbol{\beta}$ *is the associated eigenvector of the largest eigenvalue* λ *of the covariance matrix* $\boldsymbol{\Sigma}$ *and* $\boldsymbol{\beta}^T\boldsymbol{\beta} = 1$*, and* γ_1 *and* γ_2 *are the MSE-RPs of univariate symmetric distribution* $\boldsymbol{\beta}^T(\boldsymbol{X} - \boldsymbol{\mu})$*.*

8

Applications of RPs in Various Fields

This chapter collects various applications of RPs in statistics, including geometric probability problems, meta-distribution via copula theory, experimental design with mixture, and testing multivariate normality.

8.1 Simulation for Geometric Probability Problem

Statistical simulation has been widely used for problems in statistics without an analytical solution. For example, let G be a domain in $\mathbb{R}^p$ and E be the output of a stochastic process in G. We need to find the distribution of the area or volume of E. For statistics simulation, it needs a set of points that are uniformly scattered on G and also on any subarea of G including E. A set of random samples fails this job. Wang and Fang (1990a) suggested applying QMC methods to construct such a net on G when G is not a rectangle, which is the NT-net on G. Section 1.12 introduces how to build NT-nets in several regions using QMC-RPs. The following two case studies are discussed in Fang and Wang (1994). We shall show that QMC-RPs and MSE-RPs can be used as a net on G.

8.1.1 Estimation of π

The circular constant π is essential in mathematics and physics. There are various ways to estimate the famous constant π including geometric, algebraic, and Monte Carlo methods.

Monte Carlo methods: Hogg et al. (2005) in their popular textbook, "Introduction to Mathematical Statistics," introduced a way by Monte Carlo methods. Let U_1 and U_2 be *i.i.d.* as $U(-1,1)$. Define

$$X = \begin{cases} 1, & \text{if } U_1^2 + U_2^2 < 1 \\ 0, & \text{otherwise.} \end{cases}$$

It is easy to find $E(X) = \pi/4$.

Generation iid random numbers $\mathcal{U} = \{u_{i_1}, u_{i_2}, i = 1, \ldots, n\}$ and we have the correspondent $x_1, \ldots, x_n$ and $\overline{x} = \sum_{i=1}^n x_i/n$. Then $4\overline{x} \approx \pi$. However, due to the randomness of the Monte Carlo sample, this estimator of π varies. A confidence interval of 95% for π is given by

$$\left(4\overline{x} - 1.96 \cdot 4\sqrt{\overline{x}(1-\overline{x})/n},\ 4\overline{x} + 1.96 \cdot 4\sqrt{\overline{x}(1-\overline{x})/n}\right).$$

Furthermore, the convergence rate of this estimator $O_p(1/\sqrt{n})$ is slow.

DOI: 10.1201/9781003589389-8

TABLE 8.1
Absolute error between $\hat{\pi}$ and π by the four ways

n	Method	MC	ELP-net	*glp*-net	MSE-RPs
13	$\hat{\pi}$	3.07692308	4.00000000 ($n' = 16$)	3.69230769	3.07692308
	$\lvert\hat{\pi} - \pi\rvert$	0.06466958	0.85840735	0.55071504	**0.06466958**
55	$\hat{\pi}$	3.12727273	3.75000000 ($n' = 64$)	3.05454545	3.12727273
	$\lvert\hat{\pi} - \pi\rvert$	0.01431993	0.60840735	0.08704720	**0.01431993**
89	$\hat{\pi}$	3.14606742	3.52000000 ($n' = 100$)	3.19101124	3.14606742
	$\lvert\hat{\pi} - \pi\rvert$	0.00447476	0.37840735	0.04941858	**0.00447476**
144	$\hat{\pi}$	3.13888889	3.44444444 ($n' = 144$)	3.02777778	3.13888889
	$\lvert\hat{\pi} - \pi\rvert$	0.00270376	0.30285179	0.11381488	**0.00270376**
233	$\hat{\pi}$	3.14163090	3.37500000 ($n' = 256$)	3.14163090	3.19313305
	$\lvert\hat{\pi} - \pi\rvert$	0.00003825	0.23340735	**0.00003825**	0.05154039
377	$\hat{\pi}$	3.14058355	3.32000000 ($n' = 400$)	3.16180371	3.18302387
	$\lvert\hat{\pi} - \pi\rvert$	0.00100910	0.17840735	**0.02021106**	0.04143122
610	$\hat{\pi}$	3.14098361	3.30880000 ($n' = 625$)	3.16721311	3.20000000
	$\lvert\hat{\pi} - \pi\rvert$	0.00060905	0.16720735	**0.02562046**	0.05840735
987	$\hat{\pi}$	3.14083080	3.31250000 ($n' = 1024$)	3.12867275	3.12867275
	$\lvert\hat{\pi} - \pi\rvert$	0.00076185	0.17090735	**0.01291991**	**0.01291991**
1597	$\hat{\pi}$	3.13838447	3.26000000 ($n' = 1600$)	3.16593613	3.18096431
	$\lvert\hat{\pi} - \pi\rvert$	0.00320818	0.11840735	**0.02434348**	0.03937165
4181	$\hat{\pi}$	3.14183210	3.22366864 ($n' = 4225$)	3.13417843	3.13513514
	$\lvert\hat{\pi} - \pi\rvert$	0.00023944	0.08207599	0.00741423	**0.00645752**
6765	$\hat{\pi}$	3.14146341	3.20569023 ($n' = 6889$)	3.13732446	3.13909830
	$\lvert\hat{\pi} - \pi\rvert$	0.00012924	0.06409758	0.00426819	**0.00249435**
10946	$\hat{\pi}$	3.14196967	3.19165533 ($n' = 11025$)	3.13721908	3.14306596
	$\lvert\hat{\pi} - \pi\rvert$	0.00037702	0.05006268	0.00437358	**0.00147331**
17711	$\hat{\pi}$	3.14222799	3.18467365 ($n' = 17956$)	3.13929197	3.14064706
	$\lvert\hat{\pi} - \pi\rvert$	0.00063534	0.04308099	0.00230069	**0.00094560**

Let M be the number of points of $\mathcal{U}$ falling in the circle $C_2 = \{(x_1, x_2) : x_1^2 + x_2^2 \leq 1\}$, that is, we use this $\overline{x} = M/n$ as an estimator of π. However, there are different ways of building a net on a square $(0,1)^2$. The following are some methods.

(A) Equal lattice point (ELP-net): Generate the set n of points

$$\mathcal{U}_{ELP} = \left\{ \left(-1 + \frac{2i}{m}, -1 + \frac{2j}{m} \right), 1 \leq i, j \leq m - 1 \right\},$$

where $n = m^2$. Then, we can account for how many $\mathcal{U}_{ELP}$ points fall in the circle C_2. Fang and Wang (1994) pointed out that the convergence rate is also slow. Therefore, we need more efficient methods.

(B) Qusai-Monte Carlo method: Generate a set of points on $(-1,1)^2$ by Qusai Monte Carlo methods, for example, by the good lattice point method (glp) (see Definition 1.9). The convergence rate of the estimation can reach $O(\log(n)^p/n)$. For this example of estimation of π, $p = 2$. More details can refer to Fang and Wang (1994).

(C) MSE-RPs method: We employ n MSE-RPs of $U((-1,1)^2)$ to form a net $\mathcal{U}_{MSE}$ for estimation.

Generate training sequence on $(-1,1)^2$ by glp method with $n = 75025$ and generating vector $\boldsymbol{h} = [1, 46368]$, then find the set of n MSE-RPs $\{(\xi_{i1}, \xi_{i2}), i = 1, \cdots, n\}$ by LBG algorithm (Linde, Bruzo and Gray, 1980). Let c be the RPs satisfying $\xi_{i1}^2 + \xi_{i2}^2 < 1$, then $\hat{\pi} = 4c/n$.

In Table 8.1, the value of π is estimated with the set of points generated by Monte Carlo, ELP, glp method, and MSE-RPs on $(-1,1)^2$, n is the number of points in the set, and the results marked in red are the minimum error values among ELP, Glp methods, and the way with RPs. For the ELP-net, since $n = m^2$ and sometimes $\sqrt{n}$ is not an integer, so a number $n' > n$ whose gives an integer after taking the square root is chosen as the number of points on ELP-net. For the Monte Carlo method, π is estimated for $T = 2000$ with 2000 set of random samples and obtain $\hat{\pi}_j$, $j = 1, \cdots, T$, the median of these 2000 estimated value is listed in the table. Comparing the corresponding absolute error between π and $\hat{\pi}$, $|\hat{\pi} - \pi|$, GLP-net and MSE-RPs seem to have better performance than the training set generated by Monte Carlo and ELP methods. However, based on the results in the table, it is hard to say which way to generate the set is better for estimating π between glp method and MSE-RPs. It seems that the MSE-RPs are slit better than the glp method.

8.1.2 Area of intersection between a fixed circle and several random circles

Let K be a unit circle with center at the origin $\mathbf{o} = (0, 0)^\top$. Suppose that $K_1, \ldots, K_m$ be the m random circle centered at $\mathbf{o}_1, \ldots, \mathbf{o}_m$ with given radii $r_1, \ldots, r_m$, respectively. The centers $\mathbf{o}_1, \ldots, \mathbf{o}_m$ are independent and $\mathbf{o}_j \sim N_2(\mathbf{0}, \sigma_j^2 \mathbf{I}_2)$, where $\sigma_j > 0$ is known. Let $S_m = K \cap (K_1 \cup \ldots \cup K_m)$ be the overlapping region of K and the union of the m random circles $K_1, \ldots, K_m$, and denote $A(S_m)$ as the area of S_m. The goal of the problem is to find the distribution of $A(S_m)$. Finding the distribution of the overlapping area with a single random circle is not difficult. However, there is no analytic solution for the cases when $m \geq 2$. Zhou and Fang (2011) studied the problem in detail and provided the analytic formulas to calculate $A(S_m)$ for $m \leq 2$, but the analytic solution to the overlapping area with $m > 2$ is complicated. Hence, accurately estimating the area $A(S_m)$ is crucial to solving the problem. Fang and Wang (1994) suggested to apply the NT-net to estimate the overlapping area $A(S_m)$ so that we can obtain a random sample $A_1(S_m), \ldots, A_k(S_m)$ of $A(S_m)$, and then the empirical distribution of the sample can be regarded as an approximation of the distribution of $A(S_m)$. Yang et al. (2021) focused the estimation of $A(S_m)$ with a net generated by QMC-RPs and MSE-RPs.

We need to generate a sample uniformly distributed on the unit circle to estimate the area of S_m. Yang et al. (2021) gave a comparisons for estimation of $A(S_2)$ under QMC-RPs and MSE-RPs. They concluded:

Based on the numerical results, the QMC-RPs and MSE-RPs estimate $A(S_2)$ well. Compared with MSE-RPs, QMC-RPs perform slightly better in this problem, though their estimation bias values are very close.

8.1.3 Estimation of the area of belt on unit sphere

Fang and Wei (1992) and Fang and Wang (1994) studied a real-life example of the life distribution of a rotary ball roller in the production line of steel rolling. Its statistical model can be expressed in the following example.

Example 8.1 *Let $S \in \mathbb{R}^3$ be the spherical roller in the steel rolling production line. In the steel process, S is covered by a sequence of belts, $\mathcal{B} = \{G_h(\boldsymbol{x}_1), G_h(\boldsymbol{x}_2), \cdots\}$, successively, where $G_h(\boldsymbol{x})$ is the belt with thickness $2h$ and the normal direction $\boldsymbol{x}$ on S. Each belt is symmetric about a great circle of S. Let $K_n(\boldsymbol{y})$ be the number of belts in $\mathcal{B}$ which cover a specific $\boldsymbol{y} \in S$ in the first n belts of $\mathcal{B}$. For given integer m, let $T_m(h, \mathcal{B})$ be the minimum of n such that $K_n(\boldsymbol{y}) \geq m$ for some $\boldsymbol{y} \in S$. The $T_m(h, \mathcal{B})$ is called life of the roller. We want to design a sequence of belts, $\mathcal{B}$, such that the corresponding $T_m(h, \mathcal{B})$ has the longest life. Given the thickness $2h$, the belt $G_h(\boldsymbol{x})$ is determined by its normal direction $\boldsymbol{x}$. The set $\mathcal{B}$ can be denoted by a sequence of points, $\boldsymbol{x}_1, \boldsymbol{x}_2, \cdots$.*

It isn't easy to find analytic results for the life distribution of the rotary ball roller S. Therefore, statistical simulation is a good technique for finding the life distribution. It needs a Net on the spherical surface of a ball S with radius r. Without loss of generality, assume $r = 1$. Fang and Wei (1992) pointed out that it can't use the Monte Carlo method to construct a Net on the surface of S, denoted by S-net, and they propose to use Quasi-Monte Carlo methods to build a S-net. However, we can employ a set of MSE-RPs of the uniform distribution on the unit sphere.

The belt on the surface of $\mathbb{R}^3$ with thickness $2h$ can be written as

$$G_h(\mathbf{n}) = \{\mathbf{w} : |\mathbf{w}^\top \mathbf{n}| \leq h\},$$

and the area of the belt can be computed by $A(G_h(\mathbf{n})) = 4\pi h$.

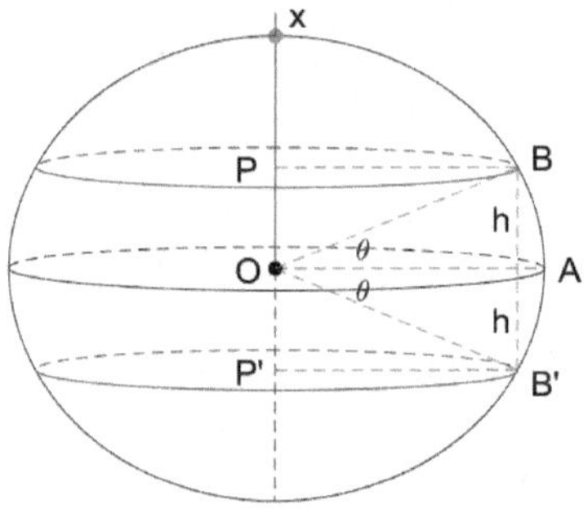

FIGURE 8.1
Belt on unit sphere with normal direction $\boldsymbol{x} = [0, 0, 1]^T$.

Given the radius of unit sphere r, the thickness of random belt $2h$, we have the area of the unit sphere is

$$S_W = 4\pi r^2, \tag{8.1}$$

and the area of a spherical cap is

$$S_{SC} = 2\pi r(1 - \cos(\pi/2 - \theta)) = 2\pi r^2(1 - \sin(\theta)), \tag{8.2}$$

where θ is the angle between $\overrightarrow{OB}$ and $\overrightarrow{OB'}$. Hence, the area of the belt on the unit sphere is

$$\begin{aligned} S_G(h) = S_W - S_{SC} &= 4\pi r^2 - 2[2\pi r(1 - \sin(\theta))] \\ &= 4\pi r^2 \sin(\theta) = 4\pi h R, \end{aligned} \tag{8.3}$$

where $\sin(\theta) = h/r$.

The area of the random belt on the unit sphere could also be obtained by simulation. Let $\boldsymbol{u}$ be the random vector of uniform distribution on unit sphere $\boldsymbol{w}$, denoted as $U(\boldsymbol{w})$, and $\boldsymbol{a}$ be a random sample of $\boldsymbol{u}$, then the distance from $\boldsymbol{a}$ to the circle K_A is $|\boldsymbol{a}^T\boldsymbol{x}|$. Hence, if $|\boldsymbol{a}^T\boldsymbol{x}| \leq h$, the point $\boldsymbol{a}$ is on the random belt $G_h(\boldsymbol{x})$, and

$$G_h(\boldsymbol{x}) = \{\boldsymbol{a} : |\boldsymbol{a}^T\boldsymbol{x}| \leq h\}, \tag{8.4}$$

where $\boldsymbol{x} \sim U(W)$. Let $N(\mathcal{X})$ be the number of elements in the set $\mathcal{X}$, the area of random belt can be estimated by

$$\hat{S}_G(h) = \frac{N(G_h(\boldsymbol{x}))}{N}, \tag{8.5}$$

where N is the number of points in training sequence $\mathcal{Y}$ on the unit sphere $\boldsymbol{W}$.

Let $\mathbf{S}^3$ be a unit sphere in $\mathbb{R}^3$ centering at the origin $(0,0,0)^T$, and random belts can cover it with a fixed width. Each belt is symmetric about a unit circle in $\mathbf{S}^3$ centered

$$G_h(\mathbf{n}) = \{\mathbf{w} : |\mathbf{w}^T\mathbf{n}| \le h\},$$

and the area of the belt can be computed by $A(G_h(\mathbf{n})) = 4\pi h$. The original problem arises in steel rolling and aims to find the distribution of a roller's lifespan and find some way to increase the lifespan (see (Fang and Wang, 1994) for details). Similar to the problem of finding distribution. Yang et al. (2021) focus on one random belt, estimate its area with QMC-RPs and MSE-RPs, and compare their performance. Suppose $\mathcal{U}^{(3)} = \{\mathbf{u}_1^{(3)}, \ldots, \mathbf{u}_k^{(3)}\}$ be a set of points uniformly distributed on the surface of S, the area of $G_h(\mathbf{n})$ can be estimated by

$$\hat{A}(G_h(\mathbf{n})) = \frac{4\pi}{k}\sum_{j=1}^{k} \mathbf{1}\{|(\mathbf{u}^{(3)})^T\mathbf{n}| \le h\} \tag{8.6}$$

with the estimation bias evaluated by $d_{G_h} = |A(G_h(\mathbf{n})) - \hat{A}(G_h(\mathbf{n}))|$ The area of the random belt on the surface of S with a given h remains the same with any direction vector, we will assume that $\mathbf{n} = (0,0,1)^T$ for convenience purpose and denote $G_h = G_h((0,0,1)^T)$ as the belt with thickness $2h$ and $A(G_h)$ as its area.

Section 6.3 mentions how to generate QMC-RPs of the uniform distribution $U(S^p)$. Let $\boldsymbol{U} \sim U(S^p)$ and $\boldsymbol{Z} \sim N_p(\mathbf{0}, \boldsymbol{I}_p)$. We have the following stochastic representation

$$\boldsymbol{U} \stackrel{d}{=} \boldsymbol{Z}/\|\boldsymbol{Z}\|. \tag{8.7}$$

It is easy to generate QMC-RPs of $N_p(\mathbf{0}, \boldsymbol{I}_p)$ and then QMC-RPs of $\boldsymbol{U}$. Yang et al. (2021) gave some numerical comparisons on estimation of areas $A(G_{0.05}), A(G_{0.15})$. They concluded that both the two types of RPs have good performance when estimating $A(G_{0.05}), A(G_{0.15})$, but the MSE-RPs are more competitive than QMC-RPs for the estimation of $A(G_{0.1})$.

8.2 RPs of Meta-Distributions via Copula Technique

Data analysis often requires a multivariate distribution from one-dimensional marginal distributions with a known dependence structure. Multivariate distributions with given margins arise from statistical theory and practical problems, especially in the context of risk and decision analysis for complex multivariate systems. Copula is a technique for constructing multivariate distributions, called *meta-distributions*. The copula is useful for estimating parameters when the dimension is large since the parameters of a multivariate distribution are estimated separately by each marginal distribution and a copula function. Using copulas, we can construct multivariate distributions with different marginal distributions. The following definition is based on the theory of probability. More general definitions can refer to Nelsen (2006) and Durante and Sempi (2010).

Definition 8.1 *If $C(z_1, \ldots, z_p)$ is the cumulative distribution function (cdf) on the unit hypercube $C^p = [0,1]^p$ with uniform $U(0,1)$ margins, then C is called a copular function or copula.*

It is well known that for given p marginal distributions $F_i(x_i), i = 1, \ldots, p$, there are infinity p-dimensional distributions with these given margins. If we know the concordance structure of the joint distribution, we might find few or unique such joint distributions $H(x_1, \ldots, x_p)$, called as *Meta-distribution* based on the given copula C.

Suppose that there are p random variables $X_i \sim F_i(x_i), i = 1, \ldots, p$, we are requested to construct a multivariate distribution H with margins $F_1, \ldots, F_p$ under a given structure of $C(z_1, \ldots, z_p)$. This kind of technique is called *copula technique.* The copula technique is an essential tool for this purpose. There are two important questions to the copula theory.

1. How can we find useful copulas?
2. For a given multivariate distribution $H(x_1, \ldots, x_p)$ how to find its corresponding copula function?

The following lemma shows the existence of the copular function.

Lemma 8.1 *Let $H(x_1, \ldots, x_p)$ be a continuous distribution with p continuous margins F_i on the domain D_i for $i = 1, \ldots, p$. Then, there exists a unique copula C such that*

$$C(z_1, \ldots, z_p) = H(F_1^{-1}(z_1), \ldots, F_p^{-1}(z_p)). \tag{8.8}$$

Sklar (1959) proposed the following theorem: Sklar's theorem for the multivariate distributions.

Theorem 8.1 *Let continuous random vector $\boldsymbol{X}$ follow $H(x_1, \ldots, x_p)$ with p continuous margins F_i on the domain D_i for $i = 1, \ldots, p$, Then there exists a unique copula $C(z_1, \ldots, z_p)$ such that*

$$H(x_1, \ldots, x_p) = C(F_1(x_1), \ldots, F_p(x_p)). \tag{8.9}$$

If $F_i, i = 1, \ldots, p$ are continuous, then C is unique on the domain $D = D_1 \otimes \cdots \otimes D_p$. Conversely, if C is a copula and $F_i, i = 1, \ldots, p$ are distribution functions, then the function C defined by (8.8) is a p-dimensional distribution function with margins $F_i, i = 1, \ldots, p$. The density function of $\boldsymbol{X}$ from (8.9) is given

$$h(x_1, \ldots, x_p) = C(F_1(x_1), \ldots, F_p(x_p)) \prod_{i=1}^{p} f_i(x_i), \tag{8.10}$$

where f_i is the pdf of X_i.

We call H is a *meta-distribution based on* C and denote $H \sim H(C; F_1, \ldots, F_p)$.

The proof of Sklar's theorem was not given in Sklar (1959), but a sketch of it was provided in Sklar (1973) and Schweizer and Sklar (1974). It should also be mentioned that some "indirect" proofs of Sklar's theorem (without mentioning copula) were later discovered by Moore and Spruill (1975). A detailed proof was given in Úbeda Flores and Sánchez (2017). Kotz and Seeger (1991) introduced the pdf of the meta-distribution, which can be decomposed into the *density weighting function* and the *product of marginal densities.*

Sklar's theorem can also be extended to apply to conditional distributions (see Patton (2006)), which is helpful for forecasting and time series applications. Let $\{(X_t, Y_t)\}_t$ denote a stochastic process and $\mathcal{F}_t$ denote an information set available at time t, and let the conditional distribution of $(X_t, Y_t)|\mathcal{F}_t$ be H_t, with conditional marginal distributions F_t and G_t.

$$H_t(x, y|\mathcal{F}_{t-1}) = C_t(F_t(x|\mathcal{F}_{t-1}), G_t(y|\mathcal{F}_{t-1}|\mathcal{F}_{t-1})).$$

The complication that arises when applying Sklar's theorem to conditional distributions is that the information set used for the margins and the copula must be the same; if different information sets are used, then the resulting function H_t can no longer be interpreted as a multivariate conditional distribution.

8.2.1 Properties of copula

Nelsen (2006) pointed out that a two-dimensional copula function $C = (u, v)$ must satisfy the following conditions.

1. *The independence copula.* Let X and Y be continuous random variables. Then X and Y are independent if and only if (X, Y) has the copula:

$$C(u, v) = uv, \ 0 \leq u, v \leq 1.$$

   See Theorem 2.4.3 of Nelsen (2006). In the literature, the independent copula is denoted by $\Pi(\boldsymbol{u})$.

2. *Fréchet-Hoeffding bounds*

$$W(u, v) \equiv \max(u + v - 1, 0) \leq C(u, v) \leq \min(u, v) \equiv M(u, v), \qquad (8.11)$$

   where M is the Fréchet-Hoeffding upper bound copula and W is the Fréchet-Hoeffding lower bound copula. Let two margins are F and G, then,

$$M(F(x), G(y)) \leq H(x, y) \leq W(F(x), G(y)).$$

3. Fréchet-Mardia copulas are defined as convex combinations of copulas

$$C^{FM}(\boldsymbol{u}) = \alpha\Pi(\boldsymbol{u}) + (1 - \alpha)M(\boldsymbol{u}), 0 < \alpha < 1,$$

   where $\Pi(\boldsymbol{u})$ is the independent copula and $M(\boldsymbol{u})$ is the Fréchet-Hoeffding upper bound copula.

4. $0 \leq \frac{\partial}{\partial u}C(u, v) \leq 1$, $0 \leq \frac{\partial}{\partial v}C(u, v) \leq 1$.

5. A copular C is a Lipschitz function, i.e.,

$$C(\boldsymbol{u}) - C(\boldsymbol{v}) \leq |\boldsymbol{u} - \boldsymbol{v}|.$$

6. The marginal rth moments of $E(X_i^r)$ are known from F_i it $E(|X_i|^r) < \infty$.

The following theorem is important in the theory of meta-distributions.

Theorem 8.2 *Let X and Y be continuous random variables with copula $C_{X,Y}$. If $\alpha(\cdot)$ and $\beta(\cdot)$ are strictly increasing on the domains D_x and D_y, respectively, then $C_{\alpha(X),\ \beta(Y)} = C_{X,Y}$, i.e., $C_{X,Y}$ is invariant under strictly increasing transformations. Furthermore,*

1. *If $\alpha(\cdot)$ is strictly increasing and $\beta(\cdot)$ is strictly decreasing, then*

$$C_{\alpha(X),\ \beta(Y)}(u, v) = u - C_{X,Y}(u, 1 - v);$$

2. *If $\alpha(\cdot)$ is strictly decreasing and $\beta(\cdot)$ is strictly increasing, then*

$$C_{\alpha(X),\ \beta(Y)}(u,v) = v - C_{X,Y}(1-u,v);$$

3. *If $\alpha(\cdot)$ and $\beta(\cdot)$ are both strictly decreasing, then*

$$C_{\alpha(X),\ \beta(Y)}(u,v) = u + v - 1 + C_{X,Y}(1-u,1-v).$$

The above results can be extended to the multivariate case. Refer to Durante and Sempi (2010) or Mai and Scherer (2017).

Theorem 8.3 *Let $\boldsymbol{X} = (X_1, \dots, X_p)$ follow a continuous distribution $H(\boldsymbol{x})$.*

1. *The copula of $\boldsymbol{X}$ is $\Pi(\boldsymbol{u}) = \prod_{i=1}^{p} u_i$ if and only if $X_1, \dots, X_p$ are independent.*

2. *The comonotonicity copula $M(\boldsymbol{u}) = \min\{u_1, \dots, u_p\}$ associated with a vector $\boldsymbol{U} = (U_1, \dots, U_p)$ uniformly distributed on $C^p = [0,1]^p$ and such that $U_1 = \cdots = U_p$ almost surely.*

3. *The countermonotonicity copula $W = (u_1, u_2) = \max(u_1+u_2-1, 0)$ associated with a vector $\boldsymbol{U} = (U_1, U_2)$ of r.v.'s uniformly distributed on $[0,1]^2$ and such that $U_1 = 1 - U_2$ almost surely. Note that $W(\boldsymbol{u})$ is not a copula for $p \geq 3$.*

4. *The Fréchet-Hoeffding bounds (refer to Lemma 1.3 of Mai and Scherer (2017))*

$$W(\boldsymbol{u}) = \max(\sum_{i=1}^{p} u_i - p + 1, 0) \leq C(\boldsymbol{u}) \leq M(\boldsymbol{u}).$$

Denote the class of all p-copulas by $\mathcal{B}_p$. Moreover, the above bounds are sharp in the following sense.

$$\inf_{C \in \mathcal{B}_p} C(\boldsymbol{u}) = W(\boldsymbol{u}), \ \sup_{C \in \mathcal{B}_p} C(\boldsymbol{u}) = M(\boldsymbol{u}).$$

5. *Although $W(\boldsymbol{u})$ is not a copula for any $p \geq 3$, but there exists an $p-$dimensional copula C for any $\boldsymbol{u} \in C^p$ such that*

$$C(\boldsymbol{u}) = W(\boldsymbol{u}),$$

see Theorem 2.10.13 of Nelsen (2006).

6. *The copula function C is a Lipschitz function that satisfies*

$$|C(\boldsymbol{u}) - C(\boldsymbol{v})| \leq \sum_{i=1}^{p} |u_i - v_i|, \ for\ \boldsymbol{u}, \boldsymbol{v} \in (0,1)^p.$$

7. *Let $T_1, \dots, T_p$ be strictly increasing functions from $\mathbb{R}$ to $\mathbb{R}$. Then C is also the copula of $(T_1(X_1), \dots, T_p(X_p))$.*

8. *For $p = 2$ the random vector (X_1, X_2) has copula $W(u_1, u_2)$ if and only if, for some strictly decreasing function T such that $X_2 = T(X_1)$. almost surely.*

Copulas are also a powerful tool for finance because the modeling problem can be split into two steps: the first step deals with the identification of the marginal distributions; and the second step consists of defining the appropriate copula in order of dependence structure in a reasonably good manner.

Definition 8.2 *A random vector* $(X,Y) \sim H(x,y)$ *is exchangeable if*

$$(X,Y) \stackrel{d}{=} (Y,X), \quad \text{or } H(x,y) = H(y,x), \ \forall (x,y) \in \mathbb{R}^2.$$

Theorem 8.4 *Let continuous random variables* $(X,Y) \sim H(x,y)$ *with margins* $F_1(x)$ *and* $F_2(y)$ *and copula* C*. Then* (X,Y) *are exchangeable if and only if* $F_1 = F_2$ *and* $C(u,v) = C(v,u)$ *for all* $(u,v) \in (0,1)^2$*.*

Definition 8.3 *If* C_1 *and* C_2 *are copulas, we say that* C_1 *is smaller than* C_2 *(or* C_2 *is larger than* C_1*), and write* $C_1 \prec C_2$ *(or* $C_2 \succ C_1$*) if* $C_1(u,v) \le C_2(u,v)$ *for all* $(u,v) \in (0,1)^2$*.*

Example 8.2 *The Cuadras-Augé family of copulas (1981).*

$$C_\theta(u,v) = [\min(u,v)]^\theta (uv)^{1-\theta} = \begin{cases} uv^{1-\theta}, & u \le v, \\ u^{1-\theta}v, & u \ge v. \end{cases}$$

For $0 \le \alpha \le \beta \le 1$ *and* $u, v \in (0,1)$ *we have*

$$\frac{C_\alpha(u,v)}{C_\beta(u,v)} = \left(\frac{uv}{\min(u,v)}\right)^{\beta-\alpha} \le 1,$$

Hence $C_\alpha \prec C_\beta$*.*

8.2.2 Copula examples

Example 8.3 *Consider*

$$H(x,y) = \frac{(x+1)(e^y - 1)}{x + 2e^y - 1}.$$

Its two margins are $U(-1,1)$ *and the standard exponential EP(1) with cdf* $1 - e^{-y}$ *and copula*

$$C(u,v) = \frac{uv}{u + v - uv}.$$

If the two margins are $F_i(x) = (1 + e^{-x})^{-1}$*, the corresponding meta-distribution is*

$$H(x,y) = (1 + e^{-x} + e^{-y})^{-1}, \ x, y \ge 0,$$

which is the Gumbel's bivariate logistic distribution Gumbel (1961).

Example 8.4 *The Gumbel's bivariate exponential distribution (cdf, Gumbel, 1960), is given by*

$$H_\theta(x,y) = 1 - e^{-x} - e^{-y} - e^{-(x+y+\theta xy)}, \ x \ge 0, y \ge 0.$$

By Lemma 1, we obtain the copula of the Gumbel's distribution

$$C_\theta(u,v) = u + v - 1 + (1-u)(1-v)e^{-\theta \ln(1-u)\ln(1-v)}.$$

Example 8.5 *Fang et al. (2000) gave a comprehensive study on the copula*

$$C(u, v; \alpha, \beta) = \frac{uv}{[1 - \beta(1 - u^{1/\alpha})(1 - v^{1/\alpha})]^{\alpha}}. \tag{8.12}$$

For different values of α and β

1. *$\beta = 0$, U and V are independent;*
2. *$\beta = 1$, it reduces into the bivariate Clayton distribution (Clayton (1978));*
3. *$\alpha = 1$, it reduces into the Ali-Mikhail-Haq copula (Ali et al. (1978)).*

Its multi-version is

$$C(u_1, \ldots, u_n; \alpha, \beta) = \frac{\prod_{i-1}^{n} u_i}{[1 - \beta \prod_{j=1}^{n}(1 - u_j^{1/\alpha})]^{\alpha}}. \tag{8.13}$$

Example 8.6 *The 3-dimension distribution*

$$H(x, y, z) = \frac{(x+1)(e^z - 1)\sin(z)}{x + 2e^y - 1}, \quad on\ [-1, 1] \times [0, \infty] \times [0, \pi/2].$$

The three 1-dimensional margins are

$$H_1(x) = \frac{x+1}{2}, H_2(y) = 1 - e^{-y}, H_3(z) = \sin(z),$$

respectively, and three 2-dimensional margins are

$$\begin{aligned} H_{12}(x, y) &= \frac{(x+1)(e^y - 1)}{x + 2e^y - 1}, \\ H_{23}(y, z) &= (1 - e^{-y})\sin(z), \\ H_{13}(x, z) &= \frac{1}{2}(x+1)\sin(z), \end{aligned}$$

respectively. The interesting reader tries to find the copula function.

8.2.3 Normal copula and meta-normal distributions

The multinormal distributions have been popularly used in data modeling. Therefore, their meta-distributions are the first group of distributions in the copula theory and applications. From the definition of the copula, the Gaussian copula of $N_p(\boldsymbol{\mu}, \boldsymbol{\Sigma})$ is the same as the copula of $\boldsymbol{Z} = (Z_1, \ldots, Z_p) \sim N_p(\boldsymbol{0}, \boldsymbol{R})$, where $\boldsymbol{R}$ is the correlation matrix of $\boldsymbol{Z}$ or $\boldsymbol{X}$. It is clear that all the marginal distributions of $\boldsymbol{Z}$ are $N(0, 1)$, but its components have a correlation matrix $\boldsymbol{R} = (r_{ij})$. Denote the cdf of $\boldsymbol{Z}$ by $G_p(z_1, \ldots, z_p)$ and define a function C.

Definition 8.4 *Define a distribution function C as*

$$C_{normal}(u_1, \ldots, u_p) = G_p(\Phi^{-1}(u_1), \ldots, \Phi^{-1}(u_p)). \tag{8.14}$$

This function C is called normal copula.

The meta-normal distribution is a class of distributions with a set of given marginal cdfs $F_{X_1}, \ldots, F_{X_p}$ and cdf

$$F_{X_1, \ldots, X_p}(p) = G_p(\Phi^{-1}(F_{X_1}(z_1)), \ldots, \Phi^{-1}(F_{X_p}(z_p))). \tag{8.15}$$

Denote this class of distributions by $\boldsymbol{X} \sim MN_p(\boldsymbol{0}, \boldsymbol{R}; F_1, \ldots, F_p)$.

The pdf of $MN_p(\mathbf{0}, \boldsymbol{R}; F_1, \ldots, F_p)$ is

$$h(\boldsymbol{x}) = C_{normal}(F_1(x_1), \ldots, F_p(x_p)) \prod_{i=1}^{p} f_i(x_i), \tag{8.16}$$

where $f_i(x_i)$ is the pdf of $F_i(x_i)$.

Let $\boldsymbol{\xi} = (\xi_1, \ldots, \xi_p)^T = (\Phi^{-1}(u_1), \ldots, \Phi^{-1}(u_p))^T$. Zhang and Singh (2019) gave a simple formula for the density of C_{normal} as

$$c_{normal}(\boldsymbol{\xi}) = c_{normal}(\xi_1, \ldots, \xi_p) = |\boldsymbol{R}|^{-1/2} \exp\left(-\frac{1}{2}\boldsymbol{\xi}^T(\boldsymbol{R}^{-1} - \boldsymbol{I})\boldsymbol{\xi}\right). \tag{8.17}$$

They also gave examples for $p = 2, 3$.

Example 8.7 *The Binormal copular. Let $\boldsymbol{Z} \sim N_2(\mathbf{0}, \boldsymbol{R})$, where the correlation matrix with* $\text{corr}(X_1, X_2) = \rho$. *Then, the normal copula is given by*

$$C_{normal}(u, v) =$$
$$\frac{1}{2\pi\sqrt{1-\rho^2}} \int_{-\infty}^{\Phi^{-1}(u)} \int_{-\infty}^{\Phi^{-1}(v)} \exp\left\{-\frac{s^2 - 2\rho st + t^2}{2(1-\rho^2)}\right\} dsdt.$$

We choose any two continuous univariate distributions as margins, for example, the exponential distribution, the t-distribution, etc.

The meta-normal distribution $\boldsymbol{X} \sim MN_2(\mathbf{0}, \boldsymbol{R}; F_1, F_2)$ is given by

$$\begin{aligned} F(x_1, x_2) &= C_{normal}(F_1(x_1), F_2(x_2)) \\ &= \frac{1}{2\pi\sqrt{1-\rho^2}} \int_{-\infty}^{\Phi^{-1}(F_1(x_1))} \int_{-\infty}^{\Phi^{-1}(F_2(x_2))} \exp\left\{-\frac{s^2 - 2\rho st + t^2}{2(1-\rho^2)}\right\} dsdt. \end{aligned} \tag{8.18}$$

Krzysztofowicz and Kelly (1996) and Kelly and Krzysztofowicz (1997) studied the *meta-Gaussian distribution* or Meta-Normal Distribution. Joe (2015) gave a more detailed study in Section 2.6 "Multivariate Gaussian/normal".

8.2.4 Measure of concordance

In the multivariate statistical analysis, the Pearson correlation coefficient between two random variables U and V is defined by

$$\text{corr}(U, V) = \frac{E(UV) - E(U)E(V)}{\sqrt{\text{Var}(U)\text{Var}(V)}} \tag{8.19}$$

has been widely used.

Two distributions $H_1(\boldsymbol{x})$ and $H_2(\boldsymbol{x})$ in $\mathbb{R}^p$ that have the same copula $C(\boldsymbol{z})$ are called them belonging to the same class of *meta-distributions* based on the copula C. A meta-distribution $H(C, F_1, \ldots, F_p)$ involves two parts:

(1) its copula $C(u_1, \ldots, u_p)$ and

(2) given marginal distributions $F_1(x_1), \ldots, F_p(x_p)$.

The copula is sometimes called the *dependence function*, as it describes the dependence/concordance between any pair of random variables in $H(C, F_1, \ldots, F_p)$. Let random vectors $(X_1, \ldots, X_p) \sim H(C, F_1, \ldots, F_p)$ and $(U_1, \ldots, U_p) \sim C(u_1, \ldots, u_p)$. It is clear

$\text{corr}(X_i, X_j) \neq \text{corr}(U_i, U_j), i \neq j$ in general. Therefore, one wants to find some other measures of concordance.

There are a variety of ways to define a measure of concordance. The necessary property for a measure dependence is "*scale-invariant*". It remains unchanged under strictly increasing transformations of the random variables. That means that the pairs (X_i, X_j) and (U_i, U_j) have the same value of concordance measure. Any set of desirable properties for a "measure of concordance" would include those in the following definition (Scarsini, M. 1984).

A numeric measure κ of association between two continuous random variables X and Y whose copula is C is a *measure of concordance* (denote as $\kappa_{X,Y}$, or κ_C) if it satisfies the following properties:

1. $\kappa_{X,Y}$ is defined for every pair (X, Y) of continuous random variables;
2. $-1 \leq \kappa_{X,Y} \leq 1$, $\kappa_{X,X} = 1, \kappa_{X,-X} = -1$;
3. $\kappa_{X,Y}$ is exchangeable, i.e., $\kappa_{X,Y} = \kappa_{Y,X}$;
4. If X and Y are independent, $\kappa_{X,Y} = 0$;
5. $\kappa_{-X,Y} = \kappa_{X,-Y} = -\kappa_{X,Y}$;
6. If C_1 and C_2 are copulas such that $C_1 \prec C_2$, then $\kappa_{C_1} \leq \kappa_{C_2}$;
7. If $C_n(X_n, Y_n) \to C$ as $n \to \infty$, then $\kappa_{C_n} \to \kappa_C$.

Note that the request exists on moments of X and Y. Therefore, a measure of concordance should be "*strictly increasing transformation-invariant (SITI)*", see Theorem 8.2.

We can use group theory, especially transformation group theory, to study the measure of concordance. Section 1.3 of Fang et al. (1990) introduced the following knowledge.

Definition 8.5 *Let G be a set of transformations from a space $\mathcal{X}$ into itself which satisfy the following conditions*

1. if $g_1, g_2 \in G$, then $g_1g_2 \in G$ where g_1g_2 is defined as $(g_1g_2)x = g_1(g_2x)$ for each $x \in \mathcal{X}$;

2. if $g \in G$ there exists a $g^{-1} \in G$ such that $gg^{-1} = g^{-1}g = e$, where e is the identity transformation in G.

Then G is called as a transformation group (TG).

Definition 8.6 *Two points x_1 and x_2 in $\mathcal{X}$ are said to be equivalent under G if there exists a $g \in G$ such that $x_2 = gx_1$. We write $x_1 \sim x_2$ (mod G). This equivalence relation satisfies:*
(i) $x \sim x$ (mod G);
(ii) $x \sim y$ (mod G) implies $y \sim x$ (mod G);
(iii) $x \sim y$ (mod G) and $y \sim z$ (mod G) implies $x \sim z$ (mod G).

Definition 8.7 *A function $f(x)$ on $\mathcal{X}$ is said to be invariant under G if*

$$f(gx) = f(x), \forall x \in \mathcal{X}, \forall g \in G.$$

A function $f(x)$ on $\mathcal{X}$ is said to be maximal invariant under G if it is invariant under G and if

$$f(x_1) = f(x_2) \;\; \textit{implies} \;\; x_1 \sim x_2 \; (\textit{mod } G).$$

Theorem 8.5 *Assume function $f(x)$ on $\mathcal{X}$ is maximal invariant under G. Then a function $h(x), x \in \mathcal{X}$ is invariant under G if and only if h is a function of $f(x)$.*

The proof can refer to Theorem 1.1 of Fang et al. (1990). Note that the maximal invariant under G is not unique.

The Group theory in copula

Denote $MD(C; F_1, \dots, F_p)$ for a meta-distribution with the copula $C(u_1, \dots, u_p)$ and marginal continuous distributions $F_i \in \mathcal{F}, i = 1, \dots, p$. where $\mathcal{F}$ is the set of univariate continuous distributions. Denote a class of random vectors following meta-distributions by

$$MD(C) = \{\boldsymbol{X} = (X_1, \dots, X_p) \sim MD(C; F_1, \dots, F_p), F_i \in \mathcal{F}, i = 1. \dots, p\}.$$

Let G be the set of $g(x)$, where g is a strictly increasing function on $\mathbb{R}$ and the operator $g_1g_2(X) = g_1(g_2(X))$ for any $g_1, g_2 \in G$. Define the transformation group in a given copula: Let $\mathcal{X} = MD(C)$ be the space and G be the transformation on $\mathcal{X}$ by the action

$$\begin{aligned} &\boldsymbol{X} = (X_1, \dots, X_p) \sim MD(C; F_1, \dots, F_p) \to \\ &g\boldsymbol{X} \sim MD(C; g_1(X_1), \dots, g_p(X_p)), g_i \in G, i = 1, \dots, p. \end{aligned}$$

The identical transformation is denoted by $e : e(X) = X$. It is easy to find that this is a transformation group. Now we can define a transformation group in a group of meta-distributions for a given copula $C(u_1, \dots, u_p)$.

Definition 8.8 *Let $\mathcal{X} = MD(C)$; $\mathcal{G} = \{(g_1, \dots, g_p) : g_i \in G, i = 1, \dots, p\}$; the operator $g_1g_2(X) = g_1(g_2(X))$ for any $g_1, g_2 \in G$; and the action*

$$\begin{aligned} &\boldsymbol{X} \sim MD(C; F_1, \dots, F_p) \to \\ &g\boldsymbol{X} \sim MD(C; g_1(X_1), \dots, g_p(X_p)), (g_1, \dots, g_p) \in \mathcal{G}\}. \end{aligned}$$

Then $\mathcal{X} = MD(C)$ forms a transformation group (or meta-copula C distributions).

If C is the normal copula defined in Subsection 8.2.3, the corresponding $MD(C)$ is the set of meta-normal distributions.

Definition 8.9 *A function $f(x)$ on $\mathcal{X} = MD(C)$ is said to be invariant under $\mathcal{G}$*

$$f(gx) = f(x), \forall x \in \mathcal{X}, \forall g \in \mathcal{G}.$$

A function $f(x)$ on $\mathcal{X}$ is said to be maximal invariant under $\mathcal{G}$ if it is invariant under $\mathcal{G}$ and if $f(x_1) = f(x_2)$ implies $x_1 \sim x_2$ (mod $\mathcal{G}$).

We can prove the following facts: Kendall's tau and Spearman's rho will be introduced in the next subsection.

1. The copula function C is a maximal invariant under $\mathcal{G}$.
2. A function $\kappa_{X,Y}$ is qualified to be a measure of depend. It must be a function of C.
3. The Kendall's tau $\tau_{X_i,X_j}, 1 \le i < j \le p$, where $\boldsymbol{X} = (X_1, \dots, X_p) \in \mathcal{X}$ is a function of C.
4. The Spearman's rho $\rho_{X_i,X_j}, 1 \le i < j \le p$, where $\boldsymbol{X} = (X_1, \dots, X_p) \in \mathcal{X}$ is a function of C.
5. Meta-elliptical distributions $\boldsymbol{X} \sim EC_2(\boldsymbol{0}, \boldsymbol{R}, h)$ studied by Fang et al. (2002) have the same value of the Kendall's tau $\tau = \frac{2}{\pi} arcsin(\rho)$ and Spearman's rho $\rho = \frac{6}{\pi} arcsin(\rho/2)$ if they have the same correlation matrix $\boldsymbol{R}$.

8.2.5 Some measures of concordance

Some concordance measures, such as Kendall's τ, Spearman's ρ, and tail dependence coefficients.

A. Kendall's τ

Definition 8.10 *Let (X_1, Y_1) and (X_2, Y_2) be random samples from $H(x, y)$. Then the Kendall's tau is defined as the difference of concordant and discordant probabilities*

$$\tau = \tau_{X,Y} = P((X_1 - X_2)(Y_1 - Y_2) > 0) - P((X_1 - X_2)(Y_1 - Y_2) < 0).$$

From Theorem 5.1.1 Nelsen (2006) we have:

Theorem 8.6 *Let X and Y be continuous random variables. Then the population version of Kendall's τ for X and Y is given by*

$$\tau = \tau_{X,Y} = \tau_C = 4\int_0^1\int_0^1 C(u,v)dC(u,v) - 1 = 4E(C(U,V)) - 1,$$

where $C(U, V)$ is the copula of (X, Y). Therefore, the Kendall's τ is a function of the copula C.

Example 8.8 *The Farlie-Gumbel-Morgenstern family of copulas C are symmetric*

$$C_\theta(u,v) = uv + \theta uv(1-u)(1-v).$$

$$dC_\theta(u,v) = \frac{\partial^2 C_\theta(u,v)}{\partial u \partial v} dudv = [1 + \theta(1-2u)(1-2v)]dudv,$$

and

$$\int_0^1\int_0^1 C_\theta(u,v)dC_\theta(u,v) = \frac{1}{4} + \frac{\theta}{18}, \quad \tau_\theta = \frac{2\theta}{9}.$$

B. Spearman's ρ

Definition 8.11 *Let $(X_1, Y_1), (X_2, Y_2)$, and (X_3, Y_3) be random samples from $H(x, y)$. Then the measure*

$$\rho_{X,Y} = P((X_1 - X_2)(Y_1 - Y_3) > 0) - P((X_1 - X_2)(Y_1 - Y_3) < 0)$$

is called Spearman's rho.

Theorem 5.1.3 of Nelsen (2006) shows that

$$\rho_{X,Y} = \rho_C = 12\int_0^1\int_0^1 uvdC(u,v) - 3 = 12\int_0^1\int_0^1 C(u,v)dudv - 3.$$

In the literature, the strictly increasing transformation-invariant (SITI) property for Kendall's τ, Spearman's ρ has been proved (Chapter 5, Nelsen (2006)).

8.2.6 RPs of some meta-distributions

In the previous chapters, we introduced applications of the several RPs in statistical inference. As the meta-distributions are multivariate continuous distributions, the theory and algorithms in Chapters 6 and 7 can be applied to the meta-distributions. Two meta-distributions are studied as follows:

Example 8.9 *Let us study the copula in Example 8.5 discussed by Fang et al. (2000).*

$$C(u, v; \alpha, \beta) = \frac{uv}{[1-\beta(1-u^{1/\alpha})(1-v^{1/\alpha})]^\alpha}.$$

Consider the meta-distributions $C(u, v; 1, 1)$ *with* $\alpha = \beta = 1$ *with two sets of margins by the exponential distribution* $EP(1)$ *or the standard normal distribution* $N(0, 1)$*. The meta-distribution* ED *has the density function*

$$h(x_1, x_2; 1, 1) = \frac{2F_1(x_1)F_2(x_2)}{[1-(1-F_1(x_1))(1-F_2(x_2))]^3} f_{X_1}(x_1) f_{X_2}(x_2),$$

where F_1 *and* F_2 *are margins with pdfs* f_1 *and* f_2*, respectively. When* F_1 *and* F_2 *are* $EP(1)$*, the corresponding pdf of the meta-distribution is*

$$\frac{2(1-e^{-x_1})(1-e^{-x_2})}{[1-e^{-x_1}e^{-x_2}]^3} e^{-x_1} e^{-x_2},$$

and the covariance $Cov(X_1, X_2)$ *is*

$$\int_0^\infty \int_0^\infty x_1 x_2 \frac{2(1-e^{-x_1})(1-e^{-x_2})}{[1-e^{-x_1}e^{-x_2}]^3} e^{-x_1} e^{-x_2} dx_1 dx_2 - 1.$$

Similarly, when F_1 *and* F_2 *are* $N(0, 1)$*, the corresponding pdf of the meta-distribution is*

$$\frac{2\int_{-\infty}^{x_1} \frac{1}{\sqrt{2\pi}} e^{-\frac{t^2}{2}} dt \int_{-\infty}^{x_2} \frac{1}{\sqrt{2\pi}} e^{-\frac{t^2}{2}} dt}{\left[1-\left(1-\int_{-\infty}^{x_1} \frac{1}{\sqrt{2\pi}} e^{-\frac{t^2}{2}} dt\right)\left(1-\int_{-\infty}^{x_2} \frac{1}{\sqrt{2\pi}} e^{-\frac{t^2}{2}} dt\right)\right]^3} \frac{1}{\sqrt{2\pi}} e^{-\frac{x_1^2}{2}} \frac{1}{\sqrt{2\pi}} e^{-\frac{x_2^2}{2}}.$$

All the marginal variances are equal to one for the cases above, and the covariance matrices with margins $EP(1)$ *or* $N(0, 1)$ *can be obtained numerically. They are*

$$\boldsymbol{\Sigma}_1 = \begin{bmatrix} 1.0000 & 0.2899 \\ 0.2899 & 1.0000 \end{bmatrix}, \quad \boldsymbol{\Sigma}_2 = \begin{bmatrix} 1.0000 & 0.4983 \\ 0.4983 & 1.0000 \end{bmatrix},$$

respectively. We can see that the covariance $Cov(X_1, X_2)$ *depends on the margins. Table 8.2 shows the estimation bias of the mean vectors* $\|\hat{\boldsymbol{\mu}} - \boldsymbol{\mu}\|_2$ *with* L_2*-norm and estimation bias of the covariance matrices* $\|\hat{\boldsymbol{\Sigma}} - \boldsymbol{\Sigma}\|_F$ *with the Frobenius distance between the two matrices defined in (1.56). The several types of RPs for the meta-distribution with* $EP(1)$ *and* $N(0, 1)$ *margins, RPs-QMC' means RPS-QMC are generated by the generalized glp method (see Subsection 1.11.1). We observe that (1) MC-RPs are worst; (2) the* QMC'*-RPs are better than MSE-RPs; and (3) MSE-RPs and* QMC'*-RPs are equally best.*

TABLE 8.2
The estimation bias of the mean and covariance matrix of the meta distribution for different margin and parameter settings

		$\|\hat{\boldsymbol{\mu}} - \boldsymbol{\mu}\|_2$				$\|\hat{\boldsymbol{\Sigma}} - \boldsymbol{\Sigma}\|_F$			
Distributions	k	MC	QMC	QMC'	MSE	MC	QMC	QMC'	MSE
$x_1 \sim EP(1)$,	8	0.626	0.043	0.043	**0**	0.259	0.733	0.733	**0.022**
$x_2 \sim EP(1)$	13	0.408	0.030	0.039	**0**	0.167	0.673	**0.002**	0.003
	21	0.224	0.017	0.023	**0**	0.103	0.539	0.008	**0.003**
	377	0.124	0.001	0.001	**0**	0.063	0.111	**0.003**	0.009
	610	0.067	0.001	0.001	**0**	0.072	0.082	**0.002**	0.009
$x_1 \sim N(0,1)$,	8	0.477	0.003	0.003	**0**	0.846	0.372	0.372	**0.227**
$x_2 \sim N(0,1)$	13	0.002	0.010	0.002	**0**	0.513	0.360	**0.054**	0.149
	21	0.086	**0**	0.005	**0**	0.508	0.219	**0.095**	0.096
	377	0.084	**0**	**0**	**0**	0.106	0.027	**0.003**	0.007
	610	0.042	**0**	**0**	**0**	0.114	0.020	**0.001**	0.004

Let us study the mete-normal distributions introduced in Subsection 8.2.3.

Example 8.10 *Let's consider two meta-Gaussian distributions. One is based on* $N_3(\mathbf{0}, \boldsymbol{R}_3)$ *with the correlation matrix*

$$\boldsymbol{R}_3 = \begin{bmatrix} 1 & 0.2 & 0.9 \\ 0.2 & 1 & -0.1 \\ 0.9 & -0.1 & 1 \end{bmatrix},$$

and another is based on $N_4(\mathbf{0}, \boldsymbol{R}_4)$ with the correlation matrix

$$\boldsymbol{R}_4 = \begin{bmatrix} 1.00 & 0.73 & -0.04 & 0.56 \\ 0.73 & 1.00 & -0.02 & 0.47 \\ -0.04 & -0.02 & 1.00 & -0.44 \\ 0.56 & 0.47 & -0.44 & 1.00 \end{bmatrix}.$$

Consider the margins $\{N(0,1), N(0,3^2), EP(5)\}$ and $\{N(0,1), N(0,3^2), N(0,5^2)\}$ in the first meta-normal distribution; and $\{N(0,1), EP(3), t(5), N(0,7^2)\}$ and $\{t(3), t(5), t(7), t(9)\}$ for the second meta-normal distribution, where $t(\nu)$ is the t-distribution with ν degrees of freedom. Tables 8.3 and 8.4 give comparisons for estimating the mean vector and covariance matrix with different margins and correlation matrices. We can see that the MSE method yields the smallest error in mean vector estimation and outperforms other methods in covariance estimation in most instances. Overall, the MSE method demonstrates the highest estimation accuracy.

We can similarly study the meta-elliptical distributions; some references can refer to Fang et al. (2002).

8.3 MSE-RPs in Uniform Design of Experiments with Mixtures

Many products are formed by mixing several ingredients; for example, the building construction concrete consists of sand, water, and one or more types of cement. The manufacturer or the experimenter who is responsible for mixing the ingredients may be interested in the hardness or compression strength of the concrete, where the hardness is a function of the percentages of cement, sand, and water in the mix. Designs for deciding how to mix

TABLE 8.3

The estimation bias of the mean and covariance matrix of the trivariate meta Gaussian distribution for different margin and parameter settings

		$\|\hat{\boldsymbol{\mu}}-\boldsymbol{\mu}\|_2$				$\|\hat{\boldsymbol{R}}_3-\boldsymbol{R}_3\|_F$			
Distributions	N	MC	QMC	QMC'	MSE	MC	QMC	QMC'	MSE
$x_1 \sim N(0,1)$,	35	0.151	0.028	0.028	**0.0001**	0.550	0.154	0.154	**0.031**
$x_2 \sim N(0,3^2)$,	101	0.389	0.008	0.018	**0.0001**	0.497	0.126	0.139	**0.038**
$x_3 \sim EP(5)$,	135	0.192	0.043	0.015	**0.0001**	0.389	0.157	0.125	**0.040**
	266	0.128	0.040	0.019	**0.0001**	0.688	0.141	0.144	**0.044**
	597	0.108	0.005	0.007	**0.0001**	0.501	0.127	0.123	**0.047**
$x_1 \sim EP(1)$,	35	0.932	0.126	0.126	**0.0004**	0.400	0.149	0.149	**0.045**
$x_2 \sim N(0,3^2)$,	101	0.296	0.012	0.029	**0.0004**	0.589	0.132	0.126	**0.050**
$x_3 \sim N(0,5^2)$,	135	0.580	0.044	0.047	**0.0004**	0.613	0.151	0.126	**0.052**
	266	0.356	0.143	0.035	**0.0004**	0.563	0.123	0.153	**0.069**
	597	0.220	0.019	0.013	**0.0004**	0.601	0.124	0.126	**0.090**

TABLE 8.4

The estimation bias of the mean and covariance matrix of the four-dimensional meta-Gaussian distribution for different margin and parameter settings

		$\|\hat{\boldsymbol{\mu}}-\boldsymbol{\mu}\|_2$				$\|\hat{\boldsymbol{R}}_4-\boldsymbol{R}_4\|_F$			
Distributions	N	MC	QMC	QMC'	MSE	MC	QMC	QMC'	MSE
$x_1 \sim N(0,1)$,	307	0.251	0.154	0.245	**0.214**	0.607	0.156	0.171	**0.062**
$x_2 \sim EP(3)$,	562	0.320	0.216	0.219	**0.214**	0.521	0.144	0.142	**0.043**
$x_3 \sim t(5)$,	701	0.314	0.221	0.223	**0.214**	0.693	0.152	0.156	**0.055**
$x_4 \sim N(0,7^2)$	1019	0.120	0.175	0.202	**0.214**	0.593	0.189	0.163	**0.069**
$x_1 \sim t(3)$,	307	0.278	0.261	0.288	**0.283**	0.692	0.106	0.081	**0.031**
$x_2 \sim t(5)$,	562	0.273	0.278	0.284	**0.283**	0.763	0.079	0.053	**0.033**
$x_3 \sim t(7)$,	701	1.272	0.276	0.284	**0.283**	1.056	0.077	0.057	**0.034**
$x_4 \sim t(9)$	1019	0.201	0.276	0.283	**0.283**	0.718	0.146	0.071	**0.038**

the ingredients are called *design of experiments with mixtures (DEM)* that have played an important role in various fields, such as chemical engineering, rubber industry, material and pharmaceutical engineering. Cornell (2011) gave a comprehensive review on DEM. New developments in DBM can refer to Fang and Wang (1994) and Fang et al. (2018). It is clear that the domain of experiment with mixture is

$$T^p = \{\boldsymbol{x} : \boldsymbol{x} \in R^p_+, x_1 + \ldots + x_p = 1\}. \tag{8.20}$$

Section 1.12 introduces an algorithm for generating an NT-net on T^p or QMC-RPs of $U(T^p)$. Fang and Wang (1994) suggested to use QMC-RPs of $U(T^p)$ for experimental designs with mixture and call as *the uniform design of experiments with the mixture (UDEM)*. In this section, we also propose to use RPs of the uniform distribution on T^p under another criterion ML1D for UDEM that will be defined in Subsection 8.3.2. Let us review some traditional DEM.

8.3.1 Designs for experimental with mixtures

There are many existing designs for DEM.

1. **Simplex-lattice design**
   Scheffé (1958) proposed the simplex-lattice design and the corresponding canonical polynomial models, laying the foundation for mixture designs. To represent

the response surface on the entire simplex region, the natural choice for a design is to spread the design points evenly on the whole simplex factor space.

Suppose that the experiments with mixtures have p components. Let m be a positive integer and suppose that each component takes $(m+1)$ equally spaced places from 0 to 1, i.e.,

$$x_i = 0, 1/m, 2/m, \cdots, 1, \quad for\ i = 1, \cdots, p. \tag{8.21}$$

A p-ingredient simplex-lattice is denoted by $\{p, m\}$ which has $\binom{p+m-1}{m}$ design points, i.e., it consists of all possible combinations of the components under the constraint $x_1 + \cdots + x_p = 1$, and each component is an element from the set $\{0, 1/m, 2/m, \ldots, 1\}$. For example, when $p = 3$, the simplex-lattice $\{s, m\}$ is as follows:

$m = 1$:3 design points (1, 0, 0), (0, 1, 0), (0, 0, 1);

$m = 2$:6 design points (1, 0, 0), (0, 1, 0), (0, 0, 1), (1/2, 1/2, 0), (1/2, 0, 1/2), (0, 1/2, 1/2);

$m = 3$:10 design points (1, 0, 0), (0, 1, 0), (0, 0, 1), (1/3, 2/3, 0), (1/3, 0, 2/3), (0, 1/3, 2/3), (2/3, 1/3, 0), (2/3, 0, 1/3), (0, 2/3, 1/3) (1/3, 1/3, 1/3).

2. **Simplex-centroid design** In a s-component simplex-centroid design, the design points form Scheffé (1963) suggested the simplex-centroid design as an alternative to the $\{p, m\}$-simplex-lattice design. as follows:

   p points correspond to the permutations of $(1, 0, 0, \cdots, 0)$,

   $\binom{p}{2}$ points correspond to the permutations of $(1/2, 1/2, 0, \cdots, 0)$,

   $\binom{p}{3}$ points correspond to the permutations of $(1/3, 1/3, 1/3, 0, \cdots, 0)$, $\cdots$,

   the centroid point $(1/p, \cdots, 1/p)$.

   Then, the total number of design points is $2^p - 1$. For example, when $p = 3$, the design points are (1, 0, 0), (0, 1, 0), (0, 0, 1), (1/2, 1/2, 0), (1/2, 0, 1/2), (0, 1/2, 1/2), and (1/3, 1/3, 1/3). The centroid design has one more design point, i.e., the centroid point. It can be seen that most of the design points of simplex-centroid design are positioned on the boundaries (vertices, edges, faces, etc.) of the simplex factor space.

3. **Axial design**
   Cornell (1975) introduced a design known as an axial design. The line segment joining a vertex of the simplex T_p with its centroid $(1/p, \cdots, 1/p)$ is called an *axis*. The distance between the centroid point and each vertex is $\sqrt{(p-1)/p}$. Let d be a positive number such that $0 < d < \sqrt{(p-1)/p}$. The experimental points of the axial design are s points on the s axes such that each point to the centroid has the same distance d. Then, each point of the axial design is located in the inner region of the simplex factor space.

4. **Uniform mixture design**
   There are two major weaknesses in the designs above.

   (1) The majority of experimental points in the simplex-lattice designs and the simplex-centroid designs are located on the boundaries (vertices, edges, or faces) of the simplex-factor space. We typically require a complete mixture experiment with non-zero proportions for each component.

(2) Second, all three designs are constrained by a limited number of points. Figure 8.2 shows the first three kinds with three components. For a three-component mixture, the axial design yields 3 points, the simplex-centroid design points produce points, and the simplex-lattice design can only provide points based on the integer parameter m. Therefore, users cannot arbitrarily set the number of design points they need.

To overcome the weakness of boundary designs, Fang and Wang (1994) proposed a uniform design with a mixture. A uniform design of experiments with p-ingredient mixtures is a set of uniformly scattered points on the domain T^p. A systematic introduction to this direction can refer to Fang and Wang (1994) and Fang et al. (2018).

5. **Scheffé-type designs**
Fang and Wang (1994) also suggested the contracted designs as Scheffé-type designs. Table 8.5 presents 10 points of the $\{3, 3\}$-simplex-lattice design and its associated Scheffé-type design. The objective is to identify a suitable value for the parameter a such that the Scheffé-type design points are scattered uniformly on T^p without any boundary points. In the literature, some criteria under distance and uniformity are given.

TABLE 8.5
The $\{3, 3\}$-simplex-lattice design and its contracted design

	Simplex-lattice			Scheffé-type design ($a > 0$)		
No.	y_1	y_2	y_3	y_1'	y_2'	y_3'
1	1	0	0	$1 - 1/a$	$1/(2a)$	$1/(2a)$
2	0	1	0	$1/(2a)$	$1 - 1/a$	$1/(2a)$
3	0	0	1	$1/(2a)$	$1/(2a)$	$1 - 1/a$
4	2/3	1/3	0	$2/3 - 1/(2a)$	$1/3$	$1/(2a)$
5	1/3	2/3	0	$1/3$	$2/3 - 1/(2a)$	$1/(2a)$
6	2/3	0	1/3	$2/3 - 1/(2a)$	$1/(2a)$	$1/3$
7	1/3	0	2/3	$1/3$	$1/(2a)$	$2/3 - 1/(2a)$
8	0	2/3	1/3	$1/(2a)$	$2/3 - 1/(2a)$	$1/3$
9	0	1/3	2/3	$1/(2a)$	$1/3$	$2/3 - 1/(2a)$
10	1/3	1/3	1/3	$1/3$	$1/3$	$1/3$

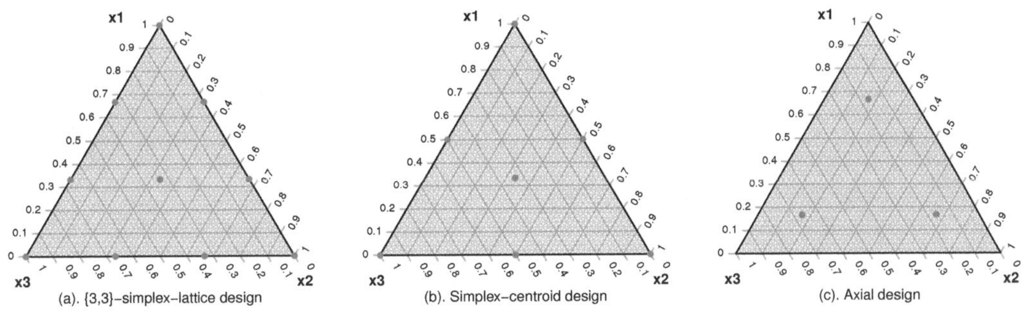

FIGURE 8.2
Some traditional mixture designs with three components ($p = 3$).

8.3.2 Representative points for mixture designs

Recently, Li et al. (2024b) considered uniform mixture design and Scheffé-type designs under the criterion the mean L_1-distance (ML1D) that is defined below.

Definition 8.12 *Let two points $\boldsymbol{x} = (x_1, \ldots, x_p)$ and $\boldsymbol{y} = (y_1, \ldots, y_p)$ in T^p. Their l_1-distance is defined as*

$$D_{l_1}(\boldsymbol{x}, \boldsymbol{y}) = \sum_{i=1}^{p} |x_i - y_i| .$$

Definition 8.13 *Let set of points $\mathcal{Y} = \{\boldsymbol{y}_1, \ldots, \boldsymbol{y}_n\} \subset T^p$ and random vector $\boldsymbol{Z} \sim U(T^p)$. Define a measure of uniformity of $\mathcal{Y}$ by*

$$ML1D(\mathcal{Y}) = E_{\boldsymbol{z}} \left[\min_{1 \le j \le k} D_{l_1}(\boldsymbol{x}, \boldsymbol{y}_j) \right],$$

where the expectation is under the distribution of $\boldsymbol{Z}$. The set with a minimum value of ML1D is called ML1D-RPs on the distribution $U(T^p)$ with size n.

It is a natural idea to consider a set of ML1D-RPs of the uniform distribution $U(T^p)$ as a design for experiments with mixtures. Note that n denotes the number of experiments in the literature. Therefore, this section uses n for the number of ML1D-RPs of $U(T^p)$.

It is known that a continuous univariate distribution has its median as its unique ML1D-RP ($n = 1$). Let $\boldsymbol{x} = (X_1, \ldots, X_p)$ represent a random vector in $\mathbb{R}^p$, and $\text{Med}(X_i)$ denote the median of $\{X_i\}$. Under the L_1-norm, the spatial median $\boldsymbol{m} = (\text{Med}(X_1), \ldots, \text{Med}(X_p))$ satisfies

$$E|\boldsymbol{x} - \boldsymbol{m}| = \min_{\boldsymbol{b} \in \mathbb{R}^p} E|\boldsymbol{x} - \boldsymbol{b}|.$$

Hence, the spatial median $\boldsymbol{m}$ is equivalent to the ML1D-RP with a size of $n = 1$ for the random variable $\boldsymbol{x}$ in $\mathbb{R}^p$.

Let $(Z_1, \ldots, Z_p) \sim U(T^p)$, From Lemma 6.1 its marginal distribution Z_i follows a beta distribution with pdf $f(z) = (p-1)(1-z)^{p-2}$ for $0 \le z < 1$. The spatial median of $\boldsymbol{z}$ is given by

$$\text{Med}(Z_i) = 2^{-\frac{1}{p-1}} \left(2^{\frac{1}{p-1}} - 1 \right), \quad \text{for } i = 1, \ldots, p.$$

When $p = 2$, the spatial median is (1/2, 1/2) that is ML1D-RP from $U(T^p)$ with $p = 2$ and $n = 1$. However, for $p > 2$, the spatial median of $\boldsymbol{Z} = (Z_1, \ldots, Z_p) \sim U(T^p)$ does not lie within T^p. For instance, we have $\text{Med}(Z_i) = 1 - 1/\sqrt{2}$ when $p = 3$. While the point $(1 - 1/\sqrt{2}, 1 - 1/\sqrt{2}, 1 - 1/\sqrt{2})$ minimizes $E|\boldsymbol{z} - \boldsymbol{b}|$ for all $\boldsymbol{b} \in \mathbb{R}^3$, its components do not sum up to one, which is not desired for mixture experiments. Li et al. (2024b) gave the following results. Its proof can be found in Subsection 8.7.1.

Theorem 8.7 *For any $p \ge 2$ and $n = 1$, the one-point ML1D-RP on T^p is*

$$\mathcal{Y}^* = \left\{ \left(\frac{1}{p}, \ldots, \frac{1}{p} \right) \right\},$$

and the corresponding

$$ML1D(\mathcal{Y}^*) = 2 \left(1 - \frac{1}{p} \right)^p.$$

Example 8.11 *When $p = 3$, we have $\mathcal{Y}^* = \{(1/3, 1/3, 1/3)\}$ and $ML1D(\mathcal{Y}^*) = 16/27$. When $p = 4$, we have $\mathcal{Y}^* = \{(1/4, 1/4, 1/4, 1/4)\}$ and $ML1D(\mathcal{Y}^*) = 81/128$.*

ML1D-RPs on T^2

The simplex T^2 is essentiaegment connecting the points $(0,1)$ and $(1,0)$. For uniform designs on the interval $[0,1]$, its RPs in QMC or MSE criteria are $\{\frac{1}{2n}, \frac{3}{2n}, \ldots, \frac{2n-1}{2n}\}$. It is easy to guess the following theorem, and its proof can be found in Subsection 8.7.2.

Theorem 8.8 *For $p=2$ and any $n \in \mathbb{N}^+$, the unique set of ML1D-RPs on T^2 of size n is*

$$\mathcal{Y}^* = \left\{ \left(\frac{1}{2n}, \frac{2n-1}{2n}\right), \left(\frac{3}{2n}, \frac{2n-3}{2n}\right), \ldots, \left(\frac{2n-1}{2n}, \frac{1}{2n}\right) \right\},$$

with the corresponding $ML1D(\mathcal{Y}^*) = 1/2n$.

8.3.3 Optimal Scheffé-type simplex-lattice design under the ML1D criterion

To adjust the boundary points of Scheffé's designs, Fang and Wang (1994) suggested contracting the experimental points toward the centroid of T^p while maintaining the pattern of the original design, and called Scheffé-type simplex-lattice design.

Table 8.5 provides an example of the Scheffé-type $\{3,3\}$-simplex-lattice design, where a is a contraction parameter that needs to be optimized. We choose the ML1D criterion to optimize the parameter a with a contraction ratio denoted by ρ for the Scheffé-type simplex-lattice design.

Let $\mathcal{D} \subset T^p$ be a simplex-lattice design on T^p and $\boldsymbol{y} = (y_1, \ldots, y_p)$ be a point of $\mathcal{D}$. Then a point $\boldsymbol{y}^* = (y_1^*, \ldots, y_p^*)$ in the Scheffé-type simplex-lattice design corresponding to $\boldsymbol{y} \in \mathcal{D}$ can be determined by the following expression:

$$\boldsymbol{y}^* - \frac{1}{p}\mathbf{1}_p = \rho\left(\boldsymbol{y} - \frac{1}{p}\mathbf{1}_p\right), \tag{8.22}$$

where $0 < \rho \leq 1$ is a parameter we refer to as the contraction ratio. By (8.22), we have

$$y_i^* = \rho y_i + \frac{1-\rho}{p}, i = 1, \ldots, p,$$

and denote the contracted design of $\mathcal{D}$ as $\mathcal{D}_\rho$. Optimizing ρ is equivalent to optimizing a in a Scheffé-type simplex-lattice design. For instance, the relationship $\rho = 1 - 3/(2a)$ is valid for the Scheffé-type $\{3,3\}$-simplex-lattice design.

Different from the Scheffé-type simplex-lattice designs, the uniform mixture designs under the ML1D criterion (called ML1D-RPs) are developed by assessing the uniformity of experimental points. We discovered exciting connections between the ML1D-RPs and the Scheffé-type simplex-lattice designs, even though these two types of designs are derived from different considerations.

The ML1D-RPs on T^2 derived in Theorem 8.8 for two-component mixture designs are considered as special cases of the Scheffé-type simplex-lattice designs. This is evident from the fact that a $\{2,m\}$-simplex-lattice design consists of $\binom{m+1}{m} = m+1$ design points, which can be expressed as $\mathcal{D} = \{(0,1), (1/m, 1-1/m), \ldots, (1,0)\}$. Theorem 8.8 shows that the unique set of ML1D-RPs on T^2 is

$$\mathcal{Y}^* = \left\{ \left(\frac{1}{2m+2}, \frac{2m+1}{2m+2}\right), \left(\frac{3}{2m+2}, \frac{2m-1}{2m+2}\right), \ldots, \left(\frac{2m+1}{2m+2}, \frac{1}{2m+2}\right) \right\}.$$

In this case, the set of ML1D-RPs $\mathcal{Y}^*$ is the contracted design $\mathcal{D}_\rho$ with the contraction ratio $\rho = m/(m+1)$. We summarize this observation in the following theorem.

Theorem 8.9 *For $p = 2$ and any $n \geq 2$, the set of ML1D-RPs on T^2 of size n is a Scheffé-type $\{2, n-1\}$-simplex-lattice design with the contraction ratio $\rho^* = (n-1)/n$, where ρ^* is optimized under ML1D criterion.*

For $p > 2$ Li et al. (2024b), the following result and its proof can refer to Subsection 8.7.3.

Theorem 8.10 *Let $p = 3$ and $m \geq 1$. The optimal Scheffé-type $\{3, m\}$-simplex-lattice design under the ML1D criterion is $\mathcal{D}_{\rho^*}$ where*

$$\rho^* = \frac{m^2 - \sqrt{21m^2 + 102m + 81} + 10m + 9}{m^2 + 7m + 6}.$$

The corresponding ML1D is

$$ML1D(\mathcal{D}_{\rho^*}) = \frac{7m^3 + 145m^2 - (14m + 54)\sqrt{21m^2 + 102m + 81} + 624m + 486}{9m(m+1)(m+6)^2}.$$

Example 8.12 *Consider $p = 3$ and $m = 1, 2, 3, 4$. The following optimal Scheffé-type $\{3, m\}$-simplex-lattice designs are obtained by Theorem 8.10.*

For $m = 1$, we have $\rho^* = \left(10 - \sqrt{51}\right)/7$. The three design points are

$$\begin{aligned}\mathcal{D}_{\rho^*} \quad = \quad & \left\{ \left(\frac{27 - 2\sqrt{51}}{21}, \frac{\sqrt{51} - 3}{21}, \frac{\sqrt{51} - 3}{21} \right), \left(\frac{\sqrt{51} - 3}{21}, \frac{27 - 2\sqrt{51}}{21}, \frac{\sqrt{51} - 3}{21} \right), \right. \\ & \left. \left(\frac{\sqrt{51} - 3}{21}, \frac{\sqrt{51} - 3}{21}, \frac{27 - 2\sqrt{51}}{21} \right) \right\}\end{aligned}$$

with the corresponding $ML1D(\mathcal{D}_{\rho^*}) = \left(631 - 68\sqrt{51}\right)/441 \approx 0.3296664$.

For $m = 2$, we have $\rho^* = \left(11 - \sqrt{41}\right)/8$. The six design points are

$$\begin{aligned}\mathcal{D}_{\rho^*} \quad = \quad & \left\{ \left(\frac{5}{4} - \frac{\sqrt{41}}{12}, \frac{\sqrt{41} - 3}{24}, \frac{\sqrt{41} - 3}{24} \right), \left(\frac{27 - \sqrt{41}}{48}, \frac{27 - \sqrt{41}}{48}, \frac{\sqrt{41} - 3}{24} \right), \right. \\ & \left(\frac{\sqrt{41} - 3}{24}, \frac{5}{4} - \frac{\sqrt{41}}{12}, \frac{\sqrt{41} - 3}{24} \right), \left(\frac{\sqrt{41} - 3}{24}, \frac{27 - \sqrt{41}}{48}, \frac{27 - \sqrt{41}}{48} \right), \\ & \left. \left(\frac{\sqrt{41} - 3}{24}, \frac{\sqrt{41} - 3}{24}, \frac{5}{4} - \frac{\sqrt{41}}{12} \right), \left(\frac{27 - \sqrt{41}}{48}, \frac{\sqrt{41} - 3}{24}, \frac{27 - \sqrt{41}}{48} \right) \right\}\end{aligned}$$

with the corresponding $ML1D(\mathcal{D}_{\rho^*}) = \left(395 - 41\sqrt{41}\right)/576 \approx 0.2299859$. Please refer to Table 8.6 and Figure 8.3. For the cases of $m = 3, 4$, refer to Li et al. (2024b).

8.4 Representative Points in Goodness-of-Fit Test

The goodness-of-fit test tells you if your sample data represents the distribution you expect to find in the population. More specifically, it is used to test if sample data fits a distribution from a specific population F_0, where F_0 is a univariate distribution. (for example, a

TABLE 8.6
Comparisons of the optimal Scheffé-type $\{p, m\}$-simplex-lattice designs ($\mathcal{D}_{\rho^*}$) and the ML1D-RPs ($\mathcal{Y}^*$) on ML1D values, where $p = 3$, $n = \binom{p+m-1}{m}$, and ρ^* is the optimal contraction ratio under the ML1D criterion

m	n	ρ^*	$ML1D(\mathcal{D}_{\rho^*})$	$ML1D(\mathcal{Y}^*)$
1	3	0.4083674	0.3297	0.3295
2	6	0.5746095	0.2230	0.2230
3	10	0.6666667	0.1770	0.1769
4	15	0.7255437	0.1439	0.1438
5	21	0.7665669	0.1213	0.1222
6	28	0.7968365	0.1049	0.1049
7	36	0.8201124	0.0924	0.0929
8	45	0.8385778	0.0825	0.0829
9	55	0.8535898	0.0746	0.0750
10	66	0.8660376	0.0681	0.0681

normal distribution or a Weibull distribution) oo multivariate distribution (for example, a multinormal or an elliptical distribution). The goodness of fit test is to test the hypothesis:

$$H_0 : F(x) = F_0(x), \quad \text{against } F(x) \neq F_0(x),$$

where F_0 is the population distribution conjectured by the user. This section considers the goodness of fit test for a univariate distribution. There are many goodness-of-fit tests commonly used in statistics:

1. **Chi-square test:** Suppose the domain of the sorted data $x_1 \leq x_2 \leq \ldots \leq x_n$ can be sprit into k groups: $(-\infty, a_1], (a_1, a_2], \ldots, (a_{k-1}, \infty)$. Denote the number of observations in each interval by $n_1, \ldots, n_k$ satisfying $n_i \geq 5, i = 1, \ldots, k$. Let random variable $X \sim F_0(x)$ and

$$p_i = P(X \in I_i), \text{ where } I_i = (a_{i-1}, a_i], i = 1, \ldots, k. \tag{8.23}$$

The test statistic is calculated by

$$T_{\chi^2} = \sum_{i=1}^{k} \frac{(n_i - np_i)^2}{np_i}. \tag{8.24}$$

and call the Pearson-Fisher chi-square test (PF chi-square test for short). It is known that the distribution of T_{χ^2} converges to $\chi^2(k-1)$, the χ^2-distribution with degrees of freedom $k - 1$. If there are m unknown parameters in $F_0(x)$, the probabilities p_i are estimated, denoted by $\hat{p}_i$, and the distribution of T_{χ^2} converges to $\chi^2(k - m - 1)$. The chi-square test is robust and can be applied to all the distributions: univariate or multivariate, discrete or continuous. However, there is a potential to increase the power of the test in applications, especially since the critical values for the chi-square test are obtained from the limiting distribution (the chi-square distribution), which may have a big bias when the n is not large enough.

2. **Kolmogorov-Smirnov test:** Let $F_n(x)$ be the empirical distribution function. The Kolmogorov-Smirnov (KS) statistic is defined by

$$T_{ks} = \sup_{-\infty < x < \infty} |F_n(x) - F_0(x)|. \tag{8.25}$$

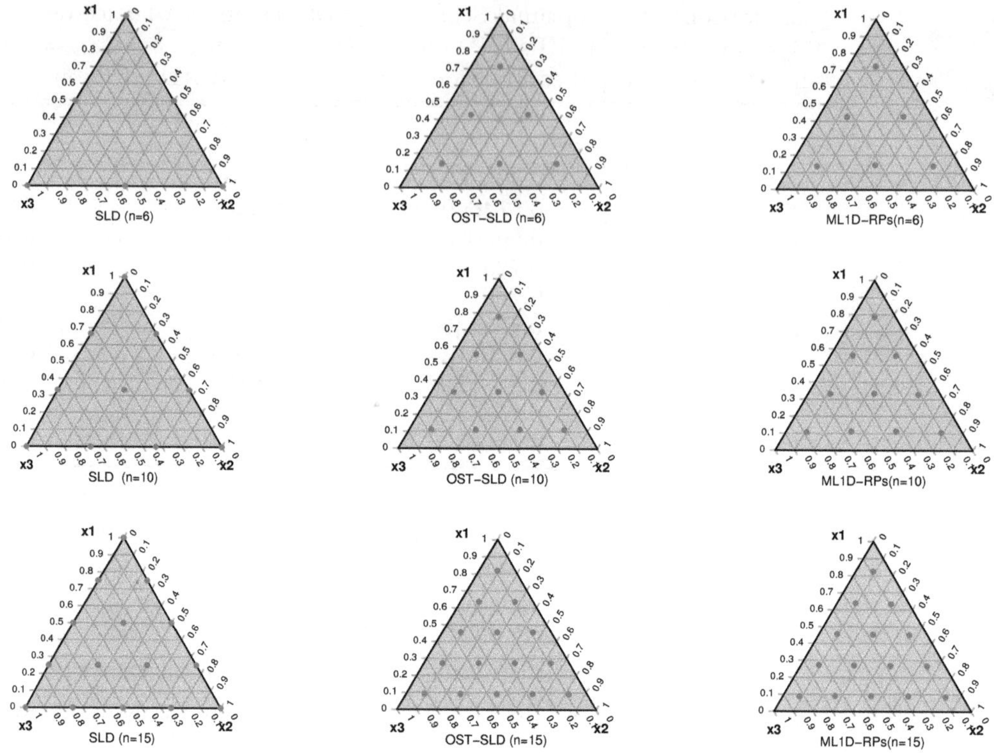

FIGURE 8.3
Comparisons of the simplex-lattice designs (SLD), the Scheffé-type simplex-lattice designs (OST-SLD) and the uniform mixture designs under ML1D criterion (ML1D-RPs), where $p = 3$ and $n = 6, 10, 15$).

The KS test is distribution-free, and T_{ks} converges to 0 almost surely as the sample size n approaches infinity according to the Glivenko-Cantelli theorem. The KS test was suggested by Kolmogorov (1933) and Smirnov (1949) obtained the limiting distribution of $Q_n = \sqrt{n}T_{ks}$ as $n \to \infty$. That is

$$Q_n(y) = \begin{cases} \sum_{k=-\infty}^{\infty}(-1)^k e^{-2k^2y^2}, & y > 0, \\ 0, & y \leq o, \end{cases}$$

which is independent of F_0.

3. **Anderson-Darling test (A^2 and W^2 statistics):** When F_0 is a continuous distribution, consider

$$Z = n\int_{-\infty}^{\infty} |F_n(x) - F_0(x)|^2 w(x) f(x)dx, \tag{8.26}$$

where $f(x)$ is the pdf of F_0 and $w(x)$ is a weighted function. Taking $w(x) = 1$, we have W^2-statistic

$$W^2 = n\int_{-\infty}^{\infty} |F_n(x) - F_0(x)|^2 f(x)dx.$$

When taking $w(x) = [F_0(x)(1 - F(x)]^{-1}$, the A^2 statistic is given by

$$A^2 = n \int_{-\infty}^{\infty} |F_n(x) - F_0(x)|^2 [F_0(x)(1 - F(x)]^{-1} f(x) dx.$$

4. **More tests:** Zhang et al. (2024b) consider more tests, like the Cramér-von Mises (CM), Anderson-Darling (AD), and Watson tests (WA).

8.4.1 A new PF chi-square test based on representative points

The classical PF chi-square test is a general approach to testing goodness-of-fit for univariate data. There is considerable discussion in the literature on effectively applying this test to practical goodness-of-fit problems. However, choosing optimal grouping intervals in constructing the chi-square statistic remains arguable and uncertain. Recently, Li et al. (2024a) suggested to use QMC-RPs or MSE-RPs to form I_i in (8.23) and called the RP-aided chi-square goodness-of-fit test. Three popular location-scale distributions (refer to Subsection 1.1.2), including the normal, logistic, and Laplace distributions, are employed as the null hypothesis of the goodness-of-fit tests. The power of RP approach was compared with the traditional PF chi-square test for three types of alternative distributions: symmetric fat-tailed distributions (student's t-distribution with degrees of freedom 5, the standard Laplace distribution, and standard Cauchy distribution), symmetric thin-tailed distributions (Kotz-type distribution and β-generalized normal distribution), and asymmetric distributions (shifted chi-square, shifted log-normal, and shifted Weibull distributions). They showed that the chi-square test outperforms the traditional construction of the same test by using equiprobable points for the grouping intervals in the sense of type I error control and power against some general alternative distributions.

8.4.2 A new Kolmogorov-Smirnov-type test based on representative points

It is known that the KS test is better than the chi-square test in most cases. Still, there are two weaknesses:

(1) The performance of the test is highly depend on the quality of the random sample, i.e., the empirical distribution $F_n(x)$;

(2) The limiting distribution determines the critical values, which results in less accuracy. Birnbaum (1952) gave a numerical tabulation of the distribution of Kolmogorov's statistic for finite sample size.

Recently, Wang et al. (2024) and Zhang et al. (2024b) proposed a new Kolmogorov-Smirnov test based on representative points for the Weibull distribution and the exponential family of distributions, respectively. The new KL test considers the following:

(1) using a QMC-data in Subsection 1.7.4 to reduce the randomness of the data; and

(2) random sampling was used to find approximate critical values for the test.

They showed that the new KS test performs better. The new KS test is introduced as follows:

Let $\mathcal{Y}_{MSE}^{(k)} = \{\xi_1^{(k)}, \xi_2^{(k)}, \ldots, \xi_k^{(k)}\}$ be the ordered MSE-RPs of F_0 and $F_{mse}^{(k)}$ be the empirical distribution of $\mathcal{Y}_{MSE}^{(k)}$, i.e.,

$$F_{mse}^{(k)}(x) = \begin{cases} 0 & x < \xi_1^{(k)} \\ p_i & \xi_i^{(k)} \leq x < \xi_{i+1}^{(k)} \quad i = 1, \ldots, k-1 \\ 1 & x \geq \xi_k^{(k)}. \end{cases}$$

where the respective probabilities are

$$p_1 = \int_{-\infty}^{\left(\xi_1^{(k)}+\xi_2^{(k)}\right)/2} f(x)dx,$$

$$p_i = \int_{\left(\xi_{i-1}^{(k)}+\xi_i^{(k)}\right)/2}^{\left(\xi_i^{(k)}+\xi_{i+1}^{(k)}\right)/2} f(x)dx \quad i = 1, \ldots, k-1,$$

$$p_k = \int_{\left(\xi_{k-1}^{(k)}+\xi_k^{(k)}\right)/2}^{\infty} f(x)dx.$$

In this sense, a new KS-type statistic is defined by

$$T_{ks-rp} = \sup_{-\infty<x<\infty} \left| F_n(x) - F_{mse}^{(k)}(x) \right|,$$

where T_{ks-rp} is the first-order distance between two empirical distribution functions. This testing statistic is derived by comparing functions: one from the observed data and another from selected representative points.

According to the Glivenko-Cantelli theorem, T_{ks} converges to 0 almost surely as the sample size n approaches infinity. Furthermore, the following theorem guarantees the asymptotic convergence of T_{ks-rp}.

Theorem 8.11 *T_{ks-rp} converges to 0 as $n \to \infty$ and $k \to \infty$.*

Proof . *According to the triangle inequality,*

$$\begin{aligned}\left| F_n(x) - F_{mse}^{(k)}(x) \right| &= \left| F_n(x) - F_{mse}^{(k)}(x) + F_0(x) - F_0(x) \right| \\ &\le |F_n(x) - F_0(x)| + \left| F_0(x) - F_{mse}^{(k)}(x)(x) \right|.\end{aligned}$$

From Corollary 3.1 We have

$$\begin{aligned}T_{ks-rp} &= \sup_{-\infty<x<\infty} \left| F_n(x) - F_{mse}^{(k)}(x) \right| \\ &\le \sup_{-\infty<x<\infty} |F_n(x) - F_0(x)| + \sup_{-\infty<x<\infty} \left| F_0(x) - F_{mse}^{(k)}(x) \right|.\end{aligned}$$

The theorem is proved.

8.4.3 A new Kolmogorov-Smirnov test for the exponential family

The exponential family of distributions is a broad class of probability distributions that includes many commonly used distributions such as the normal, exponential, Poisson, gamma, and Weibull distributions. They are characterized by a probability density function (pdf) or probability mass function (pmf) of the form as follows.

$$f(x|\theta) = h(x)\exp\{(\eta(\theta)T(x) - A(\theta))\},$$

where θ is the parameter of the distribution, $T(x)$ is the sufficient statistic, $\eta(\theta)$ is the natural parameter, $A(\theta)$ is the log partition function, and $h(x)$ is the base measure (Lehmann and Casella, 1998). The goodness of fit test for the exponential family of distributions is a pivotal facet of statistical inference. Typically, the goal of such testing is to evaluate the extent to which the data distribution aligns with the assumed distribution.

To select the optimal number of MSE-RPs, we use the loss function $L(\mathcal{Y}_{MSE}^{(k)})$ defined in (2.15)

$$L\left(\mathcal{Y}_{\mathrm{MSE}}^{(k)}\right) = \frac{\mathrm{MSE}\left(\mathcal{Y}_{\mathrm{MSE}}^{(k)}\right)}{\mathrm{Var}\,(X)}.$$

It shows how much percentage of Var(X) lost by use $Y_{MSE}^{(k)}$ replacing X and $0 \leq L\left(\mathcal{Y}_{\mathrm{MSE}}^{(k)}\right) \leq 1$. Zhang et al. (2024b) gave a way to find critical values based on T_{ks-rp}. Given the need to recalibrate critical value tables when estimating distribution parameters from samples, the Monte Carlo study in this section is designed to evaluate the effectiveness of the new KS-type test. This is achieved by simulating empirical Type I error rates and power against various sets of alternative distributions. Different parameter combinations were scrutinized via Monte Carlo calculations to ascertain the critical values for the proposed T_{ks-rp} statistic over a range of sample sizes n. Hundreds of thousands of bootstrap resamplings were randomly drawn from n observations. The procedure is as outlined below.

Algorithm 8.1 (Calculation of the critical value of T_{ks-rp})

Step 1 Take a sample of size n from the population $F(x; \boldsymbol{\theta})$ with some unknown parameters $\boldsymbol{\theta}$ and perform one thousand bootstrap resamplings from the sample.

Step 2 Estimate the unknown parameters $\boldsymbol{\theta}$ in each bootstrap resampling using the HD standardized samples or biased-correction methods by Li and Fang (2022) (refer to Subsection 5.6.3).

Step 3 Calculate T_{ks-rp} with a number of representative points k in each bootstrap and order the resulting values in ascending order.

Step 4 Determine the critical value for the α significance level, which corresponds to the $(1-\alpha)\%$ percentile of the collection of the resulting values in ascending order.

Step 5 Repeat steps 1-4 with the sample from the distribution with different combinations of parameters; we find that critical values demonstrated consistency across all combinations, thereby indicating distribution-free.

Table 8.7 lists the critical values of T_{ks-rp} for $k = 15$ and different n, which are based on the gamma distributions. To guarantee the reliability, we replicated the experiment 100 times and calculated the average of the outcomes, which are documented in Table 8.7. For more results based on other distributions, please refer to Wang et al. (2024) and Zhang et al. (2024b).

TABLE 8.7
Critical Values of T_{ks-rp} (based on the gamma distributions)

n	5	6	7	8	9	10	15	20	25
$\alpha = 0.05$	0.264	0.241	0.229	0.224	0.216	0.214	0.188	0.177	0.169
$\alpha = 0.01$	0.294	0.274	0.266	0.255	0.250	0.241	0.220	0.204	0.195
n	30	35	40	45	50	60	80	100	150
$\alpha = 0.05$	0.163	0.158	0.155	0.152	0.149	0.144	0.138	0.134	0.126
$\alpha = 0.01$	0.187	0.181	0.176	0.171	0.168	0.163	0.154	0.148	0.138
n	200	n							
$\alpha = 0.05$	0.121	$f_1(n)^i$							
$\alpha = 0.01$	0.131	$f_2(n)^{ii}$							

(i) $f_1(n) = \frac{0.910}{n+2.448} + 0.137e^{-0.001n} + 0.001$
(ii) $f_2(n) = \frac{1.387}{n+4.889} + 0.150e^{-0.001n} + 0.001$

8.4.4 Simulation studies

Table 8.8 shows empirical type I error rates of the new KS test and other tailor-made tests ($\alpha = 0.05$) under the gamma distributions with different parameters. Other tests include the Kolmogorov-Smirnov (KS) test, the Cramér-von Mises (CM) test, the Anderson-Darling (AD) test, and the Watson (WA) test. The simulation results, which include thousands of replications for each case, are summarized in Table 8.8. Empirical Type I error rates are calculated based on the T_{ks-rp} count that equals or exceeds the critical values at the specified significance level ($\alpha = 0.05$). The results suggest that all statistics effectively control their empirical Type I error rates at the 0.05 significance level. In most instances, tailor-made tests are more conservative than T_{ks-rp}, especially for smaller values of n. More studies can refer to Wang et al. (2024) and Zhang et al. (2024b).

8.4.5 Empirical power of the new KS test ($\alpha = 0.05$)

To examine the properties of the new test concerning power, we selected a variety of alternative distributions: the Gompertz, the Exponential, F, Lognormal, Beta, Pareto, Gamma, chi-square, and Weibull distributions. We drew thousands of replicates from each of these alternative distributions with varying sample sizes. The critical values were derived from the preceding tables. The findings are presented as follows.

The effectiveness of the tests is gauged by the frequency with which the null hypothesis is rejected. As illustrated in Figure 8.5. In more Figures (Wang et al. (2024) and Zhang et al. (2024b)), the proposed test exhibits markedly greater power compared to the other tests. For example, the power of the proposed test is significantly elevated when the alternatives are certain distributions. It's also worth noting that the power of the tests can vary significantly from one alternative to another, primarily due to the considerable variation in the shape of the alternative distributions. Therefore, it's not feasible to anticipate that the test statistics will consistently yield high power for all possible alternatives. In conclusion, the proposed test offers substantial benefits over traditional methods, particularly for small samples, while maintaining competitive performance in large-sample scenarios. For some distributions, the T_{ks-rp} test even showcases an all-encompassing superiority over traditional methods.

TABLE 8.8
Empirical type I error rates of the new KS test and other tailor-made tests ($\alpha = 0.05$) (gamma distributions)

Gamma(a, b)	Test	$n = 20$	$n = 50$	$n = 100$	$n = 150$	$n = 200$
Gamma$(2, 1)$	$T_{\text{ks}\ -rp}$	0.055	0.060	0.061	0.055	0.054
	T_{ks}	0.056	0.052	0.054	0.055	0.048
	T_{ad}	0.048	0.051	0.042	0.053	0.044
	T_{cm}	0.046	0.046	0.040	0.046	0.044
	T_{wa}	0.047	0.049	0.044	0.047	0.042
Gamma$(3, 1)$	T_{ks-rp}	0.042	0.056	0.050	0.051	0.050
	T_{ks}	0.052	0.051	0.049	0.048	0.052
	T_{ad}	0.050	0.050	0.056	0.041	0.053
	T_{cm}	0.049	0.050	0.054	0.037	0.050
	T_{wa}	0.051	0.052	0.055	0.039	0.052
Gamma$(3, 3)$	T_{ks-rp}	0.049	0.058	0.061	0.048	0.040
	T_{hme1}	0.051	0.047	0.058	0.053	0.053
	T_{ks}	0.058	0.053	0.052	0.050	0.045
	T_{ad}	0.050	0.047	0.058	0.046	0.053
	T_{cm}	0.053	0.050	0.054	0.048	0.055
	T_{wa}	0.052	0.052	0.054	0.050	0.052
Gamma$(5, 1)$	T_{ks-rp}	0.049	0.049	0.035	0.042	0.042
	T_{ks}	0.049	0.045	0.036	0.050	0.057
	T_{ad}	0.045	0.046	0.045	0.052	0.051
	T_{cm}	0.043	0.049	0.043	0.053	0.054
	T_{wa}	0.045	0.045	0.042	0.053	0.051

8.4.6 Simulation studies

Table 8.8 shows empirical type I error rates of the new KS test and other tailor-made tests ($\alpha = 0.05$) under the gamma distributions with different parameters. Other tests include the Kolmogorov-Smirnov (KS) test, the Cramér-von Mises (CM) test, the Anderson-Darling (AD) test, and the Watson (WA) test. The simulation results, which include thousands of replications for each case, are summarized in Table 8.8. Empirical Type I error rates are calculated based on the T_{ks-rp} count that equals or exceeds the critical values at the specified significance level ($\alpha = 0.05$). The results suggest that all statistics effectively control their empirical Type I error rates at the 0.05 significance level. In most instances, tailor-made tests are more conservative than T_{ks-rp}, especially for smaller values of n. More studies can refer to Wang et al. (2024) and Zhang et al. (2024b).

8.4.7 Real data analysis

This subsection demonstrates the effectiveness of the T_{ks-rp} test using two real-world datasets. The first dataset, presented in Table 8.9 is a frequently used engineering dataset, initially introduced by Proschan (1963) and later used by Keating et al. (1990) and Pal et al. (2005). This dataset contains 29 air conditioning (A/C) equipment failures in Boeing 720 aircraft. Bhaumik et al. (2009) applied several well-known GOF tests (including Kolmogorov-Smirnov, Anderson-Darling, and Cramér-von Mises) and achieved p-values > 0.5. These tests failed to reject the null hypothesis, suggesting that the data are consistent with the gamma distribution.

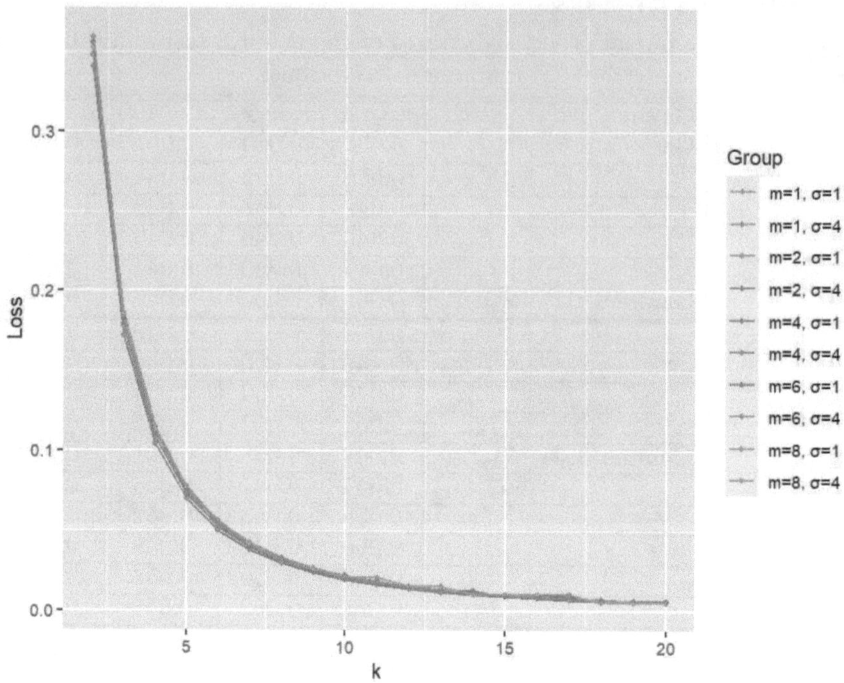

FIGURE 8.4
The power curves are compared across four different types of the Gamma distributions, with sample sizes ranging from 0 to 200.

TABLE 8.9
Failure times of air condition equipment dataset

90	10	60	186	61	49	208	44	59	62
24	56	20	79	84	44	14	29	101	23
118	25	156	310	76	26	70	208	130	

The second dataset, presented in Table 8.10, comprises data on remission times (measured in months) from 128 bladder cancer patients, as reported by Lee and Wang (2003). Ozonur and Paul (2020) analyzed this data using the likelihood ratio test and the score test. Both tests consistently concluded that the two-parameter gamma distribution does not adequately fit the remission times data in this dataset.

We can make several observations from the results presented in Table 8.11. For the first dataset, our findings align with those in the referenced studies. However, for the second dataset, the representative points-based statistic T_{ks-rp} yields p-values that challenge the assumption of a satisfactory fit for the gamma distribution at a significance level of 0.01. This suggests that the criteria to reject the null hypothesis are strict, enhancing the results' statistical significance.

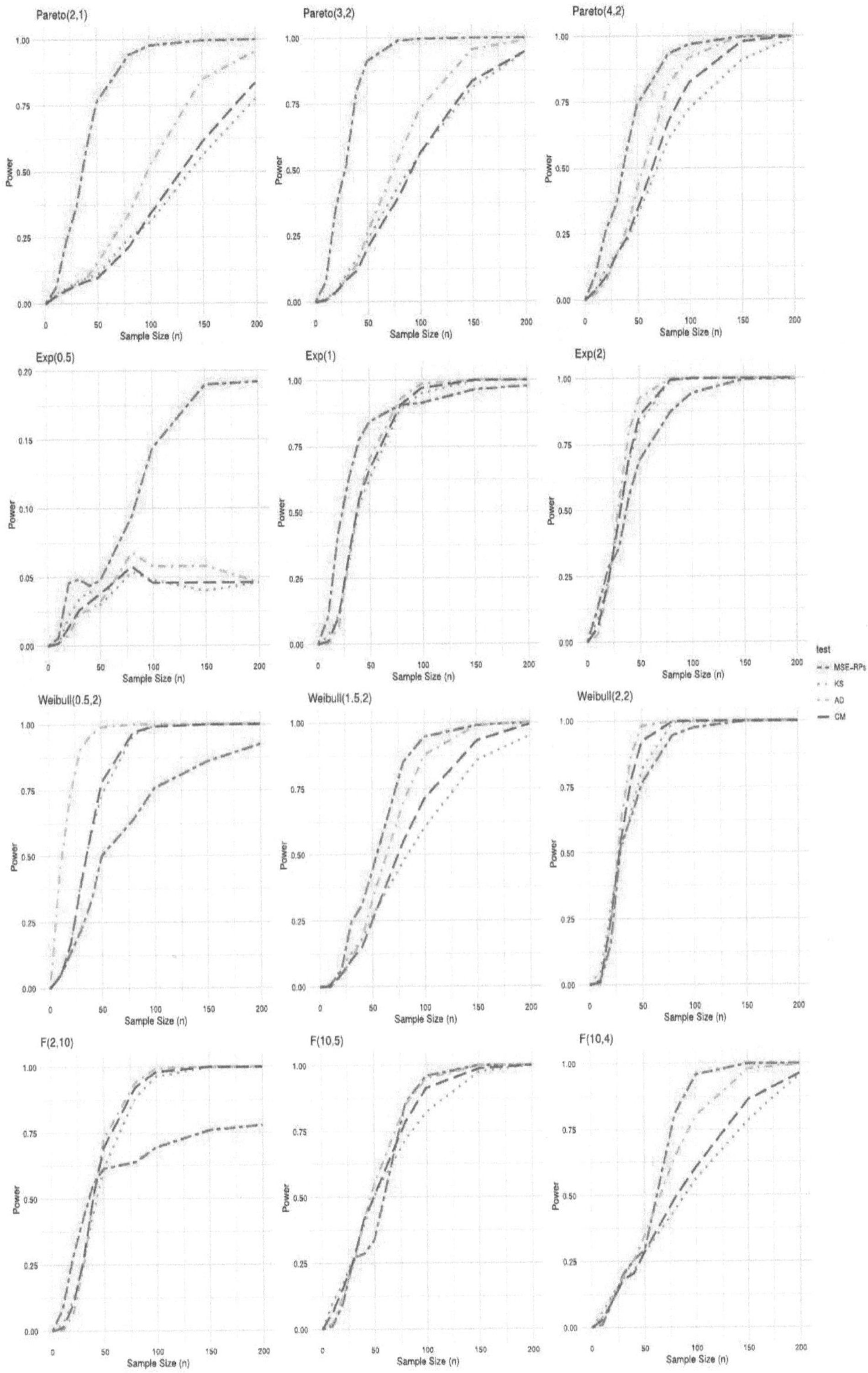

FIGURE 8.5
Statistical power is computed for sample sizes of 20 and 50 (Gamma distributions). Box plots are then generated based on the results of repeated calculations across 100 seed values.

TABLE 8.10
Bladder cancer data

0.08	6.97	2.46	9.74	3.88	15.96	4.26	79.05	11.79	8.37
12.07	2.09	9.02	3.64	14.76	5.32	36.66	5.41	1.35	18.1
12.02	21.73	3.48	13.29	5.09	6.31	7.39	1.05	7.63	2.87
1.46	2.02	2.07	4.87	0.4	7.26	0.81	10.34	2.69	17.12
5.62	4.4	3.31	3.36	6.94	2.26	9.47	2.62	14.83	4.23
46.12	7.87	5.85	4.51	6.93	8.66	3.57	14.24	3.82	34.26
5.41	1.26	11.64	8.26	6.54	8.65	13.11	5.06	25.82	5.32
0.9	7.62	2.83	17.36	11.98	8.53	12.63	23.63	7.09	0.51
7.32	2.69	10.75	4.33	1.4	19.13	12.03	22.69	0.2	9.22
2.54	10.06	4.18	16.62	5.49	3.02	1.76	20.28	2.23	13.8
3.7	14.77	5.34	43.01	7.66	4.34	3.25	2.02	3.52	25.74
5.17	32.15	7.59	1.19	11.25	5.71	4.5	3.36	4.98	0.5
7.28	2.64	10.66	2.75	17.14	7.93	6.25	6.76		

TABLE 8.11
p-values comparison for real-life studies (gamma distributions)

Dataset 1	Test	*p*-values	Dataset 2	Test	*p*-values
	$T_{\text{ks }-rp}$	> 0.05		$T_{\text{ks-rp}}$	$< \mathbf{0.01}$
	T_{ks}	> 0.05		T_{ks}	> 0.05
	T_{ad}	> 0.05		T_{ad}	< 0.05
	T_{cm}	> 0.05		T_{cm}	< 0.05
	T_{wa}	> 0.05		T_{wa}	> 0.05

8.5 Testing Multivariate Normality Based on *F*-RPs

The multivariate normal distribution, $N_d(\boldsymbol{\mu}, \boldsymbol{\Sigma})$, is a prevalent assumption in numerous statistical models and methodologies for high-dimensional data analysis. The development of methods for testing multivariate normality is an ongoing process. Given the characteristics of the multivariate normal distribution, most testing approaches exhibit varying degrees of advantage in their power performance. These approaches can be categorized into two types: multivariate and univariate. By utilizing the multivariate normal characteristic through the Mahalanobis distance (M-distance), we propose a method for testing multivariate normality. This method is based on RPs of the simple univariate F-distribution and the traditional chi-square statistic. This approach introduces a novel way to enhance the conventional chi-square test for goodness-of-fit. A Monte Carlo study revealed a substantial power improvement in the representative-point-based chi-square test compared to the traditional one. Additionally, an illustration of testing goodness-of-fit for three well-known datasets yields results that align with those obtained from classical methods.

8.5.1 Historical research

The efficacy of numerous statistical methodologies hinges on the assumption of multivariate normality (MVN) in high-dimensional data analysis. A plethora of approaches for testing MVN has been documented in the literature ((Mardia, 1970); Koziol (1982); Mudholkar et al. (1992); Liang and Bentler (1999); Liang et al. (2000); Thulin (2014); Tenreiro (2017); Kim and Park (2018); and Enomoto et al. (2020). As noted in several review papers (Andrews et al. (1973); Gnanadesikan (1977); and Looney (1995)), it is challenging for a single method to outperform all others comprehensively. To evaluate the power performance of some commonly-cited methods for testing MVN, Monte Carlo studies have been executed to compare these MVN tests (Mecklin and Mundfrom (2000); Mecklin and Mundfrom (2005)). Generally, when developing a new test, it is compared against a specific category of tests in the literature, considering a variety of alternative distributions. It would be impractical to test against every possible deviation from normality (Mecklin and Mundfrom (2005)). For instance, Ward (1988) compared the power of Mardia's skewness and kurtosis tests, the Malkovich-Afifi test, Hawkins' test, the Mardia-Foster test, and Ward's extension of the Kolmogorov-Smirnov and Anderson-Darling tests. None of these tests demonstrated effective performance against the multivariate t distribution, representing a mild deviation from normality. Ahn (1992) proposed a squared jackknife distance and advocated using an F-distribution probability plot to verify multivariate normality. Liang et al. (2000) developed projection tests for MVN based on affine invariant statistics. These tests remain effective in scenarios with high dimensions and small sample sizes. Henze (2002) posited that desirable properties such as affine invariance and consistency are typically required for MVN testing.

In this section, we introduce a novel approach to constructing MVN tests, predicated on the RPs of the univariate F-distribution and Pearson's chi-square test. The new test is based on the squared jackknife distances (Ahn, 1992) between each observation and the sample mean, which are affine invariant statistics. Consequently, we can circumvent the estimation of unknown parameters ($\boldsymbol{\mu}$, $\boldsymbol{\Sigma}$) in the d-variate normal distribution $\mathrm{N}_d(\boldsymbol{\mu}, \boldsymbol{\Sigma})$. Then, we describe an MVN test based on MSE-RPs of the F-distribution proposed by Wang et al. (2022b). A Monte Carlo study is conducted to explore the performance of the proposed test. The final part illustrates the application of the new test to three real examples, accompanied by computational results from other classical MVN tests. The concluding remarks are summarized at the end.

8.5.2 A brief review on affine invariance

A test for MVN is typically anticipated to possess the property of affine invariance. This property implies that the null distribution of the test statistic is independent of the unknown mean $\boldsymbol{\mu}$ and covariance matrix $\boldsymbol{\Sigma}$ of the multivariate normal distribution $N_d(\boldsymbol{\mu}, \boldsymbol{\Sigma})$. Henze (2002) in Proposition 2.1 highlighted that any affine invariant test for MVN is a function of the sample Mahalanobis distances (M-distance) and angles, which are defined as follows.

Let $\{\boldsymbol{x}_1, \ldots, \boldsymbol{x}_n\}$ denote a set of i.id. samples from a continuous distribution P^X. $\mathcal{N}_d$ represents the d-dimensional normal distribution $N_d(\boldsymbol{\mu}, \boldsymbol{\Sigma})$ with a mean vector $\boldsymbol{\mu} \in \mathcal{R}^d$ and a non-singular covariance matrix $\boldsymbol{\Sigma}$. The challenge of assessing MVN for the sample is to test the hypothesis.

$$H_d : P^X \in \mathcal{N}_d$$

against general alternatives. Any affine invariant statistic $T_n(\boldsymbol{x}_1, \ldots, \boldsymbol{x}_n)$ should satisfy the condition

$$T_n(\boldsymbol{A}\boldsymbol{x}_1 + \boldsymbol{b}, \ldots, \boldsymbol{A}\boldsymbol{x}_n + \boldsymbol{b}) = T_n(\boldsymbol{x}_1, \ldots, \boldsymbol{x}_n) \tag{8.17}$$

for any $\boldsymbol{b} \in \mathcal{R}^d$ and nonsingular matrix $A \in \mathcal{R}^{d\times d}$. Denote the sample mean and the sample covariance matrix by

$$\bar{\boldsymbol{x}} = \frac{1}{n}\sum_{j=1}^{n} \boldsymbol{x}_i \quad \text{and} \quad \boldsymbol{S}_n = \frac{1}{n}\sum_{j=1}^{n} (\boldsymbol{x}_j - \bar{\boldsymbol{x}})(\boldsymbol{x}_j - \bar{\boldsymbol{x}})',$$

respectively. The sample M-distance between two observations x_j and x_k is defined by

$$D_{n,jk} = (\boldsymbol{x}_j - \bar{\boldsymbol{x}})' \boldsymbol{S}_n^{-1} (\boldsymbol{x}_k - \bar{\boldsymbol{x}}), \quad j,k = 1,\ldots,n \tag{8.18}$$

In particular, when $j = k$, denote by

$$D_{n,j}^2 = D_{n,jj} = (\boldsymbol{x}_j - \bar{\boldsymbol{x}})' \boldsymbol{S}_n^{-1} (\boldsymbol{x}_j - \bar{\boldsymbol{x}}).$$

It is known that any test for MVN based on the M-distance (8.18) satisfies (8.17), so its null distribution does not depend on the normal parameters $\boldsymbol{\mu}$ and $\boldsymbol{\Sigma}$. In the following section, we will construct a test for MVN that is a function of the M-distance (8.18).

8.5.3 The jackknife distance

Define the Jackknife mean and the Jackknife covariance matrix by

$$\bar{\boldsymbol{x}}_{-i} = \frac{1}{n-1}\sum_{j\neq i} \boldsymbol{x}_j \quad \text{and} \quad \boldsymbol{S}_{-i} = \frac{1}{n-2}\sum_{j\neq i} (\boldsymbol{x}_j - \bar{\boldsymbol{x}}_{-i})(\boldsymbol{x}_j - \bar{\boldsymbol{x}}_{-i})',$$

respectively. The squared Jackknife distance from the i-th observation x_i to the sample mean is defined by

$$D_{-i}^2 = (\boldsymbol{x}_i - \bar{\boldsymbol{x}})' \boldsymbol{S}_{-i}^{-1} (\boldsymbol{x}_i - \bar{\boldsymbol{x}}), \quad i = 1,\ldots,n.$$

Ahn (1992) employed the squared Jackknife distance above to construct an F-probability plot for assessing MVN of the i.id. sample $\{\boldsymbol{x}_1,\ldots,\boldsymbol{x}_n\}$. It is pointed out that the Jackknife covariance matrix S_{-i} is related to the usual sample covariance matrix S_n by

$$\frac{1}{n-1}\boldsymbol{S}_{-i}^{-1} = \frac{1}{n-1}\boldsymbol{S}^{-1} + \frac{\gamma_i}{(n-1)^2}\boldsymbol{S}^{-1}(\boldsymbol{x}_i - \bar{\boldsymbol{x}})(\boldsymbol{x}_i - \bar{\boldsymbol{x}})'\boldsymbol{S}^{-1},$$

where

$$\boldsymbol{S} = \sum_{j=1}^{n} (\boldsymbol{x}_j - \overline{\boldsymbol{x}})(\boldsymbol{x}_j - \overline{\boldsymbol{x}})' = n\boldsymbol{S}_n,$$

$$\gamma_i = \frac{n}{n-1}\left[1 - \frac{n}{(n-1)^2}\right](x_i - \overline{\boldsymbol{x}})' \boldsymbol{S}^{-1} (\boldsymbol{x}_i - \overline{\boldsymbol{x}}).$$

The following monotonic relation between the squared Jackknife distance and the usual chi-squared distance is given in Ahn (1992):

$$D_{-i}^2 = \frac{(n-1)(n-2)D_i^2}{(n-1)^2 - nD_i^2}, \quad D_i^2 = (\boldsymbol{x}_i - \bar{\boldsymbol{x}})' S^{-1} (x_i - \bar{x}).$$

It is obvious that D_i^2 is related to the M-distance $D_{n,i}^2$ by $D_{n,i}^2 = nD_i^2$. Equations above provide a simple computational method for obtaining the squared Jackknife distances $\{D_{-i}^2 : i = 1,\ldots,n\}$ by avoiding the computation of n inverse matrices $\boldsymbol{S}_{-i}^{-1}$. The following theorem can be easily derived from Ahn (1992).

Theorem 8.12 *Under hypothesis $H_d : P^X \in \mathcal{N}_d$, the following two assertions are true:*
(1) the adjusted Jackknife distance has an exact F-distribution:

$$d_{-i}^2 = \frac{n(n-1-d)}{(n-1)(n-2)d} D_{-i}^2 \sim F(d, n-1-d)$$

(2) $\{d_{-i}^2 : i = 1,\ldots,n\}$ are asymptotically independent.

The proof is given in Subsection 8.7.4.

8.5.4 The chi-squared test based on the F-representative points

The above theorem provides an approach to testing hypothesis $H_d : P^X \in \mathcal{N}_d$. Instead of testing $H_d : P^X \in \mathcal{N}_d$, we can test

$$H_0 : \left\{d_{-i}^2 : i = 1, \ldots, n\right\} \text{ is a sample from } F(d, n-1-d), \tag{8.27}$$

against the alternative that H_0 is not true, it's clear that if hypothesis (8.27) is rejected, hypothesis $H_d : P^X \in \mathcal{N}_d$ is also dismissed. However, the converse is not necessarily true. A hypothesis test (8.27) is referred to as a necessary test (Al-Labadi et al., 2021) for the normality of the original data in the literature. Testing hypothesis (8.19) can be carried out by the classical chi-squared test for general goodness of fit. However, the classical chi-squared test faces the challenge of choosing classification cells for observed sample data Williams (1950); Dahiya and Gurland (1973); D'Agostino and Stephens (1986); Koehler and Gan (1990); and Bogdan (1995)). A classification with equiprobable cells is a common choice in the literature. However, since the RPs minimize a specific type of quadratic loss function, we conjecture that classification cells based on RPs may perform better than the simple equiprobable classification. Therefore, we propose to use the FF-RPs to construct classification cells and expect to improve the performance of the classical chi-squared test. A Monte Carlo study will be carried out in the next subsection to verify the performance of this approach.

The F-representative points are a set of points $\{F_1, \ldots, F_k\}$ (for a selected number of points k) that minimize the quadratic loss function:

$$\phi\left(x_1, \ldots, x_k\right) = \int_0^{\infty} \min_{1 \le i \le k} \left\{\left(x_i - x\right)^2\right\} f_F(x; d, n-1-d) dx,$$

where $f_F(x; d, n-1-d)$ stands for the density function of the F-distribution with degrees of freedom $(d, n-1-d)$ and $\{p_1, \ldots, p_k\}$ can be considered as a set of "representative probabilities" for the F-distribution $F(d, n-1-d)$. Based on the theorem above, a hypothesis test $H_d : P^X \in \mathcal{N}_d$ can be approximately (under large sample size n) transferred to testing (8.19) with classification intervals of RPs. The χ^2-statistic is computed by:

$$\chi_R^2 = \sum_{i=1}^{k} \frac{\left(n_i - n p_i\right)^2}{n p_i} \rightarrow \chi^2(k-1), \quad (n \rightarrow \infty \text{ under the hypothesis }),$$

where n_i is the frequency of the transformed approximately i.i.d. sample points $\{d^2{}_{-i}$ $i = 1, \ldots, n\}$ computed by (8.19) that are located in the interval. The approximate p-value is calculated by

$$P\left(\chi_R^2, v\right) = K \int_{\chi_R^2}^{\infty} z^{\frac{v}{2}-1} \exp\left(-\frac{z}{2}\right) dz, \text{ with } v = k-1, K = \left[2^{\frac{v}{2}} \Gamma\left(\frac{v}{2}\right)\right]^{-1}.$$

The Monte Carlo study in the next subsection will verify the accuracy of the chi-square approximation above by simulating the empirical type I error rates and their power against some selected sets of alternative distributions.

8.5.5 A Monte Carlo study

To compare the χ^2-test under the "representative probabilities" $\{p_1, \ldots, p_k\}$ with the traditional chi-squared test, we select the equiprobable cells for computing the traditional chi-squared test. We define the interval endpoints for a chosen number of points $k-1$ by

$$a_1 \text{ satisfies } P\left(\chi^2(k-1) < a_1\right) = \frac{1}{k}$$
$$a_2 \text{ satisfies } P\left(a_1 < \chi^2(k-1) < a_2\right) = \frac{1}{k}$$
$$\vdots$$
$$a_{k-1} \text{ satisfies } P\left(\chi^2(k-1) > a_{k-1}\right) = \frac{1}{k}$$

Denote the traditional chi-squared test based on the interval endpoints above by χ_T^2 which also has an approximate null distribution $\chi^2(k-1)$,

$$\chi_T^2 = \sum_{i=1}^{k} \frac{(N_i - n/k)^2}{n/k},$$

where N_i is the frequency of the approximate iid transformed sample points $\{d_{-i}^2 : i = 1, \ldots, n\}$ that are located in the i-th interval.

1. A Comparison between Empirical Type I Error Rates

Since the chi-squared test based on the transformed sample points $\{d_{-i}^2 : i = 1, \ldots, n\}$, as given by equation (8.19), is affine invariant under any nonsingular linear transformation of the original i.id. sample $\{\boldsymbol{x}_1, \ldots, \boldsymbol{x}_n\}$, it is only necessary to generate samples from a d-dimensional standard normal $\{\boldsymbol{x}_1, \ldots, \boldsymbol{x}_n\}$, where $\boldsymbol{I}_d$ denotes the $\boldsymbol{I}_d$ identity matrix. Based on 2000 replications for each case, the simulation results are summarized in Table 8.9. These results reveal that both statistics χ_R^2 and χ_T^2 effectively control their empirical type I error rates under the significance level of 0.05. In most cases, χ_T^2 proves to be more conservative than χ_R^2 , especially when dealing with smaller sample sizes. Both statistics exhibit similar performance at significance levels of 0.01 and 0.10.

2. A Power Comparison

The various alternative distributions are selected as follows.

(1) The multivariate t-distribution has a density function of the form

$$f_t(\|\boldsymbol{x}\|) = C_1\left(1 + \frac{\|\boldsymbol{x}\|^2}{v}\right)^{-\frac{d+v}{2}}, \quad v > 0,$$

where "$\|\cdot\|$" stands for the Euclidean norm of a vector, and let the degree of freedom $v = 5$.
(2) The β-generalized normal distribution $N_d(\boldsymbol{0}, \boldsymbol{I}_d, 1/2)$ with $\beta = 1/2$ has a density function of the form by (Goodman and Kotz, 1973):

$$f(x_1, \ldots, x_d) = \frac{\beta^d r^{d/\beta}}{2^d \Gamma^d(1/\beta)} \cdot \exp\left\{-r\sum_{i=1}^{d} |x_i|^\beta\right\}, \quad (x_1, \ldots, x_d)^T \in R^d,$$

where $r > 0$ is a parameter. Let $r = 1/2$ in the simulation and denote it by β normal;

(3) The shifted iid $\chi^2(1)$ with iid marginals, each marginal has the same distribution as that of the random variable $Y = X - E(X)$, where $X \sim \chi^2(1)$, the univariate chi-squared

TABLE 8.12
Empirical type I error rates ($\alpha = 0.05$)

n	m	Test	$d = 3$	$d = 5$	$d = 10$	$d = 15$	$d = 20$
$n = 20$	$m = 10$	χ^2_R	0.056	0.060	0.064	0.056	0.072
		χ^2_T	0.040	0.022	0.032	0.030	0.024
	$m = 20$	χ^2_R	0.050	0.070	0.061	0.064	0.062
		χ^2_T	0.038	0.026	0.038	0.038	0.022
	$m = 30$	χ^2_R	0.062	0.070	0.064	0.074	0.078
		χ^2_T	0.042	0.032	0.036	0.038	0.028
$n = 50$	$m = 10$	χ^2_R	0.032	0.035	0.048	0.046	0.035
		χ^2_T	0.027	0.030	0.039	0.038	0.033
	$m = 20$	χ^2_R	0.049	0.059	0.059	0.065	0.064
		χ^2_T	0.037	0.036	0.034	0.036	0.035
	$m = 30$	χ^2_R	0.057	0.074	0.072	0.070	0.082
		χ^2_T	0.038	0.037	0.033	0.037	0.040
$n = 100$	$m = 10$	χ^2_R	0.032	0.034	0.048	0.046	0.035
		χ^2_T	0.025	0.031	0.034	0.025	0.036
	$m = 20$	χ^2_R	0.062	0.067	0.044	0.048	0.054
		χ^2_T	0.041	0.025	0.038	0.032	0.031
	$m = 30$	χ^2_R	0.059	0.060	0.065	0.060	0.062
		χ^2_T	0.040	0.036	0.043	0.036	0.032

distribution with 1 degree of freedom and $E(X) = 1$;

(4) The distribution $N(0, 1) + \chi^2(2)$ consists of iid [d/2] normal $N(0, 1)$ marginals and $d - [d/2]$ iid $\chi^2(2)$ marginals, where $[d/2]$ stands for the integer part of $d/2$;

(5) The shifted i.i.d. exp(1) with iid marginals, each marginal has the same distribution as that of the random variable $Y = X - E(X)$, where $X \sim \exp(1)$, the univariate exponential distribution.

We selected the sample size $n = 20, \ldots, 400$ for each alternative distribution. Through simulations with 10,000 replications, the power values versus the sample size n for both statistics χ^2_R and χ^2_T are depicted in Figure 8.6. For more details, please refer to Wang et al. (2022b). As can be observed, the proposed statistic χ^2_R consistently outperforms the classical χ^2_T across different alternative distributions. This demonstrates a clear trend that the chi-squared statistic χ^2_R, based on F-representative points, can significantly enhance the performance of the traditional chi-squared statistic χ^2_T. To compare the power performance of χ^2_R with other frequently utilized tests in the literature, we selected a category of tailor-made MVN tests (as per Mardia (1980); Henze and Wagner (1997); Székely and Rizzo (2017)) against two types of alternative distributions (symmetric and non-symmetric). These are illustrated in Figure 8.7, where "hz" represents Henze and Wagner's statistic and "energy" represents Szekely and Rizzo's test. The straightforward comparison in Figure 8.7 suggests that the χ^2_R-test, based on F-representative points, can be comparable to the selected frequently used tests in the literature. Further power comparisons based on three additional alternative distributions are provided in the work of Wang et al. (2022b). These results give us confidence in using the χ^2_R -test as a supplement to enhance the application of some existing tests. More related study can refer to Cao et al. (2024).

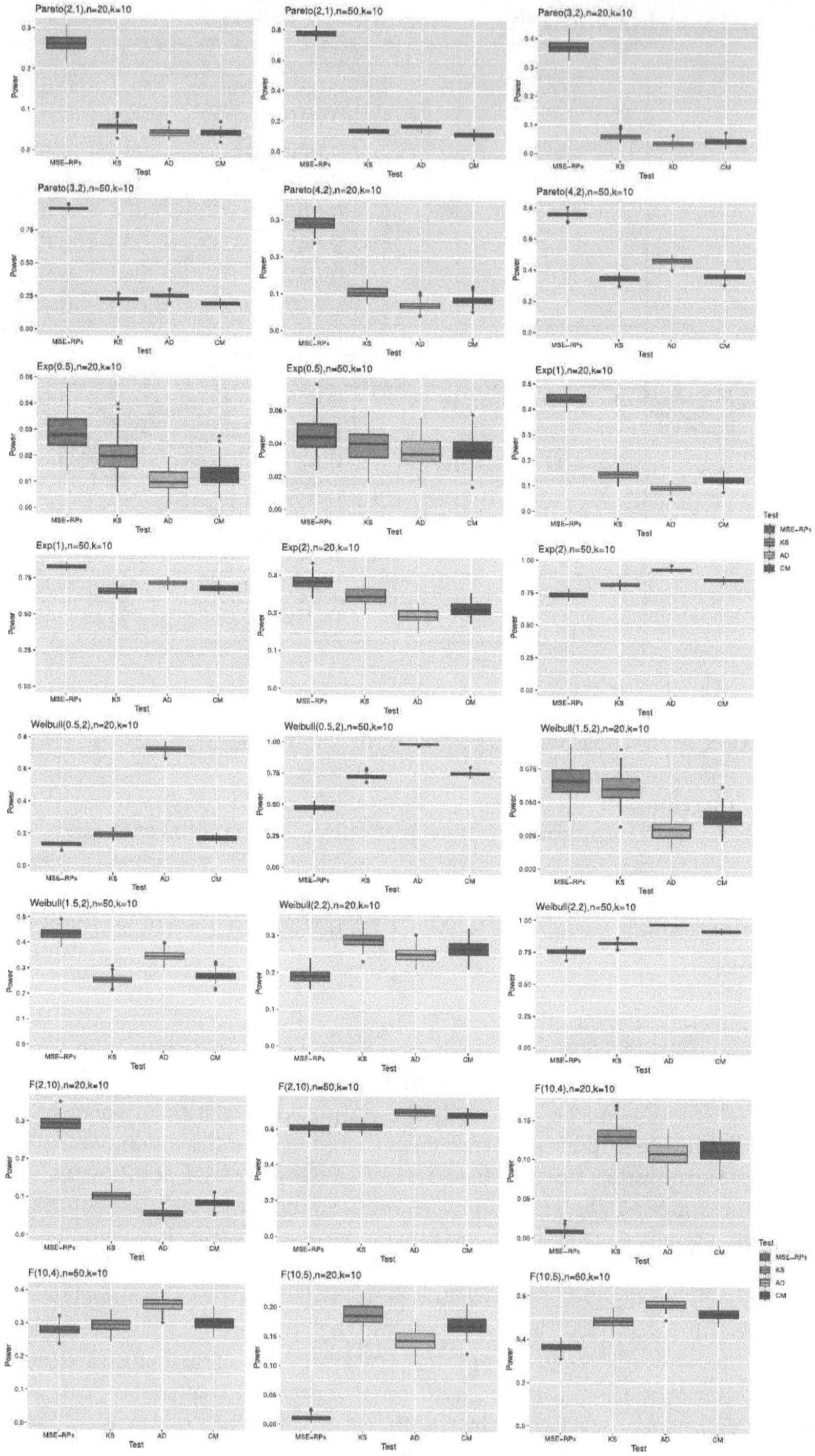

FIGURE 8.6
Power performance ($\alpha = 0.05$) of χ_R^2 and χ_T^2 in multivariate t distribution.

TABLE 8.13
Measures of jaw bones on 20 boys at four different ages

	Age in years			
	8	8.5	9	9.5
1	47.8	48.8	49.0	49.7
2	46.4	47.3	47.7	48.4
3	46.3	46.8	47.8	48.5
⋮	⋮	⋮	⋮	⋮
19	46.2	47.5	48.1	48.4
20	46.3	47.6	51.3	51.8

8.5.6 Real data analysis

Dataset 1—Ramus Bone Data in Table 8.13. Elston and Grizzle (1962) collected intriguing data on the ramus height (in millimeters) of 20 boys, each measured at ages 8, 8.5, 9, and 9.5 years old. The original study aimed to establish a standard growth curve for orthodontists' use. The data exhibits several interesting characteristics. Timm (2002) demonstrated that the recorded heights of the ramus bone appear to be normally distributed marginally but reveal some deviation from multivariate normality for $\alpha = 0.05$. To examine the multivariate normality for ages 8, 8.5, 9, and 9.5, the methods suggested by Zhou and Shao (2014) provided p-values of $0.002, 0.054, < 0.001, \text{and} < 0.001$.

Dataset 2-Rao's Cork Data in Table 8.14. The dataset gathered by Rao (1948) records the weight of cork borings taken from the north (N)(N), east (E)(E), south (S)(S), and west (W)(W) directions, comprising the thickness of bark deposits in 28 corks. The original problem was investigating whether the bark deposit varies in thickness in the four directions. Regarding the multivariate normality assumption, Srivastava and Hui (1987) rejected the hypothesis using tests with p-values of 0.01 and 0.037. Additionally, Mudholkar et al. (1992) also expressed skepticism about the multivariate normality assumption based on their p-value of 0.0302. Srivastava and Mudholkar (2003) applied the simplified Fisher combination statistic and obtained a p-value of 0.006, which also suggests a particular departure from multivariate normality.

To evaluate the assumption of multivariate normality for these data using the two chi-squared tests χ_R^2 and χ_T^2, the p-values under different m (the number of representative points) from the two chi-squared tests χ_R^2 and χ_T^2 are summarized in the following Table 8.18. Meanwhile, we employ the classic skewness, kurtosis, and univariate Shapiro-Wilk tests (Shapiro and Wilk, 1965) to assess the marginal normality of each of the four variables. Table 8.19 also presents the results of the p-values of the two statistics for skew and kurtosis from Mardia's MVN test. The variables in the following two tables are:

X_8 = boys' ′ ramus height at age 8 in Dataset 1
$X_{8.5}$ = boys' ′ ramus height at age 8.5 in Dataset 1
X_9 = boys' ′ ramus height at age 9 in Dataset 1
$X_{9.5}$ = boys' ramus height at age 9.5 in Dataset 1
X_{sl} = sepal length in Dataset 2
X_{sw} = sepal width in Dataset 2
X_{pl} = petal length in Dataset 2

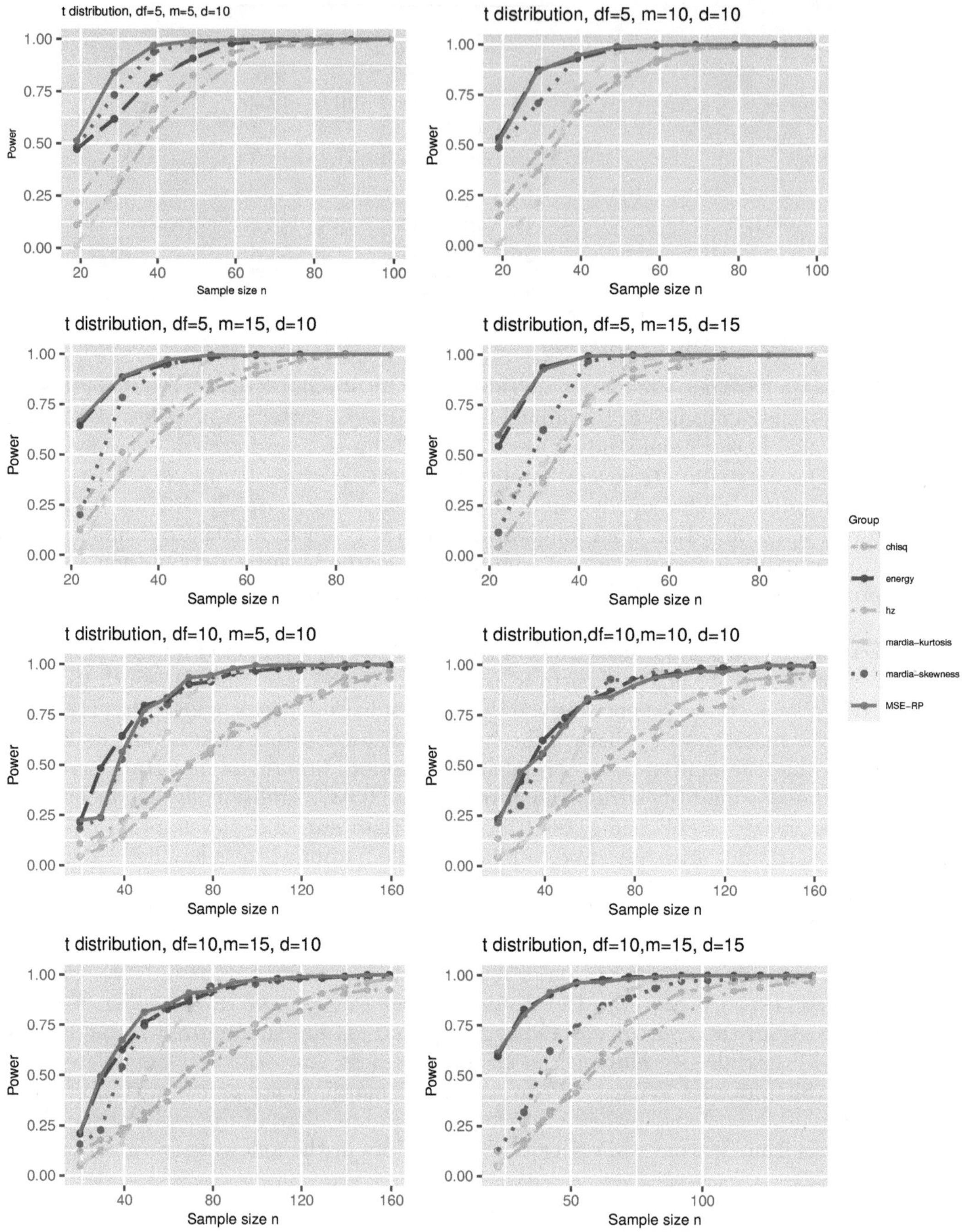

FIGURE 8.7
power ($\alpha = 0.05$) comparison with tailor-made tests in multivariate t distribution.

TABLE 8.14
Weights of cork borings (cg) in four directions for 28 trees

Tree	N	E	S	W	Tree	N	E	S	W
1	72	66	76	77	15	91	79	100	75
2	60	53	66	63	16	56	68	47	50
3	56	57	64	58	17	79	65	70	61
4	41	29	36	38	18	81	80	68	58
5	32	32	35	36	19	78	55	67	60
6	30	35	34	26	20	46	38	37	38
7	39	39	31	27	21	39	35	34	37
8	42	43	31	25	22	32	30	30	32
9	37	40	31	25	23	60	50	67	54
10	33	29	27	36	24	35	37	48	39
11	32	30	34	28	25	39	36	39	31
12	63	45	74	63	26	50	34	37	40
13	54	46	60	52	27	43	37	39	50
14	47	51	52	43	28	48	54	57	43

X_{pw} = petal width in Dataset 2
X_N = the weight of cork borings taken from north in Dataset 3
X_E = the weight of cork borings taken from east in Dataset 3
X_S = the weight of cork borings taken from south in Dataset 3
X_W = the weight of cork borings taken from west in Dataset 3

TABLE 8.15
p-values from the two chi-squared tests

Variables	χ^2-Test	$m = \mathbf{10}$	$m = \mathbf{20}$	$m = \mathbf{30}$
$(X_8, X_{8.5}, X_9, X_{9.5})$	χ_R^2	0.0016	1.661×10^{-6}	0.0005
	χ_T^2	0.0669	0.1302	0.0839
$(X_{sl}, X_{sw}, X_{pl}, X_{pw})$	χ_R^2	0.0066	0.0199	0.0076
	χ_T^2	0.0179	0.0397	0.0979
(X_N, X_E, X_S, X_W)	χ_R^2	0.0004	$< 10^{-10}$	$< 10^{-10}$
	χ_T^2	0.7110	0.3610	0.7566

The *p*-values from the representative-points chi-squared test χ_R^2 in Table 8.15 suggest a significant deviation from joint multivariate normality for the four variables in each example. In contrast, the *p*-values from the traditional chi-squared test χ_T^2 only partially imply a particular departure from joint multivariate normality. The potential deviation from joint multivariate normality is also supported by Mardia's multivariate skewness test and most of the univariate normality tests (UVN tests in Table 8.16) by Shapiro-Wilk's statistic. Since the results from the χ_R^2-test are nearly all consistent with those from the well-performed tests in the literature, the representative-point-based χ_R^2-test appears to provide more consistent results with some well-known existing tests than the traditional χ_T^2-test does.

TABLE 8.16
Skewness and kurtosis for each measurement and feature

Variables	Skewness	Kurtosis	UVN Test [1]	Mardia's MVN Test [2]
X_8	0.3069	-1.0682	0.3360	$Sk: 0.0093$
$X_{8.5}$	0.3111	-0.8932	0.6020	$Ku: 0.1125$
X_9	0.0645	-1.2076	0.5016	
$X_{9.5}$	0.0648	-1.4332	0.0905	
X_{sl}	0.3118	-0.5736	0.0102	$Sk: 4.7570 \times 10^{-7}$
X_{sw}	0.3158	0.1810	0.1012	$Ku: 0.8180$
X_{pl}	-0.2721	-1.3955	< 0.0001	
X_{pw}	-0.1019	-1.3361	< 0.0001	
X_N	0.8088	-0.3535	0.0179	$Sk: 7.6249 \times 10^{-9}$
X_E	0.8322	-0.2680	0.0135	$Ku: 0.0076$
X_S	-0.1289	-0.3767	0.0361	
X_W	0.4417	-0.7800	0.1185	

[1] *p*-values from univariate Shapiro-Wilk tests.
[2] *p*-values from Mardia's multivariate skewness and kurtosis tests.

8.6 Space Partitioning Based on MSE-RPs and Its Application

In the process of solving for the MSS-RPs, a natural partition of the domain of a continuous distribution can be obtained, where each region in the partition is represented by a representative point, and the cumulative probability of that region corresponds to the probability of the representative point. In this section, we have invited Dr. Xiaoling Peng to introduce her work (Peng et al., 2024), where she and her collaborators have combined representative point theory with big data for the evaluation of hospital specialty influence. She will introduce how to use the MSE-RPs-based partition and associated probability distribution to construct the patient regional index (PRI), reflecting the influence of the hospital by weighting the distances.

8.6.1 Space partitioning based on MSE-RPs

Continuous data grouping using the mean square error (MSE) was first proposed by Cox (1957). Let X defined on $[c, d]$ be a continuous random variable with a probability density function (p.d.f.) $f(x)$, and let $\{y; P\}$ denote the set of k representative points and their probability distribution, thereinto:

$$P(Y = y_i) = p_i = \begin{cases} \int_c^{\frac{y_1+y_2}{2}} f(x)dx, & i = 1, \\ \int_{\frac{y_{i-1}+y_i}{2}}^{\frac{y_i+y_{i+1}}{2}} f(x)dx, & i = 2, \ldots, k-1, \\ \int_{\frac{y_{k-1}+y_k}{2}}^{d} f(x)dx, & i = k. \end{cases}$$

The MSE between the representative points and the continuous distribution is defined as:

$$MSE(y_1, \ldots, y_k) = E\left(\min_{i=1,\ldots,k} (X - y_i)^2\right) = \int_c^d \min_{i=1,\ldots,k} (x - y_i)^2 f(x)dx,$$

where

$$MSE(y_1,\dots,y_k)=\int_c^{\frac{y_1+y_2}{2}}(x-y_1)^2 f(x)dx+\int_{\frac{y_1+y_2}{2}}^{\frac{y_2+y_3}{2}}(x-y_2)^2 f(x)dx$$
$$+\cdots+\int_{\frac{y_{k-1}+y_k}{2}}^{d}(x-y_k)^2 f(x)dx.$$

When the MSE is minimized, the discrete distribution $\{y; P\}$ becomes the MSE-RPs. Through the calculation process for these representative points, we can ascertain the probability distribution of the k MSE-RPs and the corresponding partition Ω:

$$\begin{aligned} y &= \{y_1, y_2, \dots, y_k\},\\ P &= \{p_1, p_2, \dots, p_k\},\\ \Omega &= \left\{[c, \frac{y_1+y_2}{2}], [\frac{y_1+y_2}{2}, \frac{y_2+y_3}{2}], \dots, [\frac{y_{k-1}+y_k}{2}, d]\right\}. \end{aligned}$$

8.6.2 Case study: patient regional index

Utilizing big data from outpatient clinics, this study aims to introduce a new metric, the Patient Regional Index (PRI). The PRI is designed to evaluate specialty influence using a partition based on Mean Squared Error-based Representative Points (MSE-RPs). The process involves segmenting the distances from patient origins to the healthcare institution into various intervals using MSE-RPs criteria. Subsequently, the PRI is constructed by weighting the number (or proportion) of patients within each interval, with the weight being inversely proportional to the distance.

(1) **Data Selection and Preprocessing**

In this study, data was obtained from a large, comprehensive hospital in South China, covering 33 departments over the period 2014–2021. The dataset included 10,098,024 outpatient visit records from 16 selected specialties, such as Pediatrics and Urology. Patients' origins were converted into latitude and longitude coordinates using Baidu Open Platform's geocoding API. The Python library `geopy` was employed to calculate distances from patients' origins to the hospital.

Most patients originate from nearby areas, so the probability of a patient journeying from a distant location is relatively low. The Gamma distribution aptly describes the distribution of distances from the patient's region to the hospital. Figure 8.8 depicts this distance distribution as a histogram overlaid with the probability density curve for $Ga(0.1954, 0.0014)$.

(2) **Patient Regional Index Calculation**

Following the principle that larger distances should be attributed with higher weights, we define the weight w_i of the ith distance interval to be inversely proportional to the probability p_i, that is, $w_i \propto 1/p_i$, $i = 1, 2, \cdots, k$. To guarantee that the sum of all weights is equal to 1, i.e., $\sum w_i = 1$, the weight of the ith distance interval is assigned as follows:

$$w_i = \frac{\frac{1}{p_i}}{\sum_{j=1}^{k}\frac{1}{p_j}}.$$

The Patient Regional Index (PRI) is then defined as:

$$PRI = \sum_{i=1}^{k} w_i r_i,$$

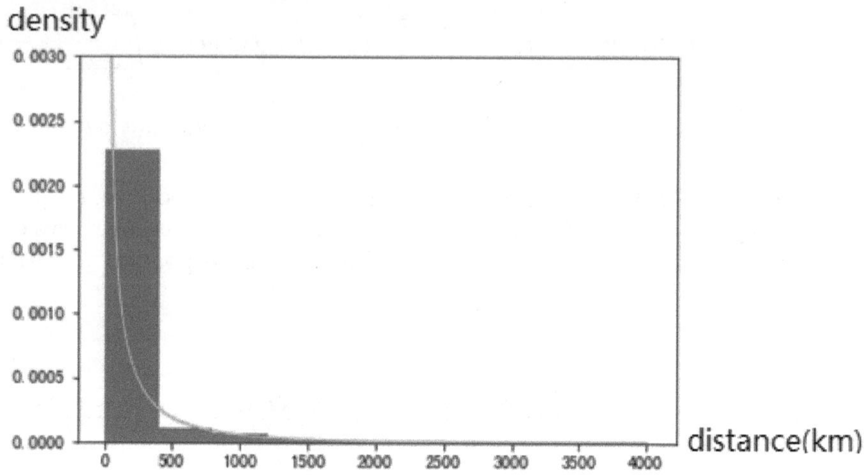

FIGURE 8.8
Distribution of distance from the patient's region to the hospital.

where r_i represents the proportion of patients in the i-th interval; as the distribution of patients' origins fluctuates between departments, the PRI scores calculated will also vary. Typically, a department will register a higher PRI score if it draws a greater percentage of patients from more remote areas. Noting that the PRI scores obtained in the previously described method were numerically small, we used the average PRI scores from 2017 as a reference point, setting this score at 100. Following this, each PRI was adjusted as follows:

$$PRI_{\text{adjusted}} = \frac{PRI}{PRI_{2017}} \times 100.$$

(3) **Results and Discussion**

Figure 8.9 compares the change in the specialty PRI of nonsurgical and surgical departments from 2014 to 2021. For comparison, specialties were divided into two groups: nonsurgical (e.g., Pediatrics, Nuclear Medicine) and surgical (e.g., Gynecology, Urology). Surgical departments exhibited greater increases in PRI, reflecting broader regional patient origins and stronger patient loyalty.

Notably, the PRI for most specialties experienced a decline post-2017, attributable to the healthcare reform initiated by the Chinese government in that year. This reform aimed to establish a hierarchical diagnosis and treatment service system, thereby encouraging patients with common diseases to initially seek help from primary healthcare institutions. Consequently, there was a notable reduction in out-of-town patients at this large comprehensive hospital (Yip et al., 2019; Zhang et al., 2017). Further, the emergence of the COVID-2019 pandemic toward the end of 2019 restricted people's mobility, as evidenced by a distinct inflection point in 2020 in Figure 8.9 for the PRI of certain specialties at this hospital, particularly the surgical ones. However, with the Chinese government's control of the epidemic, patient mobility across regions was reinstated, leading to a significant rebound in the PRIs of all specialties. The fluctuation in the PRI effectively mirrors the influence of China's healthcare reform and the COVID-2019 pandemic on specialty outpatient clinics.

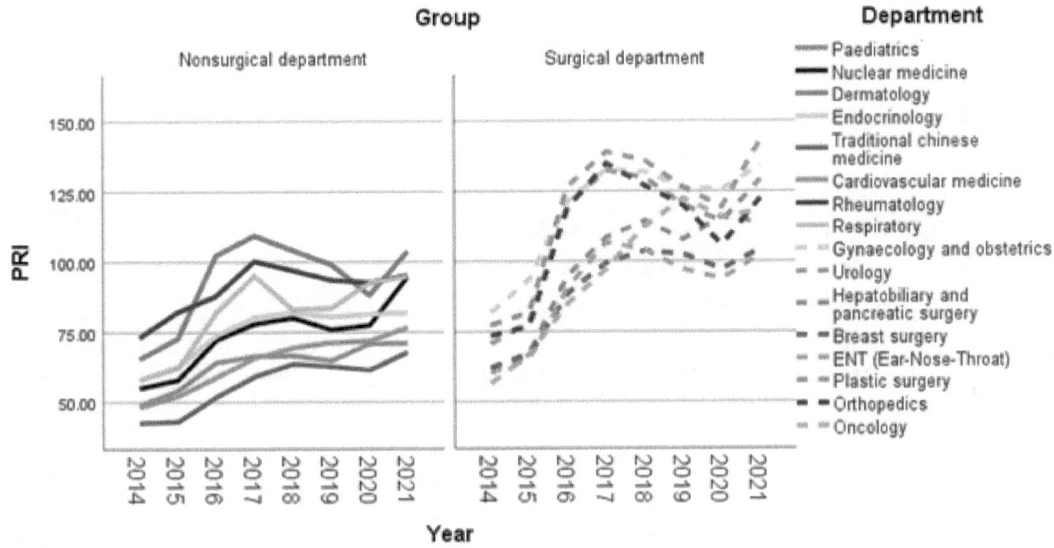

FIGURE 8.9
Comparison of PRI across specialties in a comprehensive hospital (2014–2021).

In addition, a heat map was created for selected specialties, as shown in Figure 8.10. It visualizes the distribution of patient origins and illustrates the varying influences of different specialties.

As depicted in Figure 8.10, the Department of Gynecology and Obstetrics was highly esteemed in the community, attracting patients not only from southern China but also in significant numbers from central and eastern China and even some from as far as the northeast. In comparison to pediatrics, the urology department, despite having a smaller total number of patient visits, had a substantially wider geographical coverage of its patient source. As a result, urology's PRI was considerably higher than that of pediatrics. On the other hand, in 2014, the oncology department was still in its nascent stages of development within the hospital, primarily serving patients from the nearby areas. Consequently, both its PRI and outpatient volume were relatively low.

8.7 Proofs of Theorems in Chapter 8

8.7.1 Proof of Theorem 8.7

Let $\boldsymbol{y} = (y_1, \ldots, y_p) \in T^p$, $\mathcal{Y} = \{\boldsymbol{y}\}$, $\boldsymbol{y}^* = (1/p, \ldots, 1/p)$, and $\mathcal{Y}^* = \{\boldsymbol{y}^*\}$. Let $\boldsymbol{Z} = (Z_1, \ldots, Z_p) \sim U(T^p)$ and μ denote the uniform distribution measure on T^p, then

$$ML1D(\mathcal{Y}) = E_{\boldsymbol{Z}}[D_{l_1}(\boldsymbol{z}, \boldsymbol{y})] = \int_{(z_1,\ldots,z_p)\in T^p} \left(\sum_{i=1}^{p} |z_i - y_i|\right) d\mu(\boldsymbol{z}).$$

Let τ denote a permutation of $(1, 2, \ldots, p)$, then

$$\int_{(z_1,\ldots,z_p)\in T^p} \left(\sum_{i=1}^{p} |z_i - y_i|\right) d\mu(\boldsymbol{z}) = \int_{(z_1,\ldots,z_p)\in T^p} \left(\sum_{i=1}^{p} |z_i - y_{\tau(i)}|\right) d\mu(\boldsymbol{z})$$

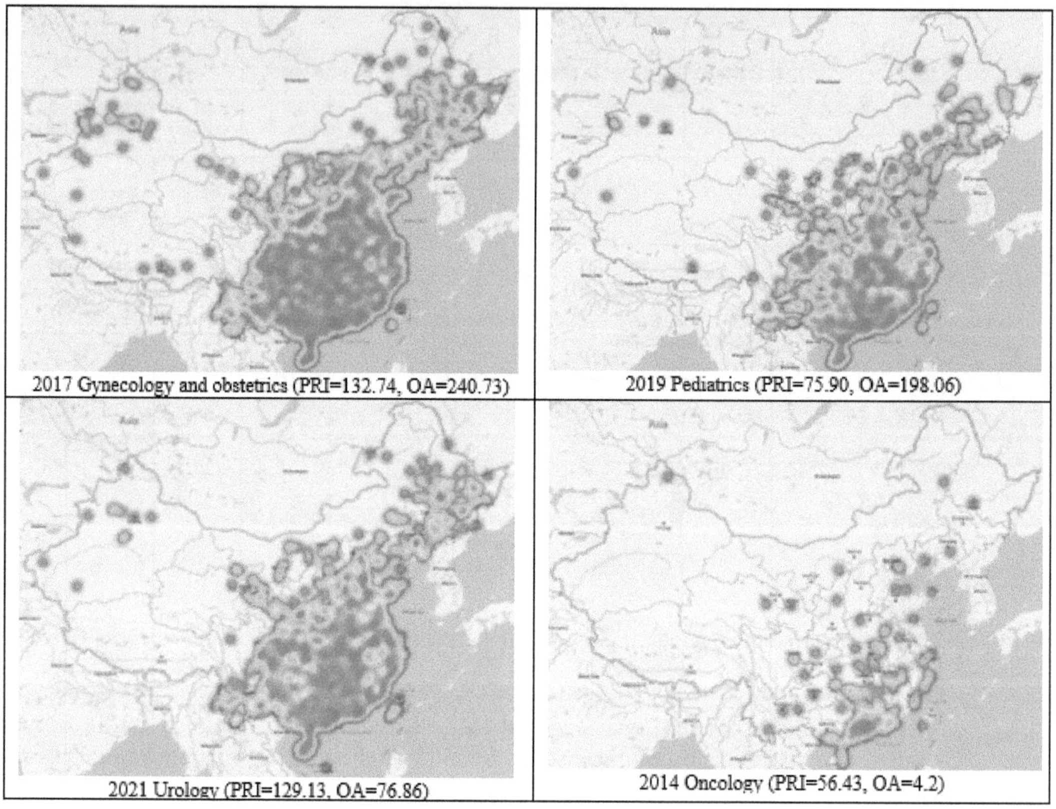

FIGURE 8.10
Heat map showing patient origin distribution for selected specialties.

because of the symmetry of T^p. Thus,

$$\begin{aligned} p!ML1D(\mathcal{Y}) &= \sum_{\tau} \int_{(z_1,\dots,z_p)\in T^p} \left(\sum_{i=1}^{p} |z_i - y_{\tau(i)}| \right) d\mu(\boldsymbol{z}) \\ &= \int_{(z_1,\dots,z_p)\in T^p} \sum_{\tau} \left(\sum_{i=1}^{p} |z_i - y_{\tau(i)}| \right) d\mu(\boldsymbol{z}) \\ &= \int_{(z_1,\dots,z_p)\in T^p} \left(\sum_{i=1}^{p} \sum_{\tau} |z_i - y_{\tau(i)}| \right) d\mu(\boldsymbol{z}) \\ &= \int_{(z_1,\dots,z_p)\in T^p} \left(\sum_{i=1}^{p} (p-1)! \sum_{j=1}^{p} |z_i - y_j| \right) d\mu(\boldsymbol{z}), \end{aligned}$$

where $\sum_\tau$ means summing over by taking all possible $p!$ permutations of $(1,\dots,p)$. Note that $\sum_j y_j = 1$, we have

$$\sum_{j=1}^{p} |z_i - y_j| \geq |\sum_{j=1}^{p} z_i - \sum_{j=1}^{p} y_j| = p|z_i - 1/p|.$$

Therefore

$$ML1D(\mathcal{Y}) \geq \int_{(z_1,\ldots,z_p)\in T^p} \left(\sum_{i=1}^{p} |z_i - 1/p|\right) d\mu(\boldsymbol{z}) = ML1D(\mathcal{Y}^*).$$

As $Z_i \sim \text{Beta}(1, p-1)$ for $i = 1, \ldots, p$, direct calculation gives

$$ML1D(\mathcal{Y}^*) = p\int_0^1 |z - 1/p| \times (p-1)(1-z)^{p-2} dz = 2(p-1)^p p^{-p},$$

which completes the proof.

8.7.2 Proof of Theorem 8.8

Suppose $\mathcal{Y} = \{\boldsymbol{y}_i = (y_i, 1-y_i), i = 1, \ldots, n\}$ and $0 \leq y_1 < \ldots < y_n \leq 1$. We prove the theorem by mathematical induction on n. We have shown the result for $n = 1$ (see Theorem 8.7). Suppose the conclusion holds for design size of $n-1, n \geq 2$. For size n,

$$\begin{aligned} ML1D(\mathcal{Y}) &= E_{\boldsymbol{Z}}[\min_{1\leq j\leq n} D_{l_1}(\boldsymbol{z}, \boldsymbol{y}_j)] \\ &= 2\Bigg(\int_0^{(y_1+y_2)/2} |z - y_1| dz + \int_{(y_1+y_2)/2}^{(y_2+y_3)/2} |z - y_2| dz \\ &\quad + \cdots + \int_{(y_{n-2}+y_{n-1})/2}^{(y_{n-1}+y_n)/2} |z - y_{n-1}| dz + \int_{(y_{n-1}+y_n)/2}^{1} |z - y_n| dz\Bigg). \end{aligned}$$

Let us first suppose that $(y_{n-1} + y_n)/2 = c \in (0,1)$ is fixed. Define

$$g_c(y_1, \ldots, y_{n-1}) = 2\Bigg(\int_0^{(y_1+y_2)/2} |z-y_1| dz + \int_{(y_1+y_2)/2}^{(y_2+y_3)/2} |z-y_2| dz + \cdots + \int_{(y_{n-2}+y_{n-1})/2}^{c} |z-y_{n-1}| dz\Bigg).$$

We have

$$\begin{aligned} g_c(y_1, \ldots, y_{n-1})/c &= 2\Bigg(\int_0^{(y_1+y_2)/2} |z/c - y_1/c| dz + \int_{(y_1+y_2)/2}^{(y_2+y_3)/2} |z/c - y_2/c| dz + \cdots \\ &\quad + \int_{(y_{n-2}+y_{n-1})/2}^{c} |z/c - y_{n-1}/c| dz\Bigg) \\ &= 2c\Bigg(\int_0^{(y_1/c+y_2/c)/2} |z' - y_1/c| dz' + \int_{(y_1/c+y_2/c)/(2)}^{(y_2/c+y_3/c)/2} |z' - y_2/c| dz' \\ &\quad + \cdots + \int_{(y_{n-2}/c+y_{n-1}/c)/2}^{1} |z' - y_{n-1}/c| dz'\Bigg), \end{aligned}$$

where in the last equality, we let $z' = z/c$.

Obviously, $g_c(y_1, \ldots, y_{n-1})/c^2$ can be viewed as the $ML1D$ of the set of points $\{(y_i/c, 1 - y_i/c), i = 1, \ldots, n-1\}$ on T^2 for the distribution $U(T^2)$. By the assumption, $g_c(y_1, \ldots, y_{n-1})/c^2$ is minimized if and only if $y_1/c = 1/[2(n-1)]$, $y_2/c = 3/[2(n-1)], \ldots$, $y_{n-1}/c = (2n-3)/[2(n-1)]$, and the corresponding minimum value of $g_c(y_1, \ldots, y_{n-1})$ is $\frac{c^2}{2(n-1)}$.

Now we optimize $ML1D(\mathcal{Y})$ with respect to c and y_n. We have $0 < c < x_n < 1$ and

$$ML1D(\mathcal{Y}) = \frac{c^2}{2(n-1)} + 2\int_c^1 |z - y_n| dz = \frac{c^2}{2(n-1)} + c^2 - 2cy_n + 2z_n^2 - 2y_n + 1.$$

Taking the partial derivative of y_n and c, we obtain that the minimum of $ML1D(\mathcal{Y})$ is attained at $y_n = (2n-1)/(2n)$ and $c = 1 - 1/n$. The corresponding

$$\mathcal{Y}^* = \left\{ \left(\frac{1}{2n}, \frac{2n-1}{2n} \right), \left(\frac{3}{2n}, \frac{2n-3}{2n} \right), \ldots, \left(\frac{2n-1}{2n}, \frac{1}{2n} \right) \right\}$$

and $ML1D(\mathcal{Y}^*) = 1/2n$. Therefore, the desired conclusion holds for size n.

8.7.3 Proof of Theorem 8.8

For $p = 3$ and $m \geq 1$, the simplex-lattice design $\mathcal{D}$ has $n = \binom{p+m-1}{m} = (m+2)(m+1)/2$ points, denoted as $\{\boldsymbol{y}_1, \ldots, \boldsymbol{y}_n\}$, and these points are all possible combinations of the components under the constraint $x_1 + x_2 + x_3 = 1$, and each component is an element from the set $\{0, 1/m, 2/m, \ldots, 1\}$. For a given contraction ratio $\rho \in (0, 1]$ and a vector component $x = i/m, 0 \leq i \leq m$, the contracted component is

$$x' = \rho x + (1-\rho)/3 = 1/3 + (i/m - 1/3)\rho.$$

By this formula, we can easily obtain the coordinates of the Scheffé type simplex-lattice design $\mathcal{D}_\rho = \{\boldsymbol{y}_1, \ldots, \boldsymbol{y}_n\}$.

The region of T^3 can be decomposed into n distinct parts (Voronoi cells) such that each contains all points closer to $\boldsymbol{y}_i$, $i = 1, \ldots, n$. In other words, each Voronoi cell contains a single point $\boldsymbol{y}_i$, $i = 1, \ldots, n$, and we denote it as T_i^3. For any point $\boldsymbol{x} \in T_i^3$, $\min_{1 \leq j \leq k} D_{l_1}(\boldsymbol{x}, \boldsymbol{y}_j) = D_{l_1}(\boldsymbol{x}, \boldsymbol{y}_i)$.

As an illustration, Figure 8.11 shows the 10 points of a Scheffé type $\{3,3\}$-simplex-lattice design and the corresponding Voronoi cells.

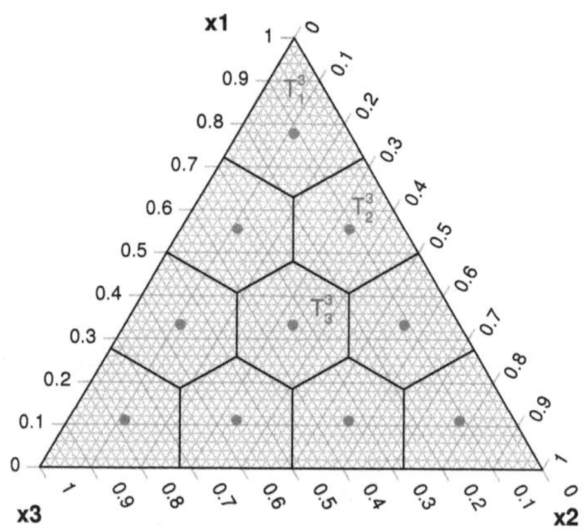

FIGURE 8.11
A contracted Scheffé type $\{3,3\}$-simplex-lattice design.

Under the L_1-metric on T^3, we can deduce that the regions (Voronoi cells) $V_i, i = 1, \ldots, n$ are polygons and can be classified into three categories by symmetry of $\mathcal{D}$, as illustrated in Figure 8.11 for the case of $m = 3$. Consequently, to compute $ML1D(\mathcal{D}_\rho)$, we only need to consider three points of $\mathcal{D}_\rho$:

$$\boldsymbol{y}_1 = \left(\frac{1}{3} + \frac{2\rho}{3}, \frac{1}{3} - \frac{\rho}{3}, \frac{1}{3} - \frac{\rho}{3}\right),$$
$$\boldsymbol{y}_2 = \left(\frac{1}{3} + \left(\frac{2}{3} - \frac{1}{m}\right)\rho, \frac{1}{3} - \frac{\rho}{3}, \frac{1}{3} + \left(\frac{1}{m} - \frac{1}{3}\right)\rho\right),$$
$$\boldsymbol{y}_3 = \left(\frac{1}{3} + \left(\frac{2}{3} - \frac{2}{m}\right)\rho, \frac{1}{3} + \left(\frac{1}{m} - \frac{1}{3}\right)\rho, \frac{1}{3} + \left(\frac{1}{m} - \frac{1}{3}\right)\rho\right),$$

and their corresponding regions T_i^3, as labelled in Figure 8.11 for the case of $m = 3$. We have $MSE_{l_1}(\mathcal{D}_\rho) = E_{\boldsymbol{X}}[\min_{1\le j\le (m+2)(m+1)/2} D_{l_1}(\boldsymbol{x}, \boldsymbol{y}_j)$ is equal to

$$3\int_{\boldsymbol{x}\in T_1^3} D_{l_1}(\boldsymbol{x}, \boldsymbol{y}_1)d\mu(\boldsymbol{x}) + (3m-3)\int_{\boldsymbol{x}\in T_2^3} D_{l_1}(\boldsymbol{x}, \boldsymbol{y}_2)d\mu(\boldsymbol{x})$$
$$+\frac{(m-1)(m-2)}{2}\int_{\boldsymbol{x}\in T_3^3} D_{l_1}(\boldsymbol{x}, \boldsymbol{y}_3)d\mu(\boldsymbol{x}),$$

where $\boldsymbol{X} \sim U(T^3)$ and μ is the uniform probability measure on T^3. By a careful check, we can obtain that the region

$$T_1^3 = \left\{(x_1, x_2, x_3) : x_1 = 1 - x_2 - x_3, 0 \le x_2, x_3 \le 1, x_1 - x_2 \ge \frac{m-1}{m}\rho, x_1 - x_3 \ge \frac{m-1}{m}\rho\right\},$$

and is equivalent to

$$V_1^2 := \left\{(x_2, x_3) : 0 \le x_2 \le \frac{1}{2} - \frac{m-1}{2m}\rho, 0 \le x_3 \le \frac{1}{2} - \frac{m-1}{2m}\rho,\right.$$
$$\left. 2x_2 + x_3 \le 1 - \frac{m-1}{m}\rho, x_2 + 2x_3 \le 1 - \frac{m-1}{m}\rho\right\}.$$

The region

$$T_2^3 = \left\{(x_1, x_2, x_3) : x_1 = 1 - x_2 - x_3, 0 \le x_2, x_3 \le 1,\right.$$
$$\left.\frac{m-3}{m}\rho \le x_1 - x_3 \le \frac{m-1}{m}\rho, x_1 - x_2 \ge \frac{m-2}{m}\rho, x_3 - x_2 \ge 0\right\},$$

and is equivalent to

$$V_2^2 = \left\{(x_2, x_3) : x_3 \ge x_2 \ge 0, 1 - \frac{m-1}{m}\rho \le x_2 + 2x_3 \le 1 - \frac{m-3}{m}\rho, 2x_2 + x_3 \le 1 - \frac{m-2}{m}\rho\right\}.$$

The region

$$T_3^3 = \left\{(x_1, x_2, x_3) : x_1 = 1 - x_2 - x_3, 0 \le x_2, x_3 \le 1,\right.$$
$$\left.\frac{m-4}{m}\rho \le x_1 - x_2 \le \frac{m-2}{m}\rho, \frac{m-4}{m}\rho \le x_1 - x_3 \le \frac{m-2}{m}\rho, -\frac{\rho}{m} \le x_2 - x_3 \le \frac{\rho}{m}\right\},$$

and is equivalent to

$$V_3^3 = \left\{(x_2, x_3) : 1 - \frac{m-2}{m}\rho \le 2x_2 + x_3 \le 1 - \frac{m-4}{m}\rho, 1 - \frac{m-2}{m}\rho \le x_2 + 2x_3\right.$$
$$\left.\le 1 - \frac{m-4}{m}\rho, -\frac{\rho}{m} \le x_2 - x_3 \le \frac{\rho}{m}\right\}.$$

Since $(X_1, X_2, X_3) \sim U(T^3)$, $(X_2, X_3) \sim U(V^2)$, and (X_2, X_3) has a probability density function $2!1_{(x_2,x_3)\in V^2}$. We have $ML1D(\mathcal{D}_\rho)$ is equal to

$$6\int_{(x_2,x_3)\in V_1^2}\left(\left|x_2-\frac{1-\rho}{3}\right|+\left|x_3-\frac{1-\rho}{3}\right|+\left|x_2+x_3-\frac{2-2\rho}{3}\right|\right)dx_2dx_3$$
$$+(6m-6)\int_{(x_2,x_3)\in V_2^2}\left(\left|x_2-\frac{1-\rho}{3}\right|+\left|x_3-\frac{1}{3}-\left(\frac{1}{m}-\frac{1}{3}\right)\rho\right|\right.$$
$$\left.+\left|x_2+x_3-\frac{2}{3}+\left(\frac{2}{3}-\frac{1}{m}\right)\rho\right|\right)dx_2dx_3+(m-1)(m-2)\int_{(x_2,x_3)\in V_3^2}\left(\left|x_2-\frac{1}{3}-\left(\frac{1}{m}-\frac{1}{3}\right)\rho\right|\right.$$
$$\left.+\left|x_3-\frac{1}{3}-\left(\frac{1}{m}-\frac{1}{3}\right)\rho\right|+\left|x_2+x_3-\frac{2}{3}+\left(\frac{2}{3}-\frac{2}{m}\right)\rho\right|\right)dx_2dx_3,$$

Tedious calculation shows

$$ML1D(\mathcal{D}_\rho)=\begin{cases}\frac{\rho^3}{m^2}+\frac{7\rho^3}{9m}+\frac{2}{27}(\rho-1)^2(\rho+8), & \text{if } 0\le\rho\le\frac{m}{m+3};\\ -\frac{m^2(\rho-1)^3+m(7\rho-16)(\rho-1)^2+3\rho^2(2\rho-9)}{27m}, & \text{if } \frac{m}{m+3}\le\rho\le 1.\end{cases}$$

It can be checked that when $\rho = 0$, $\mathcal{D}_\rho$ reduces to $\{(1/3, 1/3, 1/3)\}$, i.e., the center point of T^3, and $MSE_{l_1}(\mathcal{D}_\rho) = 16/27$, which is the same as Example 8.11. Taking derivative of $ML1D(\mathcal{D}_\rho)$ with respect to ρ we have the local minimum is attained at

$$\rho^*=\frac{m^2-\sqrt{21m^2+102m+81}+10m+9}{m^2+7m+6}$$

and the corresponding

$$ML1D(\mathcal{D}_{\rho^*})=\frac{7m^3+145m^2-(14m+54)\sqrt{21m^2+102m+81}+624m+486}{9m(m+1)(m+6)^2}.$$

It can be verified that this is a global minimum value of $ML1D(\mathcal{D}_\rho)$ over $\rho \in (0, 1]$, which completes the proof.

8.7.4 Proof of Theorem 8.12

The exact F-distribution for each d_{-i} in (1) is given in Theorem 1 of Ahn (1992). The asymptotic independence of $\{d_{-i}^2 : i = 1, \ldots, n\}$ can be verified as follows. According to the strong law of convergence,

$$\bar{x} \xrightarrow{\text{a.s}} \mu, \quad (n \to \infty)$$

where $\xrightarrow{\text{a.s}}$ means "converge almost surely". S_n can be written as

$$S_n=\frac{1}{n}\sum_{i=1}^{n}(x_i-\mu)(x_i-\mu)'+(\bar{x}-\mu)(\bar{x}-\mu)'$$

Because

$$\frac{1}{n}\sum_{i=1}^{n}(x_i-\boldsymbol{\mu})(x_i-\boldsymbol{\mu})'$$

is the sample mean of the i.id. terms $(\boldsymbol{x_i}-\boldsymbol{\mu})(\boldsymbol{x_i}-\boldsymbol{\mu})'$, the strong law of convergence (page 238, Theorem 1 of Feller (1991)) gives

$$\frac{1}{n}\sum_{i=1}^{n}(x_i-\mu)(x_i-\mu)' \xrightarrow{\text{a.s}} E\{(x_i-\mu)(x_i-\mu)'\}=\Sigma$$

According to the continuous mapping theorem (page 7, van der Vaart (1998)), it follows that

$$(\bar{x}-\mu)(\bar{x}-\mu)' \xrightarrow{\text{a.s}} 0, \quad S_n=\frac{1}{n}\sum_{i=1}^{n}(x_i-\boldsymbol{\mu})(x_i-\boldsymbol{\mu})'+(\bar{x}-\boldsymbol{\mu})(\bar{x}-\boldsymbol{\mu})' \xrightarrow{\text{a.s}} \boldsymbol{\Sigma}, \quad (n\to\infty)$$

and

$$nD_i^2 \xrightarrow{\text{a.s}} (x_i-\boldsymbol{\mu})'\boldsymbol{\Sigma}^{-1}(x_i-\boldsymbol{\mu}), \quad nD_j^2 \xrightarrow{\text{a.s}} (x_j-\boldsymbol{\mu})'\boldsymbol{\Sigma}^{-1}(x_j-\boldsymbol{\mu}) \quad (n\to\infty)$$

for any given $i\neq j$. The independence between x_i and x_j results in the asymptotic independence between nD_i^2 and nD_j^2 for $i\neq j$. Because d_{-i}^2 is a continuous function of nD_i^2 for $i=1,\ldots,n$, using the continuous mapping theorem again, we can obtain the asymptotic independence of $\{d_{-i}^2 : i=1,\ldots,n\}$. This completes the proof.

9

Representative Points for Discrete Massive Data

Previous chapters always considered representative points for continuous distributions. However, representative points have many applications for discrete data, especially for discrete massive data. In this chapter, we consider cases where the sample size N is hugely larger than the number of variables p and introduce some model-based and model-free subsampling methods.

9.1 Introduction

With the rapid advances in science and technology, the big data era brings extraordinary data. For example, the on-time airline dataset from the 2009 ASA Data Expo contains $N = 123534969$ observations on 29 variables about flight arrival and departure information for all commercial flights within the USA from October 1987 to April 2008. When the sample size N is hugely larger than the number of variables p, we can call the data "big data". Since the advance of computing resources still lags far behind the exponential growth of databases, and effective and efficient statistical tools for analyzing massive datasets are still lacking, a critical step in extensive data analysis is data reduction. Subsampling methods are potential techniques to be used for data reduction. Traditionally, subsampling has been used to refer to bootstrap, which often uses the random sampling technique. It has been proved that the performance of the random sampling method can be improved.

Some deterministic subsampling methods have been proposed based on parametric models, such as linear regression, logistic regression, generalized linear regression, quantile regression, and misspecified linear models. They assume that the underlying model between the variables and the response is a special parametric model, then give some algorithm to obtain an optimal subsample under some optimality criteria, such as the D-, A-optimality. For example, Ting and Brochu (2018) investigated optimal subsampling with influence functions. Wang et al. (2019) proposed the information-based optimal subdata selection (IBOSS) for linear regression models under the D-optimality criterion. Wang et al. (2021) proposed an orthogonal subsampling approach for big data with a focus on linear regression models, which improves the performance of IBOSS. Wang and Ma (2021) considered the optimal subsampling procedure that minimizes the asymptotic variance of the subsampling estimator for quantile regression. Moreover, Yu et al. (2022) derived the optimal Poisson subsampling method for generalized linear regression models under the A- and L-optimality criteria. Meng et al. (2021) proposed a subsampling approach in a misspecified linear model using orthogonal Latin hypercube designs.

When the model is correct, such methods provide excellent compression of big data; otherwise, poor subsampling performance may result. In practical problems, a situation

DOI: 10.1201/9781003589389-9

is that the underlying model is unknown before analyzing the data. Hence, with limited computational resources, a robust subsampling method against model misspecification may be a desirable choice. In other words, some model-free subsampling methods are necessary for representing the entire dataset, such that the subsample is statistically similar to the entire dataset. Then, we do not need a special underlying model. In a mathematical form, we assume $\mathcal{X} = \{\boldsymbol{x}_i, i = 1, \ldots, N\}$ is an observed large-scale data in R^s, and consider the problem of selecting n points from $\mathcal{X}$ to form a subdata $\mathcal{P}$ that preserves the distribution of $\mathcal{X}$. Therefore, the subdata $\mathcal{P}$ can be considered the representative points of the discrete full dataset $\mathcal{X}$. Usually, the procedure of obtaining $\mathcal{P}$ is nonuniform subsampling.

This chapter is organized as follows. Section 9.2 introduces some model-based subsampling methods, such as IBOSS and its extension. Section 9.3 gives the method of Split, which Joseph and Vakayil (2022) proposed, and its improved method, data twinning proposed by Vakayil and Joseph (2022). Section 9.4 introduces the data-driven subsampling method proposed by Zhang et al. (2024a). Section 9.5 shows the global likelihood subsampling based on the global likelihood sampler.

9.2 Model-Based Subsampling Methods

Let $(\boldsymbol{x}_1, y_1), \ldots, (\boldsymbol{x}_N, y_N)$ denote the full dataset, where $\boldsymbol{x}_i = (x_{i1}, \ldots, x_{ip})$ is the p predictor variables and y_i is the response, $i = 1, \ldots, N$. A model-based subsampling method uses a parametric model

$$y_i = g(\boldsymbol{x}_i, \boldsymbol{\beta}) + \varepsilon_i, \; i = 1, \ldots, N, \tag{9.1}$$

to link the predictor variables and the response, where g is a known function, $\boldsymbol{\beta}$ is the model parameters, ε_i is the random error with mean 0 and variance σ^2. For the model (9.1), one can use some parameter estimation method to estimate the parameters in $\boldsymbol{\beta}$, such as the least square estimation method. A model-based subsampling method assumes that the relationship between the predictor variables and the response follows the known model (9.1), then the subsampling method uses some technique to obtain the n-point subdata $(\boldsymbol{x}_1^*, y_1^*), \ldots, (\boldsymbol{x}_n^*, y_n^*)$ deterministically from the full data, such that the model parameters $\boldsymbol{\beta}$ can be estimated as accurate as possible by the subdata. In (9.1), the response can be quantitative or qualitative.

9.2.1 Subsampling for linear models

When the response is quantitative, the simplest parametric model is the linear model

$$y_i = \beta_0 + \boldsymbol{x}_i\boldsymbol{\beta}_1 + \varepsilon_i = \beta_0 + \sum_{j=1}^{p} x_{ij}\beta_j + \varepsilon_i, \; i = 1, \ldots, N, \tag{9.2}$$

with $E(\varepsilon_i) = 0, \mathrm{Var}(\varepsilon_i) = \sigma^2$. Denote $\boldsymbol{\beta} = (\beta_0, \boldsymbol{\beta}_1^T)^T = (\beta_0, \beta_1, \ldots, \beta_p)^T$, $\boldsymbol{z}_i = (1, \boldsymbol{x}_i)$, and $Z = (\boldsymbol{z}_1^T, \ldots, \boldsymbol{z}_N^T)^T$ as the coefficient matrix, $\boldsymbol{y} = (y_1, \ldots, y_N)^T$. Then, the least square estimator of $\boldsymbol{\beta}$ is as follows,

$$\hat{\boldsymbol{\beta}} = (Z^T Z)^{-1} Z^T \boldsymbol{y} = \left(\sum_{i=1}^{N} \boldsymbol{z}_i^T \boldsymbol{z}_i \right)^{-1} \sum_{i=1}^{N} \boldsymbol{z}_i^T y_i, \tag{9.3}$$

which is known as the unbiased estimator. The covariance matrix of this unbiased estimator is equal to the inverse of $\frac{1}{\sigma^2}\sum_{i=1}^{N} \boldsymbol{z}_i^T \boldsymbol{z}_i$, which is often called as the information matrix. Since the variance of the random error σ^2 is a constant, the term $1/\sigma^2$ can be omitted, and the information matrix becomes

$$M_f = Z^T Z = \sum_{i=1}^{N} \boldsymbol{z}_i^T \boldsymbol{z}_i. \tag{9.4}$$

Moreover, for the selected subdata $(\boldsymbol{x}_1^*, y_1^*), \ldots, (\boldsymbol{x}_n^*, y_n^*)$, we also assume that it follows the linear regression model

$$y_i^* = \beta_0 + \boldsymbol{x}_i^* \boldsymbol{\beta}_1 + \varepsilon_i^*,\ i = 1, \ldots, n. \tag{9.5}$$

Then, the least square estimator of $\boldsymbol{\beta} = (\beta_0, \boldsymbol{\beta}_1^T)^T$ is

$$\hat{\boldsymbol{\beta}}_s = \left(\sum_{i=1}^{n} \boldsymbol{z}_i^{*T} \boldsymbol{z}_i^*\right)^{-1} \sum_{i=1}^{n} \boldsymbol{z}_i^{*T} y_i^*, \tag{9.6}$$

which is also the best linear unbiased estimator. Here $\boldsymbol{z}_i^* = (1, \boldsymbol{x}_i^*)$. The covariance matrix of $\hat{\boldsymbol{\beta}}_s$ is

$$\text{Cov}(\hat{\boldsymbol{\beta}}_s) = \sigma^2 \left(\sum_{i=1}^{n} \boldsymbol{z}_i^{*T} \boldsymbol{z}_i^*\right)^{-1}.$$

Let δ_i be the indicator variable that signifies whether $(\boldsymbol{x}_i, y_i)$ is included in the subdata, that is, $\delta_i = 1$ if $(\boldsymbol{x}_i, y_i)$ is included and $\delta_i = 0$ otherwise. Then

$$\text{Cov}(\hat{\boldsymbol{\beta}}_s) = \sigma^2 \left(\sum_{i=1}^{N} \delta_i \boldsymbol{z}_i{}^T \boldsymbol{z}_i\right)^{-1},$$

and the corresponding information matrix becomes

$$M_s(\boldsymbol{\delta}) = \sum_{i=1}^{N} \delta_i \boldsymbol{z}_i{}^T \boldsymbol{z}_i, \tag{9.7}$$

where $\boldsymbol{\delta} = (\delta_1, \ldots, \delta_N)$ such that $\sum_{i=1}^{N} \delta_i = n$. Obtaining the best estimation of $\boldsymbol{\beta}$ naturally requires minimizing some criterion based on the covariance matrix $\text{Cov}(\hat{\boldsymbol{\beta}}_s)$. From the theory of optimality design, one can choose the D-, A-, or E-optimality criterion, see Kiefer (1959). Among them, the D-optimality criterion is used popularly, maximizing the determinant of the information matrix M_s. Therefore, the problem can be described as an optimization problem that finds an optimal $\boldsymbol{\delta}$ to maximize $|M_s(\boldsymbol{\delta})|$, that is,

$$\boldsymbol{\delta}^{opt} = \arg\max_{\boldsymbol{\delta}} \left|\sum_{i=1}^{N} \delta_i \boldsymbol{z}_i{}^T \boldsymbol{z}_i\right|,\ \ s.t.\ \sum_{i=1}^{N} \delta_i = n. \tag{9.8}$$

For calculating $|M_s(\boldsymbol{\delta})|$, Wang et al. (2019) derived the following upper bound.

Theorem 9.1 *For subdata of size n represented by $\boldsymbol{\delta}$,*

$$|M_s(\boldsymbol{\delta})| \leq \frac{n^{p+1}}{4^p \sigma^{2(p+1)}} \prod_{j=1}^{p} \left(x_{(N)j} - x_{(1)j}\right)^2 \tag{9.9}$$

where $x_{(N)j} = \max\{x_{1j}, x_{2j}, \ldots, x_{Nj}\}$ *and* $x_{(1)j} = \min\{x_{1j}, x_{2j}, \ldots, x_{Nj}\}$ *are the* N*th and first order statistics of* $x_{1j}, x_{2j}, \ldots, x_{Nj}$*. If the subdata consists of the* 2^p *points* $(a_1, \ldots, a_p)^T$ *where* $a_j = z_{(N)j}$ *or* $z_{(1)j}, j = 1, 2, \ldots, p$*, each occurring equally often, then the equality holds in* (9.9).

Usually, the number n is much smaller than 2^p, then the equality condition for the subdata in Theorem 9.1 may not exist. Then, Wang et al. (2019) suggested collecting subdata with extreme covariate values, both small and large, occurring with the same frequency, and proposed an algorithm, information-based optimal subdata selection, to select the subdata.

Algorithm 9.1 (IBOSS for linear model)

Suppose that $r = n/(2p)$ *is an integer.*

Step 1. For $x_{i1}, 1 \leq i \leq N$*, include* r *data points with the* r *smallest* x_{i1} *values and* r *data points with the* r *largest* x_{i1} *values.*

Step 2. For $j = 2, \ldots, p$*, exclude data points that were previously selected, and from the remainder, select* r *data points with the smallest* x_{ij} *values and* r *data points with the largest* x_{ij} *values.*

Step 3. Return $\hat{\boldsymbol{\beta}}^D = \left(Z_D^{*T} Z_D^*\right)^{-1} Z_D^{*T} \boldsymbol{y}_D^*$ *and the estimated covariance matrix* $\hat{\sigma}_D^2 \left(Z_D^{*T} Z_D^*\right)^{-1}$*, where* $Z_D^* = (\mathbf{1}, X_D^*), X_D^*$ *is the covariate matrix of the subdata selected in the previous steps,* $\boldsymbol{y}_D^*$ *is the response vector of the subdata and* $\hat{\sigma}_D^2 = \left\|\boldsymbol{y}_D^* - \mathrm{X}_D^* \hat{\boldsymbol{\beta}}^D\right\|^2 /(n - p - 1)$.

Compared to other subdata methods, the IBOSS method has the following advantages:

(i) It is significantly faster; for the scenario that $N > np$, it can be verified that the computational complexity of Algorithm 9.1 is $O(Np)$. Then, this algorithm is faster than the algorithmic leveraging proposed by Drineas et al. (2012), which has a computing time of $O(pNlogN)$.

(ii) It is suitable for distributed parallel computing since one can simultaneously process each covariate and find the indexes of its extreme values.

(iii) It was shown that the IBOSS method does not induce sampling error and the variances of estimators converge to 0 as the total data size N becomes large even if the subdata size n is fixed.

(iv) Data analysis for IBOSS subdata is straightforward, and the sampling distribution of an IBOSS estimator is easy to assess.

Based on those good properties, the IBOSS method has often been used for subsampling.

Example 9.1 *Suppose the total sample is randomly generated from a bivariate uniform distribution on* $[-1, 1]^2$*, its size* $N = 1000$*. We want to select an informative subsample of size* $n = 20$*. Figure 9.1 shows the total sample and subsamples selected by the uniform subsampling and IBOSS. It can be seen that the subsample by IBOSS is located at the boundary of each covariate, which is much better than the subsample by the uniform subsampling method for estimating the linear regression model.*

According to the optimality design theory, the 2^p complete design with design points at all the vertex is D-optimal for estimating the parameters in the model (9.2). From Figure 9.1(c), the points obtained by the IBOSS method are not always located at the vertices, then the estimation efficiency of the parameters in (9.2) can be improved by other methods.

Since each covariate is bounded by its minimum and maximum in the full sample, we can always scale each covariate to $[-1, 1]$. Moreover, for a scaled subsample with size n forms an orthogonal array on the two levels $\{-1, 1\}$, its combinatorial orthogonality ensures that the corresponding coefficient matrix Z satisfies $Z^T Z = nI_{p+1}$ where I_{p+1} is the identity matrix of order $p+1$, then the subsample would be D- and A-optimal for the model in (9.2).

Wang et al. (2021) proposed a new subsampling approach for big data called orthogonal subsampling (OSS). The approach is inspired by an orthogonal array of two levels, providing the best experimental design for linear regression based on the optimality design theory. OSS aims to find a subset of the full sample that forms an orthogonal array. However, the number of rows of a two-level orthogonal array of strength 2 has to be a multiple of 4, so the subsample size is not flexible for any n. Moreover, even if n is a multiple of 4, a full sample generally does not contain a subset of n points, which forms an orthogonal array. Therefore, the OSS approach finds a subsample that best approximates an orthogonal array to achieve near optimality for the problem in (9.8). The reader can refer to Wang et al. (2021) for the detailed procedure of OSS.

9.2.2 Subsampling for generalized linear model

In many datasets, the response is qualitative. For example, one can use the residents' information from the U.S. Census, such as income, age, work class, education, and the number of working hours per week, to predict whether the residents are high-income residents. Then, each response can be denoted as 1 or -1. One often uses the logistic regression model, a generalized linear model for the qualitative response. However, the whole census data set is superlarge, and the computational complexity of statistical analysis is too high. Therefore, it is natural to obtain sub-data to estimate the parameters in the logistic model as accurately as possible.

For a dataset $\{(\boldsymbol{x}_i, y_i), i = 1, \ldots, N\}$, where $\boldsymbol{x}_i = (x_{i1}, \ldots, x_{ip})$, $y_i \in \{-1, 1\}$, a logistic regression model is as follows,

$$P(y_i = 1|\boldsymbol{x}_i) = p_i(\boldsymbol{\beta}) = \frac{\exp(\boldsymbol{x}_i\boldsymbol{\beta})}{1 + \exp(\boldsymbol{x}_i\boldsymbol{\beta})}, i = 1, 2, \ldots, N. \tag{9.10}$$

The maximum likelihood estimator (MLE) of $\boldsymbol{\beta}$ is to maximize the log-likelihood function

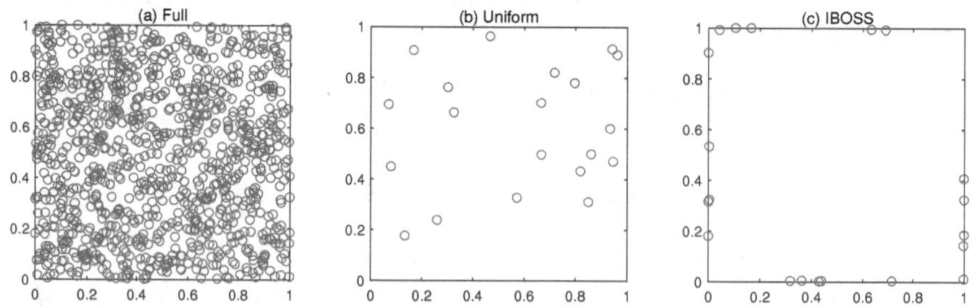

FIGURE 9.1
The selected subsample by different subsampling methods.

with respect to $\boldsymbol{\beta}$, that is,

$$\hat{\boldsymbol{\beta}}_{mle} = \arg\max_{\boldsymbol{\beta}} l(\boldsymbol{\beta}) = \arg\max_{\boldsymbol{\beta}} \sum_{i=1}^{N} [y_i \log p_i(\boldsymbol{\beta}) + (1 - y_i) \log\{1 - p_i(\boldsymbol{\beta})\}].$$

It is known that there is no general closed-form solution for $\hat{\boldsymbol{\beta}}_{mle}$, and the iterative procedure by the Newton method is as follows,

$$\hat{\boldsymbol{\beta}}^{(t+1)} = \hat{\boldsymbol{\beta}}^{(t)} + \left\{ \sum_{i=1}^{N} w_i(\hat{\boldsymbol{\beta}}^{(t)}) x_i^T x_i \right\}^{-1} \frac{\partial l(\hat{\boldsymbol{\beta}}^{(t)})}{\partial \boldsymbol{\beta}}, \tag{9.11}$$

until $\hat{\boldsymbol{\beta}}^{(t+1)}$ converges, where $w_i(\boldsymbol{\beta}) = p_i(\boldsymbol{\beta})(1 - p_i(\boldsymbol{\beta}))$. Suppose it needs ξ iterations for the optimization procedure to converge; then the computational complexity of the full data MLE typically requires $O(\xi n p^2)$.

For decreasing the running time, Wang et al. (2018) proposed an algorithm for obtaining the optimal subsample based on the A-optimality criterion (OSMAC) in the theory of optimal experimental design to approximate the full data MLE. In OSMAC, we determine the importance score of each data point first, then define the nonuniform subsampling probabilities for sampling from the full dataset by using the importance scores. The computational complexity of OSMAC is $O(Np)$, which is much smaller than that of the full data. The iteration process of OSMAC is shown in Algorithm 9.2.

Algorithm 9.2 (OSMAC for logistic model)

Step 1. Random sample r_0 data points from the full data $\{(\boldsymbol{x}_i, y_i), i = 1, \ldots, N\}$. Then, use the iteration formula (9.11) *to obtain the initial estimator of $\boldsymbol{\beta}$, denoted as $\tilde{\boldsymbol{\beta}}_{mle}$.*

Step 2. Draw a random subsample of size $r(\ll N)$, according to the probabilities $\{\pi_i\}_{i=1}^N$, from the full data, where

$$\pi_i = \frac{|y_i - p_i(\tilde{\boldsymbol{\beta}}_{mle})| \parallel x_i \parallel}{\sum_{j=1}^N |y_j - p_j(\tilde{\boldsymbol{\beta}}_{mle})| \parallel x_j \parallel}, i = 1, \ldots, N, \tag{9.12}$$

and $p_i(\cdot)$ is defined in (9.10).

Step 3. Combine the subsamples in Steps 1 and 2 to obtain the final subsample with size $n = r_0 + r$. Denote the covariates, responses, and corresponding probabilities in the subsample as x_i^*, y_i^* and π_i^*, respective, $i = 1, \ldots, n$. Denote the weighted likelihood function

$$l^*(\boldsymbol{\beta}) = \frac{1}{r} \sum_{i=1}^{r} \frac{1}{\pi_i^*} \left[y_i^* \log p_i^*(\boldsymbol{\beta}) + (1 - y_i^*) \log\left\{1 - p_i^*(\boldsymbol{\beta})\right\}\right],$$

where $p_i^*(\boldsymbol{\beta}) = \exp(\mathbf{x}_i^* \boldsymbol{\beta}) / \{1 + \exp(\mathbf{x}_i^* \boldsymbol{\beta})\}$.

Step 4. Maximize $l^*(\boldsymbol{\beta})$ to get the estimator $\tilde{\boldsymbol{\beta}}$ of $\boldsymbol{\beta}$ based on the n-point subsample, by the following iterative formula until $\tilde{\boldsymbol{\beta}}^{(t+1)}$ converges,

$$\tilde{\boldsymbol{\beta}}^{(t+1)} = \tilde{\boldsymbol{\beta}}^{(t)} + \left\{ \sum_{i=1}^{r} \frac{w_i^*\left(\tilde{\boldsymbol{\beta}}^{(t)}\right) (\mathbf{x}_i^*)^T \mathbf{x}_i^*}{\pi_i^*} \right\}^{-1} \sum_{i=1}^{r} \frac{\left\{y_i^* - p_i^*\left(\tilde{\boldsymbol{\beta}}^{(t)}\right)\right\} \mathbf{x}_i^*}{\pi_i^*},$$

where $w_i^*(\boldsymbol{\beta}) = p_i^*(\boldsymbol{\beta})\{1 - p_i^*(\boldsymbol{\beta})\}$.

In Step 2 of Algorithm 9.2, the weight π_i is asymptotic optimal in some criterion such that $\tilde{\boldsymbol{\beta}}$ converges to $\hat{\boldsymbol{\beta}}_{mle}$. In Step 1 of Algorithm 9.2, when r_0 is not too small, $\tilde{\boldsymbol{\beta}}_{mle}$ accurately approximates $\hat{\boldsymbol{\beta}}_{mle}$ obtained by the full data. However, if r_0 is large and r in Step 2 is relatively small, the efficiency of Algorithm 9.2 decreases since the subsample is similar to the random subsample. Wang et al. (2018) suggested that r_0 can be chosen as a small term compared with $r^{1/2}$, that is, $r_0 = o(r^{-1/2})$, to prove the consistency and asymptotically optimality of Algorithm 9.2.

Ai et al. (2021) extended the logistic model into the exponential family, and proposed an optimal subsampling method under the A-optimality criterion for those generalized linear models, and proved the consistency and asymptotic normality of the estimator. Moreover, Yu et al. (2022) derived optimal Poisson subsampling probabilities for generalized linear models in the context of quasi-likelihood estimation under the A- and L-optimality criteria. Readers can refer to those references for the detailed procedure.

9.3 Data Splitting

In the last section, we obtain a subsample from the full data based on some parametric models, such as linear and generalized linear models. The objective of the subdata is to accurately estimate the parameters in the given parametric model. However, if the model is not accurate, the obtained subdata may not be robust enough for different models. For example, when the actual model is quadratic, and the pre-defined model is a linear model, then the subdata based on the linear model, the parameters in the quadratic model may not be well estimated. Moreover, the obtained subdata depends on a given parametric model, and the similarity between the subdata and the complete data is always low. Then, those subdata may have poor performance in the sense of representation points. Some algorithms for obtaining subdata highly similar to the complete data are welcome. In other words, those subdata should represent the full data very well. From now, we focus on those algorithms.

Splitting the complete data into training and testing data is typical for developing statistical and machine-learning models. Randomly sampling a fraction of the full data is the simplest and probably the most common strategy. For example, 80% of the rows of the complete data can be randomly chosen for training, and the remaining 20% can be used for testing. It is well known that the randomly chosen subdata is similar to the complete data when the total data size is enormous, and the random method is speedy to operate. However, if the size of the subdata is only several hundred or thousands, the similarity can be improved by other deterministic methods. In other words, we need to propose some subsampling techniques such that the subsample is a better representation point than that from the uniform sampling method, and the subsample is the representative point of the original dataset. This section introduces two splitting methods, SPlit (Joseph and Vakayil, 2022) and Data twinning (Vakayil and Joseph, 2022).

9.3.1 SPlit

Joseph and Vakayil (2022) proposed an optimal data splitting method called Support Points-based Split (SPlit) for model validation. It splits the full dataset into training and testing sets so that the full dataset and the testing set follow the same distribution. Split is also a model-free subsampling method.

Let $\mathcal{X} = \{\boldsymbol{z}_i = (\boldsymbol{x}_i, y_i), i = 1, \ldots, N\}$ be the i.i.d dataset, that is,

$$(\boldsymbol{x}_i, y_i) \stackrel{iid}{\sim} F, \quad i = 1, \ldots, N,$$

where $\boldsymbol{x}_i = (x_{i1}, \ldots, x_{ip})$ is the p predictor variables and y_i is the response. Let $g(\boldsymbol{x}, \theta)$ be the prediction model for fitting the data, where θ is a set of unknown parameters in the model, which can be estimated by minimizing a loss function $L(y, g(\boldsymbol{x}, \theta))$, such as the squared or absolute error loss. To avoid overfitting, we do not use the entire data for estimation and divide the dataset $\mathcal{X}$ into two disjoint and mutually exclusive sets: the training set $\mathcal{X}_1$ and the testing set $\mathcal{X}_2$, which respectively contain $N-n$ and n points. The SPlit algorithm aims to retain both distributions of $\mathcal{X}_1$ and $\mathcal{X}_2$ as the same F as possible. Then, $\mathcal{X}_1$ or $\mathcal{X}_2$ can be denoted as the representation points of $\mathcal{X}$ since they have almost the same distribution.

Let $\hat{\theta}$ be the estimator of the parameter θ from the training set by minimizing a loss $L(y, g(\boldsymbol{x}, \theta))$. Define the generalization error

$$\varepsilon = E_{\boldsymbol{X},Y}\left\{L(Y, g(\boldsymbol{X}, \hat{\theta})) \mid \mathcal{X}_1\right\},$$

to quantify the model's performance. Based on the testing data $\mathcal{X}_2$ whose distribution is also F, ε can be estimated by

$$\hat{\varepsilon} = \frac{1}{n}\sum_{i=1}^{n} L(y_i^2, g(\boldsymbol{x}_i^2, \hat{\theta})), \tag{9.13}$$

where $\boldsymbol{u}_i = (\boldsymbol{x}_i^2, y_i^2)$ is the ith sample in $\mathcal{X}_2$. The estimation error $\hat{\varepsilon} - \varepsilon$ converges to 0 with the ratio $O(n^{-1/2})$ if the random sample is considered. As such, the ratio can be improved by quasi-Monte Carlo methods, whose convergence ratio is $O(n^{-1+\epsilon})$ with ϵ closing to 0.

Support points proposed by Mak and Joseph (2018) are a quasi-Monte Carlo method. It depends on the energy distance (ED, Székely and Rizzo, 2013) between the testing data $\mathcal{X}_2 = \{\boldsymbol{u}_1, \ldots, \boldsymbol{u}_n\}$ and the distribution F, which can be estimated by the following formula through the given dataset,

$$\begin{aligned}\overline{\mathrm{ED}}_{n,N} =& \frac{2}{nN}\sum_{i=1}^{n}\sum_{j=1}^{N} \| \boldsymbol{u}_i - \boldsymbol{z}_j \|_2 - \frac{1}{n^2}\sum_{i=1}^{n}\sum_{j=1}^{n} \| \boldsymbol{u}_i - \boldsymbol{u}_j \|_2 \\ & - \frac{1}{N^2}\sum_{i=1}^{N}\sum_{j=1}^{N} \| \boldsymbol{z}_i - \boldsymbol{z}_j \|_2,\end{aligned} \tag{9.14}$$

where $\| \cdot \|_2$ is the L_2-norm. From (9.14), ED does not depend on the model $g(\cdot, \theta)$ nor the loss function $L(\cdot, \cdot)$, then minimizing ED obtains a model-free testing set, that is,

$$\begin{aligned}\{\boldsymbol{u}_i^*\}_{i=1}^n \in & \mathop{\arg\min}_{\{\boldsymbol{u}_i\}_{i=1}^n \subset \mathcal{X}} \overline{\mathrm{ED}}_{n,N} \\ = & \mathop{\arg\min}_{\{\boldsymbol{u}_i\}_{i=1}^n \subset \mathcal{X}} \left\{\frac{2}{nN}\sum_{i=1}^{n}\sum_{j=1}^{N} \| \boldsymbol{u}_i - \boldsymbol{z}_j \|_2 - \frac{1}{n^2}\sum_{i=1}^{n}\sum_{j=1}^{n} \| \boldsymbol{u}_i - \boldsymbol{u}_j \|_2\right\},\end{aligned} \tag{9.15}$$

since $\frac{1}{N^2}\sum_{i=1}^{N}\sum_{j=1}^{N} \| \boldsymbol{z}_i - \boldsymbol{z}_j \|_2$ can be omitted in the optimization problem. Then, the obtained testing data in (9.15) can be viewed as the representative points of the entire dataset since it is the best set of n points to represent $\mathcal{X}$ according to the energy distance criterion.

Solving the discrete optimization problem (9.15) is difficult since the candidate set is the discrete set $\mathcal{X}$. To fastly solve this optimization problem, one can first find the solution

$\{z_i^*\}_{i=1}^n$ in a continuous space, which is very fast, then choose the closest points in $\mathcal{X}$ to $\{z_i^*\}_{i=1}^n$ by using k-d tree based nearest neighbor algorithm. However, a naive nearest neighbor assignment can lead to duplicates, such that the number of the obtained points is less than n. Joseph and Vakayil (2022) modified the naive nearest neighbor assignment so that the obtained set has exact n points. Its main idea is to avoid duplicates. However, the average case complexity of the overall procedure of SPlit is $O(pN^2)$. The quadratic growth in complexity of the dataset size for SPlit is the major computational bottleneck.

9.3.2 Twinning

The computational cost of SPlit is fairly high, especially when the data size or dimension is large. To decrease the complexity, Vakayil and Joseph (2022) further studied a much quicker algorithm, called Twinning, to make it applicable to big data problems such as data compression. Like five SPlit, Twinning also aims to partition a dataset into two disjoint sets that have similar statistical properties, and the two sets are called twins, which need not have the same size. Moreover, it was theoretically shown that the objectives of Twinning and SPlit are equivalent.

Let the partition of $\mathcal{X}$, $\mathcal{X}_1 = \{\boldsymbol{u}_1, \ldots, \boldsymbol{u}_n\}$ and $\mathcal{X}_2 = \{\boldsymbol{v}_1, \ldots, \boldsymbol{v}_{N-n}\}$ be the twins. Then $\mathcal{X}_1 \bigcap \mathcal{X}_2 =$, $\mathcal{X}_1 \bigcup \mathcal{X}_2 = \mathcal{X}$. The energy distance between $\mathcal{X}_1$ and $\mathcal{X}_2$ is

$$\begin{aligned}\overline{\mathrm{ED}}_{n,N-n} =& \frac{2}{n(N-n)} \sum_{i=1}^{n} \sum_{j=1}^{N-n} \| \boldsymbol{u}_i - \boldsymbol{v}_j \|_2 - \frac{1}{n^2} \sum_{i=1}^{n} \sum_{j=1}^{n} \| \boldsymbol{u}_i - \boldsymbol{u}_j \|_2 \\ & - \frac{1}{(N-n)^2} \sum_{i=1}^{N-n} \sum_{j=1}^{N-n} \| \boldsymbol{v}_i - \boldsymbol{v}_j \|_2 .\end{aligned}$$

Since $\mathcal{X}_1 = \mathcal{X} \setminus \mathcal{X}_2$, the twins are obtained by minimizing the ED between $\mathcal{X}_1$ and $\mathcal{X}_2$, $\overline{\mathrm{ED}}_{n,N-n}$, that is, we obtain the optimal $\mathcal{X}_1$ by the following optimization problem,

$$\mathcal{X}_1^* = \arg\min_{\mathcal{X}_1 \subset \mathcal{X}} \overline{\mathrm{ED}}_{n,N-n}, \tag{9.16}$$

then $\mathcal{X}_2^* = \mathcal{X} \setminus \mathcal{X}_1^*$. Vakayil and Joseph (2022) proved that the optimization problem (9.16) is an NP-hard problem and proposed a sequential approach to solve the problem approximately.

Assume $N = rn$. The mean idea of Twinning is to partition the N points into disjoint n closets, such that the r points in each closet are close to each other. The procedure of Twinning is as follows. First, select a starting point $\boldsymbol{u}_1$ and denote $\mathcal{V}_1^{(1)} = \{\boldsymbol{u}_1\}$, then determine the first $r-1$ closest points $\{\boldsymbol{v}_1^1, \ldots, \boldsymbol{v}_{r-1}^1\}$ to $\boldsymbol{u}_1$ in $\mathcal{D} = \{\mathcal{X} \setminus \mathcal{V}_1^{(1)}\}$, and $\boldsymbol{v}_1^1$ is nearest point to $\boldsymbol{u}_1$ and $\boldsymbol{v}_{r-1}^1$ is the farthest one to $\boldsymbol{u}_1$ among those $r-1$ points; Denote $\mathcal{V}_2^{(1)} = \{\boldsymbol{v}_1^1, \ldots, \boldsymbol{v}_{r-1}^1\}$ and update $\mathcal{D} = \{\mathcal{X} \setminus (\mathcal{V}_1^{(1)} \cup \mathcal{V}_2^{(1)})\}$. Then, $\{\mathcal{V}_1^{(1)}, \mathcal{V}_2^{(1)}\}$ forms a closet. Second, find a nearest point to $\boldsymbol{v}_{r-1}^1$ in $\mathcal{D}$ and denote as $\boldsymbol{u}_2$, and $\mathcal{V}_1^{(2)} = \{\mathcal{V}_1^{(1)}, \boldsymbol{u}_2\}$; Similarly, find the first $r-1$ closest points $\{\boldsymbol{v}_1^2, \ldots, \boldsymbol{v}_{r-1}^2\}$ to $\boldsymbol{u}_2$ in $\mathcal{D}$ and $\boldsymbol{v}_1^2$ is nearest point to $\boldsymbol{u}_2$ and $\boldsymbol{v}_{r-1}^2$ is the farthest one to $\boldsymbol{u}_2$ among those $r-1$ points; Denote $\mathcal{V}_2^{(2)} = \{\boldsymbol{v}_1^2, \ldots, \boldsymbol{v}_{r-1}^2\}$ and update $\mathcal{D} = \{\mathcal{X} \setminus (\mathcal{V}_1^{(2)} \cup \mathcal{V}_2^{(2)})\}$. Repeat the procedure until all the N points are divided into n closets. Then, $\mathcal{V}_1^{(n)}$ is the approximate solution of the representation points $\mathcal{X}_1^*$. For illustration, Figure 9.2 considers $N = 50$ and $n = 10$. The first point $\boldsymbol{u}_1$ is chosen as the point located at the upper right corner, which is marked as "1" and is the first point in $\mathcal{X}_1$. Then, the 4 nearest points to $\boldsymbol{u}_1$ are found, and the point "d" is farthest to $\boldsymbol{u}_1$. Among the remaining points of $\mathcal{X}$, the nearest point to the point "d" is the second point in $\mathcal{X}_1$, marked as "2". Continuing this procedure, we can obtain the 10 points.

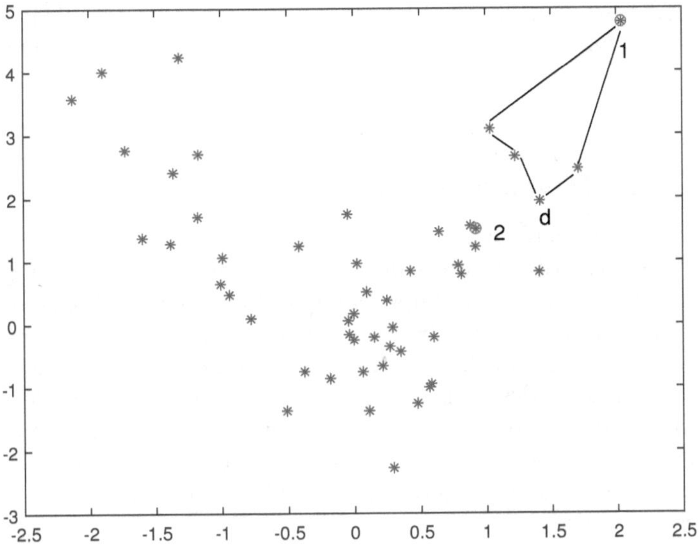

FIGURE 9.2
The illustration of twinning.

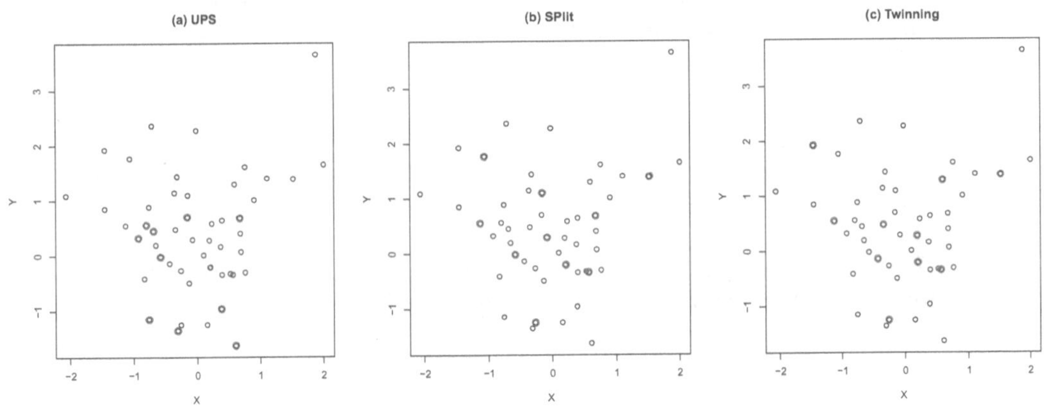

FIGURE 9.3
The performances of URS, SPlit, and twinning.

Example 9.2 *Consider* $N = 50$, $n = 10$. *Let* $x_i \overset{iid}{\sim} N(0, 1)$, *and* $y_i \overset{iid}{\sim} N(x_i, 1)$, *for* $i = 1, \ldots, N$. *Then* $r = 5$, *we choose one point from each closest point in the Twinning method. Figure 9.3 compares the performances of SPlit, Twinning, and uniform random sampling (URS). It can be easily seen that the subsample by URS is more random than the other two methods. Usually, SPlit provides better representative points than Twinning when the computational complexity is affordable. At the same time, the difference in the representative points between SPlit and Twinning is non-significant.*

Vakayil and Joseph (2022) showed that the computational complexity of Twinning is $O(pN\log N)$, which is much quicker than that of SPlit, $O(pN^2)$. Therefore, for the cases of larger N and higher dimensions, Twinning is a better choice than SPlit.

9.4 Data-Driven Subsampling

Nevertheless, the performance of both SPlit and Twinning may become bad for data compression when the dimension is high. The complexity of Twinning would also be relatively high when the data size is large enough. In addition, Zhang et al. (2024a) proposed a data-driven subsampling (DDS) method, which utilizes the rotation-inversion transformation based on a uniform design and takes the nearest neighbor substitution to obtain the subsample. Uniform design is a type of space-filling design under a given uniformity criterion, such as star discrepancy or mixture discrepancy. It aims to scatter the design points uniformly on the experimental domain, see Fang et al. (2018) for details. DDS can generate subsamples quickly and is a deterministic subsampling method. In this section, we introduce the DDS procedure.

For a given data $\mathcal{X} = \{\boldsymbol{x}_i, i = 1, \ldots, N\} \subset R^s$ with large N, denote its empirical cumulative distribution function (ECDF) as $F_{\mathcal{X}}$. To find a small data to represent $\mathcal{X}$, a natural idea is to find a subdata $\mathcal{P} \subseteq \mathcal{X}$ that has *low discrepancy* with respect to the ECDF $F_{\mathcal{X}}$. Following the F-discrepancy by Fang and Wang (1994) and the generalized ℓ_2-discrepancy by Hickernell (1998), Zhang et al. (2024a) defined the generalized empirical F-discrepancy (GEFD) to measure the difference between $\mathcal{P}$ and $\mathcal{X}$.

First, we need to transform the data $\mathcal{X}$ into the hypercube $C^s = [0,1]^s$ by $T_{\mathcal{X}} : R^s \to C^s$ of the form

$$T_{\mathcal{X}}(\boldsymbol{x}) = (F_{\mathcal{X}_{(1)}}(x_1), \ldots, F_{\mathcal{X}_{(s)}}(x_s))^T, \tag{9.17}$$

where $F_{\mathcal{X}_{(j)}}$ is the marginal ECDF of the jth component of $\mathcal{X}$, $j = 1, \ldots, s$. Clearly, $T_{\mathcal{X}}$ in (9.17) can transform the data $\mathcal{X}$ in R^s into the unit hypercube C^s component by component. Moreover, based on the theory of reproducing kernel Hilbert space, we need to give a reproducing kernel $\mathcal{K}(T_{\mathcal{X}}(\boldsymbol{x}), T_{\mathcal{X}}(\boldsymbol{z}))$ which is symmetric and non-negative definite on $C^s \times C^s$. The commonly used kernel $\mathcal{K}$ can be chosen as the kernel in wrap-around ℓ_2-discrepancy or mixture discrepancy by the theory of uniform design, and those kernels are

$$\begin{aligned} \mathcal{K}^{\mathrm{W}}(\boldsymbol{u},\boldsymbol{v}) &= \prod_{j=1}^{s}\left[\frac{3}{2} - |u_j - v_j| + |u_j - v_j|^2\right], \\ \mathcal{K}^{\mathrm{m}}(\boldsymbol{u},\boldsymbol{v}) &= \prod_{j=1}^{s}[\frac{15}{8} - \frac{1}{4}\left|u_j - \frac{1}{2}\right| - \frac{1}{4}\left|v_j - \frac{1}{2}\right| - \frac{3}{4}|u_j - v_j| + \frac{1}{2}|u_j - v_j|^2], \end{aligned} \tag{9.18}$$

respectively. Zhou et al. (2013) showed that the mixture discrepancy may be better for measuring the uniformity of point sets in C^s. Then, the GEFD for measuring the representation property of the subsample $\mathcal{P}$ is defined as follows,

Definition 9.1 *Given a data $\mathcal{X} = \{\boldsymbol{x}_1, \ldots, \boldsymbol{x}_N\} \subset R^s$, the squared generalized empirical F-discrepancy for a point set $\mathcal{P} = \{\xi_1, \cdots, \xi_n\} \subset R^s$ is defined by*

$$\begin{aligned} D^2(\mathcal{P}; \mathcal{X}, \mathcal{K}) = \int_{R^{2s}} & \mathcal{K}(T_{\mathcal{X}}(\boldsymbol{x}), T_{\mathcal{X}}(\boldsymbol{z}))\mathrm{d}(F_{\mathcal{X}}(\boldsymbol{x}) - F_{\mathcal{P}}(\boldsymbol{x})) \\ & \times \mathrm{d}(F_{\mathcal{X}}(\boldsymbol{z}) - F_{\mathcal{P}}(\boldsymbol{z})), \end{aligned} \tag{9.19}$$

where $\mathcal{K}(\cdot,\cdot)$ is a reproducing kernel function on $C^s \times C^s$ and $T_{\mathcal{X}}(\cdot)$ is defined in (9.17).

Denote $\widetilde{\mathcal{K}}(\boldsymbol{x}, \boldsymbol{z}) = \mathcal{K}(T_{\mathcal{X}}(\boldsymbol{x}), T_{\mathcal{X}}(\boldsymbol{z}))$, the GEFD can be equivalently expressed by the following explicit form,

$$\begin{aligned} D^2(\mathcal{P}; \mathcal{X}, \mathcal{K}) =& \frac{1}{N^2} \sum_{i,k=1}^{N} \widetilde{\mathcal{K}}(\boldsymbol{x}_i, \boldsymbol{x}_k) - \frac{2}{Nn} \sum_{i=1}^{N} \sum_{k=1}^{n} \widetilde{\mathcal{K}}(\boldsymbol{x}_i, \xi_k) \\ &+ \frac{1}{n^2} \sum_{i,k=1}^{n} \widetilde{\mathcal{K}}(\xi_i, \xi_k). \end{aligned} \tag{9.20}$$

Therefore, one can evaluate the GEFD criterion quickly without calculating the distribution of $\mathcal{X}$, $F_{\mathcal{X}}$, since the calculation of $T_{\mathcal{X}}$ only depends on the marginal ECDFs of the s components of $\mathcal{X}$. The proposed GEFD is reasonable for measuring the representation of small data with respect to the original big data.

Example 9.3 *Suppose the original dataset $\mathcal{X}^{(1)} = \{\boldsymbol{x}_i^{(1)}, i = 1, \ldots, 5000\}$ are generated from a binormal distribution with independent components, shown as the background grey points in Figure 9.4(a). We obtain $\mathcal{P}_1^{(1)} = \{\xi_k^{(1,1)}, k = 1, \ldots, 54\} \subset \mathcal{X}^{(1)}$ by the URS method, and $\mathcal{P}_2^{(1)} = \{\xi_k^{(1,2)}, k = 1, \ldots, n\} \subset \mathcal{X}^{(1)}$ by the subsampling method in this section. Let us choose the kernel function $\mathcal{K}^m$ of mixture discrepancy in (9.18). According to the analytical expression (9.20), it can be obtained that $D^2(\mathcal{P}_1^{(1)}; \mathcal{X}^{(1)}, \mathcal{K}^M) = 1.1212 \times 10^{-2}$ and $D^2(\mathcal{P}_2^{(1)}; \mathcal{X}^{(1)}, \mathcal{K}^M) = 4.0424 \times 10^{-4}$. Thus, $\mathcal{P}_2^{(1)}$ is better than $\mathcal{P}_1^{(1)}$ by the criterion of GEFD, which is consistent with the intuitive view in Figure 9.4(b).*

It is worth mentioning that the GEFD by Definition 9.1 is also well-defined when the components of $\mathcal{X}$ are correlated, although the transformation $T_{\mathcal{X}}$ in (9.17) translates each coordinate of $\mathcal{X}$ into $[0, 1]$ independently. The right-hand side of (9.19) is a norm of the function $F_{\mathcal{X}} - F_{\mathcal{P}}$, where the ECDF $F_{\mathcal{X}}$ is general enough to cover any joint distributions of $\mathcal{X}$.

Example 9.4 *Consider the dataset $\mathcal{X}^{(2)}$ with $N = 5000$ generated from a binormal distribution with correlated components, see Figure 9.4(c). We obtain two different 54-point subsamples $\mathcal{P}_1^{(2)}$ and $\mathcal{P}_2^{(2)}$ respectively by the URS method and the DDS method in this Section, and their GEFD values are $D^2(\mathcal{P}_1^{(2)}; \mathcal{X}^{(2)}, \mathcal{K}^M) = 7.7165 \times 10^{-3}$ and $D^2(\mathcal{P}_2^{(2)}; \mathcal{X}^{(2)}, \mathcal{K}^M) = 1.2544 \times 10^{-3}$, respectively. It is implied that $\mathcal{P}_2^{(2)}$ is better than $\mathcal{P}_1^{(2)}$ for representing the full data, which agrees with the intuitive view in Figure 9.4(d).*

Given $\mathcal{X}$, to find a small data $\mathcal{P}$ which minimizes the GEFD value leads to an NP-hard problem. Alternatively, we establish the connections between GEFD and the uniformity criterion for constructing uniform designs (Fang and Wang, 1994; Fang et al., 2018), then convert the original GEFD-optimization problem to direct utilization of existing uniform design tables. For a point set $\mathcal{P}_0 = \{\boldsymbol{x}_1, \ldots, \boldsymbol{x}_n\}$ in C^s, denote $F_{\rm u}$ as the uniform distribution on C^s, $D^2(\mathcal{P}_0; F_{\rm u}, \mathcal{K})$ as the uniformity criterion for $\mathcal{P}_0$, whose expression is as follow.

$$\begin{aligned} &D^2(\mathcal{P}_0; F_{\rm u}, \mathcal{K}) \\ &= \int_{C^{2s}} \mathcal{K}(\boldsymbol{u}, \boldsymbol{v}) {\rm d}\boldsymbol{u}{\rm d}\boldsymbol{v} - \frac{2}{n} \sum_{i=1}^{n} \int_{C^s} \mathcal{K}(\boldsymbol{u}, \boldsymbol{x}_k) {\rm d}\boldsymbol{u} + \frac{1}{n^2} \sum_{i=1}^{n} \sum_{k=1}^{n} \mathcal{K}(\boldsymbol{x}_i, \boldsymbol{x}_k), \end{aligned} \tag{9.21}$$

where $\mathcal{K}$ is the given kernel. As we know, when $N \to \infty$, the ECDF $F_{\mathcal{X}}(\boldsymbol{x})$ of $\mathcal{X}$ converges to the CDF $F(\boldsymbol{x})$ for $\boldsymbol{x} \in R^s$, and the marginal ECDF $F_{\mathcal{X}_{(j)}}(x_j)$ of the jth component

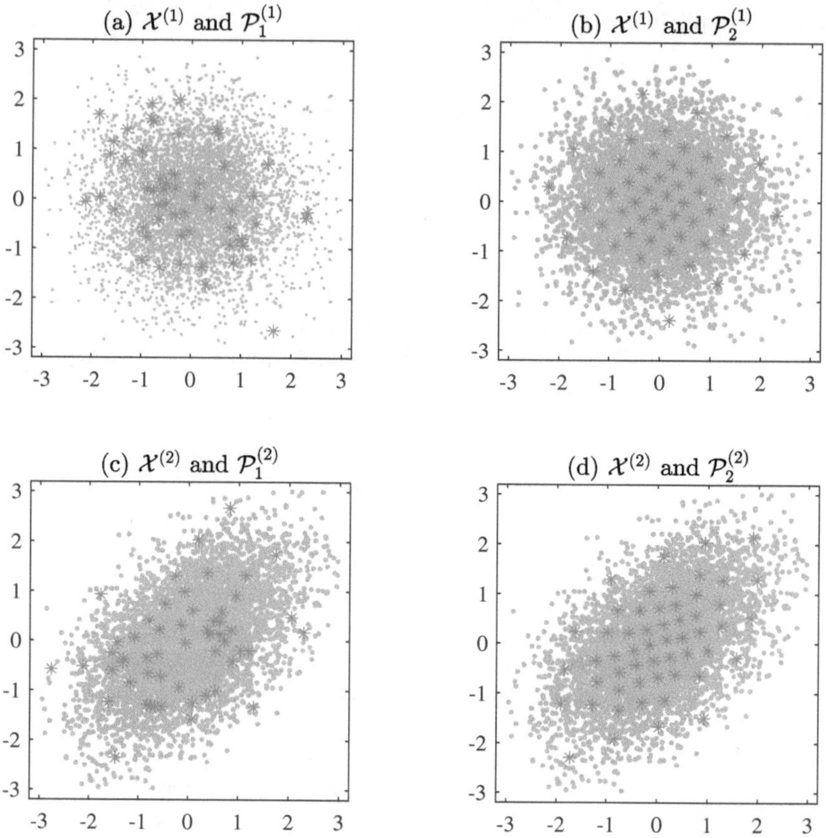

FIGURE 9.4
The scatter plots of the original binormal data $\mathcal{X}$ (grey points), the subsampled $\mathcal{P}_1$ and $\mathcal{P}_2$. Upper: independent components; Lower: correlated components.

converges to the jth marginal CDF of $F(\boldsymbol{x})$ for $j = 1, \ldots, s$. Under the joint independence condition, we can derive the asymptotic equivalence between the GEFD defined in (9.19) and the generalized ℓ_2-discrepancy in (9.21) upon the transformation $T_{\mathcal{X}}$ in (9.17).

Theorem 9.2 *Given a reference data $\mathcal{X} \subset R^s$ satisfying the joint independence*

$$F_{\mathcal{X}}(\boldsymbol{x}) = \prod_{j=1}^{s} F_{\mathcal{X}_{(j)}}(x_j), \ \forall \ \boldsymbol{x} = (x_1, \ldots, x_s)^T \in R^s, \tag{9.22}$$

suppose the number of repeated points within $\mathcal{X}$ be upper bounded by a constant c_1, and the kernel $\mathcal{K}(u, v) = \prod_{j=1}^{s} K(u_j, v_j)$ be Lipschitz continuous in the sense of $|K(u_1, v) - K(u_2, v)| \leq c_2|u_1 - u_2|$, for any $u_1, u_2, v \in [0, 1]$, and a constant c_2. Then, for any small size data $\mathcal{P} \subset R^s$,

$$D^2(\mathcal{P}; \mathcal{X}, \mathcal{K}) = D^2(T_{\mathcal{X}}(\mathcal{P}); F_{\mathrm{u}}, \mathcal{K}) + O(1/N^*),$$

where $N^* = \min_{j \in \{1:s\}} N_j$ and N_j is the number of marginal distinct values $\mathcal{X}_{(j)} = \{x_{ij}, i = 1, \ldots, N\}$ along the jth coordinate.

Theorem 9.2 translates the GEFD of $\mathcal{P} \subset R^s$ to the discrepancy of $T_{\mathcal{X}}(\mathcal{P})$ in C^s, subject to an approximation error of order $1/N^*$. Therefore, if $\mathcal{X}$ satisfies the joint independence assumption (9.22), an n-point design $\mathcal{D}$ in C^s with low discrepancy could lead to a point set $\mathcal{P}$ with low GEFD by the inverse transformation of $T_{\mathcal{X}}$, i.e. $\mathcal{P} = T_{\mathcal{X}}^{-1}(\mathcal{D})$. Theorem 1 requires the dataset $\mathcal{X}$ to satisfy the joint independence assumption (9.22). For some datasets that the assumption (9.22) may not be satisfied, we can apply the principal component analysis (PCA) method to transform the reference data. Then, for correlated data sets, such as $\mathcal{X}^{(2)}$ given in Figure 9.4, Theorem 1 can be applied to the transformed data set given by PCA. PCA is based on the singular value decomposition (SVD), which converts the data to linearly uncorrelated coordinates. Denote the rotated data as $\mathcal{Z}$. Based on the principal scores on the latent space, one can use the inversion method to find a small representation $\mathcal{Q}_{\mathcal{P}}$ with low GEFD. Then, such $\mathcal{Q}_{\mathcal{P}}$ can be converted back to the original space as the desired small data representation of $\mathcal{X}$ on R^s.

For the subsampling purpose, it is required that the obtained small data $\mathcal{P}$ should be a subset of the original data $\mathcal{X}$. Since the rotated data $\mathcal{Z}$ is only nearly independent, there are some points in $\mathcal{Q}_{\mathcal{P}}$ not belonging to $\mathcal{Z}$. For each data point, we suggest finding its nearest neighbor in $\mathcal{Z}$ as a replacement. To find the nearest neighbor of each point in $\mathcal{Q}_{\mathcal{P}} \backslash \mathcal{Z}$, k-d tree is the conventional method. Since the final subsample is directly driven by the original data, we call this procedure the data-driven subsampling method, and its detailed procedure is in Algorithm 9.3.

Algorithm 9.3 (Data-driven Subsampling)

Input *The original data* $\mathcal{X} = \{\boldsymbol{x}_1, \ldots, \boldsymbol{x}_N\} \subset R^s$, *the neighboring radius* τ *and a uniform design* $\mathcal{D} = \{\zeta_1, \ldots, \zeta_n\} \subset C^s$.

1. *Perform SVD* $\mathcal{X} = \mathcal{Z}\Lambda \boldsymbol{V}^T$ *for* $\mathcal{X}$ *to obtain the rotated data* $\mathcal{Z} = \{\boldsymbol{z}_1, \ldots, \boldsymbol{z}_N\}$ *whose components are linearly uncorrelated.*

2. *For the uniform design* $\mathcal{D}$, *obtain the data-driven space-filling design* $\mathcal{Q}_{\mathcal{P}} = \{\boldsymbol{\eta}_1, \ldots, \boldsymbol{\eta}_n\}$ *in the rotated space by using the following inverse transformation*

$$\boldsymbol{\eta}_k = T_{\mathcal{Z}}^{-1}(\zeta_k) = \left(F_{\mathcal{Z}_{(1)}}^{-1}(\zeta_{k1}), \ldots, F_{\mathcal{Z}_{(s)}}^{-1}(\zeta_{ks})\right)^T, \ k = 1, \ldots, n.$$

3. *Build a k-d tree for the rotated data* $\mathcal{Z}$.

4. *Based on the k-d tree built in Step 3, search the nearest neighbor of each design point in* $\mathcal{Q}_{\mathcal{P}}$ *from* $\mathcal{Z}$. *Denote the nearest neighbor of* $\boldsymbol{\eta}_k$ *by* $z_{i_k^*}$ *and record the sample index* i_k^*, $k = 1, \ldots, n$.

Output *Data-driven subsample* $\mathcal{P}^\dagger = \{\boldsymbol{x}_{i_1^*}, \ldots, \boldsymbol{x}_{i_n^*}\}$.

In the input of Algorithm 9.3, the uniform design can be generated by the good lattice point method, leave-one-out good lattice point method, or the threshold accepting method; the details can refer to Fang et al. (2018). DDS is a deterministic method, Zhang et al. (2024a) showed that the average case complexity is $O(sN(s + \log N))$, which is very quick. Such a procedure is computationally efficient for large-scale data in a low-dimensional space.

For an illustration, we apply the rotation-inversion construction to a two-dimensional reference data $\mathcal{X}$ and obtain the new sampled points $\mathcal{P}$, shown in Figure 9.5 (a) and (b). The original data $\mathcal{X}$ is simulated from a truncated binormal distribution with correlation, as shown in Figure 9.4. It can be found that such a small data $\mathcal{P}$ (plotted as circles) represents the original data $\mathcal{X}$ (plotted as dots) quite well. In Figure 9.5 (c), the points of $\mathcal{Q}_{\mathcal{P}}$ are shown as the circle points, and their nearest points from $\mathcal{Z}$ are shown as the star points.

Zhang et al. (2024a) gave many simulations and real case studies for describing DDS's properties in classification and regression problems. For each kind of model, they compared

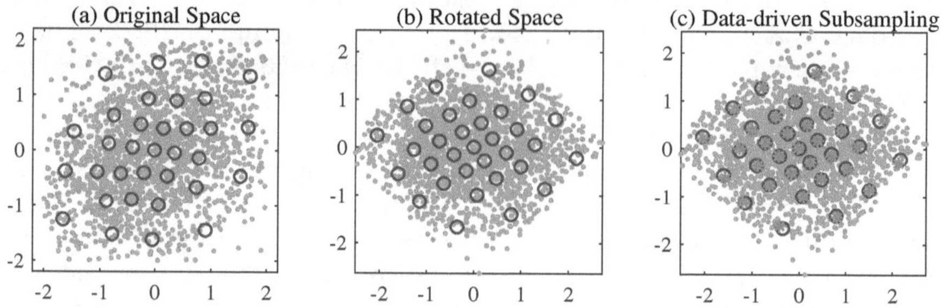

FIGURE 9.5
Illustration of data-driven subsampling for corrected data.

the prediction property of the trained models based on the subdata sets by different subsampling strategies. It was shown that DDS outperforms the URS method in various settings. If there is no a priori knowledge about the true model, model-free subsampling methods are more appropriate. They also showed that the DDS method possesses the model-free property and performs more robustly than another model-free subsampling method for different model specifications and subsample sizes. Moreover, DDS is much quicker than SPlit when the size of the full dataset is large.

Notably, independence among all the rotated variables is the key point in the DDS procedure. However, PCA only guarantees the uncorrelated property among the rotated variables. If the variables follow a normal distribution, the independence is equivalent to the non-correlated property, then DDS can perform very well. However, PCA may not convert the data to independent or approximately independent components for non-independent data generated from a more complex distribution. In those cases, Zhou et al. (2024b) proposed an efficient model-free subsampling method, which segments the original data into some regular data blocks and obtains subsamples from each data block by DDS. They showed that the proposed method has a significant speed advantage and performs more robustly for datasets with complex underlying distributions. It is an order of magnitude faster than other commonly used model-free subsampling methods when the sample size of the original dataset reaches the order of 10^7.

9.5 Global Likelihood Subsampling

A model-free subsampling method expects the subsample to be statistically similar to the full dataset. In this section, we introduce another model-free deterministic subsampling method, global likelihood subsampling (GLSS), which utilizes the global likelihood sampler (GLS, Yi et al., 2023). We also propose some variants of GLSS to accelerate the algorithm for large-scale datasets and deal with high-dimensional data with low computational complexity.

9.5.1 Global likelihood sampler

We first introduce the GLS procedure. GLS is an effective and robust sampling method from a continuous multidimensional distribution and is easy to operate. It uses randomly shifted good lattice points to generate several batches of the samples based on the scaled densities on the design points. Multiple shifts of the GLP make the exploration of the target distribution more sufficient, and hence, GLS can well avoid falling into a local mode with cheap computation.

Let $F(\theta)$ be a s-dimensional cumulative distribution function and $f(\theta)$ be the corresponding probability density function. GLS first truncates a closed and bounded hypercube $\mathcal{D}_0$ containing most of the probabilities of interest, and an M-point GLP is randomly shifted n_0 times in the truncated hypercube. In the ith shift for $i = 1, \ldots, n_0$, GLS defines the weights at the randomly-shifted GLP $\{\boldsymbol{g}_k^{(i)} \in \mathcal{D}_0 : k = 1, \ldots, M\}$ based on the scaled likelihoods $f(\boldsymbol{g}_k^{(i)})/\sum_{j=1}^{M} f(\boldsymbol{g}_j^{(i)})$ for $k = 1, \ldots, M$, then performs the multinomial sampling associated with the weighted shifted-GLP to generate an m-size sample. Based on n_0 randomly shifts on the GLP, there are n_0 multinomial samplings to generate n_0 batches of m-size samples. After running GLS, an $n = n_0 m$-size GLS sample will be produced. The detailed procedure is in Algorithm 9.4.

Algorithm 9.4 (The procedure of GLS)

Input *The target s-dimensional PDF $\pi(\theta)$ with support $\mathcal{S}$, an M-point GLP $\mathcal{Z}_0$ on $[0,1]^s$, a predefined threshold ϵ_T, the sample size N.*

1. **Truncation** *Choose $a_t < b_t$, $t = 1, \ldots, s$ and let $\mathcal{D} = [\boldsymbol{a}, \boldsymbol{b}]$ with $\boldsymbol{a} = (a_1, \ldots, a_s)^T$, $\boldsymbol{b} = (b_1, \ldots, b_s)^T$, such that $\mathcal{D} \subseteq \mathcal{S}$ is a compact subregion containing most of the probabilities of interest, i.e., $\int_{\mathcal{S}\setminus\mathcal{D}} \pi(\theta)d\theta < \epsilon_T$. Transform $\mathcal{Z}_0$ onto $\mathcal{D}$ and denote as $\mathcal{Z} = \{\boldsymbol{z}_k = (z_{k,1}, \ldots, z_{k,s}), k = 1, \ldots, M\}$. Set n, m such that $N = nm$ and let $i = 1$.*

2. **Random shift** *Generate a random shift vector $\boldsymbol{u}_i = (u_{i,1}, \ldots, u_{i,s})^T \in [\boldsymbol{0}, \boldsymbol{b} - \boldsymbol{a}]$, where $u_{i,t} \sim U(0, b_t - a_t), t = 1, \ldots, s$. Shift $\mathcal{P}$ by $\boldsymbol{z}_k \oplus \boldsymbol{u}_i$, where the tth entry of $\boldsymbol{z}_k \oplus \boldsymbol{u}_i$ is*

$$(\boldsymbol{z}_k \oplus \boldsymbol{u}_i)_t = \begin{cases} z_{k,t} + u_{i,t} & \text{if } a_t \le z_{k,t} + u_{i,t} \le b_t \\ z_{k,t} + u_{i,t} + a_t - b_t, & \text{otherwise.} \end{cases}$$

   $\mathcal{Z} \oplus \boldsymbol{u}_i = \{\boldsymbol{z}_k \oplus \boldsymbol{u}_i, k = 1, \ldots, M\}$ is the randomly-shifted GLP in the ith batch.

3. **Likelihood** *Compute $q_{i,k} = p(\boldsymbol{z}_k \oplus \boldsymbol{u}_i)/\sum_{j=1}^{M} p(\boldsymbol{z}_j \oplus \boldsymbol{u}_i)$, let $\tilde{F}_{M,i}$ be the multinomial distribution over $\{\boldsymbol{z}_k \oplus \boldsymbol{u}_i, k = 1, \ldots, M\}$ with the weight $q_{i,k}$ for $k = 1, \ldots, M$.*

4. **Batch sampling** *Sample from $\tilde{F}_{M,i}$ m times to obtain the ith batch of samples $B_i = \{\boldsymbol{x}_{i,1}, \ldots, \boldsymbol{x}_{i,m}\}$. If $i < n_0$, then set $i = i + 1$ and go to Step 2.*

Output *Drawn sample set $\mathcal{P} = \{\boldsymbol{x}_{i,j} \in \mathcal{D}, i = 1, \ldots, n_0, j = 1, \ldots, m\}$.*

In the input of Algorithm 9.4, the detailed procedure of GLP can be referred to Fang et al. (2018). Using the GLP eliminates the need for the proposal distribution in GLS; GLS can be problem-independent for different target distributions. Since the randomly-shifted GLP has better space-filling properties than random points, GLS produces more representative weighted points for the target distribution than the random method. Moreover, the multiple random shifts and multinomial samplings increase the diversity and randomness of the sampling, so GLS can be helpful and practical in exploring the structure of complicated target distribution.

Yi et al. (2023) theoretically proved that the empirical cumulative distribution function of the GLS sample uniformly converges to the target distribution in probability and with probability one under some conditions. They also showed the effectiveness, robustness, and speediness of GLS by many simulation studies and some practical applications. They found that GLS has better properties than other commonly used sampling methods such as the sampling/importance resampling (Rubin, 1987), Metropolis-Hastings algorithm (Hastings, 1970), and evolutionary Monte Carlo (Liang and Wong, 2001).

9.5.2 Global likelihood subsampling

Now, we utilize the idea of GLS to obtain an n-size subsample $\mathcal{P}$ from the N-size full dataset $\mathcal{X}$, such that $\mathcal{P}$ can statistically represent $\mathcal{X}$ for data compression. The last subsection shows that the GLS procedure gets the sample from a continuous distribution. We need some modifications for the discrete $\mathcal{X}$ for using GLS.

Assume that $\mathcal{X} = \{\boldsymbol{z}_i, i = 1, \ldots, N\}$ is the i.i.d sample from the PDF $f(\theta)$ over R^s, and $f(\theta)$ is often unknown. Then, the natural idea is to adopt the kernel density estimation (KDE, Parzen, 1962) to estimate the density function based on $\mathcal{X}$ and then to perform subsampling based on the estimated density function. For any $\boldsymbol{x} \in R^s$, the KDE of $f(\theta)$ under the uniform kernel function is given by

$$\hat{f}_K(\boldsymbol{x}) = \frac{1}{N \cdot r^s} \sum_{i=1}^{N} K\left(\frac{\boldsymbol{x} - \boldsymbol{z}_i}{r}\right), \quad K(\boldsymbol{u}) = \begin{cases} 1/v_s & ||\boldsymbol{u}|| \leq 1, \\ 0 & \text{otherwise}, \end{cases} \tag{9.23}$$

where r is the bandwidth, K is the uniform kernel, and v_s is the volume of a unit ball in R^s. Hence, the estimated density at the point $\boldsymbol{x}$ is proportional to the number of data points in the ball with θ as the center and r as the radius, denoted by $\mathcal{B}(\boldsymbol{x}, r)$. For the discrete dataset $\mathcal{X}$, we can find a bounded hypercube $\mathcal{D}$ such that $\mathcal{X} \subseteq \mathcal{D}$.

Based on $\hat{f}_K$, we wish to sample the representative points of $\mathcal{X}$ in $\mathcal{D}$. Inspired by GLS, we first give a low-discrepancy point set $\mathcal{Z} = \{\boldsymbol{z}_i, i = 1, \ldots, M\}$ in $\mathcal{D}$ such as a GLP set, and obtain the separation distance $r_{\mathcal{Z}}$ of $\mathcal{Z}$, which is also well known as the minimal pairwise distance. and estimate the densities of the design points by $\hat{f}_K$. In empirical knowledge, the radius r in $\hat{f}_K$ can be chosen as half of $r_{\mathcal{Z}}$, that is $r = r_{\mathcal{Z}}/2$. Standardizing the probabilities $\hat{f}_K(\boldsymbol{z}_k)$ into $\tilde{f}(\boldsymbol{z}_k)$, we obtain the discrete distribution $\tilde{F}_M = (\boldsymbol{z}_k, \tilde{f}(\boldsymbol{z}_k)), i = 1, \ldots, M$. Generating a random sample from $\tilde{F}_M$, we obtain one representative point. Randomly shift the GLP set $\mathcal{Z}$ in $\mathcal{D}$, continue the procedure n times, and obtain the representative points $\boldsymbol{d}_i, i = 1 \ldots, n$. Finally, we use the sequential nearest neighbor substitution for each $\boldsymbol{d}_i$ in the small region $\mathcal{B}(\boldsymbol{d}_i, r)$, such that the final subsample is from $\mathcal{X}$. The detailed procedure is shown in Algorithm 9.5.

Algorithm 9.5 (The procedure of GLSS)

Input *Full data,* $\mathcal{X} = \{\boldsymbol{x}_i \in \mathcal{D} = [\boldsymbol{a}, \boldsymbol{b}] : i = 1, \ldots, N\}$ *with* $\boldsymbol{a} = [a_1, \ldots, a_s]$ *and* $\boldsymbol{b} = [b_1, \ldots, b_s]$*; A GLP set,* $\mathcal{Z} = \{\boldsymbol{z}_i \in \mathcal{D} : i = 1, \ldots, M\}$*; The size of the final subsample is* n*.*

1. Random shift *Generate a random shift vector* $\boldsymbol{u}_i = (u_{i,1}, \ldots, u_{i,s})^T \in \mathcal{D}$*, where* $u_{i,t} \sim U(0, b_t - a_t), t = 1, \ldots, s$*. Shift* $\mathcal{Z}$ *by* $\boldsymbol{y}_k \oplus \boldsymbol{u}_i$*, where the* t*th entry of* $\boldsymbol{z}_k \oplus \boldsymbol{u}_i$ *is*

$$(\boldsymbol{z}_k \oplus \boldsymbol{u}_i)_t = \begin{cases} z_{k,t} + u_{i,t} & \textit{if } a_t \leq z_{k,t} + u_{i,t} \leq b_t \\ z_{k,t} + u_{i,t} + a_t - b_t, & \textit{otherwise.} \end{cases}$$

$\mathcal{Z} \oplus \boldsymbol{u}_i = \{\boldsymbol{z}_k \oplus \boldsymbol{u}_i : k = 1, \ldots, M\}$ *is the randomly-shifted GLP in the* i*th batch.*

2. KDE *Calculate the separation distance of* $\mathcal{Z}$, $r_{\mathcal{Z}}$. *Estimate the PDF values at the points* $\mathcal{Z} \oplus \boldsymbol{u}_i$ *by kernel density estimator in* (9.23), *where the radius* $r = r_{\mathcal{Z}}/2$.

3. Likelihood *Compute* $q_{i,k} = \hat{f}_K(\boldsymbol{z}_k \oplus \boldsymbol{u}_i)/\sum_{j=1}^{M} \hat{f}_K(\boldsymbol{z}_j \oplus \boldsymbol{u}_i)$, *let* $\tilde{F}_{M,i}$ *be the multinomial distribution over* $\{\boldsymbol{z}_k \oplus \boldsymbol{u}_i : k = 1, \dots, M\}$ *with the weight* $q_{i,k}$ *for* $k = 1, \dots, M$.

4. Batch sampling *Sample from* $\tilde{F}_{M,i}$ m *times to obtain the* i*th batch of samples* $B_i = \{\boldsymbol{x}_{i,1}, \dots, \boldsymbol{x}_{i,m}\}$. *If* $i < n_0$, *set* $i = i + 1$ *and go to Step 1.*

5. Nearest neighbor *For the* i*th point in the* $n(= mn_0)$*-point set* $\mathcal{P}_0 = \{B_i, i = 1, \dots, n_0\}$, *use the k-d tree to find the nearest neighbor in* $\mathcal{X}$, *denoted by* $\boldsymbol{x}_i^*$.

Output *The* n*-point subsample* $\mathcal{P} = \{\boldsymbol{x}_1^*, \dots, \boldsymbol{x}_n^*\}$ *in* $\mathcal{X}$.

In Algorithm 9.5, the run size of the space-filling design $\mathcal{Z}$, M, is crucial. Usually, M depends on the number of dimensions s, the full data size N, and the subsample size n. Too small M may derive poor property, while too large M results in high computational complexity. According to empirical study, we set $M = \exp(2(s + \kappa))$, where $\kappa \leq 2$. When s is larger, κ is smaller for controlling the computational complexity. Under the same s, larger M is chosen with the increase of N or n for better capturing the data structure. Moreover, other space-filling designs can also replace the GLP set under the discrepancy-based or distance-based criteria (Fang et al., 2018). Furthermore, Yi and Zhou (2023) showed that the overall time complexity of GLSS in the average case is $O(sN \log N + M \log N)$. Usually, the run size M in the GLP set is smaller than the data size of N, and then the computational complexity becomes $O(sN \log N)$, which is the same as the DDS method in the last section and is much lower than SPlit.

Example 9.5 *Consider the following two different datasets: (i) the full data* $\mathcal{Z}$ *is generated from a two-dimensional standard normal distribution; (ii)* $\mathcal{Z} = \{(x_i, y_i), i = 1, \dots, N\}$ *with* $y_i = x_i^2 + \epsilon_i, x_i \sim N(0, 1)$, $\epsilon_i \sim N(0, 1)$. *Let* $N = 2000$ *and the subsample size* $n = 60$. *In GLSS, the used GLPs in the two cases are* $M = 349$ *and 887, respectively. The dataset* $\mathcal{Z}$ *and the subsample are shown in Figure 9.6. It can be seen that for the two datasets, the subsample under GLSS better represents the full data than URS.*

Next, we establish the convergence result for GLSS based on the theoretical results for KDE and GLS. Let $f(\boldsymbol{x})$ and $F(\boldsymbol{x})$ be the underlying PDF and CDF of the data $\mathcal{X}$. Let $\hat{F}_{\mathcal{P}_0}$ be the empirical cumulative distribution function (ECDF) of $\mathcal{P}_0$ and ϱ be the mapping for the multinomial sampling. Denote $\varrho_{\boldsymbol{x}} := I_{[-\infty, \boldsymbol{x}]} \circ \varrho$, where the notation '$\circ$' represents the composition of two functions. Yi and Zhou (2023) gave the following convergence result of $\hat{F}_{\mathcal{P}_0}$.

Theorem 9.3 *Assume the following conditions to be satisfied,*

(1) f *is bounded and Hölder-continuous (i.e.* $|f(\boldsymbol{x}) - f(\boldsymbol{x}')| \leq C_\alpha ||\boldsymbol{x} - \boldsymbol{x}'||^\alpha$ *for* $\boldsymbol{x}, \boldsymbol{x}'$ *in* R^s *and* $0 < \alpha \leq 1$*), and the support of* f *is bounded,*

(2) $\hat{f}_K(\boldsymbol{x}) I_{[-\infty, \boldsymbol{x}]}(\boldsymbol{x})$ *is of uniformly bounded variation in the sense of Hardy and Krause for any* $\boldsymbol{x} \in R^s$,

(3) $\mathcal{Z} = \{\boldsymbol{z}_1, \dots, \boldsymbol{z}_M\}$ *is a low-discrepancy point set,*

(4) $\varrho_{\boldsymbol{x}}(u)$ *is of uniformly bounded variation in the sense of Hardy and Krause for any* $\boldsymbol{x} \in R^s$ *and run size* M *of* $\mathcal{Z}$,

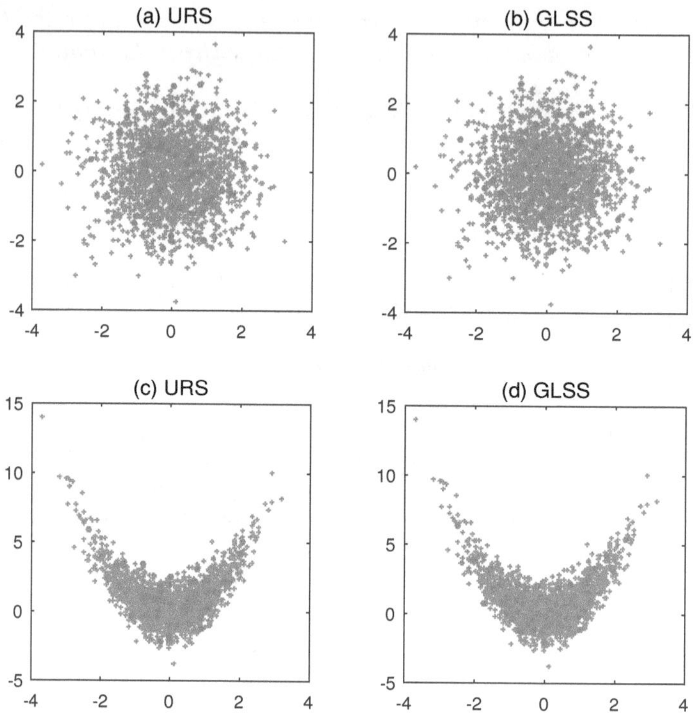

FIGURE 9.6
Compare the subsample of URS and GLSS.

then we have

$$\sup_{\boldsymbol{x}\in R^s} |\hat{F}_{\mathcal{P}_0}(\boldsymbol{x}) - F(\boldsymbol{x})| = O_{\boldsymbol{P}}(\Lambda),$$

where $\Lambda = \max\{1/M^{\alpha/d} + (M \log N/N)^{1/2}, 1/n\}$.

Under the conditions in Theorem 9.3, if $M = O((N/\log N)^{s/(s+2\alpha)})$, we have $\sup_{\boldsymbol{x}\in R^s} |\hat{F}_{\mathcal{P}_0}(\boldsymbol{x}) - F(\boldsymbol{x})| = O_{\boldsymbol{P}}(\max\{1/M^{\alpha/s}, 1/n\})$; Moreover, if $M = O((N/\log N)^{s/(s+2\alpha)})$ and $n = O(M^{\alpha/s})$, we further have $\sup_{\boldsymbol{x}\in R^s} |\hat{F}_{\mathcal{P}_0}(\boldsymbol{x}) - F(\boldsymbol{x})| = O_{\boldsymbol{P}}(1/M^{\alpha/s})$. Therefore, the convergence rate can be simplified when imposing some conditions on M and n. Moreover, from Theorem 9.3, the ECDF of the drawn design points can uniformly converge to the true CDF of the full data. Then, by the sequential nearest neighbor substitution, the final subsample $\mathcal{P}$ can also be statistically representative of the true distribution for the full data and thus of the full data.

In application, if N is not very large, e.g., N <5e5, the corresponding computational complexity may be acceptable. For larger N, we can first conduct URS on $\mathcal{X}$ to obtain an N_s-point intermediate subsample $\mathcal{X}_s$, where $n < N_s < N$, and perform GLSS on $\mathcal{X}_s$ to achieve the final subsample $\tilde{\mathcal{X}}$. The main idea is that when N is huge, the N_s-point $\mathcal{X}_s$ often have a similar statistical property with $\mathcal{X}$ when N_s is large, according to the law of large number. Denote this procedure by GLSS$_a$. Then, the computational complexity of GLSS$_a$ is $O(dN_s \log N_s + M \log N_s)$ on average, which is much smaller than that of GLSS when N is huge. Usually, N_s can be chosen as a tradeoff between the subsampling performance and computational complexity. By some empirical results, when $N >$ 5e5, $N_s = N/10$ may be an appropriate choice.

Furthermore, with the increase of the dimension s, the size M of $\mathcal{Z}$ in GLSS would increase exponentially, which increases the computational complexity significantly. Some dimension reduction techniques are needed for high-dimensional cases. Based on singular value decomposition (SVD), we propose a high-dimensional version of GLSS, denoted as GLSS_h. Its main idea is to perform SVD on the normalized full data to reduce the data dimension and conduct GLSS in the reduced latent space, finally obtaining the subsample in the original space. GLSS_h can be described in Algorithm 9.6.

Algorithm 9.6 (The procedure of GLSS_h)

Step 1 *Normalize the full data $\mathcal{X}$ to obtain $\bar{\mathcal{X}}$ and perform SVD on $\bar{\mathcal{X}}$ as $\bar{\mathcal{X}}_{N\times s} = U_{N\times s}S_{s\times s}V_{s\times s}^\top$. Denote by U' the first $s'(< s)$ columns of U;*

Step 2 *Implement the GLSS algorithm on the latent space U' and record the indexes of the selected rows with size n;*

Step 3 *Select the rows in the full data $\mathcal{X}$ corresponding to the indexes in Step 2 to obtain the final subsample.*

In practical application, s' can be chosen as the minimal number of the dimensions that can make the contribution accounting for the total variation of the full data reach 70%. In general, the number of significant variables is not large, which results in a small s'. In many cases, $d' = 2$ is a reasonable and suitable choice in many cases with good performance and low computational complexity. Moreover, in Step 2 of Algorithm 9.6, set $M = \exp(2(d' + \kappa))$ with $\kappa \leq 2$ and κ becomes large when d is large. Furthermore, for a high-dimensional huge dataset $\mathcal{X}$, we can combine the methods of GLSS_a and GLSS_h to decrease the computational complexity further.

Through many simulations and real case studies, Yi and Zhou (2023) also showed that GLSS has a significant performance improvement, compared with the uniform random sampling, regarding the prediction performance under different models. Moreover, GLSS has comparable performance with SPlit in many low-dimensional cases and outperforms Twinning in many low-dimensional cases. In many high-dimensional cases, GLSS can perform better than both SPlit and Twinning. Compared with DDS, there is no limitation on the independence of each data dimension in GLSS, so GLSS can perform much better than DDS in many cases. Here, give a simple example, and more comparisons can refer to Yi and Zhou (2023).

9.6 Proofs

9.6.1 Proof of Theorem 9.2

For ease of presentation, denote $\widetilde{\boldsymbol{x}} = T_{\mathcal{X}}(\boldsymbol{x})$ for each $\boldsymbol{x} \in \mathcal{X} \cup \mathcal{P}$, and $\widetilde{\mathcal{X}} = T_{\mathcal{X}}(\mathcal{X})$, $\widetilde{\mathcal{P}} = T_{\mathcal{X}}(\mathcal{P})$, then $\widetilde{\mathcal{X}}, \widetilde{\mathcal{P}} \subset C^s$, and the expression of $D^2(\mathcal{P};\mathcal{X},\mathcal{K})$ in (9.20) can be rewritten as,

$$
\begin{aligned}
D^2(\mathcal{P};\mathcal{X},\mathcal{K}) &= \frac{1}{N^2}\sum_{i,k=1}^{N}\mathcal{K}(\widetilde{\boldsymbol{x}}_i,\widetilde{\boldsymbol{x}_k}) - \frac{2}{Nn}\sum_{i=1}^{N}\sum_{k=1}^{n}\mathcal{K}(\widetilde{\boldsymbol{x}}_i,\widetilde{\xi}_k) + \frac{1}{n^2}\sum_{i,k=1}^{n}\mathcal{K}(\widetilde{\xi}_i,\widetilde{\xi}_k) \\
&= \int_{C^{2s}}\mathcal{K}(\boldsymbol{u},\boldsymbol{v})\mathrm{d}F_{\widetilde{\mathcal{X}}}(\boldsymbol{u})\mathrm{d}F_{\widetilde{\mathcal{X}}}(\boldsymbol{v}) - \frac{2}{n}\sum_{k=1}^{n}\int_{C^{2s}}\mathcal{K}(\boldsymbol{u},\widetilde{\xi}_k)\mathrm{d}F_{\widetilde{\mathcal{X}}}(\boldsymbol{v}) \\
&\quad + \frac{1}{n^2}\sum_{i,k=1}^{n}\mathcal{K}(\widetilde{\xi}_i,\widetilde{\xi}_k),
\end{aligned} \tag{9.24}
$$

where $F_{\widetilde{\mathcal{X}}}$ and $F_{\widetilde{\mathcal{P}}}$ are respectively the ECDFs of $\widetilde{\mathcal{X}}$ and $\widetilde{\mathcal{P}}$. Because of the joint independence assumed for $\mathcal{X}$ in (9.22), $\widetilde{\mathcal{X}}$ also satisfies the joint independence as follows,

$$
F_{\widetilde{\mathcal{X}}}(\boldsymbol{u}) = \prod_{j=1}^{s} F_{\widetilde{\mathcal{X}}_{(j)}}(u_j), \quad \forall\ \boldsymbol{u} = (u_1,\ldots,u_s)^T \in C^s,
$$

where $\widetilde{\mathcal{X}}_{(j)} = \{F_{\mathcal{X}_{(j)}}(x_{ij}), i = 1,\ldots,N\}$ is the set of values along the jth component of $\widetilde{\mathcal{X}}$, then, by using the product form of $\mathcal{K}$, the integrations over C^{2s} and C^s in (9.24) could be derived into the product form in the following

$$
\int_{C^{2s}}\mathcal{K}(\boldsymbol{u},\boldsymbol{v})\mathrm{d}F_{\widetilde{\mathcal{X}}}(\boldsymbol{u})\mathrm{d}F_{\widetilde{\mathcal{X}}}(\boldsymbol{v}) = \prod_{j=1}^{s}\int_0^1\int_0^1 K(u_j,v_j)\mathrm{d}F_{\widetilde{\mathcal{X}}_{(j)}}(u_j)\mathrm{d}F_{\widetilde{\mathcal{X}}_{(j)}}(v_j), \tag{9.25}
$$

$$
\int_{C^{s}}\mathcal{K}(\boldsymbol{u},\widetilde{\xi}_k)\mathrm{d}F_{\widetilde{\mathcal{X}}}(\boldsymbol{u}) = \prod_{j=1}^{s}\int_0^1 K(u_j,F_{\mathcal{X}_{(j)}}(\xi_{kj}))\mathrm{d}F_{\widetilde{\mathcal{X}}_{(j)}}(u_j). \tag{9.26}
$$

For each $j = 1,\ldots,s$, according to the meaning of N_j, there are also N_j different values $\hat{x}_{1j} < \cdots < \hat{x}_{N_j j} = 1$ in $\widetilde{\mathcal{X}}_{(j)}$, denote $\hat{x}_{0j} = 0$, and the replication of $\hat{x}_{ij}$ by n_{ij}, then $\hat{x}_{ij} - \hat{x}_{i-1,j} = n_{ij}/N$, $i = 1,\ldots,N_j$. Based on the mean value theorem, for any $v \in [0,1]$,

$$
\int_0^1 K(u,v)\mathrm{d}u = \sum_{i=1}^{N_j}\int_{\hat{x}_{i-1,j}}^{\hat{x}_{ij}} K(u,v)\mathrm{d}u = \sum_{i=1}^{N_j} K(u_{ij}^*,v)\cdot\frac{n_{ij}}{N},
$$

where $u_{ij}^* \in [\hat{x}_{i-1,j},\hat{x}_{ij}]$. Recall that the reproducing kernel K satisfies the Lipschitz continuous with a constant c_2. It can be easily obtained that

$$
\begin{aligned}
\left|\int_0^1 K(u,v)\mathrm{d}F_{\widetilde{\mathcal{X}}_{(j)}}(u) - \int_0^1 K(u,v)\mathrm{d}u\right| &= \left|\frac{1}{N}\sum_{i=1}^{N_j} n_{ij}\left(K(\hat{x}_{ij},v) - K(u_{ij}^*,v)\right)\right| \\
&\le \frac{c_2}{N}\sum_{i=1}^{N_j} n_{ij}\left|\hat{x}_{ij} - u_{ij}^*\right| \\
&\le \frac{c_2}{N^2}\sum_{i=1}^{N_j} n_{ij}^2.
\end{aligned} \tag{9.27}
$$

Note that, the independent condition (9.22) also yields the following constraints about N, N_j and n_{ij},

$$\begin{cases} n_{ij} \leq c_1 \prod\limits_{l \neq j}^{s} N_l, \\ N \geq \prod\limits_{l=1}^{s} N_l, \end{cases} \qquad i = 1, \ldots, N_j, \quad j = 1, \ldots, s,$$

then the upper bound in (9.27) could be shrunk as follows,

$$\frac{c_2}{N^2} \sum_{i=1}^{N_j} n_{ij}^2 \leq \frac{c_2}{\left(\prod_{l=1}^{s} N_l\right)^2} \sum_{i=1}^{N_j} \left[c_1^2 \cdot \left(\frac{\prod_{l=1}^{s} N_l}{N_j} \right)^2 \right] = \frac{c_2 c_1^2}{N_j} = \boldsymbol{O}(1/N_j).$$

which leads to

$$\int_0^1 K(u,v) \mathrm{d}F_{\widetilde{\mathcal{X}}_{(j)}}(u) = \int_0^1 K(u,v) \mathrm{d}u + \boldsymbol{O}(1/N_j), \tag{9.28}$$

thus, integrating (9.28) from 0 to 1 with reference distribution $F_{\widetilde{\mathcal{X}}_{(j)}}$ yields

$$\begin{aligned} &\int_0^1 \int_0^1 K(u,v) \mathrm{d}F_{\widetilde{\mathcal{X}}_{(j)}}(u) \mathrm{d}F_{\widetilde{\mathcal{X}}_{(j)}}(v) \\ &= \int_0^1 \left[\int_0^1 K(v,u) \mathrm{d}F_{\widetilde{\mathcal{X}}_{(j)}}(v) \right] \mathrm{d}u + \boldsymbol{O}(1/N_j) \\ &= \int_0^1 \left[\int_0^1 K(v,u) \mathrm{d}v + \boldsymbol{O}(1/N_j) \right] \mathrm{d}u + \boldsymbol{O}(1/N_j) \\ &= \int_0^1 \int_0^1 K(u,v) \mathrm{d}u \mathrm{d}v + \boldsymbol{O}(1/N_j). \end{aligned} \tag{9.29}$$

Substituting (9.29) into (9.25), and (9.28) into (9.26) respectively obtains

$$\begin{aligned} \int_{C^{2s}} \mathcal{K}(\boldsymbol{u}, \boldsymbol{v}) \mathrm{d}F_{\widetilde{\mathcal{X}}}(\boldsymbol{u}) \mathrm{d}F_{\widetilde{\mathcal{X}}}(\boldsymbol{v}) &= \prod_{j=1}^{s} \left[\int_0^1 \int_0^1 K(u,v) \mathrm{d}u \mathrm{d}v + \boldsymbol{O}(1/N_j) \right] \\ &= \int_{C^{2s}} \mathcal{K}(\boldsymbol{u}, \boldsymbol{v}) \mathrm{d}\boldsymbol{u} \mathrm{d}\boldsymbol{v} + \boldsymbol{O}(1/N^*), \end{aligned} \tag{9.30}$$

and

$$\begin{aligned} \int_{C^{s}} \mathcal{K}(\boldsymbol{u}, \widetilde{\xi}_k) \mathrm{d}F_{\widetilde{\mathcal{X}}}(\boldsymbol{u}) &= \prod_{j=1}^{s} \left[\int_0^1 K(u, F_{\mathcal{X}_{(j)}}(\xi_{kj})) \mathrm{d}u + \boldsymbol{O}(1/N_j) \right] \\ &= \int_{C^{s}} \mathcal{K}(\boldsymbol{u}, \widetilde{\xi}_k) \mathrm{d}\boldsymbol{u} + \boldsymbol{O}(1/N^*). \end{aligned} \tag{9.31}$$

Finally, substituting (9.30) and (9.31) into (9.24) completes the proof.

9.6.2 Proof of Theorem 9.3

We need the following two lemmas to prove Theorem 9.3.

Lemma 9.1 *Under the condition (1) in Theorem 9.3, then uniformly for* $r > (\log N/N)^{1/d}$, *we have*

$$\sup_{\boldsymbol{x}\in R^s} |\hat{F}_K(\boldsymbol{x}) - F(\boldsymbol{x})| = O_{\mathrm{P}}(\Delta) \quad \textit{with} \quad \Delta = r^{\alpha} + \sqrt{\frac{\log N}{Nr^d}},$$

where $\hat{F}_K$ *is the corresponding CDF of* $\hat{f}_K$.

Lemma 9.2 *Under the conditions (2)-(4) in Theorem 9.3, we have*

$$\sup_{\boldsymbol{x}\in R^s} |\hat{F}_{\mathcal{P}}(\boldsymbol{x}) - \hat{F}_K(\boldsymbol{x})| = O\left(\max\left\{\kappa(M), \frac{1}{n}\right\}\right),$$

where $\kappa(M) = 1/M$ *if* $d = 1$, *and* $(\log(M))^d/M$, *otherwise.*

The proof of Lemma 9.2 follows Lemma 1(1) and Theorem 2 in Yi et al. (2023). By Lemma 9.2, it theoretically implies that the ECDF of the drawn design points converges to the estimated CDF for $\mathcal{X}$ based on KDE. Now, we prove Theorem 9.3.

Based on Lemmas 9.1 and 9.2, by the triangle inequality, we directly have

$$\sup_{\boldsymbol{x}\in R^s} |F_{\tilde{\mathcal{Z}}_s}(\boldsymbol{x}) - F(\boldsymbol{x})| = O_{\boldsymbol{P}}\left(\max\left\{\Delta, \kappa(M), \frac{1}{n}\right\}\right).$$

Let $\mathcal{D} = \otimes_{i=1}^s [a_i, b_i]$ and δ be the volume of the union of the balls $\mathcal{B}(\boldsymbol{x}_i, r), i = 1, \ldots, M$, divided by the volume of $\otimes_{i=1}^s [a_i - r, b_i + r]$. Since r is set as half of the separation distance of $\mathcal{Z}$, we have $\delta \prod_{i=1}^s (2r + b_i - a_i) = M v_s r^s$. Without loss of generality, let $a_i = 0$ and $b_i = 1$ for $i = 1, \ldots, s$, which leads to $r = 1/[(Mv_s\delta^{-1})^{1/s} - 2]$. Then, we have $r = O(M^{-1/s})$, which is also true when a_i and $b_i (a_i < b_i)$ take any other bounded values for $i = 1, \ldots, s$. Based on the result, we have $\Delta = O(M^{-\alpha/s} + (M \log N/N)^{1/2})$. In addition, we can easily obtain that $M^{-\alpha/s} > 1/M$ if $s = 1$, and $M^{-\alpha/s} > (\log M)^s/M$ as $M \to \infty$ if $s > 1$. Thus, $O_{\boldsymbol{P}}(\max\{\Delta, \kappa(M)\}) = O_{\boldsymbol{P}}(M^{-\alpha/s} + (M \log N/N)^{1/2})$ as $M \to \infty$, which completes the proof.

Bibliography

Abaya, E.F., Wise, G.L., 1984. Convergence of vector quantizers with applications to optimal quantization. SIAM Journal on Applied Mathematics 44, 183–189.

Adcock, C.J., 2023. The linear skew-t distribution and its properties. Stats 6, 381–410.

Adelchi, A., 2003. Distributions generated by perturbation of symmetry with emphasis on a multivariate skew t-distribution. Journal of the Royal Statistical Society, Series B (Statistical Methodology) 65, 367–389.

Ahn, S.K., 1992. F-probability plot and its application to multivariate normality. Communications in Statistics-Theory and Methods 21, 997–1023.

Ai, M., Yu, J., Zhang, H., Wang, H., 2021. Optimal subsampling algorithms for big data regressions. Statistica Sinica 31, 749–772.

Al-Labadi, L., Asl, F.F., Saberi, Z., 2021. A necessary bayesian nonparametric test for assessing multivariate normality. Mathematical Methods of Statistics 30, 64–81.

Ali, M.M., Mikhal, N.N., Haq, M.S., 1978. A class of bivariate distributions including the bivariate iogist. Jouriral of Muitiijnriate Analysis into logistic. Journal of Multivariate Analysis 8, 405–412.

Altun, E., Tatlidil, H., Ozel, G., Nadarajah, S., 2018. A new generalization of skew-t distribution with volatility models. Journal of Statistical Computation and Simulation 88, 1252–1272.

Andrews, D.F., Gnanadesikan, R., Warner, J.L., 1973. Methods for assessing multivariate normality, in: Proceedings of the International Symposium on Multivariate Analysis, eds by PR.Krishnaiah. Academic Press, New York. pp. 95–116.

Antle, C.E., Bain, L.J., 1969. A property of maximum likelihood estimators of location and scale parameters. SIAM Review 11, 251–253.

Arellano-Valle, P.B., Genton, M.G., 2005. On fundamental skew distributions. Journal of Multivariate Analysis 96, 93–116.

Arthur, D., Vassilvitskii, S., 2007. k-means++, test code URL: `https://www.stanford.edu/d̃arthur/kMeansppTest.zip`.

Azzalini, A., Valle, A.D., 1996. Bias in estimation of a mixture of normal distributions. Biometrika 83, 715–726.

Barbiero, A., Hitaj, A., 2022. Discrete approximations of continuous probability distributions obtained by minimizing cramér-von mises-type distances. Statistical Papers 64, 1–29.

Bhattacharjee, M.C., 1982. The class of mean residual lives and some consequences. SIAM Journal on Algebraic Discrete Methods 3, 56–65.

Bhaumik, D.K., Kapur, K., Gibbons, R.D., 2009. Testing parameters of a gamma distribution for small samples. Technometrics 51, 326–334.

Bickel, P.J., Doksum, K.A., 1977. Mathematical Statistics. Holden-Day, San Francisco.

Bickel, P.J., Freedman, D.A., 1981. Some asymptotic theory for the bootstrap. The Annals of Statistics 9, 1196–1217.

Billingsley, P., 1995. Probability and Measure. New York: Wiley.

Birnbaum, A., 1952. Numerical tabulation of the distribution of Kolmogorov's statistic for finite sample size. Journal of the American Statistical Association 47, 425–441.

Bofinger, E., 1970. Maximizing the correlation of grouped observations. Journal of the American Statistical Association 65, 1632–1638.

Bogdan, M., 1995. Data driven versions of pearson's chi-square test for uniformity. Journal of Statistical Computation and Simulation 52, 217–237.

Branco, M.D., Dey, D.K., 2001. A general class of multivariate skew-elliptical distributions. Journal of Multivariate Analysis 79, 99–113.

Calabria, R., Pulcini, G., 1987. On the asymptotic behavior of mean residual life function. Reliability engineering 19, 165–170.

Cao, Y., Liang, J., Xu, L., Kang, J., 2024. Testing multivariate normality based on beta-representative points. Mathematics 12, 1711.

Chakraborty, S., 2015. Generating discrete analogues of continuous probability distributions–a survey of methods and constructions. Journal of Statistical Distributions and Applications 2, 1–30.

Chakraborty, S., Roychowdhury, M.K., Sifuentes, J., 2021. High precision numerical computation of principal points for univariate distributions. Sankhya B 83, 558–584.

Chen, J., 2017. Consistency of the mle under mixture models. Statistical Science 32, 47–63.

Cheng, M., Lei, S., 2022. The representative points of nonlinear transformation of random variable, in: Conference: CMAAE 2021: 2021 International Conference on Mechanical, Aerospace and Automotive Engineering, pp. 371–374.

Ciuperca, G., Ridolfi, A., Idier, J., 2003. Penalized maximum likelihood estimator for normal mixtures. Scandinavian Journal of Statistics 30, 45–59.

Clayton, U.G., 1978. A model for association in bivariate life tables and its applications in epidemiological studies of familial tendency in chronic disease incidence. Biometrika 65, 141–151.

Cornell, J.A., 1975. Some comments on designs for Cox's mixture polynomial. Technometrics 17, 25–35.

Cornell, J.A., 2011. A Primer on Experiments with Mixtures. Wiley, New Jersey.

Cox, D.R., 1957. Note on grouping. Journal of the American Statistical Association 52, 543–547.

Cox, D.R., 1962. Renewal Theory. London: Chapman & Hall.

D'Agostino, R.B., Stephens, M.A., 1986. Goodness-of-Fit Techniques. Marcel Dekker, New York.

Dahiya, R.C., Gurland, J., 1973. How many classes in the pearson chi-square test? Journal of the American Statistical Association 68, 707–712.

David, H.A., 1981. Order Statistics. New York: Wiley. Second Edition.

David, H.A., Nagaraja, H.N., 2003. Order Statistics. New York: Wiley. Third Edition.

Dhillon, I.S., Guan, Y., Kogan, J., 2002. Iterative clustering of high dimensional text data augmented by local search., in: ICDM, Citeseer. p. 131.

Dhillon, I.S., Modha, D.S., 2001. Concept decompositions for large sparse text data using clustering. Machine Learning 42, 143–175.

Diday, E., 1974. A new approach in mixed distributions detection, in: Report de Recherche, Insttut de Rechrche d'Informatique d'Automatique. p. 52.

Drineas, P., Magdon-Ismail, M., Mahoney, M., Woodru, D., 2012. Faster approximation of matrix coherence and statistical leverage. Journal of Machine Learning Research 13, 3475–3506.

Dudley, R.M., 1989. Real Analysis and Probability. The Wadsworth & Brooks Cole Math. Series, Pacific Grove, CAThe Wadsworth & Brooks Cole Math. Series, Pacific Grove, CA.

Durante, F., Sempi, C., 2010. Chapter 1: Copula theory: An introduction, in: Copula Theory and Its Applications, Proceedings of the Workshop Held in Warsaw, 25–26 September 2009, Lecture Notes in Statistics–Proceedings, 198, Springer. pp. 3–32.

Efron, B., 1967. The two sample problems with censored data, in: Proceedings of the Fifth Berkeley Symposium on Mathematical Statistics and Probability, pp. 831–853.

Efron, B., 1979. Bootstrap methods: another look at the jackknife. The Annals of Statistics 7, 1–26.

Efron, B., 2003. Second thoughts on the bootstrap. Statistical Science 18, 135–140.

Efron, B., Tibshirani, R.J., 1993. An Introduction to The Bootstrap. Chapman & Hall, New York.

Elston, R.C., Grizzle, J.E., 1962. Estimation of time-response curves and their confidence bands. Biometrics 18, 148–159.

Embrechts, P., Klüppelberg, C., Mikosch, T., 1997. Modelling Extremal Events for Insurance and Finance. Springer-Verlag.

Enomoto, R., Hanusz, Z., Hara, A., Seo, T., 2020. Multivariate normality test using normalizing transformation for Mardia's multivariate kurtosis. Communications in Statistics–Simulation and Computation 49, 684–698.

Escobar, L.A., Meeker Jr, W.Q., 1994. Fisher information matrix for the extreme value, normal and logistic distributions and censored data. Applied Statistics 43, 533–554.

Fang, B.Q., 2003. The skew elliptical distributions and their quadratic forms. Journal of Multivariate Analysis 87, 298–314.

Fang, H.B., Fang, K.T., Kotz, S., 2002. The meta-elliptical distributions with given marginalis. Journal of Multivariate Analysis 82, 1–16.

Fang, K.T., 1976. Application of the theory of the conditional distribution for the standardization of clothes. Acta Mathematicae Applicatae Sinica 2, 62–74.

Fang, K.T., Fang, H.B., von Rosen, D., 2000. A family of bivariate distributions with non-elliptical contours. Communications in Statistics: Theory and Methods 29, 1885–1898.

Fang, K.T., He, P., Yang, J., 2020. Set of representative points of statistical distributions and their applications. Scientia Sinica Mathematica 50, 1149–1168.

Fang, K.T., He, S.D., 1982. The problem of selecting a given number of representative points in a normal population and a generalized Mills' ratio. Technical Report. Department of Statistics, Stanford University, USA.

Fang, K.T., He, S.D., 1984. The problem of selecting a given number of representative points in a normal population and a generalized Mills' ratio. Acta Mathematicae Applicatae Sinica 7, 293–306. In Chinese.

Fang, K.T., Ke, X., Elsawah, A.M., 2017. Construction of uniform designs via an adjusted threshold accepting algorithm. Journal of Complexity 23, 740–751.

Fang, K.T., Kotz, S., Ng, K.W., 1990. Symmetric Multivariate and Related Distributions. Chapman and Hall, London and New York.

Fang, K.T., Liu, M.Q., Qin, H., Zhou, Y., 2018. Theory and Application of Uniform Experimental Designs. volume 221. Science Press and Springer.

Fang, K.T., Pan, J., 2023. A review of representative points of statistical distributions and their applications. Mathematics 11, 2930.

Fang, K.T., Wang, Y., 1990. A sequential algorithm for optimization and its applications to regression analysis. Lecture notes in contemporary mathematics, 17–28.

Fang, K.T., Wang, Y., 1991. A sequential algorithm for solving a system of nonlinear equations. Journal of Computational Mathematics 9, 9–16.

Fang, K.T., Wang, Y., 1994. Number-Theoretic Methods in Statistics. Chapman and Hall, London.

Fang, K.T., Wang, Y., Bentler, P.M., 1994a. Some applications of number-theoretic methods in statistics. Statistical Science 9, 416–428.

Fang, K.T., Wang, Y., Wong, H.L., 1992. A new method of generating an NT-net on the unit sphere. Technical Report. Technical report MATH-002, Hong Kong Baptist College, Hong Kong.

Fang, K.T., Wei, G., 1992. The distribution of a special first hitting time. Acta Mathematicae Applicatae Sinica 15, 460–467.

Fang, K.T., Yang, Z.H., Kotz, S., 2001. Generation of multivariate distributions by vertical density representation. Statistics 35, 381–293.

Fang, K.T., Yuan, K.H., Bentler, P.M., 1994b. Applications of number-theoretic methods to quantizers of elliptically contoured distributions. Lecture Notes-Monograph Series 24, 237–251.

Fang, K.T., Zhou, M., Wang, W.J., 2014. Applications of the representative points in statistical simulations. Science China Mathematics 57, 2609–2620.

Fei, R.C., 1990. The problem of selecting representative points in pearson distributions population. Journal of the Wuxi Institute of Light Industry 9, 74–78. In Chinese.

Fei, R.C., 1991. Statistical relationship between the representative point and the population. Journal of the Wuxi Institute of Light Industry 10, 78–81. In Chinese.

Fei, R.C., 1992. The problem of selecting representative points from population. Communications on Applied Mathematics and Computation 16, 94–96. In Chinese.

Feller, W., 1967. An Introduction to Probability Theory and Its Applications. New York: John Wiley & Sons.

Feller, W., 1991. An Introduction to Probability Theory and its Applications, Volume 2. John Wiley & Sons.

Fleisher, P.E., 1964. Sufficient conditions for achieving minimum distortion in a quantizer. IEEE Int. Conv. Rec., 104–111.

Úbeda Flores, M., Sánchez, J.F., 2017. Sklar's theorem: The cornerstone of the theory of copulas, in: Copulas and Dependence Models with Applications–Contributions in Honor of Roger B. Nelsen, Springer. pp. 241–258.

Flury, B., Riedwyl, H., 1988. Multivariate statistics: a practical approach. London: Chapman and Hall.

Flury, B.D., 1990. Principal points. Biometrika 77, 33–41.

Flury, B.D., 1993. Estimation of principal points. Journal of the Royal Statistical Society: Series C (Applied Statistics) 42, 139–151.

Foss, S., Korshunov, D., Zachary, S., 2013. An introduction to heavy-tailed and subexponential distributions. Springer.

Fu, H.H., 1985. The problem of selecting a specified number of representative points from a gama population. Journal of China University of Mining & Technology 14, 107–117. In Chinese.

Fu, H.H., 1993. The problem of the best representative points for Weibull population. Journal of China University of Mining & Technology 22, 123–134. In Chinese.

Ghitany, M., Alqallaf, F., Al-Mutairi, D.K., Husain, H., 2011. A two-parameter weighted lindley distribution and its applications to survival data. Mathematics and Computers in simulation 81, 1190–1201.

Giné, E., Zinn, J., 1989. Necessary conditions for the bootstrap of the mean. The Annals of Statistics 17, 684–691.

Gnanadesikan, R., 1977. Methods for statistical data analysis of multivariate observations, in: A Wiley publication in applied statistics.

Goodman, I., Kotz, S., 1973. Multivariate θ-generalized normal distributions. Journal of Multivariate Analysis 3, 204–219.

Govil, K.K., Aggarwal, K.K., 1983. Mean residual life function for normal, gamma and lognormal densities. Reliability Engineering 5, 47–51.

Graf, S., Luschgy, H., 2000. Foundations of Quantization for Probability Distributions. Lecture Notes in Math. 1730. Springer, Berlin.

Graf, S., Luschgy, H., 2007. Foundations of Quantization for Probability Distributions. Springer-Verlag, Berlin and Heidelberg.

Gu, X., Mathew, T., 2001. Some characterizations of symmetric two-principal points. Journal of statistical planning and inference 98, 29–37.

Gumbel, E.J., 1960. Bivariate exponential distributions. Journal of the American Statistical Association 55, 698–707.

Gumbel, E.J., 1961. Bivariate logistic distributions. Journal of the American Statistical Association 56, 335–349.

Guo, X., Jarrow, R.A., de Larrard, A., 2014. The economic default time and the arcsine law. Journal of Financial Engineering 1, 1450025.

Gupta, R.D., Kundu, D., 1999. Generalized exponential distributions. Australian & New Zealand Journal of Statistics 41, 173–188.

Hagmark, P.E., 2008. On construction and simulation of count data models. Math. Comput. Simul. 77, 72–80.

Hall, P., 1990. Asymptotic properties of the bootstrap for heavy-tailed distributions. The Annals of Probability 18, 1342–1360.

Hall, P., 1992. The Bootstrap and Edgeworth Expansion. Springcr-Verlag New York.

Hall, W.J., Wellner, J.A., 1981. Mean residual life, in: Statistics and Related Topics. New York: North Holland, pp. 169–184.

Harrell, F.E., Davis, C.E., 1982. A new distribution-free quantile estimator. Biometrika 69, 635–640.

Hartigan, J.A., 1975. Clustering Algorithms. Wiley, New York.

Hartigan, J.A., Wong, A., 1979. Algorithm as 136: A k-means clustering algorithm. Applied Statistics 28, 100–108.

Hastie, T., Stuetzle, W., 1989. Principal curves. Journal of the American Statistical Association 84, 502–516.

Hastings, W.K., 1970. Monte Carlo sampling methods using markov chains and their applications. Biometrika 57, 97–109.

Henze, N., 2002. Invariant tests for multivariate normality: a critical review. Statistical Papers 43, 467–506.

Henze, N., Wagner, T., 1997. A new approach to the BHEP tests for multivariate normality. Journal of Multivariate Analysis 62, 1–23.

Hickernell, F.J., 1998. A generalized discrepancy and quadrature error bound. Mathematics of Computation 67, 299–322.

Hickernell, F.J., 2000. What affects the accuracy of quasi-Monte Carlo quadrature?, in: Niederreiter, H., Spanier, J. (Eds.), Monte Carlo and Quasi-Monte Carlo Methods 1998, Springer-Verlag, Berlin. pp. 16–55.

Hogg, R.V., McKean, J., Craig, A.T., 2005. Introduction to Mathematical Statistics. 6th ed., Pearson Education, Boston.

Hornik, K., Feinerer, I., Kober, M., Buchta, C., 2012. Spherical k-means clustering. Journal of statistical software 50, 1–22.

Hosmer, D.W., 1973. On mle of the parameters of a mixture of two normal distributions when the sample size is small. Communications in Statistics 1, 217–227.

Hua, L.K., Wang, Y., 1981. Applications of Number Theory to Numerical Analysis. Springer and Science Press, Berlin and Beijing.

Hyndman, R.J., Fan, Y., 1996. Sample quantiles in statistical packages. The American Statistician 50, 361–365.

Inman, H.F., Bradley Jr, E.L., 1989. The overlapping coefficient as a measure of agreement between probability distributions and point estimation of the overlap of two normal densities. Communications in Statistics–theory and Methods 18, 3851–3874.

Iyengar, S., Solomon, H., 1983. Selecting representative points in normal populations, in: Recent Advances in Statistics: Papers in Honor of Herman Chernoff on his 60th Birthday. Academic Press, New York, pp. 579–591.

Jiang, J.J., He, P., Fang, K.T., 2015. An interesting property of the arcsine distribution and its applications. Statistics & Probability Letters 105, 88–95.

Joe, H., 2015. Dependence Modeling with Copulas. CRC Press, London.

Johnson, M.E., 1987. Multivariate statistical simulation. volume 192. Wiley, New York.

Jones, M.C., Faddy, M.J., 2003. A skew extension of t-distributions, with applications. Journal of the Royal Statistical Society Series B (Statistical Methodology) 65, Part 1, 159–174.

Joseph, V., Vakayil, A., 2022. Split: an optimal method for data splitting. Technometrics 64, 166–176.

Kabal, P., 1984. Quantizers for the gamma distribution and other symmetrical distributions. IEEE Transactions on Acoustics, Speech, and Signal Processing 32, 836–841.

Kalgh, W.D., Lachenbruch, P.A., 1982. A generalized quantile estimator. Communications in Statistics-Theory and Methods 11, 2217–2238.

Ke, X., Wang, S., Zhou, M., Ye, H., 2023. New approaches on parameter estimation of the gamma distribution. Mathematics 11, 927.

Keating, J.P., Glaser, R.E., Ketchum, N.S., 1990. Testing hypotheses about the shape parameter of a gamma distribution. Technometrics 32, 67–82.

Kelly, K.S., Krzysztofowicz, R., 1997. Bivariate Meta-Gaussian density for use in hydrology. Stochastic Hydrology and Hydraulics 11, 17–31.

Kiefer, J., 1959. Optimum experimental designs. Journal of the Royal Statistical Society, Series B (Statistical Methodology) 21, 272–319.

Kieffer, J.C., 1983. Uniqueness of locally optimal quantizer for log-concave density and convex error weighting function. IEEE Transactions on Information Theory 29, 42–47.

Kim, I., Park, S., 2018. Likelihood ratio tests for multivariate normality. Communications in Statistics–Theory and Methods 47, 1923–1934.

Koehler, K.J., Gan, F.F., 1990. Chi-squared goodness-of-fit tests: cell selection and power. Communications in Statistics–Simulation and Computation 19, 1265–1278.

Kolmogorov, A.N., 1933. Sulla determinazione empirica di una legge didistribuzione. Giorn Dell'inst Ital Degli Att. 4, 89–91.

Kotz, S., Fang, K.T., Liang, J.J., 1997. On multivariate vertical density representation and its application to random number generation. Statistics 30, 163–180.

Kotz, S., Seeger, J.P., 1991. A new approach to dependence in multivariate distributions, in: Advance in Probability Distributions with Given Marginals, Kluwer Academic, Dordrecht. pp. 13–50.

Kotz, S., Shanbhag, D.N., 1980. Some new approaches to probability distributions. Advances in Applied Probability 12, 903–921.

Koziol, J.A., 1982. A class of invariant procedures for assessing multivariate normality. Biometrika 69, 423–427.

Krzysztofowicz, R., Kelly, K.S., 1996. A Meta-Gaussian Distribution with Specified Marginals. Technical Report. University of Virginia.

Kullback, S., 1959. Information Theory and Statistics. Wiley, New York.

Lai, C.D., Xie, M., 2006. Stochastic Ageing and Dependence for Reliability. Springer Science & Business Media.

Lawless, J.F., 2011. Statistical Models and Methods for Lifetime Data. volume 362. John Wiley & Sons.

Lee, E., Wang, J.W., 2003. Statistical Methods for Survival Data Analysis. John Wiley & Sons.

Lévy, P., 1940. Sur certains processus stochastiques homogènes. Compositio mathematica 7, 283–339.

Lévy, P., 1948. Processus Ptochastiques et Mouvement Brownien. Gauthier-Villars, Paris.

Li, J., Liang, J., Kang, J., Peng, X., 2024a. Representative points based goodness-of-fit test for location-scale distributions. Journal of Physics: Conference Series 2890, 1–13 URL: `https://doi.org/10.1088/1742-6596/2890/1/012003`.

Li, L., Flury, B.A., 1995. Uniqueness of principal points for univariate distributions. Statistics & probability letters 25, 323–327.

Li, Y., Fang, K.T., 2022. A new approach to parameter estimation of mixture of two normal distributions. Communications in Statistics-Simulation and Computation 53, 1161–1187.

Li, Y., Fang, K.T., He, P., Peng, H., 2022. Representative points from a mixture of two normal distributions. Mathematics 10, 39–52.

Li, Y., Fang, K.T., Wang, Y., 2024b. Uniform designs of experiments with mixtures under the criterion mean l_1-distance and a new approach to Scheffé-type designs, Chinese Mathematical Society, Uniform Design Branch, 17th Academic Symposium. Invited speech.

Li, Y., Sun, Z., Fang, K.T., 2024c. Generalized inverse transformation method via representative points in statistical simulation. Communications in Statistics–Simulation and Computation, 1–27.

Liang, F., Wong, W.H., 2001. Real-parameter evolutionary Monte Carlo with applications to bayesian mixture models. Journal of the American Statistical Association 96, 653–666.

Liang, J., Bentler, P.M., 1999. A t-distribution plot to detect non-multinormality. Computational Statistics & Data Analysis 30, 31–44.

Liang, J., Li, R., Fang, H.B., Fang, K.T., 2000. Testing multinormality based on low-dimensional projection. Journal of Statistical Planning and Inference 86, 129–141.

Lin, Y., Yan, T., Fang, K.T., 2024. Some improvements on good lattice point sets. Entropy 26, 910.

Linde, Y., Buzo, A., Gray, R., 1980. An algorithm for vector quantizer design. IEEE Transactions on communications 28, 84–95.

Lindley, D.V., 1958. Fiducial distributions and Bayes' theorem. Journal of the Royal Statistical Society. Series B (Statistical Methodology) 20, 102–107.

Liu, Y., Pagès, G., 2020. Convergence rate of optimal quantization and application to the clustering performance of the empirical measure. Journal of Machine Learning Research 21, 1–36.

Lloyd, S.P., 1982. Least squares quantization in PCM. IEEE Transactions on Information Theory 28, 129–137.

Looney, S.W., 1995. How to use tests for univariate normality to assess multivariate normality. The American Statistician 49, 64–70.

Lourens, S., Zhang, Y., Long, J.D., Paulsen, J.S., 2013. Bias in estimation of a mixture of normal distributions. Journal of Biometrics & Biostatistics 4.

Luceno, A., 1999. Discrete approximations to continuous univariate distributions–an alternative to simulation. Journal of the Royal Statistical Society, Series B (Statistical Methodology) 61, 345–352.

Mai, J.F., Scherer, M., 2017. Simulating Copulas: Stochastic Models, Sampling Algorithms and Applications. World Scientific Publishing Co. Pte. Ltd. Second Edition.

Mak, S., Joseph, V.R., 2018. Support points. The Annals of Statistics 46, 2562–2592.

Mallows, C.L., 1972. A note on asymptotic joint normality. The Annals of Mathematical Statistics, 508–515.

Mardia, K.V., 1970. Measures of multivariate skewness and kurtosis with applications. Biometrika 57, 519–530.

Mardia, K.V., 1980. 9 tests of univariate and multivariate normality. Handbook of Statistics 1, 279–320.

Matsuura, S., Kurata, H., 2013. Definition and properties of m-dimensional n-principal points. Communications in Statistics-Theory and Methods 42, 267–282.

Matsuura, S., Kurata, H., Tarpey, T., 2015. Optimal estimators of principal points for minimizing expected mean squared distance. Journal of Statistical Planning and Inference 167, 102–122.

Matsuura, S., Tarpey, T., 2018. Optimal principal points estimators of multivariate distributions of location-scale and location-scale-rotation families. Statistical Papers 61, 1629–1643.

Max, J., 1960. Quantizing for minimum distortion. IRE Transactions on Information Theory 6, 7–12.

McQueen, J., 1967. Some methods for classification and analysis of multivariate observations, in: Proc. 5th Berkeley Symposium on Mathematical Statistics and Probability, 1967, pp. 281–297.

Mecklin, C.J., Mundfrom, D.J., 2000. Comparing the power of classical and newer tests of multivariate normality.

Mecklin, C.J., Mundfrom, D.J., 2005. A Monte Carlo comparison of the type I and type II error rates of tests of multivariate normality. Journal of Statistical Computation and Simulation 75, 107–93.

Meng, C., Xie, R., Mandal, A., Zhang, X., Zhong, W., Ma, P., 2021. Lowcon: a design-based subsampling approach in a misspecified linear model. Annals of Applied Statistics 30, 694–708.

Meng, L., 2016. Method for computation of the Fisher information matrix in the expectation-maximization Algorithm. Master's thesis. Thesis in Department of Applied Mathematics & Statistics, The Johns Hopkins University.

Moore, D.S., Spruill, M.C., 1975. Unified large-sample theory of general chi-squared statistics for tests of fit. The Annals of Statistics 3, 599–616.

Mudholkar, G.S., McDermott, M., Srivastava, D.K., 1992. A test of p-variate normality. Biometrika 79, 850–854.

Navruz, G., Ozdemir, F., 2020. A new quantile estimator with weights based on a subsampling approach. British Journal of Mathematical and Statistical Psychology 73, 1–16.

Nelsen, R.B., 2006. An Introduction to Copulas. Springer-Verlag, New York. 2ed edition.

Niederreiter, H., 1992. Random number generation and Quasi-Monte Carlo methods, in: SIAM CBMS-NSF Regional Conference, Philadelphia.

Ozonur, D., Paul, S., 2020. Goodness of fit tests of the two-parameter gamma distribution against the three-parameter generalized gamma distribution. Communications in Statistics–Simulation and Computation 51, 687–697.

Pagès, G., 2018. Numerical Probability, An Introduction with Applications to Finance. Springer, Universitext, Swissland.

Pagès, G., Sagna, A., 2012. Asymptotics of the maximal radius of an lr-optimal sequence of quantizers. Bernoulli 18, 360–389.

Pal, N., Jin, C., Lim, W.K., 2005. Handbook of Exponential and Related Distributions for Engineers and Scientists. London: CRC Press.

Palmitesta, P., Provasi, C., 2015. Computer generation of random vectors from continuous multivariate distributions URL: `https://www.researchgate.net/publication/268048278`.

Panaretos, V.M., Zemel, Y., 2019. Statistical aspects of wasserstein distances. Annual Review of Statistics and Its Application 6, 405–431.

Parrish, R.S., 1990. Comparison of quantile estimators in normal sampling. Biometrics 46, 247–257.

Parzen, E., 1962. On estimation of a probability density function and mode. The Annals of Mathematical Statistics 33, 1065–1076.

Patton, A.J., 2006. Estimation of multivariate models for time series of possibly different lengths. Journal of Applied Econometrics 21, 147–173.

Pearson, K., 1894. Contributions to the mathematical theory of evolution. Philosophical Transactions of the Royal Society of London. A 185, 71–110.

Pechtl, A., 1999. Distributions of occupation times of brownian motion with drift. Journal of Applied Mathematics and Decision Sciences 3, 41–62.

Peng, X., Huang, M., Li, X., Zhou, T., Lin, G., Wang, X., 2024. Patient regional index: a new way to rank clinical specialties based on outpatient clinics big data. BMC Medical Research Methodology 24, 192.

Pollard, D., 1982. Quantization and the method of k-means. IEEE Transactions on Information Theory 28, 199–205.

Proschan, F., 1963. Theoretical explanation of observed decreasing failure rate. Technometrics 5, 375–383.

Qi, Z.F., Zhang, X.R., Zhou, Y.D., 2018. Generalized good lattice point sets. Comput Stat. 33, 887–901.

Qi, Z.F., Zhou, Y.D., Fang, K.T., 2019. Representative points for location-biased datasets. Communication in Statistics–Simulation and Computation 48, 458–471.

Rao, C.R., 1948. Tests of significance in multivariate analysis. Biometrika 35 Pts 1-2, 58–79.

Ridolfi, A., Idier, J., 2001. Penalized maximum likelihood estimation for univariate normal mixture distributions, in: AIP Conference Proceedings, American Institute of Physics. pp. 229–237.

Rizzo, M.L., Székely, G.J., 2016. Energy distance. Wiley Interdiscip Rev. 8, 27–38.

Rosenblatt, M., 1956. Remarks on some nonparametric estimates of a density function. The Annals of Mathematical Statistics, 832–837.

Ross, S.M., 2013. Simulation. Academic Press, New York.

Rover, C., Friede, T., 2017. Discrete approximation of a mixture distribution via restricted divergence. Journal of Computational and Graphical Statistics 26, 217–222.

Rubin, D.B., 1987. A nonitertive sampling/importance resampling alternative to the data augmentation algorithm for creating a few imputations when fractions of missing information are modest: the sir algorithm. Journal of the American Statistical Association 82, 543–546.

Sankaran, P.G., Sunoj, S.M., 2004. Identification of models using failure rate and mean residual life of doubly truncated random variables. Statistical Papers 45, 97–109.

Sarabia, J.M., Gómez-Déniz, E., 2008. Construction of multivariate distributions: a review of some recent results (with discussion). SORT (Statistics and Operations Research Transactions) 32, 3–36.

Scheffé, H., 1958. Experiments with mixtures. Journal of the Royal Statistical Society, Series B (Statistical Methodology) 20, 344–360.

Scheffé, H., 1963. The simplex-centroid design for experiments with mixtures. Journal of the Royal Statistical Society, Series B (Statistical Methodology) 25, 235–263.

Schweizer, B., Sklar, A., 1974. Operations on distribution functions not derivable from operations on random variables. Studia Mathematica 52, 43–52.

Serfling, R.J., 1980. Approximation Theorems of Mathematical Statistics. John Willey & Sons, New York.

Sfakianakis, M.E., Verginis, D.G., 2008. A new family of nonparametric quantile estimators. Communications in Statistics–Simulation and Computation 37, 337–345.

Shao, J., Tu, D., 1995. The Jackknife and Bootstrap. Springer series in statistics, New York.

Shapiro, S.S., Wilk, M.B., 1965. An analysis of variance test for normality (complete samples). Biometrika 52, 591–611.

Shi, X., 1986. A note on bootstrapping the U-statistics. Chinese Journal of applied probability and statistics 2, 144–148.

Singh, K., 1981. On the asymptotic accuracy of Efron's bootstrap. The Annals of Statistics 9, 1187–1195.

Sklar, A., 1959. Fonctions de répartition *à* n dimensions et leurs marges. Annales de l'ISUP 8, 229–231.

Sklar, A., 1973. Random variables, joint distributions, and copulas. Kybernetica 9, 449–460.

Srivastava, D.K., Mudholkar, G.S., 2003. Ch. 24. Goodness-of-fit tests for univariate and multivariate normal models. Handbook of Statistics 22, 869–906.

Srivastava, M., Hui, T., 1987. On assessing multivariate normality based on shapiro-wilk w statistic. Statistics & Probability Letters 5, 15–18.

Stampfer, E., Stadlober, E., 2002. Methods for estimating principal points. Communications in Statistics-Simulation and Computation 31, 261–277.

Su, Y., 2000. Asmptotically optimal representative points of bivariate random vectors. Statistica Sinica 10, 559–575.

Székely, G.J., Rizzo, M.L., 2013. Energy statistics: A class of statistics based on distances. Journal of Statistical Planning and Inference 143, 1249–1272.

Székely, G.J., Rizzo, M.L., 2017. The energy of data. Annual Review of Statistics and Its Application 4, 447–479.

Tarpey, T., 1994. Two principal points of symmetric, strongly unimodal distributions. Statistics & Probability Letters 20, 253–257.

Tarpey, T., 1995. Principal points and self-consistent points of symmetrical multivariate distributions. Journal of multivariate analysis 53, 39–51.

Tarpey, T., 1997. Estimating principal points of univariate distributions. Journal of Applied Statistics 24, 499–512.

Tarpey, T., 1998. Self-consistent patterns for symmetric multivariate distributions. Journal of classification 15, 57–79.

Tarpey, T., 2000. Parallel principal axes. Journal of Multivariate Analysis 75, 295–313.

Tarpey, T., 2007. A parametric k-means algorithm. Computational statistics 22, 71–89.

Tarpey, T., Flury, B.D., 1996. Self-consistency: a fundamental concept in statistics. Statistical Science 11, 229–243.

Tarpey, T., Li, L., Flury, B.D., 1995. Principal points and self-consistent points of elliptical distributions. The Annals of Statistics 23, 103–112.

Tarpey, T., Loperfido, N., 2015. Self-consistency and a generalized principal subspace theorem. Journal of Multivariate Analysis 133, 27–37.

Tashiro, Y., 1977. On methods for generating uniform random points on the surface of a sphere. Annals of the Institute of Statistical Mathematics 29, 295–300.

Tenreiro, C., 2017. A new test for multivariate normality by combining extreme and nonextreme bhep tests. Communications in Statistics–Simulation and Computation 46, 1746–1759.

Thulin, M., 2014. Tests for multivariate normality based on canonical correlations. Statistical Methods & Applications 23, 189–208.

Timm, N.H., 2002. Applied Multivariate Analysis. Springer, New York.

Ting, D., Brochu, E., 2018. Optimal subsampling with influence functions, in: Advances in Neural Information Processing Systems, pp. 3650–3659.

Troutt, M.D., 1991. A theorem on the density of the density ordinate and an alternative interpretation of the box-muller method. Statistics 29, 463–466.

Troutt, M.D., 1993. Vertical density representation and a further remark on the box-muller method. Statistics 24, 81–83.

Trushkin, A., 1982. Sufficient conditions for uniqueness of a locally optimal quantizer for a class of convex error weighting functions. IEEE Transactions on Information Theory 28, 187–198.

Vakayil, A., Joseph, V., 2022. Data twinning. Statistical Analysis and Data Mining 15, 598–610.

van der Vaart, A.W., 1998. Stochastic Convergence. Cambridge University Press. Cambridge Series in Statistical and Probabilistic Mathematics, p. 5–24.

Wang, H., Ma, Y., 2021. Optimal subsampling for quantile regression in big data. Biometrika 108, 99–112.

Wang, H., Yang, M., Stufken, J., 2019. Information-based optimal subdata selection for big data linear regression. Journal of the American Statistical Association 114, 393–405.

Wang, H., Zhu, R., Ma, P., 2018. Optimal subsampling for large sample logistic regression. Journal of the American Statistical Association 113, 829–844.

Wang, L., Elmstedt, J., Wong, W., Xu, H., 2021. Orthogonal subsampling for big data linear regression. Annals of Applied Statistics 13, 1273–1290.

Wang, M., Wang, W., 2017. Bias-corrected maximum likelihood estimation of the parameters of the weighted lindley distribution. Communications in Statistics-Simulation and Computation 46, 530–545.

Wang, S., Fang, K.T., Ye, H., 2022a. A new bias-corrected estimator method in extreme value distributions with small sample size. Journal of Statistical Computation and Simulation 92, 3862–3884.

Wang, S., Liang, J., Peng, H., Ye, H., 2024. A new Kolmogorov-Smirnov test based on representative points in Weibull distributions. Communications in Statistics–Simulation and Computation , 1–18 URL: https://doi.org/10.1080/03610918.2024.2391871.

Wang, S., Liang, J., Zhou, M., Ye, H., 2022b. Testing multivariate normality based on F-representative points. Mathematics 10, 4300.

Wang, Y., Fang, K.T., 1990a. Number theoretic method in applied statistics. Chinese Annals of Mathematics, Series B 11, 41–55.

Wang, Y., Fang, K.T., 1990b. Number theoretic method in applied statistics (ii). Chinese Annals of Mathematics, Series B 11, 384–395.

Wang, Y., Fang, K.T., 1996. Uniform design of experiments with mixtures. Science in China (Series A) 39, 264–275.

Ward, P.J., 1988. Goodness-of-fit Tests for Multivariate Normality. The University of Alabama.

Weyl, H., 1916. Uber die gleichverteilung der zahlem mod eins. Mathematische Annalen 77, 313–352.

Williams, C.A., 1950. The choice of the number and width of classes for the chi-square test of goodness of fit. Journal of the American Statistical Association 45, 77–86.

Wolberg, W., Street, W., Mangasarian, O., 1995. The diagnostic database. Breast Cancer Wisconsin (Diagnostic). UCI Mach. Learn. Repos. URL: https://archive.ics.uci.edu/ml/datasets/Breast+Cancer+Wisconsin+.

Xiong, Z., Xiao, Y., Ning, J., Qin, H., 2022. Representative points based on power exponential kernel discrepancy. Axioms 11, 711.

Xu, L.H., Fang, K.T., He, P., 2021. Representative points of the exponential distribution. Statistical Papers 63, 197–223.

Xu, L.H., Fang, K.T., Pan, J., 2023. Representative points of the exponential distribution. Computational Statistics and Data Analysis 52, 3290–3313.

Xu, L.H., Li, Y.N., Fang, K.T., 2024. The resampling method via representative points. Statistical Papers 65, 3621–3649.

Yamamoto, W., Shinozaki, N., 2000. On uniqueness of two principal points for univariate location mixtures. Statistics & probability letters 46, 33–42.

Yan, T., Fang, K.T., Yin, H., 2024. A novel approach for parameter estimation of bimodal mixture Weibull distributions in failure data modeling. Statistics and Computing 34:221.

Yang, J., He, P., Fang, K.T., 2022. Three kinds of discrete approximations of statistical multivariate distributions and their applications. Journal of Multivariate Analysis 188, 104829.

Yang, S.S., 1985. A smooth nonparametric estimator of a quantile function. Journal of the American Statistical Association 80, 1004–1011.

Yi, S., Zhou, Y., 2023. Model-free global likelihood subsampling for massive data. Statistics and Computing 33, 9.

Yi, S.Y., Liu, Z., Liu, M.Q., Zhou, Y., 2023. Global likelihood sampler for multimodal distributions. Journal of Computational and Graphical Statistics 32, 927–937.

Yip, W., Fu, H., Chen, A., et al., 2019. 10 years of health-care reform in china: progress and gaps in universal health coverage. Lancet 394, 1192–1204.

Yu, F., 2022. Uniqueness of principal points with respect to p-order distance for a class of univariate continuous distribution. Statistics & Probability Letters 183, 109341.

Yu, J., Wang, H., Ai, M., Zhang, H., 2022. Optimal distributed subsampling for maximum quasi-likelihood estimators with massive data. Journal of the American Statistical Association 117, 265–276.

Zador, P.L., 1982. Asymptotic quantization error of continuous signals and the quantization dimension. IEEE Transactions on Information Theory 28, 139–149.

Zhang, L., Cheng, G., Song, S., et al., 2017. Efficiency performance of china's health care delivery system. International Journal of Health Planning and Management 32, 254–263.

Zhang, L., Singh, V.P., 2019. Copulas and Their Applications in Water Resources Engineering. Cambridge University Press.

Zhang, M., Zhou, Y.D., Zhou, Z., Zhang, A.J., 2024a. Model-free subsampling method based on uniform designs. IEEE Transactions on Knowledge and Data Engineering 36, 1210–1220.

Zhang, Y., Wang, S., Ke, X., Ye, H., 2024b. A new Kolmogorov-Smirnov test based on representative points in the exponential distribution family. Journal of Statistical Computation and Simulation , 1–18 URL: `https://doi.org/10.1080/00949655.2024.2385687`.

Zhang, Y.T., Fang, K.T., 1982. An Introduction to Multivariate Analysis. Science Press, Beijing. In Chinese.

Zhou, M., Shao, Y., 2014. A powerful test for multivariate normality. Journal of Applied Statistics 41, 351–363.

Zhou, M., Wang, W.J., 2016. Representative points of student's t_n distribution and applications in statistical simulation. Acta Mathematicae Applicatae Sinica 39, 620–640. In Chinese.

Zhou, Y., Fang, K.T., 2011. A note on statistics simulation for geometric probability problems. Scientia Sinica Mathematica 41, 253–264. In Chinese.

Zhou, Y., Fang, K.T., 2019. FM criterion in representative points. Scientia Sinica Mathematics 49, 1009–1020. In Chinese.

Zhou, Y., Fang, K.T., Ning, J., 2013. Mixture discrepancy for quasi-random point sets. Journal of Complexity 29, 283–301.

Zhou, Y., He, P., Ning, J.H., Fang, K.T., 2020. Methods and applications of stochastic simulation. Higher Education Press, China. In Chinese.

Zhou, Y.F., Lin, Y., Fang, K.T., Yin, H., 2024a. The representative points of generalized alpha skew-t distribution and applications. Entropy 26, 889.

Zhou, Z., Yang, Z., Zhang, A., Zhou, Y., 2024b. Efficient model-free subsampling method for massive data. Technometrics 66, 240–252.

Zoppè, A., 1995. Principal points of univariate continuous distributions. Statistics and Computing 5, 127–132.

Index

FSC
www.fsc.org
MIX
Papier | Fördert gute Waldnutzung
FSC® C083411

Kolibri 360 GmbH
Ferdinand-Jühlke-Straße 7
99095 Erfurt, Deutschland
produktsicherheit@kolibri360.de